HANDBOOK OF ANALYTICAL TECHNIQUES FOR FORENSIC SAMPLES

HANDBOOK OF ANALYTICAL TECHNIQUES FOR FORENSIC SAMPLES

Current and Emerging Developments

CHAUDHERY MUSTANSAR HUSSAIN
Adjunct Professor, Academic Advisor and Lab Director, New Jersey Institute of Technology (NJIT), Newark, NJ, United States

DEEPAK RAWTANI
Associate Professor (Nanotechnology) and Associate Dean, School of Pharmacy, National Forensic Sciences University, Gandhinagar, India

GAURAV PANDEY
Research Fellow, National Forensic Sciences University, Gandhinagar, India

MAITHRI THARMAVARAM
Research Fellow, National Forensic Sciences University, Gandhinagar, India

ELSEVIER

Elsevier
Radarweg 29, PO Box 211, 1000 AE Amsterdam, Netherlands
The Boulevard, Langford Lane, Kidlington, Oxford OX5 1GB, United Kingdom
50 Hampshire Street, 5th Floor, Cambridge, MA 02139, United States

Library of Congress Cataloging-in-Publication Data
A catalog record for this book is available from the Library of Congress

British Library Cataloguing-in-Publication Data
A catalogue record for this book is available from the British Library

ISBN: 978-0-12-822300-0

Publisher: Susan Dennis
Acquisitions Editor: Kathryn Eryilmaz
Editorial Project Manager: Ruby Smith
Production Project Manager: Debasish Ghosh
Cover Designer: Miles Hitchen

Typeset by SPi Global, India

Dedication

We would like to dedicate this handbook to a pair of forensic pioneers, the honorable Dr. J.M. Vyas, Vice-Chancellor, National Forensic Sciences University, and Prof. S.O. Junare, Campus Director, National Forensic Sciences University.

CONTENTS

Preface *xiii*

1. Introduction to chemical analysis of forensic samples **1**

1 Introduction 1
2 Chemical analytical techniques used for the analysis of evidence 1
3 Conclusion 13
References 14

2. Forensic sampling and sample preparation techniques **17**

1 Introduction 17
2 Types of forensic samples 18
3 Preparation techniques for forensic samples for instrumental analyses 22
4 Conclusion 31
References 31

3. UV-visible and fluorescence spectroscopy for forensic samples **37**

1 Introduction 37
2 Principle and theory of UV-visible and fluorescence spectroscopy 38
3 Instrumentation of UV-visible and fluorescence spectroscopy 40
4 Analysis of forensic samples using UV-visible and fluorescence spectroscopy 43
5 Conclusion 51
References 52

4. FTIR and NIR spectroscopy in forensic science **55**

1 Introduction 55
2 Principle and theory of infrared and near-infrared spectroscopy 56
3 Instrumentation of IR and NIR spectroscopy 58
4 Spectrum analysis in FTIR and NIR 61
5 FTIR and NIR analysis of forensic samples 61
6 Conclusion 69
References 71

5. Atomic absorption and emission spectrometry in forensic analysis **75**

1 Introduction 75

2 Principle and theory of atomic absorption and emission spectroscopy 75

3 Instrumentation of atomic absorption and atomic emission spectroscopy 77

4 Analysis of forensic samples via AAS and AES 80

5 Conclusion 87

References 88

6. NMR spectroscopy for forensic samples **91**

1 Introduction 91

2 Principle and instrumentation of NMR spectroscopy 92

3 NMR in the world of forensic science 96

4 Conclusion 103

References 104

7. Raman spectroscopy in forensic science **109**

1 Introduction 109

2 Principle of Raman spectroscopy 110

3 Raman spectroscopy for forensic sample analysis 113

4 Conclusion 124

References 125

8. HPLC for the toxicological analysis of forensic samples **129**

1 Introduction 129

2 High performance liquid chromatography: An insight 130

3 Applications of high-performance liquid chromatography in forensic toxicology 136

4 Conclusion 144

References 145

9. Gas chromatography in forensic science **149**

1 Introduction 149

2 Principle and theory of gas chromatography 149

3 Instrumentation of gas chromatography 152

4 Forensic sample analysis via gas chromatography 154

5 Conclusion 164

References 164

10. HPTLC in forensic science **169**

1 Introduction 169
2 Principle and theory of HPTLC 169
3 HPTLC instrumentation 172
4 HPTLC analysis of forensic samples 176
5 Conclusion 184
References 185

11. Hyphenated techniques for forensic sample analysis **189**

1 Introduction 189
2 Hyphenated techniques 190
3 Hyphenated technique-mediated analysis of forensic samples 198
4 Conclusion 207
References 207

12. Optical microscopy for forensic samples **213**

1 Introduction 213
2 Fundamentals of optical microscopy 214
3 Types of optical compound microscopes 214
4 Optical microscopic analysis of forensic samples 223
5 Conclusion 231
References 232

13. Electron microscopy for forensic samples **235**

1 Introduction 235
2 Principle and theory of electron microscopes 236
3 Instrumentation of electron microscopes 238
4 Electron microscopy-based forensic investigations 243
5 Conclusion 254
References 255

14. Atomic force microscopy for forensic samples **259**

1 Introduction 259
2 Principle, working, and instrumentation of atomic force microscopy 260
3 Atomic force microscopy for forensic investigations 263
4 Conclusion 274
References 275

15. Energy dispersive X-ray (EDX) coupled microscopy in forensic science **281**

1 Introduction 281
2 Principle and theory of EDX 281
3 Instrumentation of EDX 284
4 EDX-coupled microscopic analysis of forensic samples 285
5 Conclusion 296
References 298

16. Mass spectrometry in forensic chemistry **301**

1 Introduction 301
2 Principle and theory of mass spectrometry 301
3 Overview of mass spectrometry 302
4 Applications of mass spectrometry for forensic investigations 307
5 Conclusion 316
References 316

17. X-ray diffraction for forensic samples **321**

1 Introduction 321
2 Principle and theory of X-ray diffraction 322
3 Instrumentation of X-ray diffraction 324
4 X-ray diffraction for forensic evidence analysis 327
5 Conclusion 335
References 336

18. Lab-on-chip devices **339**

1 Introduction 339
2 Architecture of a lab-on-chip device 340
3 Forensic applications of lab-on-chip devices 344
4 Conclusion 353
References 354

19. Nanotechnology in forensic science **359**

1 Introduction 359
2 An insight into the world of nanotechnology 360
3 Nanotechnology-mediated forensic investigations 365
4 Conclusion 373
References 374

20. Ethics and legal issues of forensic analysis techniques **381**

1 Introduction 381
2 Legal issues in forensic science 382
3 Ethics and theories of ethics 388
4 Code of ethics in different fields of forensic science 391
5 Conclusion 393
References 393

21. Accreditations for forensic science laboratories **395**

1 Introduction 395
2 History and need of accreditation in forensic science 395
3 Accreditation in forensic science 397
4 Conclusion 406
References 407

22. Quality control and quality assurance in forensic science laboratories **409**

1 Introduction 409
2 Quality assurance in forensic science 409
3 Quality assurance in different fields of forensic science 412
4 Quality control for the analysis of forensic samples 415
5 Conclusion 421
References 421

23. Concluding notes **425**

1 Introduction 425
2 Analytical techniques and forensic science 425
3 Rereads of the book 426
4 Conclusion 428
References 428

Exercise 1: UV-visible spectroscopic analysis of forensic samples (chemical and biological) *431*
Exercise 2: Fluorescence spectroscopic analysis of forensic samples (biological) *433*
Exercise 3: FTIR analysis of forensic samples (chemical, physical, and biological) *435*
Exercise 4: NIR spectroscopic analysis of forensic samples (chemical and biological) *437*
Exercise 5: Atomic absorption spectrometric analysis of forensic samples (chemical and biological) *439*
Exercise 6: Optical microscopic analysis of physical and biological forensic samples *441*

Exercise 7: SEM analysis of physical and biological forensic samples 443
Exercise 8: TEM analysis of physical and biological forensic samples 445
Exercise 9: AFM analysis of physical and biological forensic samples 447
Exercise 10: Mass spectrometric analysis of forensic samples (chemical and biological) 449

Index 451

Preface

Forensic science has long served as the link between criminal activities and the justice system. The investigation consists of myriad steps, from investigating the evidence at the crime scene to processing and documenting the collected evidence to be produced as testimony in court. Analytical techniques such as spectroscopic, chromatographic, and microscopy techniques have been vastly used for the analysis of various evidence. Apart from such conventional techniques, emerging techniques such as nanotechnology and lab-on-chip devices have also been used to analyze various evidence. In order to compile the different analytical techniques used in forensic science, this handbook has been divided into two parts: part 1 has sections that contain chapters on the aforementioned techniques while part 2 gives an overview of the exercises using such techniques for different forensic samples.

Part 1 of this book has six sections. Section I sheds light on the intricacies of various types of evidence and their collection. There are various types of evidence collected at the crime scene, namely biological, physical, and chemical evidence. Biological evidence consists of samples such as hair, blood, semen, and insects while chemical evidence consists of samples such as poisons and explosives. Physical evidence consists of samples such as fingerprints, glass, fibers, and paints. The nature of every piece of evidence is vastly different and therefore it needs suitable analytical techniques for its analysis. Section II deals with spectroscopic techniques that are based on the amount of energy absorbed by the sample (molecular or atomic species) while keeping in check with certain limitations. UV-visible, infrared, near-infrared, and Raman are some of the spectroscopic techniques commonly used for the quantitative and qualitative analysis of samples. Absorption, emission, and nuclear magnetic resonance spectroscopic techniques are also used for the same. Apart from spectroscopic techniques, chromatographic techniques, which are discussed in Section III, are also used for the identification of the sample. These techniques are based on the separation of substances on the basis of their affinity toward either the stationary phase or the mobile phase. In Section IV, various microscopy techniques such as optical and electron microscopes for the analysis of various pieces of evidence have been discussed. Microscopic techniques are useful in forensic science as they can provide confirmatory identification of the samples. As mentioned before, emerging techniques such as mass spectrometry, X-ray diffraction, nanotechnology, and lab-on-chip devices are incredibly useful for the analysis of forensic evidence. Section V deals with the applications of such forensic evidence in forensic science. A very crucial aspect of the use of such analytical techniques is their standardization and the ethics of the personnel involved in their operation. The results obtained from analytical techniques need to be highly reliable and valid so that they can be produced in a court of law. In order to ensure this, various accreditation boards have been set up that are responsible for overseeing the accreditation, testimony, and interpretation of data that can be produced in court; these are discussed in Section VI. Part 2 of this book is the experimental section, which sheds light on the practical applications of various analytical techniques for the analysis of evidence.

Overall, this handbook has been written with the intent of briefing students and researchers about the use of various analytical techniques for the analysis of forensic samples. Special thanks to Ms. Ruby Smith (editorial project manager) and the editorial team at Elsevier for their dedicated support and help during this project. In the end, all thanks to Elsevier for publishing the handbook.

Chaudhery Mustansar Hussain
Deepak Rawtani
Gaurav Pandey
Maithri Tharmavaram

CHAPTER 1

Introduction to chemical analysis of forensic samples

1 Introduction

Forensic science is a field in which science is applied during a criminal investigation for the analysis of evidence obtained through the course of investigation. Because it is mainly a scientific field, several analytical techniques are used for the analysis of various evidence found at the scene of the crime. The personnel involved in the crime scene investigation include those who are present at the crime scene such as doctors, police, and other technicians responsible for the collection of evidence. Forensic scientists working in the laboratory are more experienced with the different analytical techniques used for their analysis. Biological, physical, and chemical evidence are most widely encountered during the investigation. Prior to analysis with the analytical techniques, these samples are usually prepped in order to ensure optimal analysis. In this chapter, a brief overview of the different analytical techniques used for the analysis of the evidence encountered during the investigation is given. This chapter will provide a basic outline of the various analytical techniques used during evidence analysis.

2 Chemical analytical techniques used for the analysis of evidence

In this section, spectroscopic, chromatographic, microscopic, and other emerging analytical techniques are discussed in brief. Fig. 1 gives a brief overview of the various forensic evidence and the analytical techniques used for their analysis.

2.1 Spectroscopic techniques

Spectroscopic techniques are based on the interaction of the sample with electromagnetic radiation. Electromagnetic radiation is a form of energy that consists of the electrical and the magnetic component propagating at right angles to each other. The electromagnetic radiation basically has a dual

Handbook of Analytical Techniques for Forensic Samples
https://doi.org/10.1016/B978-0-12-822300-0.00001-X

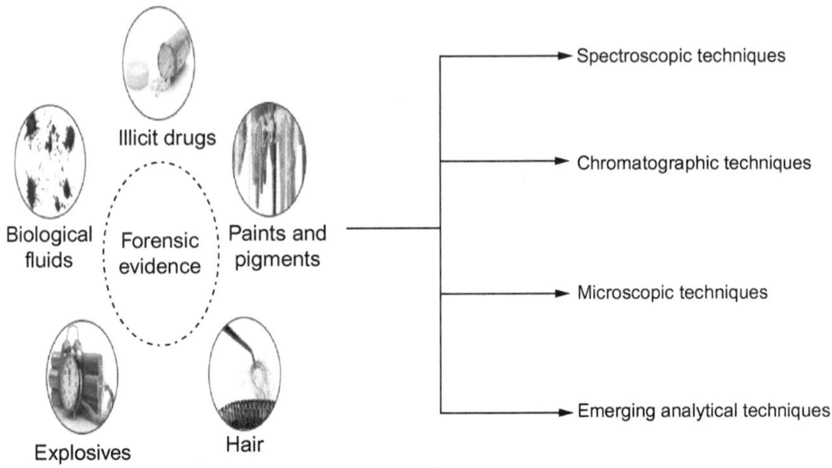

Fig. 1 Analytical techniques for the examination of forensic evidence.

nature, that is, it can travel in either a wave form or in a particulate form. The electromagnetic radiation as such consists of packets of energy known as photons. It is due to these particles that the electromagnetic radiation consists of both the wave- and particle-like behavior. Spectroscopic analytical techniques involve the interaction of matter with the electromagnetic radiation of a wavelength range from 10^{-3} to 10^{-8} nm, which basically covers the infrared to UV-visible range. In this section UV-visible and fluorescence, infrared, nuclear magnetic resonance, Raman, and atomic absorption and emission spectrometry.

2.1.1 UV-visible and fluorescence spectroscopy

UV-visible spectroscopy is a technique that measures absorption occurring in the UV-visible range. The UV region extends from 190 to 350 nm while the visible range extends from 350 to 800 nm. The main principle behind the UV-visible technique is the electronic transitions that occur from the bonding to non- and antibonding orbitals, that is, from n, π, and sigma to π^* and sigma*, respectively. Such transitions occur in compounds that consist of double bonds or are aromatic. The absorbance by such compounds in this range is determined by the Beer-Lambert law, which takes into account the concentration of the sample with the intensity of the transmitted light. Through this technique, it is possible to analyze myriad compounds. Fluorescence spectroscopy is a technique that also takes into account the electronic transitions from the ground state to a higher state. When the sample is irradiated with a high energy source, it absorbs the energy and transitions to a higher state. However, it loses its energy and falls back to its ground state while emitting fluorescence energy for a noticeably short period of time. The fluorescence emitted is characteristic to each element and is therefore crucial for the analysis of various evidence. These spectroscopic techniques are among the most ubiquitous tools used for the analysis of various forensic evidence such as blood, semen, saliva, urine, and other body fluids; counterfeit food products; drugs; paints; or petroleum products. Some of the samples can be directly analyzed through this technique while some require appropriate sample preparation techniques (Altemose, 1986; Räty et al., 2004).

2.1.2 Infrared and near-infrared techniques

Infrared and near-Infrared techniques are vibrational spectroscopic techniques that take into account the vibrational frequencies of the molecules when they are irradiated by strong infrared radiation. The infrared radiation ranges from 700 nm to 1 mm. When the radiation hits the sample, the molecule absorbs the energy, due to which the bonds start to vibrate. The behavior of the molecules at such a stage is equivalent to that of a spring and vibrations such as rocking, twisting, stretching, and bending take place. This vibrational frequency is characteristic of a particular compound and can thus be easily used for the determination of the compound. Infrared spectroscopy is usually used in conjunction with the Fourier transform technique, which uses a Michelson interferometer and is based on the constructive and destructive interference of the radiation obtained from the sample. The final spectra can be easily developed through Fourier transformation. This technique is nondestructive and is incredibly useful for the analysis of evidence such as hair, blood, and other biological evidence as well as illicit and counterfeited products, explosives, and gunshot residue. Near-infrared spectroscopy is a technique similar to infrared spectroscopy, except that it uses near-IR radiation for the analysis of compounds. The NIR range is from 0.7 to 1.4 μm. This technique is widely used due to its simplicity and nondestructive nature. It works by taking into account the overtone and combination absorption bands. The absorption here is weak and occurs as a series of overlapping bonds. Because the absorption bands are weak, they are highly useful in analyzing samples that are strong absorbers and have high light scattering (Bell and Xu, 2018; El-Azazy, 2019; Ewing and Kazarian, 2017).

2.1.3 Nuclear magnetic resonance spectroscopy

Nuclear magnetic resonance (NMR) spectroscopy is an interesting analytical technique that takes into account the magnetic moment of nuclei spin in the presence of an external magnetic field for the structural determination of a chemical compound. Typically, nuclei possess a neutron and protons that contribute to the spin of the nuclei. If an external magnetic field is applied then based on the energy level of the nuclei spin, there is an energy transfer from the ground energy level to the higher energy level. This occurs at a particular wavelength and frequency and is characteristic to the compound. Typically, H NMR is used as the hydrogen nucleus is composed of only a single proton, thus resulting in better spectra. However, apart from H NMR, phosphorous NMR is also used for the analysis of different forensic evidence. NMR in forensic science is typically used to analyze illicit drugs, explosives, poisons and toxins, and body fluids such as blood, semen, and saliva. It has also been used for the analysis of postmortem changes. NMR is a highly sensitive tool that is capable of not only elucidating the structural details of a compound, but can also easily identify any impurities, thus determining quality. These features make NMR a highly critical tool for the analysis of forensic evidence (Elipe and Victoria, 2012a,b).

2.1.4 Raman spectroscopy

Raman spectroscopy is yet another vibrational spectroscopy technique, and it is mainly used for the structural analysis of a compound. In this technique, Raman scattering serves as the main basis of analysis. Typically in IR spectroscopy, the sample is irradiated with infrared radiation. When the energy of the molecule matches that of the radiation source, absorption occurs, followed by the vibration of the molecular bonds. However, in Raman scattering, the radiation source is typically a laser that has a single wavelength and frequency. Once the radiation hits the sample, the electron cloud around the nuclei inelastically scatters the radiation, due to which the energy of the scattered photon is different from that of the incident photon. This difference corresponds to one vibrational unit; this type of scattering is known as Raman scattering. Raman scattering has two types of shifts: the stokes and the antistokes shift. The stokes shift occurs when the electron transfers from a lower energy level to a higher energy level while the antistokes shift occurs when the electron moves from a higher energy level to a lower energy level. Due to these phenomena, there is a difference in the frequency and wavelength of the molecule that is characteristic of its structure. In forensic science, Raman spectroscopy is used for its ease of use and nondestructive nature. Samples such as blood, saliva, semen, fingerprints, illicit drugs, paints, pigments, explosives, and gunshot residue are widely used. It is also used for the on-spot detection of counterfeit documents (Fikiet et al., 2018; John and George, 2017).

2.1.5 Atomic emission and absorption spectroscopy

Atomic absorption and emission spectroscopy are atomic spectra techniques that consider absorption and emission by atoms when they are subjected to a monochromatic light source. These techniques are excellent to perform the elemental mapping of a sample. In atomic absorption spectroscopy, a monochromatic radiation source emitted by the atomization of free atoms is absorbed by the sample. Because atomic absorption spectroscopy is based on the absorption of the radiation emitted by the source, it follows the Beer Lambert law, which correlates the amount of radiation absorbed with the concentration of the element in the sample. Through this, it is possible to obtain the characteristic intensities of transmitted radiation. Therefore, it can be used for the quantitative determination of the element in the sample. In atomic emission spectroscopy, when the radiation hits the sample, the electrons absorb some of the radiation and transition to a higher energy level. Once excited, they tend to come back to their ground state, which then leads to the emission of radiation of a wavelength. This emission is characteristic of the element present in the sample. Quantitative analysis can also be performed because the amount of emitted radiation is proportional to its concentration. Because these techniques are primarily used for the quantitative estimation and mapping of elements in the sample, it is used in forensic science mainly for the analysis of explosives, gunshot residue, and other ammunition. Even samples such as fibers and flammable materials can also be analyzed through this technique (Cantle, 1982; Moore, 1989).

2.2 Chromatographic techniques

Chromatography was first developed in 1900 by a Russian scientist, Mikhail Tsvett. It was initially used to separate the plant pigments in a leaf. It is basically a separation technique that employs a stationary phase and a mobile phase. The separation takes places on the basis of the affinity of the analytes toward either the stationary phase or the mobile phase. Typically, the analytes get adsorbed on the stationary phase if they have more affinity toward it, due to which they travel slower. If they have higher affinity toward the mobile phase, they tend to travel faster. Based on this basic principle, the analytes in any compound can be easily separated and identified. Substances that travel slowly have a higher retention time while those that travel faster have a lower retention time. In this section, we discuss the various types of chromatographic techniques such as high-performance liquid chromatography (HPLC), gas chromatography, and high-performance thin layer chromatography (HPTLC) as well as offer a brief overview of hyphenated techniques.

2.2.1 High performance liquid chromatography

HPLC is a type of column chromatography technique in which the stationary phase is in a column and the mobile phase is passed through it under high pressure. The entire process in HPLC is automated and currently several instruments come with autosamplers as well. The stationary phase is of many types in HPLC, including normal phase, reverse phase, ion exchange, and size exclusion. In normal phase HPLC columns, the stationary phase, which is usually silica gel, is more polar than the mobile phase. In reverse phase columns, the stationary phase is nonpolar, and the mobile phase used is polar. Ion exchange columns and size exclusion columns, use acidic/basic columns or porous stationary phases, respectively. Illicit drugs such as opioids and cannabis as well as pesticides, poisons, plant toxins, and alkaloids are analyzed using HPLC. The analysis of each sample demands its own protocol wherein factors such as the type of column to be used, the mobile phase, the pH of the mobile phase, flow rate, retention time, and pressure are considered. The detectors used in HPLC also vary according to the type of sample to be analyzed. Commonly used detectors are UV/visible, fluorescence, refractive index, electrochemical, conductivity detector, evaporative light scattering detector, and chiral detectors (Ahuja, 2005; Daldrup et al., 1986; Ghosh, 1992).

2.2.2 Gas chromatography

Gas chromatography is yet another interesting chromatography technique that is widely used in forensic laboratories. In this technique, the mobile phase is a noble or unreactive gas such as argon or helium. It is passed through the stationary phase, which is present either as a packed column or coated inside a capillary. In this technique, the sample is carried along with the carrier gas or the mobile phase and is passed through the column. There, based on the affinity of the sample toward the stationary or the mobile phase, it will be separated at different retention times. While this is a highly versatile technique, it has certain drawbacks. One is that this technique is unsuitable for thermally labile compounds, as it can get pyrolyzed when the sample is converted to the vapor state prior to analysis. However, with proper sample preparation techniques and the use of protectants to ensure sample integrity, it is now possible to accurately separate the compounds in a sample. Illicit drugs and blood alcohol content are some of the widely analyzed forensic samples. Apart from this, explosives and other flammable substances, accelerants, and even pesticides and poisons can also be detected through this technique (Blumberg, 2012; de Coning and Swinley, 2019).

2.2.3 High performance thin layer chromatography

High-performance thin-layer chromatography is probably one of the most versatile chromatographic techniques used for the analysis of forensic evidence. In this technique, a thin-layer chromatography or TLC plate coated with the adsorbent or the stationary phase is used and incubated in a chamber saturated with the mobile phase. The sample is also carefully loaded on the TLC plate prior to the chromatographic development. The retention time of the samples in this case is calculated according to the distance traveled by the analytes present in the sample. This technique is highly useful in forensic science as it offers photographic proof of the evidence that can be easily presented in court. Also, every stage in this technique is independent of each other, due to which any cumulative errors are minimized. While samples such as illicit drugs and toxins are commonly analyzed using this technique, samples such as food adulterants, ink, fabric counterfeiting, and even explosives can also be detected using this technique. This technique also offers excellent automation, optimization, and detection with minimum sample preparation (Bensakhria, 2017; Srivastava, 2011).

2.2.4 Hyphenated techniques

While chromatographic techniques are useful for the separation of analytes and their detection is found through their retention time, it is sometimes not possible to accurately determine the structure of the compound. Therefore, these techniques are often paired with other techniques that are capable of elucidating the molecular structure. GC, HPLC, and HPTLC are often paired or coupled with techniques such as a mass spectrometer (MS), FTIR, and NMR. GC-MS is a technique in which the sample is separated as usual in the GC. Once they are separated, a specific device is used to couple the two instruments, through which the samples from GC are transported to the MS. The MS then elucidates the structure of the molecule. Similarly, in the case of LC-MS, once separated by the chromatographic technique, the compounds undergo structural analysis by the MS. Similarly, HPTLC can also be paired with FTIR and MS. A crucial aspect of such hyphenated techniques is the link that connects the two. Because both techniques use different sample forms, it is important that appropriate sample preparation techniques are used (Bogusz, 1999; Cimpoiu, 2011; Kitson et al., 1996).

2.3 Microscopic techniques

Microscopic techniques are among the most widely used techniques for the preliminary and confirmatory analysis of forensic samples. While spectroscopic and chromatographic techniques are capable of obtaining accurate structural and chemical information, they do not provide an actual visualization of the compound. Microscopic techniques overcome this limitation and provide hard results that can be easily produced in court. In this section, the different electron, optical, and probe microscopes are discussed.

2.3.1 Optical microscopes

Optical microscopes are probably among the most versatile and ubiquitous tools used for the visualization of forensic evidence. Optical microscopes visualize samples by hitting the samples with a light source, which generates photons. These photons interact with the sample and the reflected and scattered light are collected by a series of lenses. There are mainly two types of optical microscopes, the simple and the compound microscope. The simple microscope consists of a single lens while the compound microscope consists of a series of objective and condenser lenses. Most microscopes used in a forensic laboratory are compound microscopes. A compound microscope consists of a light source, a condenser lens that is responsible for collecting and focusing the light obtained from the light source onto the sample, and an objective lens that focuses and collects the light from the sample. Typically, phase contrast, bright field, fluorescence, polarized, and stereomicroscopes are among the most commonly used microscopes for the analysis of various evidence. While any biological samples, explosives, and ammunition are commonly analyzed using the microscopes, samples such as fibers, glass, paints, and even questioned documents can be easily analyzed through optical microscopes (Haynes, 1984; Johnson, 2013).

2.3.2 Electron microscopes

Electron microscopes such as scanning electron microscopes and transmission electron microscopes are among the most advanced microscopes capable of visualizing a compound with extremely high resolution and magnification. SEM is used to analyze the surface morphology of a material while TEM is used to visualize its internal morphology. SEM is capable of visualizing the surface morphology by analyzing the backscattered and the secondary electrons, which are produced when the beam of the primary electrons hits the same. The backscattered and secondary electrons are generated as a result of the impingement of the incident electrons on the sample. In TEM, the primary electron beam is made to pass through an ultrathin sample and only the transmitted electrons are detected. Based on the obstructions present inside the material, some of the electrons get scattered, due to which those areas appear darker in the image. Due to such a mode of functioning, SEM requires samples that are conductive in nature so that the primary incident electrons can easily interact with the surface. Therefore, prior to usage, the samples are coated with a metal such as gold. In TEM, the samples must be ultrathin in order to ensure that the electrons can pass through easily. In forensic science, SEM has found more usage than TEM due to the nature of the evidence obtained. Biological evidence such as hair, pollen, teeth, skin, or bone cells can be easily imaged through these techniques. The microscopic analysis of such evidence itself is confirmatory in nature. Apart from this, various anthropogenic and geological samples have also been analyzed using this technique (Amelinckx et al., 1997; Fultz and Howe, 2013; Goldstein et al., 2003).

2.3.3 Atomic force microscope

Atomic force microscopy (AFM) is a type of scanning probe microscope used to produce three-dimensional (3D) images of the sample surface. In this technique, a probe or small tip moves carefully along the surface of the sample in order to get a 3D morphology of the surface. The working of this microscope is similar to that of a blind man walking on the road with his stick. The probe is usually attached to a cantilever that captures the movements of the probe. A laser light falls on the surface of the tip in order to monitor its movement. There are basically three types of working modes in AFM: contact mode, noncontact mode, and tapping mode. In contact mode, the tip or the probe is in constant contact with the sample surface and based on the forces detected by the tip on the sample surface, the 3D morphology is obtained. In noncontact mode, the tip is not in contact with the surface of the sample and again moves by considering a minimum threshold of force deflection on the sample surface. In tapping mode, raster scanning occurs while the tip is constantly oscillating to and fro from the sample surface. Depending on the type of sample to be analyzed, the different working modes can be selected. In forensic science, however, the use of this microscope is still not widespread, and it remains in the research stages. However, this is a highly promising technique as it offers a 3D surface analysis of the sample (Pandey et al., 2017b).

2.3.4 *Energy dispersive X-ray coupled microscopy in forensic science*

Energy dispersive X-ray (EDX) is a technique commonly used for the elemental analysis of a material. In this technique, the sample is irradiated with an electron beam, due to which backscattered, auger, and other secondary electrons are generated. These electrons are typical of the elements present in the sample, due to which quantitative and qualitative elemental information can be easily obtained. The EDX is usually coupled with electron microscopes where the electron gun of the microscope itself acts as the source for EDX as well. The EDX detector is usually kept close to the sample in order to efficiently capture the scattered electrons. Typically, an EDX detector consists of an Si-lithium detector or a silicon drift detector. Just like in an electron microscope, the accelerating voltage of the electron beam and the thickness of the sample play crucial roles in determining the elemental information. In forensic science, EDX has been majorly paired with SEM to analyze samples. Typical forensic evidence such as biological samples, explosives, and toxins are routinely analyzed through this technique. Apart from this, paints, glass, and gunshot residue are also analyzed through this technique (Abd Mutalib et al., 2017; Hodoroaba, 2020).

2.4 Emerging analytical techniques

Spectroscopic, microscopic, and chromatographic techniques are routinely used in forensic laboratories across the world. Currently, these techniques have seen an increase in their supplementation with emerging analytical techniques such as mass spectrometry, X-ray diffraction, lab-on-chip devices, and nanotechnology. In this section, a brief overview of such techniques will be discussed along with how they have been used for the analysis of various forensic evidence.

2.4.1 *Mass spectrometry*

Mass spectrometry, as mentioned before, is capable of analyzing organic compounds by vaporizing the samples into a gas phase, then separating the ions according to their mass/charge ratio (m/z). Typically, a mass spectrometer consists of an ion source that is responsible for the ionization of the samples, a mass analyzer where the ions are separated according to the m/z ratios, and a detector that detects the separated ions. This technique is capable of not only obtaining structural information but also performing a quantitative estimation. Both liquid and solid samples can be easily analyzed through this technique. A typical mass spectrum consists of the m/z ratio values on the *x*-axis and their relative intensities on the *y*-axis. The strongest peak corresponds to the parent ion of the sample, which is basically the ionized form of the sample with the molecular weight closest to it upon fragmentation. There are many ionization methods such as electron ionization, chemical ionization, field or matrix desorption, atmospheric pressure chemical ionization, plasma desorption, thermospray ionization, and fast atom bombardment. Mass analyzers such as a quadrupole analyzer, time of flight (TOF), and ion trap are commonly used. Detectors such as photomultiplier tubes, Faraday cups, and scintillators are also used. Typically, mass spectrometers are often coupled with liquid, gas, or thin layer chromatographic techniques in order to get an accurate analysis of the forensic evidence. Here too, this technique is used for the analysis of myriad forensic evidence such as biological samples, illicit drugs, and explosives. Gunshot residue and counterfeit inks can also be analyzed through this technique. This is a highly versatile technique and therefore both solid and liquid samples can be easily analyzed (de Hoffmann and Stroobant, 2007).

2.4.2 X-ray diffraction

X-ray diffraction (XRD) is a technique used to analyze the crystallinity of a material. In X-ray diffraction, when X-rays, preferably of higher energy, hit the atoms that are arranged in a material, they undergo two types of scattering: elastic or inelastic. In elastic scattering, the scattered rays have the same energy as the incident electron while in inelastic scattering, the energy of the scattered rays is not equal to the energy of the incident rays. Due to this, there is a constructive or destructive interference based on the crystal orientation of the sample. The constructive interference follows the Bragg law, which takes into account the angle between the radiation and the crystal planes, the wavelength of the radiation, and the spacing between the crystal planes. Through the Bragg equation, it is possible to obtain information regarding the crystallinity of the sample. Typically, the analysis takes place either by fixing or varying the angle between the incident ray and the crystalline planes. Metals such as copper, molybdenum, or iron are commonly used to generate the X-ray sources by subjecting them to a high-voltage electron beam under vacuum conditions. The samples can either by analyzed in a powdered form or as a whole crystal. Based on the type of sample to be analyzed, the type of X-ray diffraction method can also be selected. Scintillation, solid state, or charge coupled devices are often used as detectors in this technique. XRD has been commonly used to analyze the crystalline structure of drugs, minerals, and other metals. But in forensic science, it has also been used to analyze paints, pigments, bones, and even fibers (Bishnoi et al., 2017; Brügemann and Gerndt, 2004).

2.4.3 Nanotechnology

Nanotechnology is the science of nanomaterials, that is, those materials whose size is in the nanometer range. Due to their small size, they have enhanced optical, magnetic, and electrical properties along with increased reactivity contributed by a significant increase in their surface area. Due to such properties, nanomaterials have found myriad use in various applications. Nanomaterials such as nanoparticles, nanotubes, nanoshells, and nanofibers are often used in various applications. However, such nanomaterials are often functionalized, that is, the surfaces of these nanomaterials are often modified by incorporating a guest molecule or a functional group. Due to functionalization, the intrinsic properties of the nanomaterials get further enhanced. Nanomaterials are commonly synthesized through two routes: the bottom-up or the top-down approach. In the bottom-up approach, a precursor salt solution is taken and in the presence of a catalyst or a reducing agent, nanomaterials are formed. In the top-down approach, the bulk material is ground to nanomaterials through various mechanical and chemical methods. The characterization of these nanomaterials is done through various spectroscopic, electron, and probe microscopic techniques. The electron microscopes are usually paired with EDX to get elemental information. In forensic science, nanomaterials have been widely used for the development of sensors that allow the detection of explosives, gunshot residue, and other toxins and pesticides. Due to their nanosize, they have also been used for the detection of latent fingerprints as they can easily get attached to the ridges of a fingerprint, thus allowing enhanced selective and sensitive detection of the fingerprints with increased contrast (Pandey et al., 2017a; Rawtani et al., 2019; Tharmavaram et al., 2018).

2.4.4 Lab-on-chip devices

Lab-on-chip devices are among the most advanced devices created for the analysis of forensic samples. These devices are miniaturized laboratories and are developed though microfluidic technology. These are basically an amalgamation of fluidics, electronics, and optics. These devices can carry out independent reactions that usually need an entire laboratory to work. Due to this, the amount of sample required is very small (microliters to nanoliters) and the reaction is very rapid, due to which on-spot detection kits can be easily developed. A lab-on-chip device typically consists of an injector or a sample holder, and it is connected to a sample processing chamber where any contaminants are removed through centrifugation or filtration. This is followed by a reagent mixer that serves as the main component for mixing the reagents. Once this is done, it is sent to the reaction chamber where the sensors are attached to monitor the product formation to analyze the reactions. The device also consists of a transducer that is responsible for converting the chemical signals to electrical signals and converting that into a readable output. Lab-on-chip devices are highly useful for analyzing forensic evidence, as they require very small amounts of samples and allow the on-spot detection of evidence with high specificity. While the use of lab-on-chip devices for the analysis of evidence is still mostly in the research stage, it holds immense potential for use in the field. They can be used for the analysis of DNA, illicit drugs, and even explosives. They can also be used for the analysis of various biological evidence such as blood and can easily detect the difference between any red-colored liquid and actual blood. The use of lab-on-chip devices in the field will significantly reduce the time required for a criminal investigation, due to which various proactive forensic measures can also be implemented (Abgrall and Gué, 2007; Gupta et al., 2016).

3 Conclusion

The analysis of various forensic evidence is highly necessary, as this can form the crux of any criminal investigation. Apart from criminal investigation, the analysis of various samples is also required to implement security measures and laws, especially in airports. Over the years, samples such as illicit drugs; explosives; gunshot residue; biological evidence such as blood, semen, saliva, DNA, pollen, and diatoms; and other physical evidence such as paints, pigments, fabrics, and questionable documents have been analyzed using various analytical techniques. In this chapter, the various chemical analytical techniques for the analysis of various forensic samples have been discussed in brief. Spectroscopic techniques such as UV-visible, infrared, Raman, nuclear magnetic resonance and atomic emission and absorption spectroscopies have been used for the preliminary and confirmatory analysis of evidence along with chromatographic techniques such as high-performance liquid chromatography, gas chromatography, and thin-layer chromatography. These techniques are usually paired with a mass spectrometer to obtain information regarding structure. Typically, the analysis of evidence is in tandem with optical and electron microscopic techniques that provide an accurate visualization of the sample. Apart from such conventional techniques, emerging analytical techniques such as the use of nanotechnology and lab-on-chip devices also have immense potential in the analysis of various forensic evidence.

References

Abd Mutalib, M., Rahman, M.A., Othman, M.H.D., Ismail, A.F., Jaafar, J., 2017. Scanning electron microscopy (SEM) and energy-dispersive X-ray (EDX) spectroscopy. In: Membrane Characterization. Elsevier, pp. 161–179, https://doi.org/10.1016/B978-0-444-63776-5.00009-7.

Abgrall, P., Gué, A.-M., 2007. Lab-on-chip technologies: making a microfluidic network and coupling it into a complete microsystem—a review. J. Micromech. Microeng. 17, R15–R49. https://doi.org/10.1088/0960-1317/17/5/R01.

Ahuja, S. (Ed.), 2005. Handbook of Pharmaceutical Analysis by HPLC, Separation Science and Technology. Elsevier Acad. Press, Amsterdam.

Altemose, I., 1986. Evolution of instrumentation for UV-visible spectrophotometry part I. J. Chem. Educ. 63, A216. https://doi.org/10.1021/ed063pA216.

Amelinckx, S., Van Dyck, D., van Landuyt, J., van Tendeloo, G., 1997. Electron Microscopy: Principles and Fundamentals. VCH Verlagsgescllschaft mbH, Weinheim (Federal Republic of Germany).

Bell, S.E.J., Xu, Y., 2018. Infrared spectroscopy: industrial applications. In: Reference Module in Chemistry, Molecular Sciences and Chemical Engineering. Elsevier, https://doi.org/10.1016/B978-0-12-409547-2.14526-3. B9780124095472144000.

Bensakhria, A., 2017. Thin–layer chromatography (TLC) analytical toxicology. In: Analytical Toxicology. https://www.analyticaltoxicology.com/en/thin-layer-chromatography-tlc/. (Accessed 3 March 2020).

Bishnoi, A., Kumar, S., Joshi, N., 2017. Wide-angle X-ray diffraction (WXRD). In: Microscopy Methods in Nanomaterials Characterization. Elsevier, pp. 313–337, https://doi.org/10.1016/B978-0-323-46141-2.00009-2.

Blumberg, L.M., 2012. Theory of gas chromatography. In: Gas Chromatography. Elsevier, pp. 19–78, https://doi.org/10.1016/B978-0-12-385540-4.00002-X.

Bogusz, M.J., 1999. Hyphenated liquid chromatographic techniques in forensic toxicology. J. Chromatogr. B Biomed. Sci. Appl. 733, 65–91. https://doi.org/10.1016/S0378-4347(98)00520-9.

Brügemann, L., Gerndt, E.K.E., 2004. Detectors for X-ray diffraction and scattering: a user's overview. Nucl. Instrum. Methods Phys. Res. A 531, 292–301. https://doi.org/10.1016/j.nima.2004.06.019.

Cantle, J.E., 1982. Atomic Absorption Spectrometry., p. 465.

Cimpoiu, C., 2011. HPTLC hyphenated with FTIR: principles, instrumentation and qualitative analysis and quantitation. In: Srivastava, M. (Ed.), High-Performance Thin-Layer Chromatography (HPTLC). Springer Berlin Heidelberg, Berlin, Heidelberg, pp. 385–394, https://doi.org/10.1007/978-3-642-14025-9_18.

Daldrup, T., Michalke, P., Szathmary, S., 1986. HPLC in forensic chemistry. In: Engelhardt, H. (Ed.), Practice of High Performance Liquid Chromatography. Springer Berlin Heidelberg, Berlin, Heidelberg, pp. 241–285, https://doi.org/10.1007/978-3-642-69225-3_9.

de Coning, P., Swinley, J., 2019. Overview and theory. In: A Practical Guide to Gas Analysis by Gas Chromatography. Elsevier, pp. 1–43, https://doi.org/10.1016/B978-0-12-818888-0.00001-2.

de Hoffmann, E., Stroobant, V., 2007. Mass Spectrometry: Principles and Applications, third ed. J. Wiley, Chichester, West Sussex, England; Hoboken, NJ.

El-Azazy, M., 2019. Introductory chapter: infrared spectroscopy—a synopsis of the fundamentals and applications. In: El-Azazy, M. (Ed.), Infrared Spectroscopy—Principles, Advances, and Applications. IntechOpen, https://doi.org/10.5772/intechopen.82210.

Elipe, S., Victoria, M., 2012a. Basic Concepts of NMR Spectroscopy. John Wiley & Sons Ltd, Canada, pp. 1–36.

Elipe, S., Victoria, M., 2012b. Historical Development of NMR and LC-NMR. John Wiley & Sons Ltd, Canada, pp. 39–52.

Ewing, A.V., Kazarian, S.G., 2017. Infrared spectroscopy and spectroscopic imaging in forensic science. Analyst 142, 257–272. https://doi.org/10.1039/C6AN02244H.

Fikiet, M.A., Khandasammy, S.R., Mistek, E., Ahmed, Y., Halámková, L., Bueno, J., Lednev, I.K., 2018. Surface enhanced Raman spectroscopy: a review of recent applications in forensic science. Spectrochim. Acta A Mol. Biomol. Spectrosc. 197, 255–260. https://doi.org/10.1016/j.saa.2018.02.046.

Fultz, B., Howe, J., 2013. Transmission Electron Microscopy and Diffractometry of Materials 771.

Ghosh, M.K., 1992. HPLC Methods on Drug Analysis. Springer Berlin Heidelberg, Berlin, Heidelberg, https://doi.org/10.1007/978-3-642-76506-3.

Goldstein, J.I., Newbury, D.E., Echlin, P., Joy, D.C., Lyman, C.E., Lifshin, E., Sawyer, L., Michael, J.R., 2003. The SEM and its modes of operation. In: Goldstein, J.I., Newbury, D.E., Echlin, P., Joy, D.C., Lyman, C.E., Lifshin, E., Sawyer, L., Michael, J.-R. (Eds.), Scanning Electron Microscopy and X-Ray Microanalysis, third ed. Springer US, Boston, MA, pp. 21–60, https://doi.org/10.1007/978-1-4615-0215-9_2.

Gupta, S., Ramesh, K., Ahmed, S., Kakkar, V., 2016. Lab-on-chip technology: a review on design trends and future scope in biomedical applications. Int. J. Biosci. Psychiatr. Technol. IJBSPT 8, 311–322. https://doi.org/10.14257/ijbsbt.2016.8.5.28.

Haynes, R., 1984. Optical Microscopy of Materials. Springer US, Boston, MA, https://doi.org/10.1007/978-1-4757-6085-9.

Hodoroaba, V.-D., 2020. Energy-dispersive X-ray spectroscopy (EDS). In: Characterization of Nanoparticles. Elsevier, pp. 397–417, https://doi.org/10.1016/B978-0-12-814182-3.00021-3.

John, N., George, S., 2017. Raman Spectroscopy. Elsevier, https://doi.org/10.1016/B978-0-323-46140-5.00005-4.

Johnson, S.A., 2013. Phase contrast microscopy. In: Biomedical Optical Phase Microscopy and Nanoscopy. Elsevier, pp. 3–18, https://doi.org/10.1016/B978-0-12-415871-9.00001-6.

Kitson, F.G., Larsen, B.S., McEwen, C.N., 1996. Gas Chromatography and Mass Spectrometry—A Practical Guide. Academic Press, USA.

Moore, G.L., 1989. Introduction to Inductively Coupled Plasma Atomic Emission Spectrometry, Analytical Spectroscopy library. Elsevier, Amsterdam; New York.

Pandey, G., Rawtani, D., Agrawal, Y.K., 2017a. Future aspects of halloysite nanotubes in forensic investigations. J. Nanomed. Res. 6.

Pandey, G., Tharmavaram, M., Rawtani, D., Kumar, S., Agrawal, Y., 2017b. Multifarious applications of atomic force microscopy in forensic science investigations. Forensic Sci. Int. 273, 53–63. https://doi.org/10.1016/j.forsciint.2017.01.030.

Räty, J., Peiponen, K.-E., Asakura, T., 2004. UV-Visible Reflection Spectroscopy of Liquids, Springer Series in Optical Sciences. Springer Berlin Heidelberg, Berlin, Heidelberg, https://doi.org/10.1007/978-3-540-45093-1.

Rawtani, D., Tharmavaram, M., Pandey, G., Hussain, C.M., 2019. Functionalized nanomaterial for forensic sample analysis. TrAC Trends Anal. Chem. 120, 115661. https://doi.org/10.1016/j.trac.2019.115661.

Srivastava, M. (Ed.), 2011. High-Performance Thin-Layer Chromatography (HPTLC). Springer Berlin Heidelberg, Berlin, Heidelberg, https://doi.org/10.1007/978-3-642-14025-9.

Tharmavaram, M., Pandey, G., Rawtani, D., 2018. Surface modified halloysite nanotubes: a flexible interface for biological, environmental and catalytic applications. Adv. Colloid Interf. Sci. 261, 82–101. https://doi.org/10.1016/j.cis.2018.09.001.

CHAPTER 2

Forensic sampling and sample preparation techniques

1 Introduction

A forensic sample is evidence collected from the scene of a crime. This evidence provides information about the crime in many ways such as individual testimony, documented proof, or through the submission of material objects in courtroom trials. On a broader scale, evidence can be of two types: real and demonstrative. Real evidence is directly related to and collected from the crime scene, and it can assist in identifying the victim or the suspect. Most forensic samples collected from the crime scene such as biological fluids, explosives, gunshot residue (GSR), drugs, toxins, alcohol, etc., are real evidence. On the other hand, demonstrative evidence is not directly collected from the crime scene. This is developed at later stages of forensic investigation in order to assist and understand the importance of real evidence. Examples of such evidence include a digitally simulated crime scene, figures demonstrating hair characteristics, bloodstain pattern demonstration, etc. Evidence collected from the crime scene can be circumstantial, conclusive, conflicting, corroborating, derivative, exculpatory, foundational, hearsay, incriminating, presumptive, probative, rebuttal, or tainted (Houck and Siegel, 2010). As far as the transfer of evidence is concerned, it can be direct (from person to person or to a location) or indirect (involvement of intermediate persons or objects). In indirect transfer, the chances of contamination of forensic samples are very high. At a crime scene, normally three modes of searches are performed for the identification and collection of forensic samples. These search patterns can be lane or strip, spiral, and grid (Houck and Siegel, 2010). Using these search patterns, different kinds of forensic samples are properly collected, labeled, and forwarded in a suitable manner to the laboratory for further testing.

In this chapter, different types of forensic samples belonging to three different categories—physical, chemical, and biological—are initially discussed. The chapter also discusses the collection, preservation, and handling

Handbook of Analytical Techniques for Forensic Samples
https://doi.org/10.1016/B978-0-12-822300-0.00002-1

of such forensic samples. Further, the preparation techniques involved for the examination of forensic samples through various analytical instruments are also elaborated.

2 Types of forensic samples

Broadly, the forensic samples collected from a crime scene can be classified into three categories: physical, chemical, and biological (Robertson, 2016). This section discusses the types of forensic samples falling into these categories as well as their collection, preservation, and handling.

2.1 Physical forensic samples

2.1.1 Classification

Physical forensic samples can be classified mainly into five categories: questioned documents, digital samples, impression marks, ballistics samples, and trace samples. The latter three categories include a large number of forensic samples. Impression marks collected at the crime scene include lip prints, fingerprints, footprints, and tire and tool marks. Forensic samples related to ballistics include firearms, GSR, bullets, and cartridge cases. Trace forensic samples mainly include soil, paint, fiber, and glass. These are the different types of physical forensic samples that are encountered during crime scene investigations (Collett, 2005).

2.1.2 Collection, preservation, and handling

The collection, preservation, and handling of different physical evidence vary. As far as documents are concerned, they can be found either intact or in a torn or charred state. Torn documents should be collected by switching off the fan and picking them up, preferably with forceps. They should be properly arranged on a glass or plastic sheet during reconstruction. In the case of charred documents, if they are present in a container, then the entire container should be collected. If they are present in a heap with unburned papers inside, then the entire heap should be collected and sent to the laboratory. All the collected documents whether charred, torn, or intact must be preserved after photography in a cardboard or plastic box during transportation to the laboratory.

In the case of digital evidence, if evidence is present in small devices such as a cell phone or tablet, then it can be directly collected using gloves. If, however, it is present in devices that cannot be moved such as a desktop computer, then the evidence must be collected and preserved on a

CD-R or DVD. In the case of switched-off systems such as a laptop or desktop computer, the hard drive must be collected and signed for by the user and officer in charge. Also, nearby places must be searched for any paper or diary that may have the password. In the case of switched-on systems, data must be collected through digital forensic tools. If no expert is available, then the power supply should be removed from the system and the hard disk should be recovered. All the digital samples must be photographed and sent to the laboratory in antistatic packing material such as a paper bag, cardboard box, or envelope (Casey, 2004).

Impression marks include fingerprints, lip prints, footprints/shoeprints, and tire and tool marks. Latent fingerprints after development with fingerprint powder are lifted by adhesive tape and preserved on glass slides for transportation. In the case of patent or plastic fingerprints, the object is photographed and taken directly for laboratory analysis. Lip prints are also collected, preserved, and transported in a similar fashion. Photography, lifting, tracing, and casting are the methods for the collection of footprints/shoeprints or tire marks. Casting can be done by resin, plaster of Paris, wax, etc. Adhesive tape and gelatin lifters are used to lift impression marks from different surfaces. Tool marks are collected either by collecting the entire object, or by casting using plastic or rubber (softened) on the surface. The photographs in this case must be taken perpendicular to the tool marks (Robertson, 2016).

Firearms, cartridges, and GSR are the ballistic forensic samples found at the crime scene. If a firearm is recovered, then it should be safely collected after unloading and a record of the live and fired cartridges should be kept. It should be packed in a cellophane packet if any foreign material such as blood, hair, or tissue is present on the firearm. The firearm should be transported in a wooden box with sufficient padding. In the case of ammunition, parts such as live cartridges, discharged bullets, bullet from the cadaver, etc., should be safely recovered and packed separately. GSR is collected using a cast of molten wax on the suspect's hand. If GSR is on clothes, then the entire material should be packed in cellophane packets and sent to the laboratory for further testing (Saferstein, 2013).

Glass, fibers, paints, and soils are the trace forensic samples found. All glass particles present at the crime scene must be collected separately and kept in cotton padding. If glass is present in an object which can be transported such as a shoe or a window, then the entire object should be transported for analysis. In the case of fibers or threads, they should be collected using tweezers and placed in envelopes or paper, then sealed in another envelope. During paint collection in hit-and-run cases, the sample should be collected from the freshly dented regions of the vehicle. In the case of burglary, the area where the tool was hit should be used for paint collection and the tool should also be sealed with paper and tape for transportation. Soil samples can be collected by vacuuming and packing in plastic containers for transport to the laboratory (Saferstein, 2013).

2.2 Chemical forensic samples

2.2.1 Classification

Chemical forensic samples can be classified as drugs, toxins, fire and arson samples, explosives, and petroleum products. Drugs include samples such as normal ones used at high doses, injections, and narcotic and psychotropic drugs. Toxins are mainly pesticides, alcohols, or poisons. Explosive samples can be intact, residue, or a sample of soil with an explosive. In the case of petroleum products, further classification includes products such as kerosene, diesel, and biofuel. These different kinds of chemical evidence are collected and preserved in different manners (Houck and Siegel, 2010).

2.2.2 Collection, preservation, and handling

Drugs (solids and powders) are generally collected in inert dry containers and transported in customized tablet boxes. Preservatives are not added; however, refrigeration is maintained if required. Plastic vials or pill boxes are used for collecting counterfeit tablets. Abused drugs and psychoactive substances are seized in heat-sealed bags of plastic for solids and plastic or glass containers for liquid samples, which should be wrapped in clean cloth. All the containers need to be properly labeled before transport to the laboratory (Houck and Siegel, 2010).

In the case of alcohol poisoning, body fluids such as blood and urine need to be collected. Blood should be collected from the subclavian or femoral vessel in an anticoagulant coated vial, and transported in ice boxes. In the case of urine, there is no requirement of a preservative. In cases of fatal poisoning, visceral organs are collected with different preservatives. A rectified spirit in acid poisoning and a common salt or saturated solution of it in cases of alcohol, carbon monoxide, cyanide, or snake venom poisoning are used as preservatives for the transportation of viscera. Common salt with potassium oxalate is used for preserving blood in poisoning cases.

In cases of fire and arson, samples such as cigarettes, matchsticks, and containers of flammable liquids are of forensic importance. These should be carefully collected. Flammable liquids can be collected using a portable vapor detector. Soil samples with traces of flammable liquids should also be collected. The samples can be collected in metallic airtight containers. Glass containers with metal lids or caps can also be used. Burnt materials such as paper, ash, soot, wood, etc., should be packed separately in fresh glass containers in order to avoid sample loss.

Explosive samples need to be collected in the presence of bomb disposal squads. Unexploded materials should be collected by wearing safety gear, and only after a proper examination by the squad. For the chemical analysis of explosive residue, swab samples are collected from the objects, people, and dead bodies that came into contact with the explosive after the blast. Vacuum pumps, vapor detectors, and metal detectors can also be used for collecting explosive residue. Paper or polyethylene bags are used for packing samples such as soil, clothing, unused explosives, and other unnecessary objects. Glass containers for liquid samples and airtight glass containers for gas samples are used at the explosion site (Saferstein, 2013).

Petroleum products in cases of adulteration are collected from all gasoline and diesel tanks in airtight aluminum or plastic containers. In cases of counterfeiting, fake and original products must be packed separately in suitable containers. While collecting petroleum products, the container should not be filled to the brim. Some space should be left at the top for volume variations due to expansion. Containers with volatile products should be transported upside down so that the loss of low boiling point parts can be avoided (Collett, 2005).

2.3 Biological forensic samples

2.3.1 Classification

Broadly, human or animal biological forensic samples can be classified into body fluids and nonbody fluids. Body fluids include blood, semen, saliva, vomit, milk, vaginal secretion, sweat, tears, and urine. Nonbody fluids mainly include teeth, nails, feces, bone, hair, and tissue. Apart from them, there can be other biological samples as well of botanical, entomological, or microbial origin. All those collecting these samples must wear proper masks, gloves, and hair nets in order to prevent cross-contamination. For the packaging and transportation of biological samples, plastic bags and containers are avoided because they may cause microbial contamination (Houck and Siegel, 2010).

2.3.2 Collection, preservation, and handling

Body fluids are collected with or without the use of preservatives. Blood is one of the most common body fluids found at a crime scene. It can be wet or dry, or can be in the form of stains on some objects. Wet blood is soaked in cotton fabric, air dried, and packed along with a reference cotton fabric as control. Dry blood, if present on small objects, is collected along with the object, or with a swab if present on immovable objects. Swabs are then transported in paper envelopes or bags. If objects can be cut, then bloodstained parts are carefully cut and transported to the laboratory. Liquid semen samples are collected using a swab, which is dried and packed in an envelope for transport. Semen-stained objects such as clothes should be transported as such in clean paper bags after moisture removal. Condom samples with semen are transported in refrigerated condition inside the containers. Swabbing is also used for the collection of vaginal and nasal secretions as well as saliva. The swabs are air dried before transportation. Urine samples are collected in liquid form in glass vials. Vomit, if dried, is collected using scraping with a gauze piece. If present in liquid form, it is collected in a glass vial and transported in refrigerated condition (Saferstein, 2013).

In the case of nonbody fluids, collection and preservation are different. Tissues, bones, and teeth are collected for the identification of the cadaver. If tissues are present, then skeletal muscle is collected in a glass or polypropylene container. The addition of formalin is avoided. If tissues are not present, then bone is collected (preferably longer bones if available). Teeth are also collected in cases of mutilated or burned bodies. All these samples are transported in an ice box or dry ice to the laboratory.

Botanical forensic samples are collected and preserved by pressing and drying in natural conditions, thereby maintaining the physical features of the plant. In the case of fresh plant samples, plastic containers and freezing should be avoided. Entomological samples such as insects and their different developmental stages present in cadavers should be collected in glass containers. Microbial samples such as fungi should be collected and dried for preservation; this maintains the morphological features. The dried fungal samples must be transported in paper bags or pasteboard boxes to the laboratory for further analyses (Houck and Siegel, 2010; Saferstein, 2013).

3 Preparation techniques for forensic samples for instrumental analyses

3.1 Physical forensic samples

As discussed in the previous section, a number of physical forensic samples are encountered at the crime scene. These samples are collected and sent to the laboratory for the analysis of their structural and chemical profiling.

Many techniques are available for the analysis of physical forensic samples. Broadly, these techniques can be classified into spectroscopic, chromatographic, and microscopic techniques. Apart from these, other techniques that do not fall into these categories are also used. This section discusses the preparation involved for the analysis of physical forensic samples through these instrumental techniques.

3.1.1 Spectroscopic analyses

Spectroscopic techniques mainly include UV-visible, fluorescence, infrared, atomic, Raman, and nuclear magnetic resonance (NMR) spectroscopy. For physical forensic sample analysis, all these techniques except NMR are used during investigations.

UV-visible and fluorescence spectroscopy can be used to analyze physical samples such as fibers, fingerprints, paints, and questionable documents. Fibers are treated with different solvents (mostly organic) for the extraction of dyes, which can be analyzed using UV-visible and fluorescence spectroscopy (Campiglia et al., 2016). Fingerprints do not need any kind of treatment. They are directly developed using fluorescent powders (Risoluti et al., 2019). Paint chips are treated with alcohol to remove resins and the top layer before UV-visible analysis (Trzcińska et al., 2009). Questionable documents are initially treated with mixtures of aqueous and organic solvents to extract inks, which can be analyzed through UV-visible spectroscopy (Saviello et al., 2018).

IR spectroscopy can be used for the chemical profiling of physical evidence such as currency notes, GSR, questionable documents, and paints. Currency notes require no treatment for IR analysis. GSR and questioned document analysis in attenuated total reflectance (ATR) mode also need no specific preanalysis treatment (Bueno et al., 2013; Xia et al., 2019). Paint analysis, however, requires KBr for preparing pellets of paint samples via the pressed pellet technique, which can be analyzed in IR (Sciutto et al., 2014).

Atomic spectroscopy can be used for the elemental analysis of physical samples such as GSR, fibers, glass, papers, and soils. Samples with GSR as well as fibers and papers are initially treated with strong acids before atomic spectroscopic analysis (Sungur and Gülmez, 2015). Glass and soils are finely crushed before treating with strong acids. Such treated samples can be used for atomic spectroscopy (Hickman et al., 1983).

Raman spectroscopy also assists in the chemical profiling of physical evidence such as paints, GSR, fibers, questionable documents, and soils. Paint chips can be attached to glass slides using double-sided adhesive tape (De Gelder et al., 2005) while GSR on cloth or paper tissues can be directly analyzed (Karahacane et al., 2019). Fibers and documents do not require any pretreatment, but soils need to be crushed and sieved before Raman analysis (Xing et al., 2016).

3.1.2 Chromatographic analyses

High-performance liquid chromatography (HPLC), gas chromatography (GC), high-performance thin-layer chromatography (HPTLC), and other techniques hyphenated with these instruments come under chromatographic techniques. However, in the case of physical samples, only GC and related hyphenated techniques have been used by researchers.

GC has been used for the analysis of physical forensic samples such as fingerprints, paints, and documents. Fingerprints present on the surfaces need to be extracted using organic solvents before going for GC analysis (Weyermann et al., 2011). Paint samples need to undergo pyrolysis at a very high temperature before being subjected to GC analysis (Milczarek and Zięba-Palus, 2009). Inks are extracted from documents using suitable solvents, which are then subjected to GC (Koenig et al., 2015).

MS hyphenated with GC has assisted in the analysis of physical forensic samples such as fibers. Just like the paint samples, fibers also undergo pyrolysis before being subjected to GC-MS analysis (Causin et al., 2006).

3.1.3 Microscopic analyses

Optical, electron, and atomic force microscopy are leading in the forensic analysis of physical samples. Energy dispersive X-ray (EDX) spectroscopy can also be coupled with these microscopy techniques for the elemental analysis of physical forensic samples.

Optical microscopy has contributed in studying the structural details of fibers, glass, paints, documents, and soil. Fibers are mounted on slides using resins for optical microscopy (De Wael and Vanden Driessche, 2011). Glass and documents can be directly visualized while paint flakes need to be peeled off to remove each layer of paint for the analysis (Wright et al., 2011). Soil samples need to be sieved before microscopic analysis can be done.

Scanning electron microscopy (SEM) has played an important role in understanding the morphology of physical evidence such as questionable documents, GSR, tool marks, fibers, and soil. The samples are coated with inert metals such as gold or palladium through sputtering, and mounted on aluminum stubs for SEM analysis (Basu, 2006). For EDX analysis, no specific treatment is required. Samples present on the sample holder can be directly used for EDX analysis as well.

Atomic force microscopy (AFM) has played a role in understanding the surface topography of physical samples such as plastic wrapping material, fibers, GSR, fingerprints, and documents. These samples are directly mounted on the sample holder using adhesive tape, followed by their structural analysis using AFM (Pandey et al., 2017).

3.1.4 Other analyses

Mass spectrometry (MS) and X-ray diffraction (XRD) are other analytical techniques used for examining physical forensic samples. Recent advancements such as nanotechnology also aid in the physical sample investigations.

MS has helped in the analysis of physical evidence such as GSR, fibers, and documents. GSR does not require any specific preparation while fibers need to undergo an extraction process for recovering dyes, which are analyzed by MS. Similarly, documents are treated to recover inks for MS analysis (Ifa et al., 2009).

XRD has assisted in understanding the crystalline nature of physical samples such as paint, paper, fibers, GSR, and soil. Fibers, fabrics containing GSR, and papers are directly mounted on the sample holder using adhesive tape for XRD analysis. Paint and soil samples are ground into fine powders and subjected to XRD analysis (Kugler, 2003).

Nanotechnology has aided in the analysis of physical forensic samples such as currency notes (counterfeited) and fingerprints. There is no requirement of sample preparation in this type of analysis (Rawtani et al., 2019).

3.2 Chemical forensic samples

Large chunks of evidence recovered from the crime scene include chemical samples of forensic importance. Many kinds of instrumental techniques are employed for the analysis of various properties of such samples such as absorbance, molecular weight, crystallinity, molecular vibrations, structure, etc. These details help in the identification of the recovered chemical samples. For the analysis of different properties, different techniques are available. These techniques require the preparation of samples before they can be analyzed in the instrument. This section therefore sheds light on the preparation of chemical forensic samples for studying their properties using different instruments.

3.2.1 Spectroscopic analyses

Spectroscopic techniques used for the analysis of physical forensic samples are also used for the analysis of chemical forensic samples.

Drugs, counterfeit beverages, and petroleum products are analyzed using UV-visible and fluorescence spectroscopy. Beverages do not need any kind of treatment for UV-visible analysis. Drugs, however, need to be solubilized in a suitable polar or nonpolar solvent, depending on the drug's solubility for such analysis (Banister et al., 2015). Petroleum products are treated for the removal of oxygen of any kind of fluorescent impurity before fluorescence spectroscopy (Lloyd, 1980).

IR spectroscopy can be used for the analysis of chemical samples such as counterfeit and illicit drugs as well as explosives. For Fourier transform IR analysis, drug and explosive samples need to be prepared with KBr using the pressed pellet technique; if the analysis is done using the ATR mode, no pretreatment is required.

The elemental analysis of chemical samples such as drugs, explosives, and ignitable liquids can be performed with atomic spectroscopy. Drugs and explosive powders are initially treated with peroxide and strong acids for digestion, followed by washing with deionized water (French et al., 2013). For ignitable liquids such as petrol and diesel, no pretreatment is needed for atomic spectroscopy (Chan, 1981).

Raman spectroscopy can also be done for chemical samples such as drugs and explosives. Drugs need to be solubilized, preferably in an aqueous solvent, for Raman analysis (D'Elia et al., 2018) while explosives do not require any kind of pretreatment for such analysis (Elbasuney and El-Sherif, 2017).

NMR spectroscopy also assists in the analysis of chemical samples such as counterfeit beverages, drugs, explosives, and toxins. Alcoholic solutions of beverages can be prepared with a mixture of buffers and deuterated water for NMR analysis (Kuballa et al., 2018). Drug and toxin solutions are generally prepared in deuterated water with mild acid while explosives are prepared in dimethyl sulfoxide (DMSO) (Paul et al., 2019).

3.2.2 Chromatographic analyses

Most of the chromatographic techniques, including those used for the analysis of physical forensic samples, have been used for the examination of chemical forensic samples.

HPLC can be used for the analysis of chemical forensic samples such as drugs and pesticides. The drug is dissolved in a suitable solvent, depending upon its solubility and the operation mode of HPLC (Freiermuth and

Plasse, 1997). Pesticides, if present, need to be prepared in solution form, and if present in other samples need to be extracted, then used for HPLC analysis (Nasiri et al., 2020).

GC has assisted in the analysis of chemical forensic samples such as alcohols, drugs, explosives, and toxins. Alcohol can be directly analyzed while drugs and toxins need to be extracted from the sample using organic solvents and then used for GC (Papoutsis et al., 2010; Valente et al., 2015). Explosive samples need to be prepared in organic solvents for GC analysis (Calderara et al., 2003).

HPTLC has contributed in the analysis of drugs, explosives, textile dyes, inks, and toxins. Drugs, toxins, and explosives need to be prepared in solution form using a polar or nonpolar solvent, or extracted from the objects in which they are present before being spotted on HPTLC plates (Kumar Kuila and Chandra Lahiri, 2012; Turkmen et al., 2013). Dyes and inks can be either directly spotted or diluted in solvent for spotting on the HPTLC plate (Sherma, 2016).

Chromatographic techniques hyphenated mainly with MS have played a role in the separation and analysis of drugs, explosives, inks, and toxins. The sample preparation for these samples is similar for hyphenated techniques to that used for normal chromatographic techniques.

3.2.3 Microscopic analyses

Chemical forensic samples are not generally analyzed using microscopic techniques. There are very few instances in which microscopic assistance was needed for such samples. However, EDX attached to microscopes plays a role in the analysis of the elemental composition of chemical samples.

Optical microscopy has been used by forensic scientists for the analysis of explosives present in skin samples of victims of bomb blasts. Tissues are fixed in wax, followed by their sectioning and mounting on glass slides for visualization under the microscope (Turillazzi et al., 2010).

Explosives have also been analyzed using AFM. A powder sample of the explosive is mounted on glass slides using double-sided adhesive tape, and then AFM analysis is carried out (Pandey et al., 2017).

3.2.4 Other analyses

MS, XRD, nanotechnology, and lab-on-chip devices have also contributed their part in the analysis of chemical forensic samples.

MS hyphenated with other instruments has analyzed chemical samples such as drugs, explosives, and petroleum products. Solutions of drugs and explosives need to be prepared in suitable solvents before being subjected to MS analysis. Petroleum products can be directly injected into the MS instrument if present in pure form (Ifa et al., 2009).

XRD has contributed to the analysis of the crystallinity of drugs and explosives. These samples, if not present in powdered form, are finely ground and filled in the sample holder for XRD analysis (Schachel et al., 2020).

Nanotechnology-based sensors are able to detect traces of illicit drugs and explosives. These samples, if present in solution form, greatly enhance the sensitivity of detection using nanosensors (Rawtani et al., 2019).

Lab-on-chip devices have also been used for the analysis of drugs and explosives. These samples need to be diluted in different solvents before loading on the device. The solution needs to be filtered properly in order to remove any large impurities that may block channels in these miniaturized devices (Abgrall and Gué, 2007).

3.3 Biological forensic samples

Biological forensic samples are very commonly encountered at the crime scene because of the involvement of humans or animals, either as the victim or the suspect. Similar to the physical and chemical forensic samples, biological samples can also be analyzed using spectroscopic, chromatographic, microscopic, and other analytical techniques. The sample preparation is different for different techniques. This section therefore elaborates the preparation of biological samples for analysis using different instrumental techniques.

3.3.1 Spectroscopic analyses

Spectroscopic techniques for the analysis of physical and chemical forensic samples are also useful for biological forensic sample analysis.

Different types of body fluids are analyzed using UV-visible and fluorescence spectroscopy. These fluids such as saliva, semen, urine, blood, etc., can be used directly or after centrifugation and filtration for UV-visible and fluorescence analysis (Zapata et al., 2015).

IR spectroscopy can be used for the analysis of biological samples such as body fluids, hair, and entomological evidence. Body fluids need no specific treatment upon analysis using the ATR mode (Elkins, 2011). Similarly, hair samples can also be placed directly onto the ATR assembly for IR analysis (Boll et al., 2017). However, because entomological evidence is collected sometimes in alcohol, they need to be dried to remove the alcohol before the IR analysis (Barbosa et al., 2018).

Atomic spectroscopy assists in the elemental analysis of biological samples such as tissues, blood, teeth, nails, and bones. In the case of tissue and blood, the samples are initially soaked in acid, preferably nitric acid at high pressure and temperature, before the analysis (Acar et al., 2004). Teeth, nails, and bones are ground properly and mixed with nitric acid before their atomic spectroscopy (Kłys et al., 1999).

Body fluids and human remains are also analyzed via Raman spectroscopy. If body fluids are present on objects such as fabric, they can be analyzed directly; if present as a whole, a smear of fluid on a glass slide is prepared for Raman analysis (Fikiet and Lednev, 2019; Hager et al., 2018). For tissues or human remains, thin sections are prepared and kept on glass slides for the analysis (Schotsmans et al., 2020).

NMR spectroscopy has also contributed to the analysis of body fluids as well as tissues. Tissue samples are initially frozen and powdered, followed by treatment with acid to solubilize the hydrophobic components. Then, their extracts are prepared in deuterated water with trimethylsilyl propionic acid (TSP) (Hirakawa et al., 2009). Body fluids are centrifuged followed by freezing, and then reconstituted in deuterated water with TSP for NMR analysis (Scano et al., 2013).

3.3.2 Chromatographic analyses

The separation of components of biological forensic samples such as food items and plant alkaloids can be performed using chromatographic techniques such as HPLC and related hyphenated techniques.

HPLC has been used for the analysis of biological forensic samples such as plant alkaloids. These need to be extracted in organic solvents at basic pH. The sonication process also assists in this extraction of alkaloids, and the extracted sample is then used for HPLC (de Oliveira, 2014).

The hyphenation of MS with chromatographic techniques has assisted in the forensic analysis of food products. Food items are subjected to extraction procedures for extracting toxins or other metabolites of forensic importance, followed by the analysis of the extract through a hyphenated technique (Sanchis et al., 2019).

3.3.3 Microscopic analyses

Microscopic techniques used for analyzing physical forensic samples have also contributed in studying the morphological details of biological forensic samples. The EDX associated with microscopes has also assisted in knowing the elements of forensic importance present in the sample.

Optical microscopy has aided in the investigation of biological samples such as bone, teeth, hair, skin, and entomological evidence. Thin sections of these samples are cut, generally using microtome, followed by mounting on a glass slide. Then, they are stained with dyes in order to visualize the structural details under a microscope (Wilson and Wheeler, 2009).

SEM has also assisted in the analysis of biological samples such as entomological evidence, diatoms, tissues, teeth, hair, and bone. Before performing the SEM analysis, all these samples must be thoroughly dried and washed to remove moisture and impurities, and then fixed on aluminum stubs for viewing under the electron microscope (Basu, 2006). The same samples attached on sample holders can be used for elemental composition analysis via EDX coupled with SEM.

AFM has helped forensic experts in understanding the surface characteristics of biological specimens such as blood and hair. Blood can be smeared on a glass slide while hair has to be mounted on a sample holder using double-sided adhesive tape for AFM analysis (Pandey et al., 2017).

3.3.4 Other analyses

Techniques such as MS, XRD, lab-on-chip devices, and nanotechnology have assisted in the investigation of biological forensic samples as well.

MS in combination with other techniques is used for the analysis of foreign contents in body fluids such as urine and serum as well as in hair and plant extracts. Ultrasonication, centrifugation, and filtration steps are used to extract foreign materials from biological samples, which are then subjected to MS analysis (Ifa et al., 2009).

XRD has been used for the analysis of biological forensic samples such as burned cremains and human remains. The samples are subjected to heat treatments in a muffle furnace at very high temperatures before being used for XRD analysis (Piga et al., 2016).

DNA samples have been very efficiently and specifically detected using nanotechnology-based sensors. These sensors do not require any specific sample preparation if the DNA is in solution form. However, if the DNA is present inside the cells, then it needs to be extracted to be detected (Rawtani et al., 2019).

Lab-on-chip devices have also shown their potential in the analysis of DNA and serological samples. However, these samples need to be treated to make them free from any large impurities because these impurities may block the ultrasmall channels present in such devices, thereby hindering the process (Abgrall and Gué, 2007).

4 Conclusion

Evidence recovered at the crime scene can be broadly categorized into three classes: physical, chemical, and biological. Depending on the nature of the evidence, the collection, preservation, and handling of such forensic samples differ. This needs to be kept in mind while collecting the evidence because improper sampling and transportation may contaminate or destroy important information that could be obtained through analysis in the laboratory. Spectroscopic, chromatographic, and microscopic analytical techniques along with other techniques such as MS and XRD have been very well established in forensic investigations of different evidence. These techniques require some kind of sample preparation before the analysis, which was discussed in this chapter. Recent advancements such as nanotechnology and lab-on-chip devices are relatively new and require more extensive research to gain the trust of forensic investigators and come into the mainstream, thereby serving as "silver bullets" for forensic investigations.

References

Abgrall, P., Gué, A.-M., 2007. Lab-on-chip technologies: making a microfluidic network and coupling it into a complete microsystem—a review. J. Micromech. Microeng. 17, R15–R49. https://doi.org/10.1088/0960-1317/17/5/R01.

Acar, K., Boz, B., Kurtulus, A., Divrikli, U., Elci, L., 2004. Using of atomic absorption spectrometry for diagnosis of electrical injuries (an experimental rat study). Forensic Sci. Int. 146, S3–S4. https://doi.org/10.1016/j.forsciint.2004.09.007.

Banister, S.D., Stuart, J., Conroy, T., Longworth, M., Manohar, M., Beinat, C., Wilkinson, S.M., Kevin, R.C., Hibbs, D.E., Glass, M., Connor, M., McGregor, I.S., Kassiou, M., 2015. Structure–activity relationships of synthetic cannabinoid designer drug RCS-4 and its regioisomers and C4 homologues. Forensic Toxicol. 33, 355–366. https://doi.org/10.1007/s11419-015-0282-9.

Barbosa, T.M., de Lima, L.A.S., dos Santos, M.C.D., Vasconcelos, S.D., Gama, R.A., Lima, K.M.G., 2018. A novel use of infra-red spectroscopy (NIRS and ATR-FTIR) coupled with variable selection algorithms for the identification of insect species (Diptera: Sarcophagidae) of medico-legal relevance. Acta Trop. 185, 1–12. https://doi.org/10.1016/j.actatropica.2018.04.025.

Basu, S., 2006. Scanning electron microscopy in forensic science. In: Encyclopedia of Analytical Chemistry. American Cancer Society. https://doi.org/10.1002/9780470027318.a1122.

Boll, M.S., Doty, K.C., Wickenheiser, R., Lednev, I.K., 2017. Differentiation of hair using ATR FT-IR spectroscopy: a statistical classification of dyed and non-dyed hairs. Forensic Chem. 6, 1–9. https://doi.org/10.1016/j.forc.2017.08.001.

Bueno, J., Sikirzhytski, V., Lednev, I.K., 2013. Attenuated total reflectance-FT-IR spectroscopy for gunshot residue analysis: potential for ammunition determination. Anal. Chem. 85, 7287–7294. https://doi.org/10.1021/ac4011843.

Calderara, S., Gardebas, D., Martinez, F., 2003. Solid phase micro extraction coupled with on-column GC/ECD for the post-blast analysis of organic explosives. Forensic Sci. Int. 137, 6–12. https://doi.org/10.1016/S0379-0738(03)00256-1.

Campiglia, A.D., Rex, M., Muñoz de la Peña, A., Goicoechea, H.C., 2016. Excitation–emission matrix fluorescence spectroscopy combined with MCR-ALS as a tool for the forensic analysis of similar and dissimilar sets of textile fiber extracts. Anal. Methods 8, 8314–8321. https://doi.org/10.1039/C6AY02757A.

Casey, E., 2004. Digital Evidence and Computer Crime: Forensic Science, Computers and the Internet, second ed. Academic Press, London; San Diego, CA.

Causin, V., Marega, C., Schiavone, S., Guardia, V.D., Marigo, A., 2006. Forensic analysis of acrylic fibers by pyrolysis–gas chromatography/mass spectrometry. J. Anal. Appl. Pyrolysis 75, 43–48. https://doi.org/10.1016/j.jaap.2005.04.009.

Chan, L., 1981. The determination of tetraalkyl lead compounds in petrol using combined gas chromatography-atomic absorption spectrometry. Forensic Sci. Int. 18, 57–62. https://doi.org/10.1016/0379-0738(81)90139-0.

Collett, D., 2005. Forensic Evidence: Science and the Criminal Law, second ed. CRC Press. https://doi.org/10.1201/9781420038064.

D'Elia, V., Montalvo, G., Ruiz, C.G., Ermolenkov, V.V., Ahmed, Y., Lednev, I.K., 2018. Ultraviolet resonance Raman spectroscopy for the detection of cocaine in oral fluid. Spectrochim. Acta A Mol. Biomol. Spectrosc. 188, 338–340. https://doi.org/10.1016/j.saa.2017.07.010.

De Gelder, J., Vandenabeele, P., Govaert, F., Moens, L., 2005. Forensic analysis of automotive paints by Raman spectroscopy. J. Raman Spectrosc. 36, 1059–1067. https://doi.org/10.1002/jrs.1408.

de Oliveira, B.H., 2014. High pressure liquid chromatography analysis of plant alkaloids. In: Meyers, R.A. (Ed.), Encyclopedia of Analytical Chemistry. John Wiley & Sons, Ltd, Chichester, UK, pp. 1–20. https://doi.org/10.1002/9780470027318.a9921.

De Wael, K., Vanden Driessche, T., 2011. Dichroism measurements in forensic fibre examination part 1: dyed polyester fibres. Sci. Justice 51, 57–67. https://doi.org/10.1016/j.scijus.2010.10.001.

Elbasuney, S., El-Sherif, A.F., 2017. Instant detection and identification of concealed explosive-related compounds: induced stokes Raman versus infrared. Forensic Sci. Int. 270, 83–90. https://doi.org/10.1016/j.forsciint.2016.11.036.

Elkins, K.M., 2011. Rapid presumptive "fingerprinting" of body fluids and materials by ATR FT-IR spectroscopy. J. Forensic Sci. 56, 1580–1587. https://doi.org/10.1111/j.1556-4029.2011.01870.x.

Fikiet, M.A., Lednev, I.K., 2019. Raman spectroscopic method for semen identification: azoospermia. Talanta 194, 385–389. https://doi.org/10.1016/j.talanta.2018.10.034.

Freiermuth, M., Plasse, J.-C., 1997. Determination of morphine and codeine in plasma by HPLC following solid phase extraction. J. Pharm. Biomed. Anal. 15, 759–764. https://doi.org/10.1016/S0731-7085(96)01911-5.

French, H.E., Went, M.J., Gibson, S.J., 2013. Graphite furnace atomic absorption elemental analysis of ecstasy tablets. Forensic Sci. Int. 231, 88–91. https://doi.org/10.1016/j.forsciint.2013.04.021.

Hager, E., Farber, C., Kurouski, D., 2018. Forensic identification of urine on cotton and polyester fabric with a hand-held Raman spectrometer. Forensic Chem. 9, 44–49. https://doi.org/10.1016/j.forc.2018.05.001.

Hickman, D.A., Harbottle, G., Sayre, E.V., 1983. The selection of the best elemental variables for the classification of glass samples. Forensic Sci. Int. 23, 189–212. https://doi.org/10.1016/0379-0738(83)90148-2.

Hirakawa, K., Koike, K., Uekusa, K., Nihira, M., Yuta, K., Ohno, Y., 2009. Experimental estimation of postmortem interval using multivariate analysis of proton NMR metabolomic data. Legal Med. 11, S282–S285. https://doi.org/10.1016/j.legalmed.2009.02.007 Selected papers from the 7th International Symposium ADVANCES IN LEGAL MEDICINE (ISALM).

Houck, M.M., Siegel, J.A., 2010. Fundamentals of Forensic Science, second ed. Academic Press, Burlington, MA.

Ifa, D.R., Jackson, A.U., Paglia, G., Cooks, R.G., 2009. Forensic applications of ambient ionization mass spectrometry. Anal. Bioanal. Chem. 394, 1995–2008. https://doi.org/10.1007/s00216-009-2659-2.

Karahacane, D.S., Dahmani, A., Khimeche, K., 2019. Raman spectroscopy analysis and chemometric study of organic gunshot residues originating from two types of ammunition. Forensic Sci. Int. 301, 129–136. https://doi.org/10.1016/j.forsciint.2019.05.022.

Kłys, M., Lech, T., Zięba-Palus, J., Białka, J., 1999. A chemical and physicochemical study of an Egyptian mummy 'Iset Iri Hetes' from the Ptolemaic period III–I B.C. Forensic Sci. Int. 99, 217–228. https://doi.org/10.1016/S0379-0738(98)00192-3.

Koenig, A., Magnolon, S., Weyermann, C., 2015. A comparative study of ballpoint ink ageing parameters using GC/MS. Forensic Sci. Int. 252, 93–106. https://doi.org/10.1016/j.forsciint.2015.03.027.

Kuballa, T., Hausler, T., Okaru, A.O., Neufeld, M., Abuga, K.O., Kibwage, I.O., Rehm, J., Luy, B., Walch, S.G., Lachenmeier, D.W., 2018. Detection of counterfeit brand spirits using 1H NMR fingerprints in comparison to sensory analysis. Food Chem. 245, 112–118. https://doi.org/10.1016/j.foodchem.2017.10.065.

Kugler, W., 2003. X-ray diffraction analysis in the forensic science: the last resort in many criminal cases. Adv. X-ray Anal. 46, 16.

Kumar Kuila, D., Chandra Lahiri, S., 2012. Determination of the association constants of charge-transfer complexes between N,N-diethylaniline and nitro explosives using high performance thin layer chromatography with scanning video densitometry and spectrophotometry. J. Solut. Chem. 41, 36–52. https://doi.org/10.1007/s10953-011-9779-y.

Lloyd, B.F., 1980. Examination of petroleum products of high relative molecular mass for forensic purposes by synchronous fluorescence spectroscopy. Analyst 105, 97–109.

Milczarek, J.M., Zięba-Palus, J., 2009. Examination of spray paints on plasters by the use of pyrolysis-gas chromatography/mass spectrometry for forensic purposes. J. Anal. Appl. Pyrolysis 86, 252–259. https://doi.org/10.1016/j.jaap.2009.06.009.

Nasiri, M., Ahmadzadeh, H., Amiri, A., 2020. Sample preparation and extraction methods for pesticides in aquatic environments: a review. TrAC Trends Anal. Chem. 123, 115772. https://doi.org/10.1016/j.trac.2019.115772.

Pandey, G., Tharmavaram, M., Rawtani, D., Kumar, S., Agrawal, Y., 2017. Multifarious applications of atomic force microscopy in forensic science investigations. Forensic Sci. Int. 273, 53–63. https://doi.org/10.1016/j.forsciint.2017.01.030.

Papoutsis, I.I., Athanaselis, S.A., Nikolaou, P.D., Pistos, C.M., Spiliopoulou, C.A., Maravelias, C.P., 2010. Development and validation of an EI–GC–MS method for the determination of benzodiazepine drugs and their metabolites in blood: applications in clinical and forensic toxicology. J. Pharm. Biomed. Anal. 52, 609–614. https://doi.org/10.1016/j.jpba.2010.01.027.

Paul, S., Karar, M., Paul, P., Mallick, A., Majumdar, T., 2019. Dual mode nitro explosive detection under crowded condition: conceptual development of a sensing device. J. Photochem. Photobiol. A Chem. 379, 123–129. https://doi.org/10.1016/j.jphotochem.2019.04.038.

Piga, G., Gonçalves, D., Thompson, T.J.U., Brunetti, A., Malgosa, A., Enzo, S., 2016. Understanding the crystallinity indices behavior of burned bones and teeth by ATR-IR and XRD in the presence of bioapatite mixed with other phosphate and carbonate phases. Int. J. Spectrosc. 2016, 1–9. https://doi.org/10.1155/2016/4810149.

Rawtani, D., Tharmavaram, M., Pandey, G., Hussain, C.M., 2019. Functionalized nanomaterial for forensic sample analysis. TrAC Trends Anal. Chem. 120, 115661. https://doi.org/10.1016/j.trac.2019.115661.

Risoluti, R., Filetti, V., Iuliano, G., Niola, L., Schiavone, S., Arcudi, G., Materazzi, S., 2019. Updating procedures in forensic chemistry: one step cyanoacrylate method to develop latent fingermarks and subsequent DNA profiling. Microchem. J. 147, 478–486. https://doi.org/10.1016/j.microc.2019.03.056.

Robertson, B., 2016. Interpreting evidence: evaluating forensic science in the courtroom. second edJohn Wiley and Sons, Inc, Chichester, West Sussex, UK; Hoboken.

Saferstein, R., 2013. Forensic Science: From the Crime Scene to the Crime Lab, 2nd ed. Pearson/Prentice Hall, Boston.

Sanchis, Y., Coscollà, C., Yusà, V., 2019. Comprehensive analysis of photoinitiators and primary aromatic amines in food contact materials using liquid chromatography High-Resolution Mass Spectrometry. Talanta 191, 109–118. https://doi.org/10.1016/j.talanta.2018.08.047.

Saviello, D., Trabace, M., Alyami, A., Mirabile, A., Giorgi, R., Baglioni, P., Iacopino, D., 2018. A combined surface enhanced Raman spectroscopy (SERS)/UV–vis approach for the investigation of dye content in commercial felt tip pens inks. Talanta 181, 448–453. https://doi.org/10.1016/j.talanta.2018.01.049.

Scano, P., Locci, E., Noto, A., Navarra, G., Murgia, F., Lussu, M., Barberini, L., Atzori, L., De Giorgio, F., Rosa, M.F., d'Aloja, E., 2013. ^{1}H NMR metabolite fingerprinting as a new tool for body fluid identification in forensic science. Magn. Reson. Chem. 51, 454–462. https://doi.org/10.1002/mrc.3966.

Schachel, T.D., Stork, A., Schulte-Ladbeck, R., Vielhaber, T., Karst, U., 2020. Identification and differentiation of commercial and military explosives via high performance liquid chromatography—high resolution mass spectrometry (HPLC-HRMS), X-ray diffractometry (XRD) and X-ray fluorescence spectroscopy (XRF): towards a forensic substance database on explosives. Forensic Sci. Int. 308, 110180. https://doi.org/10.1016/j.forsciint.2020.110180.

Schotsmans, E.M.J., Wessling, R., McClue, W.A., Wilson, A.S., Edwards, H.G.M., Denton, J., 2020. Histology and Raman spectroscopy of limed human remains from the Rwandan genocide. J. Forensic Legal Med. 70, 101895. https://doi.org/10.1016/j.jflm.2020.101895.

Sciutto, G., Prati, S., Bonacini, I., Oliveri, P., Mazzeo, R., 2014. FT-NIR microscopy: an advanced spectroscopic approach for the characterisation of paint cross-sections. Microchem. J. 112, 87–96. https://doi.org/10.1016/j.microc.2013.09.021.

Sherma, J., 2016. Advances in the thin-layer chromatographic forensic analysis of inks. J. Liq. Chromatogr. Relat. Technol. 39, 549–557. https://doi.org/10.1080/10826076.2016.1183130.

Sungur, Ş., Gülmez, F., 2015. Determination of metal contents of various fibers used in textile industry by MP-AES. J. Spectrosc. 2015, 1–5. https://doi.org/10.1155/ 2015/640271.

Trzcińska, B., Zięba-Palus, J., Kościelniak, P., 2009. Application of microspectrometry in the visible range to differentiation of car paints for forensic purposes. J. Mol. Struct. 924–926, 393–399. https://doi.org/10.1016/j.molstruc.2008.10.037.

Turillazzi, E., Monaci, F., Neri, M., Pomara, C., Riezzo, I., Baroni, D., Fineschi, V., 2010. Collection of trace evidence of explosive residues from the skin in a death due to a disguised letter bomb. The synergy between confocal laser scanning microscope and inductively coupled plasma atomic emission spectrometer analyses. Forensic Sci. Int. 197, e7–e12. https://doi.org/10.1016/j.forsciint.2009.12.012.

Turkmen, Z., Mercan, S., Cengiz, S., 2013. An HPTLC method for the determination of oleandrin in *Nerium* plant extracts and its application to forensic toxicology. J. Plan. Chromatogr. Mod. TLC 26, 279–283. https://doi.org/10.1556/JPC.26.2013.3.13.

Valente, N.I.P., Tarelho, S., Castro, A.L., Silvestre, A., Teixeira, H.M., 2015. Analysis of organophosphorus pesticides in whole blood by GC-MS-μECD with forensic purposes. J. Forensic Legal Med. 33, 28–34. https://doi.org/10.1016/j.jflm.2015.03.006.

Weyermann, C., Roux, C., Champod, C., 2011. Initial Results on the composition of fingerprints and its evolution as a function of time by GC/MS analysis: initial results on the composition of fingerprints. J. Forensic Sci. 56, 102–108. https://doi.org/10.1111/j.1556-4029.2010.01523.x.

Wilson, L.J., Wheeler, B.P., 2009. Optical microscopy in forensic science. In: Meyers, R.A. (Ed.), Encyclopedia of Analytical Chemistry. John Wiley & Sons, Ltd, Chichester, UK, p. a9101. https://doi.org/10.1002/9780470027318.a9101.

Wright, D.M., Bradley, M.J., Mehltretter, A.H., 2011. Analysis and discrimination of architectural paint samples via a population study. Forensic Sci. Int. 209, 86–95. https://doi.org/10.1016/j.forsciint.2011.01.001.

Xia, J., Zhang, J., Zhao, Y., Huang, Y., Xiong, Y., Min, S., 2019. Fourier transform infrared spectroscopy and chemometrics for the discrimination of paper relic types. Spectrochim. Acta A Mol. Biomol. Spectrosc. 219, 8–14. https://doi.org/10.1016/j.saa.2018. 09.059.

Xing, Z., Du, C., Tian, K., Ma, F., Shen, Y., Zhou, J., 2016. Application of FTIR-PAS and Raman spectroscopies for the determination of organic matter in farmland soils. Talanta 158, 262–269. https://doi.org/10.1016/j.talanta.2016.05.076.

Zapata, F., Fernández de la Ossa, M.Á., García-Ruiz, C., 2015. Emerging spectrometric techniques for the forensic analysis of body fluids. TrAC Trends Anal. Chem. 64, 53–63. https://doi.org/10.1016/j.trac.2014.08.011.

CHAPTER 3

UV-visible and fluorescence spectroscopy for forensic samples

1 Introduction

UV-visible and fluorescence spectroscopy, just like any other spectroscopic techniques, deal with the use of the electromagnetic spectrum for the analysis of different compounds by observing and understanding the interaction of this spectrum with the compounds. These techniques are based on the absorption of radiation in the UV-visible region and the emission of energy from the electrons when they travel from the excited state to the ground state. The electromagnetic spectrum is composed of particles of energy that move with two vector components, namely the electric and magnetic fields that are perpendicular to each other. The particles of energy move in unison most of the time, due to which they follow the rules of classical physics. However, these very particles also behave like discrete packets of energy, due to which they also possess particle-like behavior. Therefore, in brief, the electromagnetic spectrum has both a wave-like and a particle-like nature, which is also known as wave-particle duality. The movement of light occurs with a frequency and a velocity. The frequency is defined as the number of oscillations made by the wave per second while the velocity of the propagation is determined in terms of the units of distance traveled per second. The frequency of the electromagnetic spectrum remains constant in every media; however, the velocity changes with the media. These parameters play a critical role in the interaction of the electromagnetic spectrum with the compounds. During interaction, the particle-like nature comes into play where every photon interacts in a very discrete manner with the electron cloud of the compounds. These mechanics serve as the basis of the UV-visible and fluorescence spectroscopy techniques.

In forensic science, these techniques play a very crucial role as they offer results within a few minutes. These techniques are also used in the preliminary and confirmatory analysis of various forensic evidence such as drugs

Handbook of Analytical Techniques for Forensic Samples
https://doi.org/10.1016/B978-0-12-822300-0.00003-3

and poisons (Pandey et al., 2017; Rawtani et al., 2019). In this chapter, the principle and theory of UV-visible and fluorescence spectroscopy are discussed in brief and the applications of these techniques in the analysis of various forensic evidence is also discussed.

2 Principle and theory of UV-visible and fluorescence spectroscopy

2.1 UV-visible spectroscopy

When a photon hits the electron cloud of a compound, there is an energy transfer from the photon to the electron, which causes the electrons to be promoted from a ground state to an excited state. The energy difference between the two molecular levels is exactly equal to that of the energy of the photon. These are the electronic transitions that play crucial roles in determining the nature of the sample. Conventionally, when the light hits the sample, the processes that take place are transmission, reflection, refraction, luminescence, and scattering. They are dependent on factors such as the solvent purity, polarity, temperature, solute concentration, ionic strength, and pH. When electronic transitions occur at a higher energy, that is, in the ultraviolet region, then the corresponding molecular spectra are observed. Here, the main transitions that occur are the sigma to sigma star (bonding to antibonding) and the pi to pi star (bonding and nonbonding to antibonding). These transitions are exclusively possible in the presence of compounds known as chromophores. Similarly, a liquid sample that is colored shows a molecular spectrum in the visible region (Penner, 2017).

When a parallel beam of radiation with an intensity I is sent through a cuvette with a thickness l containing the sample of concentration c, and travels with a particular wavelength λ, then it is assumed that the absorbance of the radiation by the sample is directly proportional to the intensity of the light that has been transmitted through the cuvette. Typically, according to the Beer Law:

$$A = alc$$

where A stands for the sample's absorbance and a is the absorptivity of the sample.

However, the instrument itself does not directly calculate the absorbance but rather takes into account the light transmitted through the sample. Here, the absorbance is considered as the negative logarithm of the fraction of transmitted light and is also equivalent to the thickness of the absorbing compounds present in the sample. This is known as the Lambert law.

Therefore, this technique uses the Beer-Lambert law, whose expression is similar to that of the Beer law. This law functions on the basis of myriad assumptions such as the radiation source, which is considered to be monochromatic; no energy loss due to phenomena such as scattering or reflection; and a constant temperature. This law also has certain limitations such as at high concentrations of the sample, there are deviations in the molar absorptivity coefficients. Also, there is increased scattering, shifts in the chemical equilibrium, and probable fluorescence and phosphorescence that may arise from the increase in concentration. However, despite the limitations, this law serves as the most crucial technique in sample analysis (Thomas and Burgess, 2007).

UV-visible spectroscopy can also be used in terms of derivative spectroscopy where the first, second, or third order of the rate of change of absorbance with respect to wavelength is used to perform appropriate quantitative and qualitative analysis of compounds. Derivative spectroscopy also allows spectral differentiation and resolution enhancement. A first-order derivative is a technique in which the plot of the rate of change of absorbance with respect to the change in wavelength shows its start and end at zero with equal spread on either side of the graph at the same wavelength. This feature is similar to all odd number derivatives. In the second-order derivative, the minimum and the maximum both are at the same wavelength but the bands are negative. In the fourth order, similar features occur but the band is positive (Gill et al., 1982). Also, only liquid and solution samples can be analyzed using this technique. Even the presence of a suspension can cause excessive scattering of light, leading to skewed results.

The absorption maximum of a compound or λ_{max} is calculated by a set of rules known as the Woodward-Fieser rules. The chromophores present, the auxophores that are basically substituents in the chromophores, and the type of solvent used are considered during the calculation of the absorption maximum. In these rules, a base value for a homoannular diene, that is, a diene with two double bonds in a single ring, and a heteroannular diene, that is, a bond with two double bonds in two rings, is predecided and any extra points are added as the number of attachments to the rings such as acetate ether or any alkyl substituent group increase. The solvents also have a specific UV absorbance cutoff that is also taken into account during the calculation of absorbance (Liljefors and Allinger, 1978).

2.2 Principle of fluorescence spectroscopy

The basis of this technique is the photonic processes that are involved in the electronic transitions from a ground state to the excited state. As mentioned in the previous section, when the energy of the photon matches the energy of each molecular level, then the electrons get excited and move to a higher level. However, it is important to consider that each molecular level or electronic state is further subdivided into a number of subvibrational levels that are separated by a distance of 100 cm^{-1}. Due to this, the actual distance between the ground state of the electron and the excited state of the electron is significantly higher, thus promoting the preference of the electron to rest in its ground state. Similar to a chromophore, a fluorophore is responsible for emitting energies with wavelengths longer than the excitation energy during the relaxation of the excited electron to the ground state. These molecules get excited when the polarization of the incident light is aligned along with the excitation dipole of the molecule, thus resulting in fluorescence.

Once excited, the electrons undergo two main types of decay to return to the ground state: radioactive and nonradioactive decay. In radioactive decay, photon emission occurs while in nonradioactive decay, photon emission does not occur. The fluorescence lifetime of a molecule is therefore measured as the combined rate of the radioactive and nonradioactive emissions (So and Dong, 2001).

3 Instrumentation of UV-visible and fluorescence spectroscopy

UV-visible spectroscopy typically consists of a radiation source, a monochromator, a sample holder, and a detector (Fig. 1). Initially, the radiation

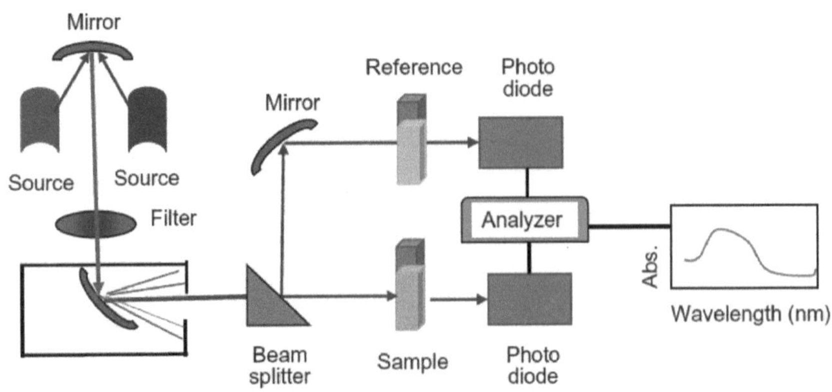

Fig. 1 Instrumentation of UV-visible spectroscopy (House, 2018).

source was a tungsten filament, which slowly evolved to deuterium arc lamps, xenon arc lamps, and light-emitting diodes. The sources emit light of wavelengths covering the UV and visible range. The light emitted from these sources is later passed through a grating slit, which is responsible for focusing the radiation on the sample. These gratings are highly polished surfaces consisting of parallel, equally spaced grooves. Also, filters are placed that eliminate radiation of varying wavelengths and focus only on the UV and visible range. Also, a series of focusing and folding mirrors is placed in order to aid the movement of the radiation on the sample. These gratings, mirrors, and filters together comprise the monochromator and serve to produce a highly paralleled monochromatic light (Räty et al., 2004).

There are usually two sample holders in this instrumentation in order to place a cuvette for the reference and one for the sample. The cuvettes in both cases must be of equal size so that there is minimal instrumental error. The cuvette is also usually made of quartz, plastic, or even glass where its opposite edges are transparent and the remaining edges are roughened.

Detectors in UV-visible spectroscopy are usually phototubes, photomultiplier tubes, photo diode array detectors, and charge coupled device detectors. The phototube detects the transmitted light by using a light-sensitive cathode and an anode. When voltage is applied to the electrodes, the photon in the transmitted light hits the electrodes, due to which current is generated; this is further amplified and processed. In photomultiplier tubes, along with the cathode and anodes, there are several dynodes placed in a series. Just like in a phototube, the impinging of the photons on the electrodes generates a current that is increased due to the multiple striking of the ejected electrons, which further increases the sensitivity of this detector toward the radiation. The diode array detector consists of multichannels of a single silicon chip capable of detecting radiation of varied wavelengths but at the cost of sensitivity. Similar to the diode array detector, charge coupled devices also have multiple channels of silicon chips. These elements, commonly known as pixels, interact with light to develop an electrical charge, due to which an electric current is developed that is further amplified through amplifiers (Altemose, 1986).

The instrumentation for fluorescence spectroscopy is also similar to that of UV-visible spectroscopy in that each contains a light source and monochromator (Fig. 2). The light sources commonly used are xenon arc lamps, mercury lamps, xenon-mercury lamps, quartz tungsten halogen lamps, light-emitting diodes, and laser diodes, which provide a uniform intensity of radiation over a broad range. The monochromators in this case also

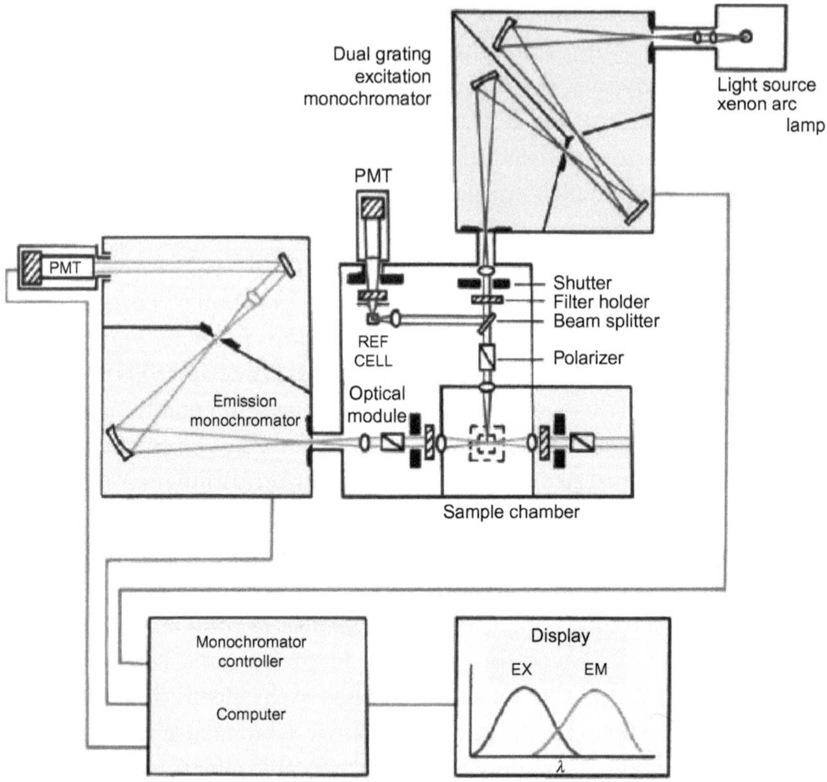

Fig. 2 Instrumentation of fluorescence spectroscopy (Lakowicz, 2006).

consist of gratings and filters that are placed in the excitation and emission paths in order to select a specific wavelength. The gratings used are either concave or convex. Beam splitters are also placed in the excitation pathway to ensure that radiation of a specific wavelength passes through. The beam splitter consists of a thin section of quartz that is responsible for the reflection of about 4% of the incident light, which is sent to a reference chamber that contains a highly fluorescent quantum counter. This is highly crucial in case any deviations from the light source need to be adjusted. This is done by subtracting the intensity of the light source with the intensity of the fluorophore in the reference chamber. Polarizers are also present in both the light paths and are used in case of analysis of polarized compounds. They ensure that the excitation light path reaching the sample is polarized in order to achieve an appropriate fluorescence response. The sample holders are also designed to hold the cuvette. However, here the cuvette is completely transparent from all sides, unlike the cuvette used for UV-visible spectroscopy where only two sides are completely transparent. Detectors, which are another crucial aspect of this instrument, are placed in an orthogonal alignment with the radiation source. Conventionally, photomultiplier tubes are used as they are capable of detecting the short-lived fluorescence span of a compound with increased sensitivity. As mentioned before, these detectors consist of a series of dynodes and electrodes that is responsible for the generation and amplification of the electric current generated due to the striking of photons. However, this detector has certain limitations such as the generation of dark current, which is produced as a result of excessive exposure of light on the detector; this usually occurs at high voltages. Apart from photomultiplier tubes, charge coupled devices are also used to detect fluorescence with increased radiation (Lakowicz, 2006).

4 Analysis of forensic samples using UV-visible and fluorescence spectroscopy

UV-visible and fluorescence spectroscopic techniques are among the basic analytical techniques for sample characterization. These techniques are widely used in different industries such as food, pharmaceuticals, textiles, dyeing, etc. Forensic science also utilizes these techniques for the analysis of different evidence such as body fluids, counterfeit beverages, drugs, fibers, fingerprints, paint, petroleum products, and questionable documents. This section discusses the research that uses UV-visible and fluorescence spectroscopy for the analysis of the aforementioned samples.

4.1 Body fluid

Body fluids such as blood, sweat, semen, saliva, urine, etc., are among the most common kinds of evidence found at a crime scene. This evidence is analyzed for different characteristics using myriad analytical techniques. Zapata et al. have shown the potential of spectroscopic techniques for the forensic investigation of body fluids (Zapata et al., 2015). This section describes the studies carried out on blood and saliva using UV-visible and fluorescence spectroscopy.

In a study by Spence and Asmussen et al., UV-visible and fluorescence spectroscopy were used for the analysis of blood collected from footwear impressions. The blood sample was initially treated with leucocrystal violet (LCV) for spectra enhancement. The UV-visible spectra was obtained in the range of 200–800 nm. For the fluorescence spectra, the excitation wavelength (λ_{ex}) was 630 nm and the emission wavelength (λ_{em}) range was 661–900 nm. The analysis was done for LCV, LCV in hydrogen peroxide, a working solution of LCV, and blood mixed with a working solution of LCV. For blood with LCV, a strong peak was observed at 630 nm, and the fluorescence spectra at this λ_{ex} showed that LCV is a weak fluorophore (Spence and Asmussen, 2003). In another study, body fluid identification was carried out by analyzing RNA and DNA collected from blood and saliva. For this purpose, DNA was isolated and quantified using a UV-visible spectrophotometer (nanovolume) (van der Meer et al., 2013). A fluorescent probe with a nanocluster of Cu/Ag with DNA was developed to sense poisoning due to hydrogen sulfide in blood. The λ_{em} for the fluorometric analysis was from 500 to 675 nm. The developed probes had a very high intensity of fluorescence at 500 nm. Fluorescence quenching was studied using the UV-visible spectra with sodium sulfide. The spectra revealed a decrease in the maxima at 450 nm for the probes after the addition of sulfide salt in the probe solution. A decreasing trend in the fluorescence intensity of the probes was observed upon increasing the concentration of sulfide salt (Ding et al., 2017). A recent study by Kaur et al. focused on the investigation of the degradation of hemoglobin and DNA in blood samples with aging and preservation using UV-visible spectroscopy. A nanodrop spectrophotometer was used to obtain UV-visible spectra in the range of 350–550 nm. It was observed that with decreasing levels of preservation quality for bloodstain samples, the hemoglobin and DNA content started degrading, which was evident with the increasing noise in the spectra (Kaur et al., 2020).

In the case of saliva as well, the fluorescence spectroscopic technique has shown its potential. In a study by Yuvaraj et al., metabolites present in saliva were characterized using fluorescence spectroscopy. The λ_{ex} for the study was 405 nm. The metabolites in the saliva of normal people and oral cancer patients were observed. A remarkable difference in the fluorescence spectra of saliva from both populations was evident. This difference was because of the increased levels of porphyrin in the saliva of cancer patients (Yuvaraj et al., 2014). Denny et al. also utilized fluorescence spectroscopy for the forensic sensing of human saliva. Both dried and liquid saliva were analyzed during the study, which showed λ_{ex} at 350 and 345 nm, respectively. It was concluded that the fluorescence intensity, peak at λ_{ex}, and area under the curve can be used to detect the existence of human saliva. Fig. 3 shows the fluorescence spectra of dried and liquid saliva (Denny et al., 2018). In another study by Li et al., a fluorescent probe was developed using SiC nanoparticles stabilized by bovine serum albumin for the detection of saliva.

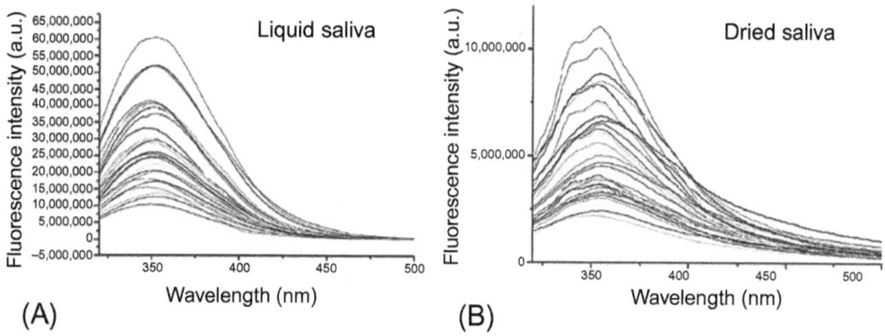

Fig. 3 Fluorescence spectra of liquid (undiluted) (A) and dried (diluted) saliva (B) (Denny et al., 2018).

The λ_{ex} and λ_{em} for the study were 320 and 410 nm, respectively. The probe targeted the bacteria *Streptococcus salivarius* in saliva for its detection. The study was able to detect saliva in mixed samples of saliva with blood, semen, and urine (Li et al., 2019).

4.2 Counterfeit beverages

Counterfeiting in food articles has remained a matter of concern for consumers. Forensic experts play a role in detecting such counterfeit products using different analytical techniques. This section discusses research utilizing the spectroscopic technique to detect counterfeit beverages.

In a study by Martins et al., UV-visible spectroscopy was used to discriminate original and counterfeit brands of whiskey. Seven different brands of this alcoholic beverage were selected for the analysis. The spectra was obtained from 190 to 1100 nm. The absorbance spectra of different brands of whiskey had different intensities in the UV region, probably because of congeners, caramel, and phenolic compounds in the drink. The study in combination with multivariate data analysis was able to differentiate between branded and counterfeit whiskey (Martins et al., 2017).

4.3 Drugs

UV-visible and fluorescence spectroscopy are among the preliminary studies for drug characterization in most industries and laboratories. This section discusses the research that employed these techniques for the forensic analysis of different drugs.

In a study by Gill et al., UV-visible spectroscopy (derivative) was used for the forensic toxicological analysis of amphetamines present in low concentrations in a liver extract. The derivative mode of analysis helped in the differentiation of compounds that had similar absorbance spectra. Methadone and carmoisine were also analyzed during the study. In the case of methadone, sharp bands were observed in the range of 250–275 nm while carmoisine had a broad absorbance region from 450 to 600 nm. Amphetamine showed peaks in the range of 240–280 nm (Gill et al., 1982). The precursors, intermediates, and byproducts formed during the synthesis of ecstasy were analyzed using UV spectroscopy in a study by Renton et al. UV spectroscopy was also assisted by other analytical techniques for this purpose. The drug was prepared by three synthetic routes. The absorbance spectra of the drug as well as its precursors, intermediates, and byproducts were obtained. Further, this data along with data from other techniques were used as reference for the identification of ecstasy in case samples (Renton et al., 1993). In another study, illicit preparations of cocaine were studied using UV-visible spectroscopy (derivative). This variant of absorption spectroscopy was useful for resolving mixtures of cocaine with lidocaine, procaine, benzocaine, and tetracaine. The zero order maxima for different samples were: 274 (cocaine); 261 (lidocaine); 271.8 (procaine); 271.9 (tetracaine); and 270 (benzocaine) nm. In case of mixtures, the maxima were different: 270.3 (lidocaine and cocaine); 273.2 (procaine and cocaine); 273 (tetracaine and cocaine); and 270.7 (benzocaine and cocaine) nm. The solvent for preparing these solutions was HCl (Arufe-Martinez and Romero-Palanco, 1988). A similar kind of study on cocaine and related adulterants such as procaine and lidocaine was done by Cruz et al. using absorption spectroscopy (derivative). However, in contrast to the previous study, a phosphate buffer of pH 7.8 was used as the solvent instead of HCl, and the spectra were obtained in the range of 220–380 nm (Cruz et al., 1994). A study from Brazil used fluorescence spectroscopy to identify cocaine present on currency notes that were in circulation in Rio de Janeiro. The λ_{ex} for the study was 315 nm while the λ_{em} was 230 nm. It was observed that around 86% of the collected banknotes had traces of cocaine. Banknotes with higher values such as 50 and 100 had very low levels of the drug on them (Almeida et al., 2015).

4.4 Fibers

Fibers are common evidence found at the crime scene, and they can be analyzed for morphological and chemical characteristics using various characterization techniques. Spectroscopic techniques such as UV-visible, Fourier transform infrared, Raman, and X-ray fluorescence have shown their potential for the analysis of textile fibers (Meleiro and García-Ruiz, 2016). In this section, studies employing UV-visible or fluorescence spectroscopy for the forensic examination of fibers are discussed.

In a study by Abbott et al., UV-visible spectroscopy was used along with Raman spectroscopy for the analysis of black dyes present in textile fibers such as cotton, nylon, polyester, viscose, and acrylic. The spectra were obtained from 200 to 800 nm. The target dye was reactive black 5. Peaks were obtained in the region of 550–650 nm, and some fibers with this dye also showed a peak around 300 nm (Fig. 4) (Abbott et al., 2010). Cotton fibers dyed with various colored compounds collected as evidence were differentiated in a study using UV-visible microspectrophotometry. The spectra were obtained in the range of 200–800 nm (transmittance mode). It was observed that fibers had different transmittance spectra when colored with

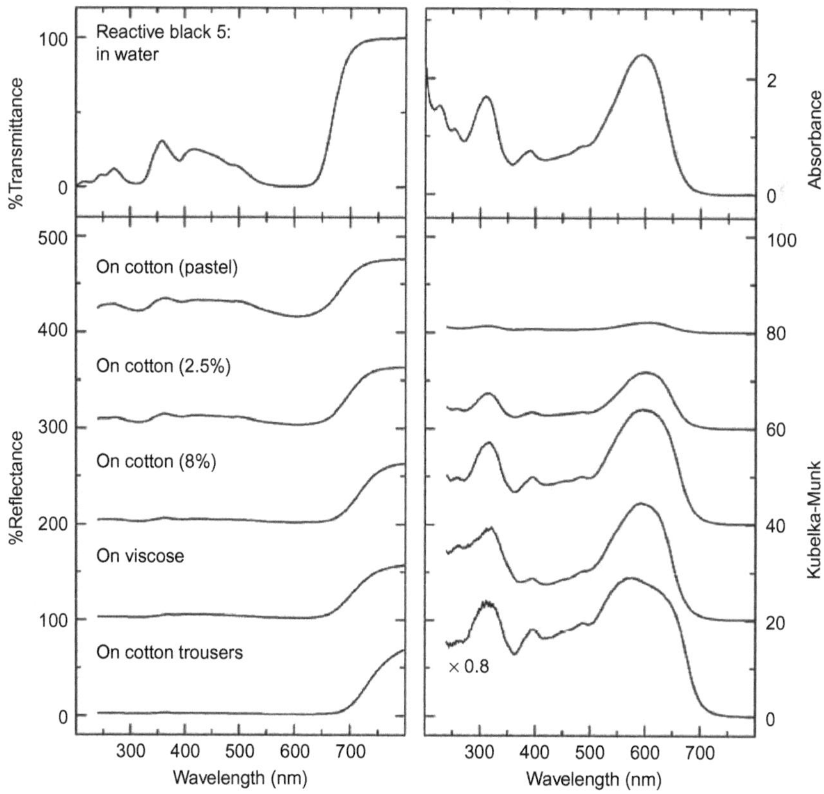

Fig. 4 Absorption spectra of reactive black 5 dye in various environments (Abbott et al., 2010).

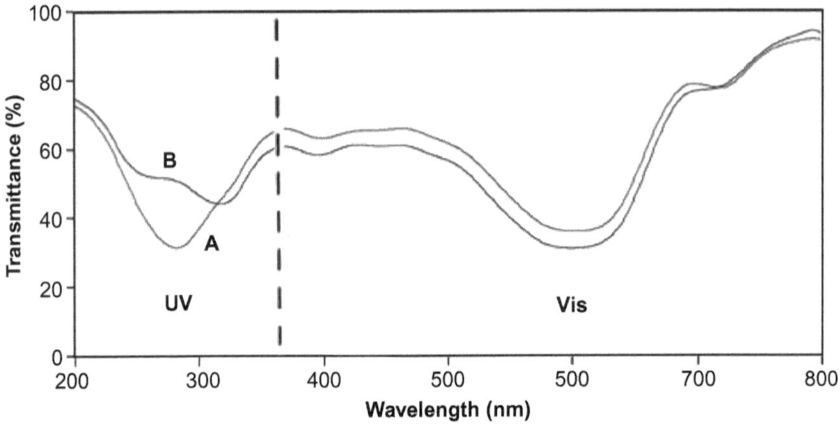

Fig. 5 Transmittance spectra of cotton fiber dyed with (A) reactive blue 184; and (B) reactive blue 238 (Was-Gubala and Starczak, 2015).

different dyes. However, the concentration of these dyes could not be found because the Beer Lambert law was not obeyed. Fig. 5 shows the UV-visible spectra of cotton fibers dyed with two different types of reactive blue dye (Was-Gubala and Starczak, 2015). In a study by Campiglia et al., extracts from textile fibers were examined with the help of fluorescence spectroscopy. The λ_{ex} was 330 nm and the λ_{em} was 500 nm. The analysis helped in differentiating visually indistinguishable fibers as well as fibers of the same material with the same dye but collected from different cloth (Campiglia et al., 2016). A recent study combined the data obtained from the UV-visible spectra of different textile dyes with multivariate statistics in order to classify them and know their concentration in the fiber samples (Rich et al., 2020).

4.5 Fingerprints

Fluorescence spectroscopy has aided in the investigation of fingerprints collected freshly or after some time. In addition, different UV lights have also been used for the visualization of fingerprints on different surfaces.

In a study by Akiba et al., latent fingerprints were excited with a laser having different excitation wavelengths in the UV region. A classic fluorescence spectra of a fingerprint shows two peaks at 440 and 330 nm. It was observed that the fluorescence intensity was the highest at 330 nm when 280 nm λ_{ex} was used (Akiba et al., 2007). In another study by Liu et al., fluorescent nanorods were used for the detection of latent fingerprints.

The nanorods were specific to human blood fingerprints, and inhibited the development of prints due to the blood of poultry or livestock. The λ_{ex} for the study was 980 nm, obtained via a laser (Liu et al., 2018). In a recent study, UV and fluorescence light were used to visualize latent fingerprints. The wavelength for UV light was 254 nm; for fluorescent light, it was in the range of 415–515 nm (Risoluti et al., 2019).

4.6 Paint

Paint analysis has forensic importance in cases related to hit and run or vandalism. Different types of techniques are employed to study the morphological and chemical features of paint and its layers. Visible spectroscopy has also been utilized in some studies.

In a study, visible spectroscopy was used for the forensic investigation of car paints. Paint chips of blue, red, and green were observed through light in a perpendicular manner at the cross-section as well as at the top layer of the paint fragment. The analysis was carried out in the transmittance and reflectance modes. It was observed that the spectra showed greater variation for different colors when the top layer analysis was done. Fig. 6 shows the

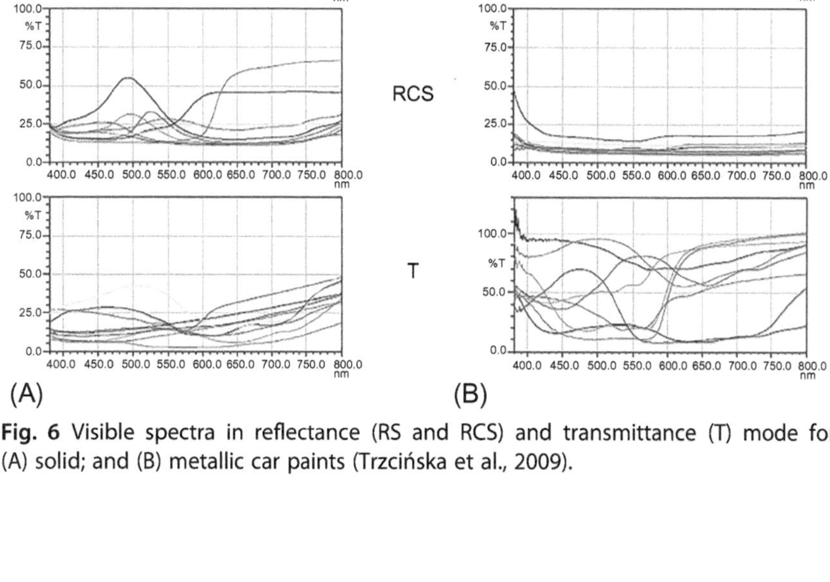

Fig. 6 Visible spectra in reflectance (RS and RCS) and transmittance (T) mode for (A) solid; and (B) metallic car paints (Trzcińska et al., 2009).

absorption spectra of car paints (Trzcińska et al., 2009). In another study, visible spectroscopy extended to the near-infrared region was used to forensically analyze and distinguish different car paints. The analysis was carried out in the range of 400–1100 nm (Ferreira et al., 2017).

4.7 Petroleum products

Petroleum products are generally studied using chromatographic techniques during forensic investigations. However, the luminescence of petroleum products is a well-known phenomenon. This property of petroleum products was utilized by Lloyd to forensically analyze them using fluorescence spectroscopy. The λ_{ex} and λ_{em} were 286 and 310 nm, respectively. It was concluded that fluorescence was produced in the petroleum products mainly because of the presence of heterocycles of sulfur and alkylated aromatic hydrocarbons (Lloyd, 1980).

4.8 Questionable documents

Documents become forensically important in cases related to forgery. The inks used in such questionable documents as well as line crossings and fibers are analyzed using different characterization tools. UV-visible and fluorescence spectroscopy have also assisted in this regard, mostly in the investigation of inks used in questionable documents.

In a study by Vogt et al., fountain pen inks were studied using capillary electrophoresis embedded with fluorescence and a UV-visible detection method. The spectra (UV-visible and fluorescence) of ink samples collected at variable λ_{ex} and λ_{em} were compared with those of standard ink compositions in order to identify them (Vogt et al., 1997). In another study, ballpoint pen inks of blue were discriminated on the basis of their visible spectra, which was recorded in the range of 400–750 nm. It was observed that inks of different brands had different overall absorbance spectra, with their maxima ranging from 550 to 700 nm (Thanasoulias et al., 2003). Causin et al. used absorption spectroscopy in the range of the UV, visible, and near-infrared region for the forensic examination of paper. The spectra were recorded in the range of 200–900 nm. Different brands of paper were differentiated on the basis of their average reflectivity in the 680–900 nm range as well as their maxima in the 272–360 nm range (Causin et al., 2012). In another study, black inkjet inks used for printing documents were analyzed using visible spectra extended to the near-infrared region. The spectra were recorded from 500 to 1000 nm. It was observed that inks of different brands had different absorption spectra in the 600–1000 nm range (Gál et al., 2015).

UV-visible spectroscopy extended to the near-infrared region has also been used for the analysis of ballpoint pen inks of blue in a nondestructive manner. The diffuse reflectance mode was also used during the analysis. The spectra were acquired in the range of 200–800 nm. Crystal violet was found as the common dye in all the ink samples (Kumar and Sharma, 2017). The same group of researchers used UV-visible spectroscopy in combination with multiple regression analysis to date the inks collected from ballpoint pens. The spectra were recorded in the range of 200–800 nm. The combined data of both the techniques helped in dating inks aged up to 267 days (Sharma and Kumar, 2017). Saviello et al. also used UV-visible spectroscopy along with surface-enhanced Raman spectroscopy to investigate the content of dyes used in the inks of felt tip pens (Saviello et al., 2018). The degradation behavior of paper and the sequences involved in the process were also examined using UV-visible spectroscopy in a recent study. The paper samples were aged at a high temperature for a period of 35 days. The absorption spectra of different paper samples differed in the UV visible range regarding the absorbance values corresponding to the 350 and 280 nm wavelengths (Zięba-Palus et al., 2020).

5 Conclusion

UV-visible spectroscopy and fluorescence spectroscopy are the most widely used techniques for the analysis of various forensic liquid evidence due to their ability to provide immediate and valid results. UV-visible spectroscopy is based on the detection of UV or visible radiation emitted by the sample due to electronic transitions and the Beer Lambert law, which calculates absorbance on the basis of the intensity of the transmitted light from the sample. While this technique is widely used for preliminary and even confirmatory analysis of forensic evidence, there are certain limitations to this technique. This technique is suitable for the analysis of liquid samples but for the detection of solid samples, a special apparatus is needed. Also, the Beer-Lambert law tends to deviate when the sample concentration is high, due to which it is necessary to perform appropriate dilution techniques in order to bring the concentration of the sample to the optimal range. Also, liquid samples such as those containing metal solutions need to be prepared through different techniques to produce a colored solution that can be detected through colorimetric analysis. Despite such flaws, this technique is still the most commonly used technique for the analysis of organic solutions.

Fluorescence spectroscopy is highly useful to determine the fluorescence properties of a sample. This technique detects the fluorescence that is emitted when an excited electron falls back to its ground state. This technique also offers high sensitivity and specificity and can correlate the concentration of the sample with the fluorescence intensity. Despite these advantages, there are certain limitations such as loss of photostability and recognition capability. If the radiation source is very strong, photobleaching may occur, due to which the sample may gradually lose its fluorescence. Despite such limitations, these techniques are indeed highly useful for the detection of forensic samples such as body fluid, counterfeit beverages, drugs, fibers, fingerprints, paint, petroleum products, and questionable documents.

References

Abbott, L.C., Batchelor, S.N., Smith, J.R.L., Moore, J.N., 2010. Resonance Raman and UV–visible spectroscopy of black dyes on textiles. Forensic Sci. Int. 202, 54–63. https://doi.org/10.1016/j.forsciint.2010.04.026.

Akiba, N., Saitoh, N., Kuroki, K., 2007. Fluorescence spectra and images of latent fingerprints excited with a tunable laser in the ultraviolet region. J. Forensic Sci. 52, 1103–1106. https://doi.org/10.1111/j.1556-4029.2007.00532.x.

Almeida, V.G.K., Cassella, R.J., Pacheco, W.F., 2015. Determination of cocaine in Real banknotes circulating at the State of Rio de Janeiro, Brazil. Forensic Sci. Int. 251, 50–55. https://doi.org/10.1016/j.forsciint.2015.02.014.

Altemose, I., 1986. Evolution of instrumentation for UV-visible spectrophotometry part I. J. Chem. Educ. 63, A216. https://doi.org/10.1021/ed063pA216.

Arufe-Martinez, M.I., Romero-Palanco, J.L., 1988. Identification of cocaine in cocaine-lidocaine mixtures ("rock cocaine") and other illicit cocaine preparations using derivative absorption spectroscopy. J. Anal. Toxicol. 12, 192–196. https://doi.org/10.1093/jat/12.4.192.

Campiglia, A.D., Rex, M., Muñoz de la Peña, A., Goicoechea, H.C., 2016. Excitation–emission matrix fluorescence spectroscopy combined with MCR-ALS as a tool for the forensic analysis of similar and dissimilar sets of textile fiber extracts. Anal. Methods 8, 8314–8321. https://doi.org/10.1039/C6AY02757A.

Causin, V., Casamassima, R., Marruncheddu, G., Lenzoni, G., Peluso, G., Ripani, L., 2012. The discrimination potential of diffuse-reflectance ultraviolet–visible–near infrared spectrophotometry for the forensic analysis of paper. Forensic Sci. Int. 216, 163–167. https://doi.org/10.1016/j.forsciint.2011.09.015.

Cruz, A., López-Rivadulla, M., Bermejo, A.M., Sánchez, I., Fernández, P., 1994. Sequential second derivative spectroscopy of cocaine and adulterants in street drug samples. Part I: cocaine, procaine, and lidocaine. Anal. Lett. 27, 2663–2675. https://doi.org/10.1080/00032719408005998.

Denny, S.E., Nazeer, S.S., Sivakumar, T.T., Nair, B.J., Jayasree, R.S., 2018. Forensic application of fluorescence spectroscopy: an efficient technique to predict the presence of human saliva. J. Lumin. 203, 696–701. https://doi.org/10.1016/j.jlumin.2018.07.022.

Ding, Y., Li, X., Chen, C., Ling, J., Li, W., Guo, Y., Yan, J., Zha, L., Cai, J., 2017. A rapid evaluation of acute hydrogen sulfide poisoning in blood based on DNA-Cu/Ag nanocluster fluorescence probe. Sci. Rep. 7, 9638. https://doi.org/10.1038/s41598-017-09960-1.

Ferreira, K.B., Oliveira, A.G.G., Gonçalves, A.S., Gomes, J.A., 2017. E valuation of hyperspectral imaging visible/near infrared spectroscopy as a forensic tool for automotive paint distinction. Forensic Chem. 5, 46–52. https://doi.org/10.1016/j.forc.2017.06.001.

Gál, L., Oravec, M., Gemeiner, P., Čeppan, M., 2015. Principal component analysis for the forensic discrimination of black inkjet inks based on the Vis–NIR fibre optics reflection spectra. Forensic Sci. Int. 257, 285–292. https://doi.org/10.1016/j.forsciint.2015.09.011.

Gill, R., Bal, T.S., Moffat, A.C., 1982. The application of derivative UV-visible spectroscopy in forensic toxicology. J. Forensic Sci. Soc. 22, 165–171. https://doi.org/10.1016/S0015-7368(82)71466-5.

House, J.E., 2018. Molecular spectroscopy. In: Fundamentals of Quantum Mechanics. Elsevier, pp. 271–296, https://doi.org/10.1016/B978-0-12-809242-2.00011-5.

Kaur, S., Saini, V., Dalal, R., 2020. UV-visible spectroscopic effect on Haemoglobin & DNA degradation: a forensic approach. Forensic Sci. Int. 307, 110078. https://doi.org/10.1016/j.forsciint.2019.110078.

Kumar, R., Sharma, V., 2017. A novel combined approach of diffuse reflectance UV–Vis-NIR spectroscopy and multivariate analysis for non-destructive examination of blue ballpoint pen inks in forensic application. Spectrochim. Acta A Mol. Biomol. Spectrosc. 175, 67–75. https://doi.org/10.1016/j.saa.2016.12.008.

Lakowicz, J.R., 2006. Instrumentation for fluorescence spectroscopy. In: Principles of Fluorescence Spectroscopy. Springer, Boston, MA, pp. 27–61.

Li, X., Ding, Y., Ling, J., Yao, W., Zha, L., Li, N., Chang, Y., Wang, Y., Cai, J., 2019. Bacteria-targeting BSA-stabilized SiC nanoparticles as a fluorescent nanoprobe for forensic identification of saliva. Microchim. Acta 186, 756. https://doi.org/10.1007/s00604-019-3890-y.

Liljefors, T., Allinger, N., 1978. Conformational analysis. 128. The Woodward-Fieser rules and .alpha.,.beta.-unsaturated ketones. J. Am. Chem. Soc. 100, 1068–1073.

Liu, C.-M., Zhang, L.-Y., Li, L., Li, B.-Y., Wang, C.-G., Wang, T.-T., 2018. Specific detection of latent human blood fingerprints using antibody modified NaYF4: Yb, Er, Gd fluorescent upconversion nanorods. Dyes Pigments 149, 822–829. https://doi.org/10.1016/j.dyepig.2017.11.050.

Lloyd, B.F., 1980. Examination of petroleum products of high relative molecular mass for forensic purposes by synchronous fluorescence spectroscopy. Analyst 105, 97–109.

Martins, A.R., Talhavini, M., Vieira, M.L., Zacca, J.J., Braga, J.W.B., 2017. Discrimination of whisky brands and counterfeit identification by UV–Vis spectroscopy and multivariate data analysis. Food Chem. 229, 142–151. https://doi.org/10.1016/j.foodchem.2017.02.024.

Meleiro, P.P., García-Ruiz, C., 2016. Spectroscopic techniques for the forensic analysis of textile fibers. Appl. Spectrosc. Rev. 51, 278–301. https://doi.org/10.1080/05704928.2015.1132720.

Pandey, G., Rawtani, D., Agrawal, Y.K., 2017. Future aspects of halloysite nanotubes in forensic investigations. J. Nanomed. Res. 6. https://doi.org/10.15406/jnmr.2017.06.00153.

Penner, M.H., 2017. Basic principles of spectroscopy. In: Nielsen, S.S. (Ed.), Food Analysis. Springer International Publishing, Cham, pp. 79–88, https://doi.org/10.1007/978-3-319-45776-5_6.

Räty, J., Peiponen, K.-E., Asakura, T., 2004. UV-visible reflection spectroscopy of liquids, springer series in optical sciences. Springer Berlin Heidelberg, Berlin, Heidelberg. https://doi.org/10.1007/978-3-540-45093-1.

Rawtani, D., Tharmavaram, M., Pandey, G., Hussain, C.M., 2019. Functionalized nanomaterial for forensic sample analysis. TrAC Trends Anal. Chem. 120, 115661. https://doi.org/10.1016/j.trac.2019.115661.

Renton, R.J., Cowie, J.S., Oon, M.C.H., 1993. A study of the precursors, intermediates and reaction by-products in the synthesis of 3,4-methylenedioxymethylamphetamin e and its application to forensic drug analysis. Forensic Sci. Int. 60, 189–202. https://doi.org/10.1016/0379-0738(93)90238-6.
Rich, D.C., Livingston, K.M., Morgan, S.L., 2020. Evaluating performance of lasso relative to PCA and LDA to classify dyes on fibers. Forensic Chem. 18, 100213. https://doi.org/10.1016/j.forc.2020.100213.
Risoluti, R., Filetti, V., Iuliano, G., Niola, L., Schiavone, S., Arcudi, G., Materazzi, S., 2019. Updating procedures in forensic chemistry: one step cyanoacrylate method to develop latent fingermarks and subsequent DNA profiling. Microchem. J. 147, 478–486. https://doi.org/10.1016/j.microc.2019.03.056.
Saviello, D., Trabace, M., Alyami, A., Mirabile, A., Giorgi, R., Baglioni, P., Iacopino, D., 2018. A combined surface enhanced Raman spectroscopy (SERS)/UV–vis approach for the investigation of dye content in commercial felt tip pens inks. Talanta 181, 448–453. https://doi.org/10.1016/j.talanta.2018.01.049.
Sharma, V., Kumar, R., 2017. Dating of ballpoint pen writing inks via spectroscopic and multiple linear regression analysis: a novel approach. Microchem. J. 134, 104–113. https://doi.org/10.1016/j.microc.2017.05.014.
So, P.T., Dong, C.Y., 2001. Fluorescence Spectrophotometry. John Wiley and Sons.
Spence, L., Asmussen, G., 2003. Spectral enhancement of leucocrystal violet treated footwear impression evidence in blood. Forensic Sci. Int. 132, 117–124. https://doi.org/10.1016/S0379-0738(03)00003-3.
Thanasoulias, N.C., Parisis, N.A., Evmiridis, N.P., 2003. Multivariate chemometrics for the forensic discrimination of blue ball-point pen inks based on their Vis spectra. Forensic Sci. Int. 138, 75–84. https://doi.org/10.1016/j.forsciint.2003.08.014.
Thomas, O., Burgess, C. (Eds.), 2007. UV-Visible Spectrophotometry of Water and Wastewater, Techniques and Instrumentation in Analytical Chemistry. Elsevier, Amsterdam; Boston.
Trzcińska, B., Zięba-Palus, J., Kościelniak, P., 2009. Application of microspectrometry in the visible range to differentiation of car paints for forensic purposes. J. Mol. Struct. 924–926, 393–399. https://doi.org/10.1016/j.molstruc.2008.10.037.
van der Meer, D., Uchimoto, M.L., Williams, G., 2013. Simultaneous analysis of micro-RNA and DNA for determining the body fluid origin of DNA profiles. J. Forensic Sci. 58, 967–971. https://doi.org/10.1111/1556-4029.12160.
Vogt, C., Vogt, J., Becker, A., Rohde, E., 1997. Separation, comparison and identification of fountain pen inks by capillary electrophoresis with UV-visible and fluorescence detection and by proton-induced X-ray emission. J. Chromatogr. A 781, 391–405. https://doi.org/10.1016/S0021-9673(97)00621-3.
Was-Gubala, J., Starczak, R., 2015. UV–Vis microspectrophotometry as a method of differentiation between cotton fibre evidence coloured with reactive dyes. Spectrochim. Acta A Mol. Biomol. Spectrosc. 142, 118–125. https://doi.org/10.1016/j.saa.2015.01.116.
Yuvaraj, M., Udayakumar, K., Jayanth, V., Prakasa Rao, A., Bharanidharan, G., Koteeswaran, D., Munusamy, B.D., Murali Krishna, C., Ganesan, S., 2014. Fluorescence spectroscopic characterization of salivary metabolites of oral cancer patients. J. Photochem. Photobiol. B Biol. 130, 153–160. https://doi.org/10.1016/j.jphotobiol.2013.11.006.
Zapata, F., Fernández de la Ossa, M.Á., García-Ruiz, C., 2015. Emerging spectrometric techniques for the forensic analysis of body fluids. TrAC Trends Anal. Chem. 64, 53–63. https://doi.org/10.1016/j.trac.2014.08.011.
Zięba-Palus, J., Trzcińska, B., Wesełucha-Birczyńska, A., Moskal, P., Sacharz, J., 2020. The sequence of changes observed during degradation process of paper by the use of UV/VIS and FTIR spectrometry with application of the PCA and 2D correlation method for forensic purposes. J. Mol. Struct. 1205, 127651. https://doi.org/10.1016/j.molstruc.2019.127651.

CHAPTER 4

FTIR and NIR spectroscopy in forensic science

1 Introduction

Infrared spectroscopy is a type of vibrational spectroscopy in which the energy of the incident electromagnetic radiation is absorbed by the molecule, gets excited, and then undergoes rotations or vibrations. Infrared spectroscopy is a widely used technique because of its versatility and rapid results. A myriad of liquids, solids, and gaseous samples can be easily analyzed using this technique. It also requires minimal sample preparation and based on the type of technique used, it can also be nondestructive in nature.

This technique involves the use of infrared rays, and the amount of rays absorbed by the sample is taken into account. The principle and theory behind this technique will be discussed further in this chapter. While IR spectroscopy takes into account the vibrational energy of the molecules, another technique known as near-infrared takes into account both the electronic and vibrational transitions. This makes NIR a very unique technique capable of analyzing a wide range of molecules. The IR region lies between 700 nm and 1 mm while NIR lies between 800 nm and 1.2 mm.

Quite often, samples are analyzed through an amalgamation of Raman, near-infrared, and IR spectroscopy in order to get a comprehensive idea about the compound. Forensic science extensively uses these techniques to analyze samples such as paints, explosives, drugs, and counterfeit compounds (Pandey et al., 2017; Rawtani et al., 2019). Because these techniques are nondestructive and reliable, they are capable of analyzing the samples. In this chapter, the use of IR and NIR in the analysis of biological samples, inks, questionable documents, explosives, gunshot residue, illicit and counterfeit drugs, bank notes, and paint is discussed. Also, the principle, theory, and instrumentation of the techniques are given.

Handbook of Analytical Techniques for Forensic Samples
https://doi.org/10.1016/B978-0-12-822300-0.00004-5

2 Principle and theory of infrared and near-infrared spectroscopy

2.1 Principle of infrared spectroscopy

Infrared spectroscopy, just like any other spectroscopy technique, is based on the use of electromagnetic radiation to analyze compounds. Electromagnetic radiation consists of an electric and magnetic component that are perpendicularly aligned with each other and travel in a sinusoidal manner. As mentioned before, IR spectroscopy is based on the absorption of the IR by the sample, due to which it gets excited and starts to vibrate. The vibrational signal sent out is characteristic of each molecule and thus serves as the basis of this spectroscopy.

However, not all molecules are capable of showing infrared absorption. A molecule should possess an electric dipole moment, due to which there is a change in vibrations when it absorbs the IR. Normally, there are two types of molecules on whose basis the ability to absorb IR radiation is decided: heteronuclear and homonuclear molecules. Heteronuclear molecules consist of two atoms, due to which an electric dipole is created that is capable of vibrating upon IR absorption; homonuclear molecules have an electric dipole of zero and therefore are said to be IR inactive (Stuart, 2004).

In IR, the molecules are often considered to be similar to two balls connected by a spring. Therefore, upon absorption of IR by these molecules, they undergo translational and rotational motions through the stretching or bending of the bonds in which either their length or angle changes. The stretching can be of two types: symmetric and asymmetric stretching. In symmetric stretching, the bond length is changed equally on both sides of the atom while in asymmetric stretching, the bond length changes unequally. These movements are commonly termed vibrations. The vibrations are described in terms of degrees of freedom. For instance, a molecule that is diatomic will undergo only stretching and bending and therefore will have only one degree of vibrational freedom. On the other hand, molecules that have more than two atoms have 3N degrees of freedom. In this case, the molecule may be distinguished either as a linear molecule or a nonlinear molecule. In a polyatomic molecule, stretching and bending also occurs. Here, the bending can be of four types: deformation, rocking, twisting, and wagging. In deformation, the position of one atom is constant while the other two molecules move toward each other. In the case of rocking, both atoms move in the same direction while in wagging, the two atoms move toward the surface of the paper. In twisting, the atoms move in opposite directions to each other with variations in their bond length.

Apart from the conventional vibrations of stretching and bending, several other vibrations also take place that may often complicate the readings. These include overtone and combination bands, Fermi resonance, coupling, and vibration-rotation bands (Mohamed et al., 2017; Stuart, 2004).

A very crucial aspect of analysis with IR spectroscopy is identifying the zones of the IR region. The IR region is divided into three main parts: the far-IR, the mid-IR, and the near-IR. The mid-IR region, which lies in the range of 400–1400 cm^{-1}, is further divided into the fingerprint region and the functional group region. This region is responsible for the characteristic signal of every molecule, thus its name as the fingerprint region (Bell and Xu, 2018).

2.2 Principle and theory of NIR spectroscopy

The NIR range as mentioned above, lies between 14,000 and 4000 cm^{-1} and was first discovered by Sir William Herschel in 1800. He carried out an experiment in which he dispersed sunlight using a prism in three thermometers with blackened carbon bulbs. He found that the part in which there was no visible light showed the highest heating. NIR spectroscopy is also similar to IR in that they both use vibrational energies. In NIR, two vibrational models are considered, namely, the harmonic oscillator and the nonharmonic oscillator. The harmonic oscillator is also known as the linear model and is applied to diatomic molecules. In this case, a spring model is considered that obeys Hooke's law while nonharmonic oscillators or nonlinear models are applied to polyatomic molecules. This model is also applied with larger amplitudes of vibrations due to repulsive and disassociation forces. NIR uses the overtone and combination bands to analyze a compound. Overtone bands arise when there are multiple fundamental absorption frequencies while combination bands are formed when two fundamental bands absorb energy simultaneously. The bands formed due to the overtone and combination phenomenon are often weak in NIR. This region is highly transparent to radiation, which makes it highly valuable. NIR can also penetrate much further into samples and is therefore suitable for the analysis of bulk materials. It is also suitable for identifying the overtones, combination modes, hydrogen bonding, and intramolecular interactions (Ahmed et al., 2010; Manley and Baeten, 2018; Ozaki et al., 2017).

3 Instrumentation of IR and NIR spectroscopy

3.1 Instrumentation of IR spectroscopy

Currently, Fourier transform IR (FTIR) is a widely used IR technique. This instrumentation is based on dispersive infrared spectrometers in which a grating monochromator or a beam splitter is used to split the radiation from the source into different optical paths and through multiple reflectance, the light enters the detector after which an interferogram is generated. In FTIR spectrometers, an interferometer is used, which is basically a signal produced as a result of a change in path length between two different beams (Fig. 1).

A Michelson interferometer is most commonly used for FTIR; however, interferometers such as Savart-Plate (Kawata et al., 1989), stop-scan (Biggs et al., 1990), and JPL MkIV (Toon, 1991) have also been developed. In this chapter, however, only the Michelson interferometer will be discussed. The Michelson interferometer was developed by Michelson in 1891 and functions by dividing a beam of IR radiation into two beams with distinct paths with a beam splitter. The separate beams are later combined and the changes within their path length and frequency are detected. Also, mirrors are used that deflect the line at sharp angles. Here, one mirror is generally kept fixed while the other is kept movable. When the light strikes a mirror, it travels to a certain distance until it reaches the other mirror. If the distance between the two mirrors is the same, then the distance traveled by the beams will also be the same, and therefore there will be a zero-path difference. However, if the distance between the mirrors varies, then there will be an optical path difference. This difference can be either constructive or destructive based on the number of crests and troughs formed. Also, a cosine wave is formed

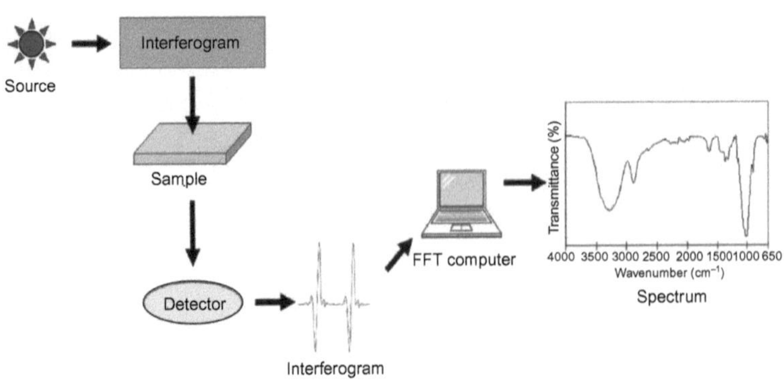

Fig. 1 Instrumentation of Fourier transform infrared spectroscopy (Mohamed et al., 2017).

that is known as the interferogram, and it is represented in terms of the function of time and signal intensity. Using Fourier transform, this interferogram is converted into an FTIR spectrum (Bell and Xu, 2018; Mohamed et al., 2017; Stuart, 2004).

The beam splitters used are also selected on the type of IR analysis to be performed. For instance, if mid- and near-IR analysis is to be performed, then potassium bromide and caesium iodide are used.

For analysis of samples in FTIR, different sample handling techniques are used such as transmission, attenuated total reflectance (ATR), diffuse reflectance (DRIFTS), and true specular reflectance or absorption. The FTIR instrumentation varies with the type of sample handling. In the transmission technique, the sample is mixed with potassium bromide in a ratio of 1:100 to form a pellet through a hydraulic press. In the instrumentation of this type of sampling, a special slot to keep the sample-containing sample holder is made and the analysis happens in a horizontal linear fashion. This technique, however, is very tedious and cannot be performed for liquid samples. ATR is another technique in which liquid and solid samples can be analyzed without any sample preparation. In this technique, the IR beam is sent directly to a dense crystal at an angle. The sample is kept on top of this crystal. When the beam is sent through this crystal, it hits the sample and undergoes multiple reflectance, which is further detected. In DRIFTS, the radiation sent to the sample gets scattered within the sample matrix through a process known as true diffuse reflectance. These reflected rays are further detected. In the case of true reflectance or absorption, the refractive index of the sample is taken into account. Here, the energy that is transmitted through the sample is not considered, but rather the rays that are reflected off the samples are considered and detected.

FTIR detectors have evolved over the years. Initially, thermocouple detectors were used followed by bolometers, Golay cells, and pyroelectric and photoconductive cells. Pyroelectric and photoconductive are fast and sensitive detectors used in the mid-IR and near-IR ranges, respectively (Mohamed et al., 2017).

3.2 Instrumentation of NIR spectroscopy

The NIR spectroscopy instrumentation is composed of four main parts: the radiation source, the wavelength selectors, the sample holders, and the detectors (Fig. 2). Thermal or nonthermal sources are often used to generate IR radiation of both a narrow and a wide range. Thermal sources such as

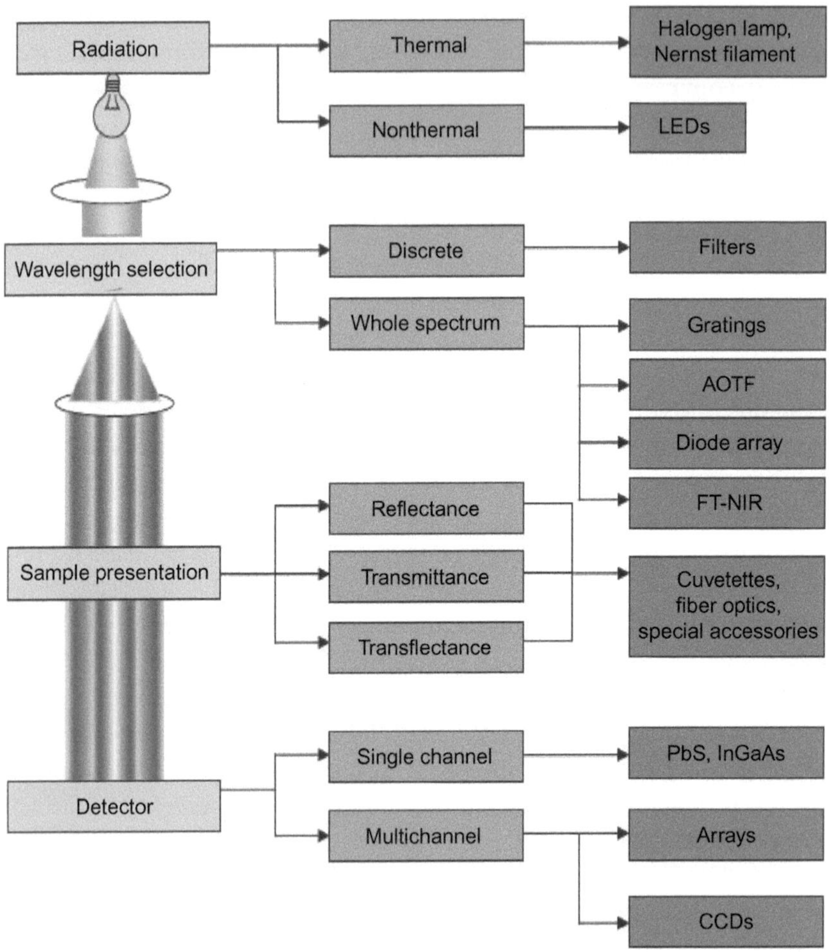

Fig. 2 Instrumentation of a near-infrared spectrometer (Manley and Baeten, 2018).

tungsten-halogen lamps and nonthermal sources such as light-emitting diodes are used.

The role of wavelength selectors is to ensure that the wavelengths are either discrete or continuous. Spectrometers containing discrete wavelengths have very limited wavelengths, which are controlled by special filters. In continuous wavelength spectrometers, a grating or diffraction monochromator is used that is similar to the one in FTIR. Diode arrays are also used to maintain a continuous wavelength; they are also capable of performing simultaneous detections of every wavelength. Apart from these, acoustic-optic tunable filters are also used, which generate a continuous wavelength by using radio frequency signals that are capable of changing the refractive index of any crystal. A Michelson interferometer is also used in this spectrometer.

Similar to FTIR, the instrumentation of NIR is also dependent on the sampling technique. Here too, detachable models for samples are used. This spectroscopy technique can be used in the reflectance, interactance, diffuse, and transmittance modes. The detectors used in NIR are most commonly photon detectors such as diode arrays, charge-coupled devices, and multichannel detectors (Ahmed et al., 2010; Manley and Baeten, 2018).

4 Spectrum analysis in FTIR and NIR

In FTIR analysis, the spectrum obtained contains the wavenumber in the *x*-axis and the intensity in the *y*-axis. The intensity of the peak can be of either the transmittance or the absorbance of the samples. Based on the functional groups present in the sample, different peaks of varying intensity, broadness, or sharpness are formed at different wavenumbers. While these wavenumbers are specific to the groups present in the sample, similar functional groups also tend to come in a similar wavenumber range and therefore need additional tests to confirm the identity of the sample. There are a vast number of FTIR frequency libraries that contain information about the frequency of each functional group. Through these libraries, it is possible to easily interpret the data with high certainty.

In the case of NIR, the plot is formed between the wavelength that is on the *x*-axis and the absorbance that is on the *y*-axis. With increases in the light-scattering efficiency and absorptivity, there is a shift in the wavelength toward 2500 nm. With a decrease in the band overlap and penetration depth of the molecule, there is a decrease in the wavelength range; it moves toward 700 nm. Just like FTIR, there are several NIR frequency tables as well that can be easily obtained and used to interpret the data.

5 FTIR and NIR analysis of forensic samples

FTIR and NIR techniques have been very commonly practiced by researchers for the qualitative and quantitative analysis of different kinds of analytes. In the field of forensic science, different research works have employed these techniques for the analysis of forensic samples such as samples of biological origin; inks and questionable documents; explosives and gunshot residue (GSR); illicit and counterfeit drugs; bank notes; and paint. Such research works are discussed in this section.

5.1 Samples of biological origin

Myriad biological samples such as hair, body fluids, insects and their larvae in cadavers, etc., are encountered at a crime scene. This section discusses the biological samples (blood stains, body fluids, insects, hairs, and fingerprints) whose analysis requires the utilization of FTIR and NIR techniques.

In a study by Lin et al., ATR-FTIR spectroscopy was used for the determination of the age of bloodstains that were collected from indoor and outdoor simulated crime scenes. Bloodstain samples that were around 107 days old were taken and analyzed with ATR-FTIR in combination with chemometrics. The ATR assembly was equipped with a diamond crystal, and the analysis was done in the spectral range of 1800–900 cm^{-1}. It was observed that with time, the peaks at 1649 (which correspond to the helical structure of hemoglobin) and 1533 (which correspond to the amide II) cm^{-1} changed with the age of the bloodstain (Fig. 3) (Lin et al., 2017). In another study, ATR FTIR was also used for the preliminary fingerprinting of body fluids, along with some other biological materials. The samples taken were semen, blood, feces, fingerprints, hair, cerumen, fingernails, saliva, tears, vaginal fluid, nasal mucus, and urine. The spectral range for the study was

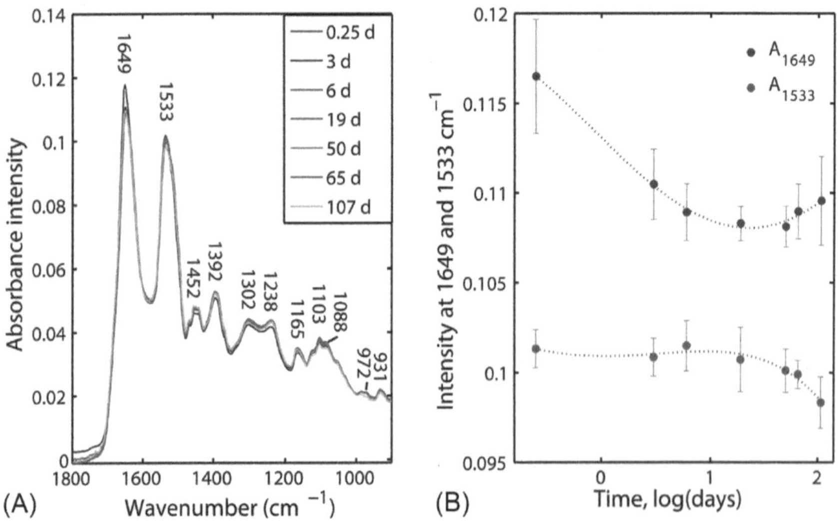

Fig. 3 (A) Averaged FTIR spectra of bloodstains; (B) time related trend in the 1649 and 1533 peaks of bloodstains (Lin et al., 2017).

4000–400 cm^{-1}. The methyl and ethyl stretches as well as specific proteins in all samples were responsible for differentiating between each sample in the fingerprint region of the FTIR spectra (Elkins, 2011). Fingerprints have also been analyzed using this technique. In a study, chemical imaging through FTIR was used to detect fingerprints that were not visible through conventional techniques. The optimum spectral resolution for obtaining the FTIR images of fingerprints was 32 or 16 cm^{-1}. It was observed that FTIR images of fingerprints fumed with cyanoacrylate were of high quality (Tahtouh et al., 2007). In another study, NIR was used for the efficient imaging of fingerprints present on different surfaces. The luminescence and absorption properties of fingerprints treated with NIR dyes were investigated in the range of 1100–650 nm. Absorbance for NIR was found in samples treated with a physical developer, iodine, ninhydrin, and powdering. Luminescence for NIR was found for samples treated with cyanoacrylate, genipin, indanedione, diazafluorenone (DFO), and ninhydrin (Maynard et al., 2009).

Hair analysis has also been done using these techniques. In a study by Signori and Lewis, the damaging effects of bleaching and weathering on hair were studied using ATR-FTIR. The crystal in the ATR assembly was diamond. It was found that the amount of cysteic acid increased from the middle of the hair toward its tip region (Signori and Lewis, 1997). In another study, compartments in hair follicles were imaged through FTIR. The analysis was done in the spectral range of 4000–700 cm^{-1}. The images taken were very distinctive for every layer of tissue in the hair follicle. The study provided a less-destructive and label-free spectroscopic technique for distinguishing between tissue layers in hair follicles (Lau et al., 2011). In another study, dyed and nondyed hair were differentiated using ATR-FTIR spectroscopy. The crystal in the ATR assembly was diamond/ZnSe with a single bounce reflection. The spectral range for the study was from 4000 to 650 cm^{-1}. The volunteers for the hair sampling had variable sex, age, and race. The dyed hairs were even differentiated among themselves based on the brand of dye, its manufacturer, and color (Boll et al., 2017). In addition to these samples, entomological evidence such as insects and their different developmental stages were also studied using FTIR/NIR. In a research work by Pickering et al., ATR-FTIR spectroscopy was used for the differentiation of maggots. Four species were taken for the species identification study: *C. vomitoria, M. domesticus, L. sericata* while for the life cycle study, *C. vomitoria* was selected. The study proved to be helpful in understanding the identity as well as examining the life cycle of commonly found insects on cadavers (Pickering et al., 2015). In another study, NIR and ATR-FTIR were used for the identification of insects belonging to the Sarcophagidae family of the Diptera order. The NIR analysis was done in the spectral range of 2500–1000 nm. The study showed the potential of IR spectroscopy in developing a database for different medicolegal important insect species to be used by forensic investigators (Barbosa et al., 2018).

5.2 Inks and questionable documents

The examination of inks has been extensively carried out by forensic researchers and investigators using FTIR/NIR spectroscopy. In a study, inks on questionable documents were examined using micro-FTIR spectroscopy. Around 70 samples of blue and black inks were taken from gel and ballpoint pens. Chemicals such as ethylene glycol, styrene, and luxol fast blue were observed in the IR spectra of dyes (Fig. 4) (Zięba-Palus and Kunicki, 2006). In another study, black inkjet inks were differentiated using fiberoptic reflection spectra visible NIR spectroscopy. Nineteen samples of six different brands were analyzed during the study, for which the spectral range

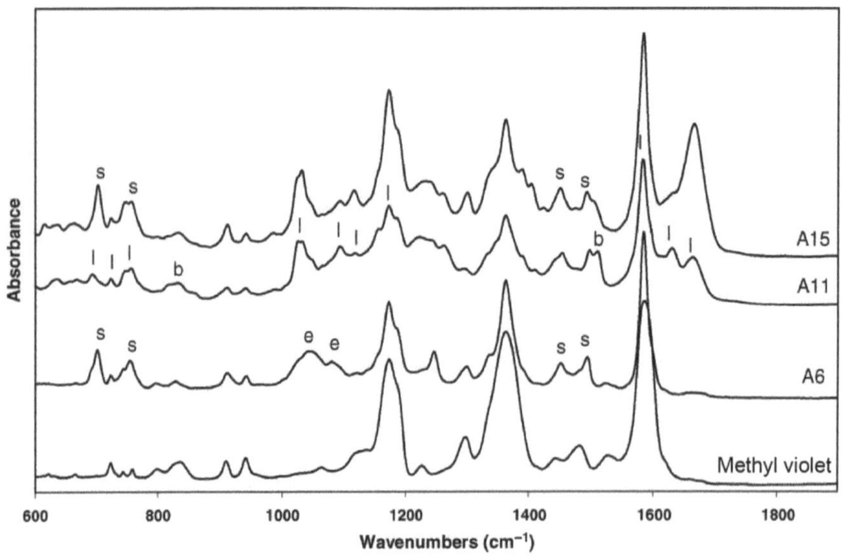

e – Ethylene glycol

l – Luxol fast blue

s – Styrene

Fig. 4 IR spectra of blue inks (A15, A11, A6) from ballpoint pens and the main dye (methyl violet) (Zięba-Palus and Kunicki, 2006).

was 1100–200 nm. It was observed that the spectra of different brands varied significantly, mostly in the region of 600–1000 nm (Gál et al., 2015). Another research work by Kumar and Sharma focused on utilizing NIR along with UV and visible spectroscopy for the analysis of blue ink from ballpoint pens. It was observed that all 57 samples were differentiated on the basis of their spectra (Kumar and Sharma, 2017). Oravec et al. used FT-NIR spectroscopy to classify black inks from inkjet printers. The samples were collected from three widely sold brands of printers. The spectral region for the study was 10,000–4000 cm^{-1}. The use of chemometrics along with this technique helped in the classification of ink samples from different printers (Oravec et al., 2019). In another study from Pakistan, FTIR spectroscopy was used to discriminate between fountain pen inks of red, green, blue, and black (Sharif et al., 2019).

Diffuse reflectance UV-visible NIR spectroscopy has also been used for the examination of questionable documents. However, the study mainly focused on variations in the UV and visible region, and not in the NIR region of the spectra (Causin et al., 2012). The same characterization technique was used by Kumar et al. as well for the classification of different types of paper for office use, writing, and photocopying. Like the previous study, the UV-visible region helped more in paper discrimination than the NIR region (Kumar et al., 2015). In another study, ATR-FTIR spectroscopy was applied to classify white copy papers, which are used to prepare formal and informal documents. It was observed that the region between 1500 and 1200 cm^{-1} was the most likely region for the discrimination of papers (Chuen Lee et al., 2016). Toners printed on questionable documents have also been analyzed using NIR spectroscopy in combination with chemometrics (Materazzi et al., 2017). Relics of papers of cultural heritage have also been analyzed using the combination of FTIR and chemometrics (Xia et al., 2019).

5.3 Explosives and gunshot residue

Infrared-based techniques such as FTIR and NIR have been widely used in forensic laboratories for the identification and chemical characterization of explosive materials and their residue. The detection of explosives through IR techniques for forensic, international, and homeland security applications has been compiled in a past review (López-López and García-Ruiz, 2014). In a research study, residues collected after consumer fireworks blasts were examined using the ATR-FTIR technique. Initially, the IR spectra of

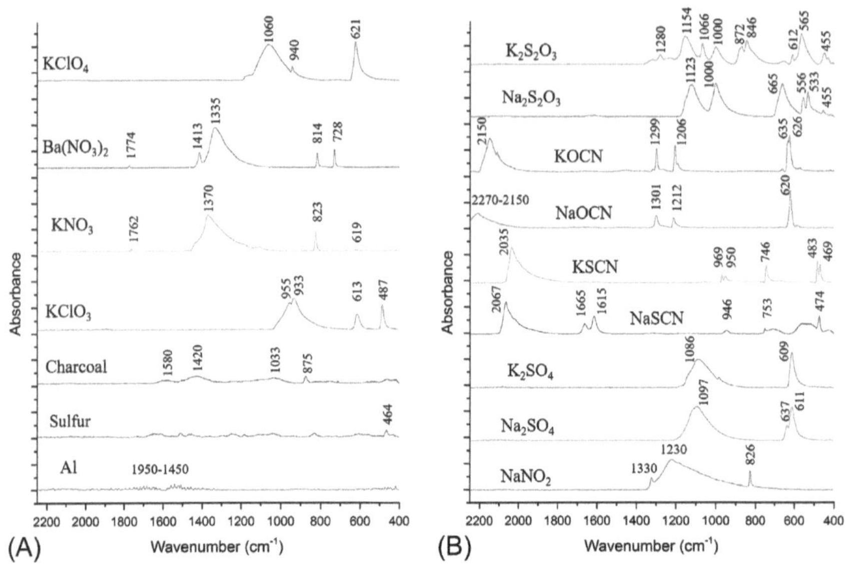

Fig. 5 IR spectra of (A) chemicals used in fireworks; and (B) chemicals found after blast residue of fireworks (Martín-Alberca et al., 2016).

the chemicals that comprised these fireworks were recorded and further, the residues of these fireworks after blasting were analyzed (Fig. 5). The spectral range for the analysis was 4000–400 cm^{-1}. Some of the chemical compounds found in the residues through IR analysis were charcoal, potassium nitrate, potassium chlorate, potassium perchlorate, and barium nitrate (Martín-Alberca et al., 2016). Propellants used in improvised explosive devices have also been discriminated using photoacoustic spectroscopy based on FTIR. In a study, propellants from four different brands were differentiated based on the aforementioned technique in combination with chemometrics. The analysis was done in the range of 3500–500 cm^{-1}. The spectra obtained showed peaks regarding the typical chemical components of any propellant such as nitroglycerin, nitrocellulose, diphenylamine, and guanidine (Álvarez et al., 2017).

In addition to explosives, GSR, which is commonly collected from crime scenes with firearm-related incidents, has also been analyzed through FTIR/NIR techniques. In a review that focused on developing luminescent markers using metal organic frameworks for GSR, the potential of IR techniques was shown for GSR analysis (Harshey et al., 2020). In a research work by Bueno et al., ATR-FTIR spectroscopy was utilized for the investigation of GSR. The study was focused on the determination of ammunition through GSR analysis. The spectra were obtained in the range of 1800–600 cm^{-1}. It was observed that most of the important peaks were observed in this region, and outside this region, only peaks specific to C—H stretching were observed (Bueno et al., 2013).

5.4 Illicit and counterfeit drugs

Illicit materials such as drugs and liquor have been widely examined through the use of IR techniques such as FTIR and NIR. These illicit materials are extensively consumed at rave parties. Many of the forensic researchers have utilized FTIR/NIR in their research work for the detection of these materials. In a study, ATR-FTIR was used for the chemical profiling of cocaine, a widely consumed illicit drug. The ATR assembly was equipped with a diamond crystal having a single reflection. The spectral range for the study was 1800–650 cm^{-1} because this region was the fingerprint region for the cocaine analysis (Marcelo et al., 2015). In another study by Pereira et al., mid-IR spectroscopy was used for the detection and classification of ecstasy tablets containing psychoactive drugs. The scanning range was 4000–400 cm^{-1}. The IR spectra revealed characteristic peaks for the drugs present in the seized tablets of ecstasy (Pereira et al., 2016). Illicit liquors collected from different locations in Punjab, India, were also examined using a combination of ATR-FTIR with chemometrics (Yadav and Sharma, 2019).

In addition to illicit materials, counterfeit drugs have also been analyzed through these techniques. In a recent study, counterfeit Viagra and Cialis were detected using ATR-FTIR spectroscopy in combination with chemometry. A total of 53 commercial and 104 counterfeit drug samples were analyzed during the study. The spectral range for the study was from 1800 to 525 cm^{-1}. The spectral region between 1800 and 525 cm^{-1} proved to be the fingerprint region for differentiating between the drug samples (Ortiz et al., 2013). A similar group of researchers carried out a research work to develop a framework in order to select characterization techniques to detect counterfeiting in the same drugs, that is, Cialis and Viagra. Along with other techniques, ATR-FTIR spectroscopy was used with a diamond crystal equipped single reflection ATR assembly. The spectral range for scanning the samples was 4000–525 cm^{-1} (Anzanello et al., 2014). In another study focusing on the pharmaceutical market of Brazil, tablet forms of sildenafil and tadalafil were analyzed through ATR-FTIR spectroscopy. The spectra were obtained in the range of 4000–400 cm^{-1}. The specific peaks for drugs were obtained in the IR spectra. Fig. 6 shows the overlayed and averaged IR spectra of sildenafil (Coelho Neto and Lisboa, 2017).

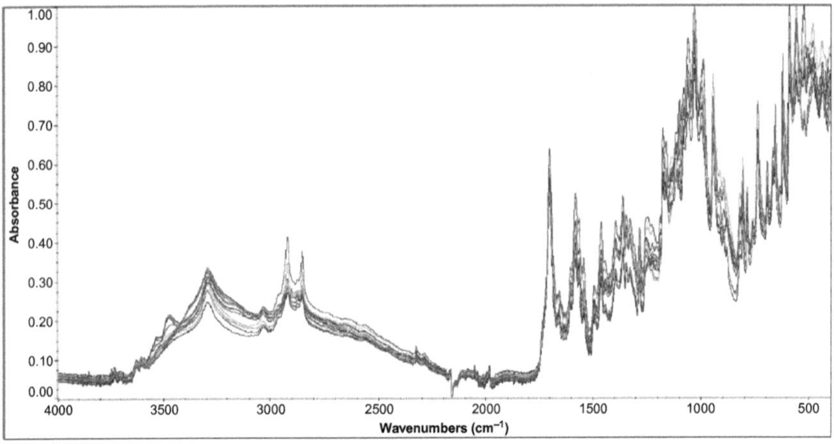

Fig. 6 IR spectra (overlayed and averaged) of sildenafil showing characteristic peaks at 734, 939, 1171, and 1698 cm^{-1} (Coelho Neto and Lisboa, 2017).

5.5 Bank notes

Counterfeiting currency notes is a major issue for any country. It directly affects the growth and economy of the country and brings it down to very low levels if these notes are not detected in a proper time frame. Various types of analytical techniques can be employed to check for counterfeiting in bank notes. Forensic experts have also utilized IR technology for this purpose.

In a study by Correia et al., portable NIR spectroscopy was utilized for the analysis of authentic and counterfeit bank notes of Brazil. The counterfeit notes were made from laserJet and deskjet printers. The spectral range for the analysis was 1676–908 nm. The spectral regions of 1600–1400 and 1100–1000 nm were of relevance to the discrimination between authentic and counterfeit notes. The region below 1200 nm had a distinction between authentic notes. Hence, it was proposed that this region could be used to differentiate between authentic notes of different values as well (Correia et al., 2018).

5.6 Paint

Paint analysis is of great forensic importance when it comes to investigations related to hit and run or vandalism. As per the Locard exchange principle, paint traces are exchanged whenever any vehicle hits an object or person, or any tool is smashed on any wall or object. Such cases require the analysis of paint by which the origin of that paint source, its manufacturer, etc., can be identified. This section focuses on the research works that have utilized FTIR/NIR spectroscopic techniques for the examination of paints.

In a study by Muehlethaler et al., FTIR in combination with Raman spectroscopy and chemometrics was used to forensically analyze paint samples. A total of 34 red paint specimens were collected and analyzed through FTIR spectroscopy, in which the spectra was recorded for the spectral range of 4000–650 cm^{-1} with a resolution of 4 cm^{-1}. The composition of paints as per the IR spectra of seven groups formed was comprised of binders such as acrylic and styrene as well as extenders such as magnesium silicate and calcium carbonate (Muehlethaler et al., 2011). Another group led by Muehlethaler published a research work focusing on the use of FTIR spectroscopy for the discrimination of spray paints of different colors such as green, blue, and red. A total of 74 cans of spray paint was used for the study. The IR spectra were obtained from 4000 to 650 cm^{-1} at 4 cm^{-1} resolution. It was observed that alkyd binders (orthophthalic acid) were commonly used in paint cans collected from the market as compared to the acrylic binders. The study proposed developing a database of the FTIR analysis of paint samples that can assist crime investigators in solving cases (Muehlethaler et al., 2014). In another study by Sciutto et al., FT-NIR was used as an advanced spectroscopic technique for the examination of cross-sections of paint samples. The IR spectra were recorded in the NIR and MIR region in the range of 8000–675 cm^{-1}, having a mode of total reflection and a resolution of 4 cm^{-1}. Fig. 7 shows the NIR spectra of the first layer of paint sample and reference NIR spectra for animal glue and egg protein. The spectra of paint matched with that of animal glue at 4896 cm^{-1}, suggesting the presence of animal glue in the paint sample (Sciutto et al., 2014).

6 Conclusion

In this chapter, the theory and principle and a brief overview about the instrumentation of infrared and near-infrared spectroscopies were given. These techniques are highly useful in the analysis of forensic samples as they are nondestructive in nature and require minimal sample preparation. FTIR is the most commonly used technique all over the world. This technique is capable of analyzing the chemical composition of solids, liquids, and even gaseous samples through the detection of fundamental vibrations, especially rotational and translational. NIR is capable of providing both the structural and chemical composition of any sample, as it is depends on the overtone

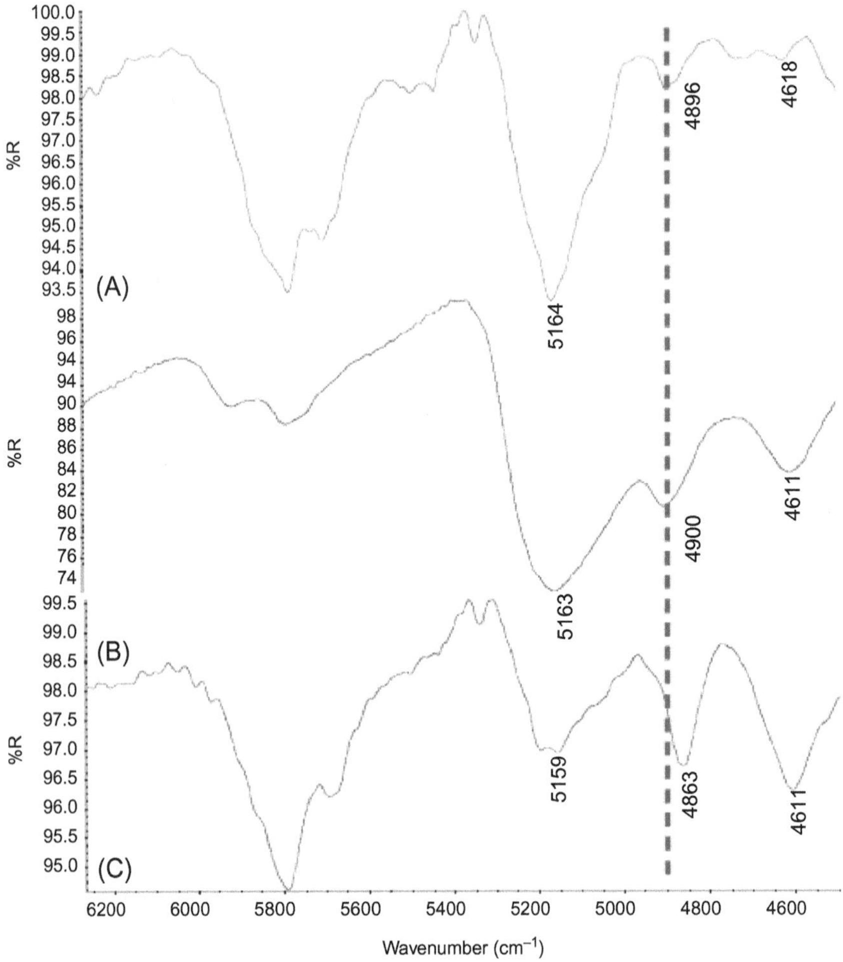

Fig. 7 NIR spectra of (A) first layer of paint sample; (B) animal glue; (C) white egg protein (Sciutto et al., 2014).

and combination bands produced by the sample. It can also provide fast and reliable results and separate the main sample from any contaminant.

In forensic science, these techniques have been used for the analysis of biological samples, inks, questionable documents, explosives, gunshot residue, illicit and counterfeit drugs, bank notes, and paints. These techniques can also be paired with microscopes in order to not only get the chemical and structural composition of a sample, but also to get a detailed visual look. These techniques are also widely used in forensic science laboratories across the world. While these techniques provide excellent analysis of the sample, further research into the sampling techniques and the development of novel detectors will significantly enhance the current mode of analysis in forensic science.

References

Ahmed, M.K., McLeod, M.P., Nézivar, J., Giuliani, A.W., 2010. Fourier transform infrared and near-infrared spectroscopic methods for the detection of toxic diethylene glycol (DEG) contaminant in glycerin based cough syrup. Spectroscopy 24, 601–608. https://doi.org/10.1155/2010/608749.

Álvarez, Á., Yáñez, J., Contreras, D., Saavedra, R., Sáez, P., Amarasiriwardena, D., 2017. Propellant's differentiation using FTIR-photoacoustic detection for forensic studies of improvised explosive devices. Forensic Sci. Int. 280, 169–175. https://doi.org/10.1016/j.forsciint.2017.09.018.

Anzanello, M.J., Ortiz, R.S., Limberger, R., Mariotti, K., 2014. A framework for selecting analytical techniques in profiling authentic and counterfeit Viagra and Cialis. Forensic Sci. Int. 235, 1–7. https://doi.org/10.1016/j.forsciint.2013.12.005.

Barbosa, T.M., de Lima, L.A.S., dos Santos, M.C.D., Vasconcelos, S.D., Gama, R.A., Lima, K.M.G., 2018. A novel use of infra-red spectroscopy (NIRS and ATR-FTIR) coupled with variable selection algorithms for the identification of insect species (Diptera: Sarcophagidae) of medico-legal relevance. Acta Trop. 185, 1–12. https://doi.org/10.1016/j.actatropica.2018.04.025.

Bell, S.E.J., Xu, Y., 2018. Infrared spectroscopy: industrial applications. In: Reference Module in Chemistry, Molecular Sciences and Chemical Engineering. Elsevier, https://doi.org/10.1016/B978-0-12-409547-2.14526-3. B9780124095472144000.

Biggs, P., Hancock, G., Heard, D.E., Wayne, R.P., 1990. A stop-scan interferometer used for time-resolved FTIR emission spectroscopy. Meas. Sci. Technol. 1, 630–636. https://doi.org/10.1088/0957-0233/1/7/016.

Boll, M.S., Doty, K.C., Wickenheiser, R., Lednev, I.K., 2017. Differentiation of hair using ATR FT-IR spectroscopy: a statistical classification of dyed and non-dyed hairs. Forensic Chem. 6, 1–9. https://doi.org/10.1016/j.forc.2017.08.001.

Bueno, J., Sikirzhytski, V., Lednev, I.K., 2013. Attenuated total reflectance-FT-IR spectroscopy for gunshot residue analysis: potential for ammunition determination. Anal. Chem. 85, 7287–7294. https://doi.org/10.1021/ac4011843.

Causin, V., Casamassima, R., Marruncheddu, G., Lenzoni, G., Peluso, G., Ripani, L., 2012. The discrimination potential of diffuse-reflectance ultraviolet–visible–near infrared spectrophotometry for the forensic analysis of paper. Forensic Sci. Int. 216, 163–167. https://doi.org/10.1016/j.forsciint.2011.09.015.

Chuen Lee, L., Liong, C.-Y., Aziz Jemain, A., 2016. Applying Fourier-transform infrared spectroscopy and self-organizing maps for forensic classification of white-copy papers. Int. J. Adv. Sci. Eng. Inf. Technol. 6, 1033. https://doi.org/10.18517/ijaseit.6.6.1425.

Coelho Neto, J., Lisboa, F.L.C., 2017. ATR-FTIR characterization of generic brand-named and counterfeit sildenafil- and tadalafil-based tablets found on the Brazilian market. Sci. Justice 57, 283–295. https://doi.org/10.1016/j.scijus.2017.04.009.

Correia, R.M., Domingos, E., Tosato, F., Aquino, L.F.M., Fontes, A.M., Cáo, V.M., Filgueiras, P.R., Romão, W., 2018. Banknote analysis by portable near infrared spectroscopy. Forensic Chem. 8, 57–63. https://doi.org/10.1016/j.forc.2018.02.003.

Elkins, K.M., 2011. Rapid presumptive "fingerprinting" of body fluids and materials by ATR FT-IR spectroscopy. J. Forensic Sci. 56, 1580–1587. https://doi.org/10.1111/j.1556-4029.2011.01870.x.

Gál, L., Oravec, M., Gemeiner, P., Čeppan, M., 2015. Principal component analysis for the forensic discrimination of black inkjet inks based on the Vis–NIR fibre optics reflection spectra. Forensic Sci. Int. 257, 285–292. https://doi.org/10.1016/j.forsciint.2015.09.011.

Harshey, A., Das, T., Srivastava, A., 2020. Analytical contributions of lanthanide based metal-organic frame works as luminescent markers: recent trends in gunshot residue analysis. Microchem. J. 154, 104597. https://doi.org/10.1016/j.microc.2020.104597.

Kawata, S., Inouye, Y., Minami, S., 1989. Compact multichannel FTIR-sensor with a savart-plate interferometer. In: Cameron, D.G. (Ed.), Presented at the Intl Conf on Fourier and Computerized Infrared Spectroscopy, Fairfax, VA, p. 567, https://doi.org/10.1117/12.969605.

Kumar, R., Sharma, V., 2017. A novel combined approach of diffuse reflectance UV–vis-NIR spectroscopy and multivariate analysis for non-destructive examination of blue ballpoint pen inks in forensic application. Spectrochim. Acta A Mol. Biomol. Spectrosc. 175, 67–75. https://doi.org/10.1016/j.saa.2016.12.008.

Kumar, R., Kumar, V., Sharma, V., 2015. Discrimination of various paper types using diffuse reflectance ultraviolet–visible near-infrared (UV-vis-NIR) spectroscopy: forensic application to questioned documents. Appl. Spectrosc. 69, 714–720. https://doi.org/10.1366/14-07663.

Lau, K., Hedegaard, M.A.B., Kloepper, J.E., Paus, R., Wood, B.R., Deckert, V., 2011. Visualization and characterisation of defined hair follicle compartments by Fourier transform infrared (FTIR) imaging without labelling. J. Dermatol. Sci. 63, 191–198. https://doi.org/10.1016/j.jdermsci.2011.05.002.

Lin, H., Zhang, Y., Wang, Q., Li, B., Huang, P., Wang, Z., 2017. Estimation of the age of human bloodstains under the simulated indoor and outdoor crime scene conditions by ATR-FTIR spectroscopy. Sci. Rep. 7, 13254. https://doi.org/10.1038/s41598-017-13725-1.

López-López, M., García-Ruiz, C., 2014. Infrared and Raman spectroscopy techniques applied to identification of explosives. TrAC Trends Anal. Chem. 54, 36–44. https://doi.org/10.1016/j.trac.2013.10.011.

Manley, M., Baeten, V., 2018. Spectroscopic technique: near infrared (NIR) spectroscopy. In: Modern Techniques for Food Authentication. Elsevier, pp. 51–102, https://doi.org/10.1016/B978-0-12-814264-6.00003-7.

Marcelo, M.C.A., Mariotti, K.C., Ferrão, M.F., Ortiz, R.S., 2015. Profiling cocaine by ATR–FTIR. Forensic Sci. Int. 246, 65–71. https://doi.org/10.1016/j.forsciint.2014.11.011.

Martín-Alberca, C., Zapata, F., Carrascosa, H., Ortega-Ojeda, F.E., García-Ruiz, C., 2016. Study of consumer fireworks post-blast residues by ATR-FTIR. Talanta 149, 257–265. https://doi.org/10.1016/j.talanta.2015.11.070.

Materazzi, S., Risoluti, R., Pinci, S., Saverio Romolo, F., 2017. New insights in forensic chemistry: NIR/Chemometrics analysis of toners for questioned documents examination. Talanta 174, 673–678. https://doi.org/10.1016/j.talanta.2017.06.044.

Maynard, P., Jenkins, J., Edey, C., Payne, G., Lennard, C., McDonagh, A., Roux, C., 2009. Near infrared imaging for the improved detection of fingermarks on difficult surfaces. Aust. J. Forensic Sci. 41, 43–62. https://doi.org/10.1080/00450610802172248.

Mohamed, M.A., Jaafar, J., Ismail, A.F., Othman, M.H.D., Rahman, M.A., 2017. Fourier transform infrared (FTIR) spectroscopy. In: Membrane Characterization. Elsevier, pp. 3–29, https://doi.org/10.1016/B978-0-444-63776-5.00001-2.

Muehlethaler, C., Massonnet, G., Esseiva, P., 2011. The application of chemometrics on infrared and Raman spectra as a tool for the forensic analysis of paints. Forensic Sci. Int. 209, 173–182. https://doi.org/10.1016/j.forsciint.2011.01.025.

Muehlethaler, C., Massonnet, G., Esseiva, P., 2014. Discrimination and classification of FTIR spectra of red, blue and green spray paints using a multivariate statistical approach. Forensic Sci. Int. 244, 170–178. https://doi.org/10.1016/j.forsciint.2014.08.038.

Oravec, M., Beganović, A., Gál, L., Čeppan, M., Huck, C.W., 2019. Forensic classification of black inkjet prints using Fourier transform near-infrared spectroscopy and linear discriminant analysis. Forensic Sci. Int. 299, 128–134. https://doi.org/10.1016/j.forsciint.2019.03.041.

Ortiz, R.S., de CássiaMariotti, K., Fank, B., Limberger, R.P., Anzanello, M.J., Mayorga, P., 2013. Counterfeit Cialis and Viagra fingerprinting by ATR-FTIR spectroscopy with chemometry: can the same pharmaceutical powder mixture be used to falsify two medicines? Forensic Sci. Int. 226, 282–289. https://doi.org/10.1016/j.forsciint.2013.01.043.

Ozaki, Y., Genkawa, T., Futami, Y., 2017. Near-infrared spectroscopy. In: Encyclopedia of Spectroscopy and Spectrometry. Elsevier, pp. 40–49, https://doi.org/10.1016/B978-0-12-409547-2.12164-X.

Pandey, G., Rawtani, D., Agrawal, Y.K., 2017. Future aspects of halloysite nanotubes in forensic investigations. J. Nanomed. Res. 6. https://doi.org/10.15406/jnmr.2017.06.00153.

Pereira, H.V., Amador, V.S., Sena, M.M., Augusti, R., Piccin, E., 2016. Paper spray mass spectrometry and PLS-DA improved by variable selection for the forensic discrimination of beers. Anal. Chim. Acta 940, 104–112. https://doi.org/10.1016/j.aca.2016.08.002.

Pickering, C.L., Hands, J.R., Fullwood, L.M., Smith, J.A., Baker, M.J., 2015. Rapid discrimination of maggots utilising ATR-FTIR spectroscopy. Forensic Sci. Int. 249, 189–196. https://doi.org/10.1016/j.forsciint.2015.01.036.

Rawtani, D., Tharmavaram, M., Pandey, G., Hussain, C.M., 2019. Functionalized nanomaterial for forensic sample analysis. TrAC Trends Anal. Chem. 120, 115661. https://doi.org/10.1016/j.trac.2019.115661.

Sciutto, G., Prati, S., Bonacini, I., Oliveri, P., Mazzeo, R., 2014. FT-NIR microscopy: an advanced spectroscopic approach for the characterisation of paint cross-sections. Microchem. J. 112, 87–96. https://doi.org/10.1016/j.microc.2013.09.021.

Sharif, M., Batool, M., Chand, S., Farooqi, Z.H., Tirmazi, S.A.A.S., Athar, M., 2019. Forensic discrimination potential of blue, black, green, and red colored fountain pen inks commercially used in Pakistan, by UV/visible spectroscopy, thin layer chromatography, and Fourier transform infrared spectroscopy. Int. J. Anal. Chem. 2019, 1–10. https://doi.org/10.1155/2019/5980967.

Signori, V., Lewis, D.M., 1997. FTIR investigation of the damage produced on human hair by weathering and bleaching processes: implementation of different sampling techniques and data processing. Int. J. Cosmet. Sci. 19, 1–13. https://doi.org/10.1111/j.1467-2494.1997.tb00161.x.

Stuart, B., 2004. Infrared Spectroscopy: Fundamentals and Applications. Wiley.

Tahtouh, M., Despland, P., Shimmon, R., Kalman, J.R., Reedy, B.J., 2007. The application of infrared chemical imaging to the detection and enhancement of latent fingerprints: method optimization and further findings. J. Forensic Sci. 52, 1089–1096. https://doi.org/10.1111/j.1556-4029.2007.00517.x.

Toon, G.C., 1991. The JPL MkIV interferometer. Opt. Photonics News 2, 19. https://doi.org/10.1364/OPN.2.10.000019.

Xia, J., Zhang, J., Zhao, Y., Huang, Y., Xiong, Y., Min, S., 2019. Fourier transform infrared spectroscopy and chemometrics for the discrimination of paper relic types. Spectrochim. Acta A Mol. Biomol. Spectrosc. 219, 8–14. https://doi.org/10.1016/j.saa.2018.09.059.

Yadav, P.K., Sharma, R.M., 2019. Classification of illicit liquors based on their geographic origin using attenuated total reflectance (ATR)—Fourier transform infrared (FT-IR) spectroscopy and chemometrics. Forensic Sci. Int. 295, e1–e5. https://doi.org/10.1016/j.forsciint.2018.12.017.

Zięba-Palus, J., Kunicki, M., 2006. Application of the micro-FTIR spectroscopy, Raman spectroscopy and XRF method examination of inks. Forensic Sci. Int. 158, 164–172. https://doi.org/10.1016/j.forsciint.2005.04.044.

CHAPTER 5

Atomic absorption and emission spectrometry in forensic analysis

1 Introduction

Atomic spectroscopy is an excellent analytical tool used for the detection and measurement of elements in a sample with high precision and confidence. Atomic spectroscopy mainly consists of two types of techniques: atomic absorption spectroscopy and atomic emission spectroscopy. In forensic science, these techniques are of utmost importance because of their ability to detect even trace amounts of elements in very small amounts of a sample. Through this, it is possible to associate the available material with the victim or the perpetrator. It can also be used for the sampling of varied samples with high sensitivity and quantification.

Atomic spectroscopy is typically based on the analysis of the electromagnetic radiation emitted by the atoms in an element. This electromagnetic radiation is highly unique to the particular atom; therefore the detection is very accurate even for small sample amounts. Typically, in forensic science, samples such as glass, paint, soil, drugs, and gunshot residue are commonly encountered. Therefore, through atomic spectroscopy it is possible to detect these elements and obtain a confirmatory analysis of the same. In this chapter, the principle and theory of atomic absorption spectroscopy and atomic emission spectroscopy are discussed along with their instrumentations. Also, their applications in the analysis of various forensic evidence are also discussed.

2 Principle and theory of atomic absorption and emission spectroscopy

The electromagnetic radiation that is composed of two components—electric and magnetic fields oscillating perpendicularly with each other—is often represented in the form of a line spectrum where each line is composed of monochromatic radiation, that is, radiation of a single wavelength. In this spectrum, a particular region can be found exclusively for the

Handbook of Analytical Techniques for Forensic Samples
https://doi.org/10.1016/B978-0-12-822300-0.00005-7

analysis of atomic spectrometry ranging from 190 to 850 nm covering the entire range from the UV to infrared. Every atom in an element has a number of energy bands. When these atoms are radiated with an external source, the electrons in these levels get excited, absorb energy, and transition to a higher energy level. This absorption is specific to the quantum theory in which the energy absorbed is equivalent to the energy difference between the energy levels of the ground state and the excited state of the atom.

In order to obtain further insight into the ground and excited states of the electron, it is important to understand the atomic theory of the atom, which describes the different orbitals of the atom in which the electrons are present. These orbitals get filled progressively as the atomic number increases. The electron is considered to be in a ground state if it is present in the first orbital, which is the s orbital. It is considered to be in an excited state when it transitions to a higher level such as the p, d, or f orbital. The transition of the electrons from the ground state to the excited state also depends on the wavelength of the light responsible for the excitation of the electron. In atomic absorption spectroscopy, the absorption of light by electrons is taken into account. The principle of atomic emission theory will be explained further in the section.

The absorption of light by an atom is governed by a very crucial law known as the Beer Lambert law, which takes into account the intensity of the transmitted light from the electrons and relates it to its concentration. The Beer Lambert law is represented as follows:

$$A = abc = \log (I_o/I)$$

where I_o is the intensity of the initial radiation while I is the intensity of the transmitted light. a denotes the absorptivity coefficient while b and c correspond to the path length and the concentration, respectively. However, this law holds certain limitations. One limitation is that this law is valid as long as the sample is homogenous and does not hold for higher or extremely low concentrations. Therefore, a modification of this law takes into account the charge, mass, speed, and number of atoms that are free to absorb the light. The ability of the atom to efficiently absorb light is expressed in terms of the oscillator strength, which also serves as the basis for performing the quantitative analysis of the samples.

When the sample is irradiated with a light source, it absorbs energy and transitions to a higher energy level from its ground state. However, once it reaches the higher energy level, it quickly loses its energy to come back to the ground state. When it transitions back to the ground state, it emits a certain

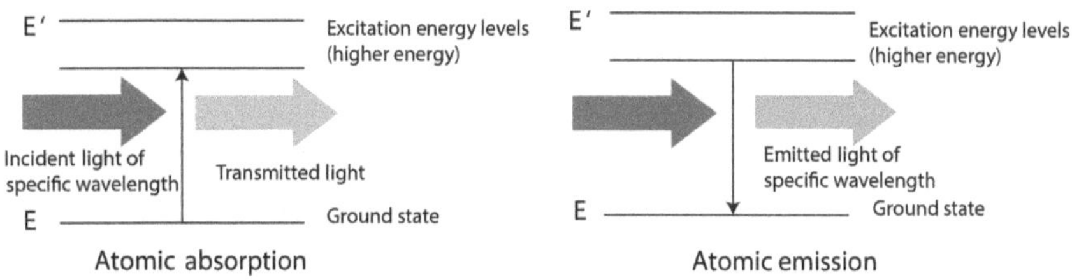

Fig. 1 Principles of atomic spectroscopy.

amount of energy, which lasts for a very short time. This emitted energy serves as the basis of the detection of elements in atomic emission spectroscopy. Similar to the amount of energy absorbed by the atoms that is determined by the energy difference between the two energy levels, the emission is also characterized by the energy difference between the two levels that corresponds to the excited and ground states of the electron. Also, as the number of electrons increases, the intensity of the emitted radiation also increases. Therefore, under appropriate conditions, it is possible to correlate the intensity of the emitted radiation with the concentration of the atoms present in the sample. The emission is characteristic to each atom due to the unique electronic structure of each atom and is radiated in the form of discrete wavelength lines instead of a continuous spectrum. These discrete lines, as mentioned before, are monochromatic in nature and therefore accurately represent the atom (Cantle, 1982; Van Loon, 1980; Welz and Sperling, 1999). Fig. 1 highlights the various techniques of atomic spectroscopy.

3 Instrumentation of atomic absorption and atomic emission spectroscopy

In this section, the instrumentation of atomic absorption and atomic emission spectroscopy will be briefly discussed.

3.1 Atomic absorption spectroscopy

This technique consists of five main components: mainly the radiation source, a sample cell that also consists of an atomizer to split the sample into its atomic counterparts, a monochromator, a grating slit, and a detector. Apart from this, a computer is also used to display and process the results obtained.

Typically, the atomizer used in AAS is a chemical flame that is formed through a combination of fuel gas and oxidants. Once, the sample is prepared which is usually in the liquid form, it is passed through a nebulizer through a capillary and is dispersed in the form of small droplets into the flame. Here, the portion where the measurement radiation passes through is known as the absorption volume. However, many techniques have now replaced flames with arcs and plasma, which provide even more energy to convert the liquid samples into gaseous forms. Apart from this conversion, the atomizer is also responsible for providing an appropriate chemical environment to allow both the oxidation and reduction reactions. It should also not hamper the incident measurement radiation in any way and should be nonturbulent. The transport of the sample through the sample usually occurs in three phases in which the sample droplets are formed, followed by the separation of larger droplets. Once they are separated, the droplets are transported into the flame. Also, the samples are introduced into the instrument either through injection or through a pump. Several techniques have also allowed atomization without flames. In these techniques, electrothermal atomization occurs in which a furnace that is electrically heated is used as an apparatus to hold the sample volume that is later atomized due to the heat. This technique is useful in cases where the solid samples need to be directly measured or when the chemical changes occurring during the atomization need to be directly measured. However, this technique is less sensitive and precise than the flame technique.

The radiation sources used for the production of the measurement radiation are hydrogen or a deuterium-filled hollow cathode lamp (HCL) that is responsible for emitting the light of the continuum radiation. Later, an HCL line source was used that emitted light of discrete wavelengths, which turned out to be more suitable for analysis as this technique required a very narrow range of the spectrum. This lamp typically consists of a cathode that is usually a metallic cup or cylinder and an anode encased in a glass shell. Because AAS allows the detection of only one atom at a time, the cathode is made of the metal that is supposed to be analyzed. To these electrodes, high-energy electric voltage is applied through which the emission spectra are generated. Apart from this, multielement sources hat consist of cathodes made of two or more elements and electrodeless discharge lamps are also used. Another major component is the monochromator, which is responsible for the isolation and control of the transmitted light and is used in the form of a diffraction grating slit. These slits are placed between the source and the sample and also between the sample and the detector. Due to this, it is possible to create a light source of a very narrow range and a specific wavelength reaches the sample. The transmitted radiation is also carefully separated in

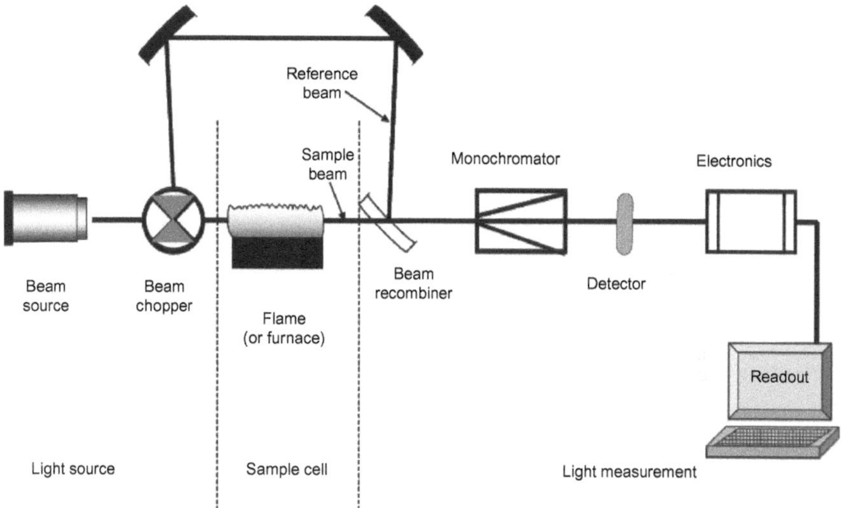

Fig. 2 Instrumentation of atomic absorption spectroscopy (Cheremisinoff, 1996).

order to allow the transmission of the characteristic radiation only. Typically, photomultiplier tubes are used as detectors in which an electric current is generated when the photons from the transmitted radiation hit the tube. This electric current is further amplified and displayed on the computer for visualization and processing, wherein the intensity of the transmitted light is compared with the intensity of the initial radiation and the absorbance is determined (Fig. 2). Various types of samples such as aqueous, organic and inorganic solutions, organic solids and gaseous samples can be analyzed through this technique. Aqueous solutions are usually degassed, and any impurities are removed either through filtration or precipitation. Organic and inorganic samples such as petroleum products, fertilizers, or soil samples that are commonly encountered in forensic science can be analyzed by selecting a suitable solvent either for their extraction or dissolution. Gaseous samples are not frequently encountered in forensic science and if encountered they can be analyzed either by dissolving them in a liquid solvent or by converting them into a solution (Cantle, 1982; Moore, 1989a,b; Van Loon, 1980; Welz and Sperling, 1999).

3.2 Instrumentation of atomic emission spectroscopy

The instrumentation of flame AES is similar to that of flame AAS in that it also contains a radiation source, an atomizer, a monochromator or a filter, and a detector. However, in this section, emphasis is given on the use of inductively coupled plasma AES, which uses plasma as the ionization source. Plasma is generated when a gaseous mixture of cations and anions consisting of extremely high energies interact with each other. They are typically generated from argon and are therefore inert. Typically, in an ICP-AES, an argon plasma source, a monochromator, a detector, and a computer are present. The plasma source is composed of three quartz tubes that are concentrically arranged around a copper coil. This coil acts as the load coil because a radiofrequency power is applied to this power. Argon gas is pumped in this system and it is used to initiate the plasma through the application of a spark that produces the cations and anions. Due to the application of the RF power to the copper coil, a magnetic field is generated that is responsible for maintaining the path of the ions generated. The argon ions generated collide with free electrons until plasma that is self-sustaining and of high energy is produced. This technique is called "inductively coupled" because of the use of a magnetic field to generate electromotive forces. Once the plasma is generated, the samples are aerosolized and carried to the base of the coil where the plasma is being generated. Once the sample hits the sample, it undergoes a series of processes that ends in the excitation of the electrons present in the atoms. Once the electrons are excited, the emission radiation is emitted, which is further detected by detectors. However, prior to detection, there is a series of prism and diffraction grating monochromators that isolate the emission radiation produced by the atoms. This optical system is also present between the plasma and the sample. The detectors used in this case are usually photomultiplier tubes or solid-state detectors such as charge coupled devices (CCD), complementary metal oxide semiconductors (CMOS), and charge injection devices (CID). This entire instrument is paired with a computer that consists of software wherein the parameters can be easily entered. Similar to AAS, this technique can also be used to analyze organic and inorganic solids as well as liquid samples. Every sample has a specific sample preparation and extraction technique. Significant research has been done on the sample preparation of commonly encountered forensic samples; any new user can consult research papers for further information on the sample preparation of forensic evidence. Fig. 3 shows the instrumental setup of ICP-AES (Lajunen, 1992; Moore, 1989a; Yeung et al., 2017).

4 Analysis of forensic samples via AAS and AES

Atomic spectroscopy such as AAS and AES has assisted forensic investigations for many years in the elemental analysis of different samples. However,

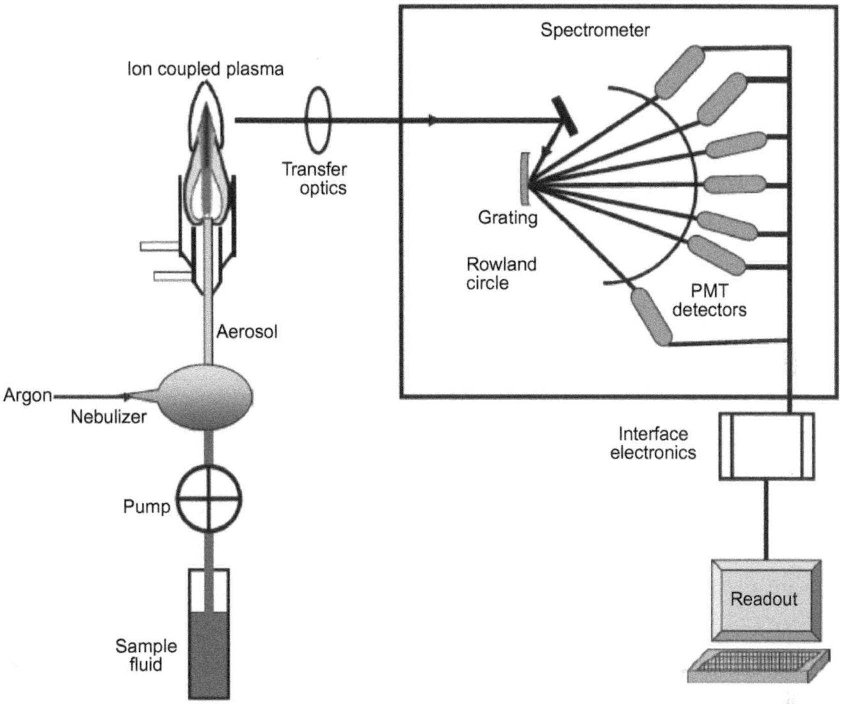

Fig. 3 Instrumental setup of ICP-atomic emission spectroscopy (Yeung et al., 2017).

with the emergence of newer techniques for elemental analysis, the use of AAS and AES has somewhat faded away from forensic laboratories. In this section, we discuss research works from the past that used AAS and AES for the analysis of evidence such as biological samples, drugs, explosives, gunshot residue (GSR), ammunition, textile fibers, glass, ignitable materials, inks, paper, and soil.

4.1 Biological samples

Biological samples are often found at a crime scene. These samples need to undergo elemental analysis in order to check for any foreign elements that may have caused harm to the victim, or give a clue about the crime committed. AAS has assisted in detecting metals in biological samples in the past (Klotz et al., 2013). Research works on biological forensic samples using AAS and AES are presented here.

In a study, fragments of a mummy were analyzed for the elements present using AAS and AES. AAS was used to detect elements such as Na, Mg, K,

Ca, Zn, Fe, Tl, Pb, Mn, and Cu while AES was used to detect Na. Samples of teeth, bones, nails, and resin from a mummy were ashed and treated with acid for AAS analysis. In bone fragments, elements were similar to those present in today's people while in nails, the percentage of elements was higher except for Pb and Cu (Kłys et al., 1999). In another study, inductively coupled plasma AES (ICP-AES) was used for the analysis of tissues from the brain, heart, pancreas, liver, kidney, ovary, and spleen during autopsy. The instrument was operated in axial mode. In most of the tissues, Fe was found in the highest concentration while Co was the lowest. Mn was majorly present in the spleen while Cu was in the brain. Al was found in the kidney while Cd was found in the cerebellum of the brain (Rahil-Khazen et al., 2002). Electrical injuries have also been analyzed by AAS using a rat as the animal model. Cu and Fe were detected in the skin samples, which were initially soaked in nitric acid for AAS analysis. Because the electrical wire used for electrocution was made of Cu, its content was higher in the skin samples while the difference in study and control rats for Fe was not significant (Acar et al., 2004). Fish and mussels were analyzed for the presence of Cd, Zn, and Cu using the AAS technique. This study could be useful in cases of metal poisoning via food (Manutsewee et al., 2007). AAS has also assisted in cases related to drowning, mainly through the analysis of Sr concentrations (Pérez-Cárceles et al., 2008). In a study, pleural effusion and cardiac blood were analyzed for Sr concentration in dead bodies drowned in seawater or freshwater. Samples were treated with nitric acid followed by AAS analysis at 460.7 nm. It was found that concentrations of Sr were higher in bodies recovered from seawater as compared to those found in freshwater (Gurler et al., 2015).

4.2 Drugs

Drugs, especially psychotropic drugs, are often abused by youngsters. Other drugs if given in excess can have adverse health effects, and can also be fatal in several cases. AAS and AES have aided in forensic research for the elemental analysis of drug samples.

In a study, conventional Chinese medicines were analyzed for the presence of Pb, Cu, and Hg using AAS. The samples were digested with dilute nitric acid prior to use. The wavelengths for flame AAS for Pb, Cu, and Hg were 283.3, 324.8, and 253.6 nm, respectively. The limit of detection (LOD) values for Pb, Cu, and Hg were 4, 0.2, and 0.03 μg/g, respectively. The developed method could be used for the forensic analysis of metals in such drugs (Chow et al., 1995). Mohammadzai and Hinze developed a method for the indirect estimation of cocaine via AAS. The cocaine sample was complexed with a salt of Fe(III), which was analyzed via AAS. The wavelength for the analysis was 248.3 nm, with a spectral bandwidth of 0.2 nm. Argon was flowing through the system at a rate of 300 mL/min. The method developed had an LOD of 0.01 ng/mL of cocaine (Mohammadzai and Hinze, 2004). A Research has also been done related to the elemental analysis of ecstasy tablets. Fierro et al. performed an analysis of Hg and As in ecstasy tablets. The research team used AAS for the analysis of As at a wavelength of 193.7 nm. Upon comparing the results with other electrochemical methods such as differential pulse anodic stripping voltammetry (DPASV) and the flow-injection mercury system (FIMS), AAS lagged behind in the analysis of As (Fierro et al., 2006). French et al. differentiated two ecstasy seizures on the basis of Ba content, analyzed via AAS (French et al., 2013).

4.3 Explosives, gunshot residue, and ammunition

Explosives, GSR, and ammunition come under the ballistics division of forensic science. Their elemental analysis has been carried out in the past using atomic spectroscopic techniques. Some such research works are discussed in this section.

Regarding the elemental analysis of explosives using AAS, a study was published in 2008 for industrial explosives. Identification particles, which help in postblast analysis for the identification of explosives, were analyzed. Flame AAS was used to detect elements such as Mg, Zn, Cu, and Pb. The LOD values for Mg, Zn, Pb, and Cu were 0.2, 1.9, 2.4, and 1.3 mg/g, respectively, of the explosive sample. The study can be helpful in determining the explosives used in terrorist activities as well (Husáková et al., 2008). GSR is the most common type of forensic evidence analyzed by atomic spectroscopy. Yourd et al. used AES coupled with ICP for the analysis of trace elements present in bullet samples. Argon gas was flowing at a rate of 15 mL/min while the temperature of the spray chamber was ambient. The method developed could be helpful for the analysis of microelements such as Ag, Bi, Cd, As, Cu, Sn, and Sb in lead collected from bullet samples (Yourd et al., 2001). Molina et al. used ICP-AES for the elemental analysis of GSR in suicides. The GSR was analyzed for elements such as Pb, Sb, and Ba. The study showed the ability of ICP-AES to differentiate between non-self-inflicted as well as self-inflicted wounds based on the elements in the GSR while also proving that this technique has merit over energy dispersive X-ray spectroscopy in elemental analysis at lower concentrations (Molina et al., 2007). The shooting distance was also estimated by studying the Pb content in the bullet holes via AAS. It was concluded that with increasing distance, the Pb content in the GSR around the bullet hole keeps decreasing (Gagliano-Candela et al., 2008). A similar kind of study was performed by Turillazzi et al., but they included the analysis of Sb and Ba along with Pb. The instrument used in this study was ICP-AES rather than AAS (Turillazzi et al., 2013). In the case of ammunition, atomic spectroscopy has also played a role in that comparison. In a study, firearm cartridge case brass was compared for elements such as Ni, Fe, and Pb via graphite furnace AAS. For the analysis, hollow cathode lamps for Ni, Fe, and Pb were used in order to provide the characteristic radiation. The estimated content of these elements helped in classifying the brass cartridges according to their origin (Heye and Thornton, 1994).

4.4 Textile fibers

Fibers usually encountered at crime scenes are analyzed in forensic laboratories to get information about their structure and composition. Microscopic techniques assist in the structural profiling of fibers. Spectroscopic techniques play a role in the chemical profiling of these fibers, which can help in their identification (Houck, 2009). Elemental analysis can also help in comparing different fibers collected from the crime scene. Atomic spectroscopic techniques have also assisted forensic scientists in this regard, and such research works are discussed here.

Saracoglu et al. used AAS to determine the trace elements (heavy metals) present in textile materials. Flame AAS as well as graphite furnace AAS were used for the study. Elements such as Fe, Pb, Ni, Cu, and Zn were detected in different textile samples taken. In most of the samples, Pb was found to be below the detection limit while Ni was present in an ample amount in most samples (Saracoglu et al., 2003). Tuzen et al. performed an elemental analysis of textiles produced in Turkey for the presence of heavy metals such as Cu, Cd, Zn, Mn, Fe, and Ni using flame AAS and graphite furnace AAS. Oeko-Tex regulates the metal content in the textile materials. The concentrations of Cu, Cd, Zn, Mn, Fe, and Ni were in the range of 0.1–341 μg/g of the textile sample. As per Oeko-Tex guidelines, the Cu and Cd concentrations were found to exceed the limit (Tuzen et al., 2008). Tonetti and Innocenti also checked the presence of heavy metals (Cr VI and total Cr) in textile materials and further verified the method developed for the analysis. AAS was used for the estimation of total Cr. The wavelength for the Cr was 357.9 nm, and Ar was flowing as an inert gas at a rate of 250 mL/min. The method developed for the extraction and estimation of Cr from textiles could assist forensic experts in comparing different fibers on the basis of their elemental composition (Tonetti and Innocenti, 2009). Another research work analyzed metals such as Cd, Al, Co, Zn, Tl, Pb, Ni, Mn, Fe, Cu, and Cr in textile fibers of different types such as acrylic, cotton, nylon, polyester, polypropylene, and viscose. These different-colored fibers were taken for the analysis. The metal content in the fibers was found to be higher upon extraction via wet digestion while lower as per Oeko-Tex guidelines when estimated in sweat extract (Sungur and Gülmez, 2015).

4.5 Glass

Glass analysis is required in cases related to vehicular accidents or burglary where broken glass fragments of windows are found as evidence. The elemental analysis of such glass samples via atomic spectroscopic techniques can help in linking the fragments to their original source (Buscaglia, 1994). The research regarding the use of such techniques for the forensic analysis of glass is discussed in this section.

In a study by Howden et al., Fe and Mg were analyzed in the fragments of glass with the help of flameless AAS. The samples were digested by polytetrafluoroethylene prior to AAS analysis. The wavelength for Fe was 248.5 nm while for Mg, it was 202.5 nm; the source was a hollow cathode lamp. It was found that the results were replicable in the window glass samples collected for the analysis (Howden et al., 1977). Hickman performed the elemental analysis of glass samples in the form of sheets and used the data to discriminate them. ICP-AES was used for elements such as Mn, Mg, Fe, Al, Ba, and Sr; flame atomization AES was used for elements such as Li, Rb, and K; and AAS was used for As. It was concluded that glasses with higher refractive index values (>1.519) can be discriminated using more than four elements while glasses with a refractive index lower than 1.519 can be discriminated on the basis of two or fewer elements. Sr helped in discriminating glasses with a lower refractive index (1.5165) while Ba helped in discriminating glasses with a higher refractive index (Hickman, 1983). Hickman and his coresearchers again carried out a similar kind of study that was focused on selecting variable elements in the glass fragments in order to classify them. Tableware, containers, and sheet glasses were used for the analysis (Hickman et al., 1983). In a relatively recent study, Pb was determined in glass samples using graphite furnace AAS. The wavelengths for the Pb analysis were 205.328 and 283.06 nm, among which the LOD value was higher at the previous wavelength (Kelestemur and Özcan, 2015).

4.6 Ignitable materials

Ignitable materials and their traces are often encountered in cases of fire and arson. Elemental analyses of such materials through AAS and AES coupled with other techniques in research works for forensic purposes are discussed in this section.

Chan used AAS combined with gas chromatography (GC) for the analysis of tetraalkyl Pb compounds present in petrol. AAS acted as the detector in the GC instrument. The flame used for AAS was air-acetylene while the wavelength selected was 217 nm, with a slit width of 1 nm. The advantage with this analysis was that there was no requirement of sample preparation. Upon analysis of petrol coming from two refineries, it was found that petrol from one refinery had tetraethyl and tetramethyl Pb in the mixture while the other refinery's petrol had only tetraethyl Pb (Chan, 1981). A similar kind of study was performed on gasoline (petrol) using AES in a combination with GC. AES was also combined with microwave-induced plasma (MIP). The LOD value for different kinds of tetraalkyl lead compounds was less than 1 μg/L of the petrol sample. The technique (GC-MIP-AES) had the advantage of fast analysis (in seconds) rather than minutes as in the case of AAS or ICP-AES (Rodriguez Pereiro and Łobiński, 1997).

4.7 Inks and paper

Inks and papers are readily investigated in questionable documents in forgeries. The elemental analysis of such evidence can help in linking it to the source from which it originated, thereby going a step toward solving the cases.

In a study from late 1970s, papers were categorized into different types on the basis of their elemental profile. Elements such as Mn, Cd, Sb, Cu, Cr, Mg, Ag, Co, Pb, and Fe were analyzed in the paper samples from different manufacturers using electrothermal AAS. The lowest wavelength was for Mg (202.5 nm) while the highest was for Cr (425.5 nm). The LOD values for these metals varied from 0.001 to 520 μg/g of the paper sample (Simon et al., 1977). Lee et al. used AAS and ICP-AES for the elemental analysis of red sealing inks collected from the paper samples. Five different types of samples were taken for the study. Elements such as Zn, Pb, and Ca were determined in the study. However, the study showed demerits over mass spectrometry, mainly because of the destructive nature of sample analysis (Lee et al., 2008).

4.8 Soil

The elemental analysis of soil samples collected from crime scenes can give an idea about its mineralogy, which can further help in identifying the location where this soil belongs. This assists in knowing the probable location of a crime.

In a research work, AAS and AES in combination with other advanced techniques were used for the analysis of soils. In addition to soils, other samples such as sediments, rocks, and ceramic materials were also analyzed. Elements such as Cr, Pb, Cu, Zn, and Ni were analyzed during the study. The study compared the results with the certified reference materials (CRMs) as well as reference materials (RMs). The method developed could be helpful in identifying the type of soil encountered during the analysis (Schrön et al., 2000). ICP-AES was used by Pye et al. for the elemental composition analysis of soil samples. Many elements were analyzed in the soil samples, among which W showed maximum variability. Among others, metal oxides as well as Sr, Nd, Sm, La, and Rb showed less variability, and therefore were considered good for differentiating among different soil samples. The study recommended carrying out the sample analysis of soils belonging to the same case in a single or sequential run in order to get reproducible results (Pye et al., 2006). ICP-AES was also used by Cheshire et al. for the differentiation of soil samples that were collected from the close proximity. The samples had single as well as mixed soils from different geographically similar locations. Elements such as Al, Ca, P, Zn, Be, Fe, Ni, Ti, and Sc were analyzed through ICP-AES. The analysis results when combined with X-ray fluorescence spectroscopy (XRF) and ICP-MS helped in differentiating the soils based on their elemental composition (Cheshire et al., 2017).

5 Conclusion

Atomic spectroscopy techniques, especially atomic absorption and emission spectroscopies, are widely used in forensic science labs for the determination of elements in the forensic evidence. These techniques are based on the excitation of atoms via a radiation source, due to which the electrons absorb energy and transit to a higher energy level. The absorption difference in the transmitted light serves as the principle for the AAS technique.

However, once it transitions to a higher energy level, it loses its energy and comes back to the ground state, during which the loss of energy is emitted in the form of radiation of the characteristic wavelength. This serves as the principle of the AES technique. The instrumental setup of both techniques are very similar in the fact that they both contain a light source, a sample holder or atomizer, a monochromator, and a detector. In the case of AAS, a chemical flame is used as the ionization source. This is also similar in the case of flame AES. However, recently, ICP-AES is more frequently used as it offers better, sensitive, and accurate results. ICP-AES uses plasma as the source of ionization of the atoms, after which they emit their characteristic radiation. In this chapter, a brief overview of the principle and theory of these techniques and their instrumentation has been given. Also, their application in the analysis of various forensic evidence such as biological samples, drugs, explosives, gunshot residue, ammunition, fibers, glass, ignitable materials, inks, papers, and soils have also been discussed.

References

Acar, K., Boz, B., Kurtulus, A., Divrikli, U., Elci, L., 2004. Using of atomic absorption spectrometry for diagnosis of electrical injuries (an experimental rat study). Forensic Sci. Int. 146, S3–S4. https://doi.org/10.1016/j.forsciint.2004.09.007.

Buscaglia, J.A., 1994. Elemental analysis of small glass fragments in forensic science. Anal. Chim. Acta 288, 17–24. https://doi.org/10.1016/0003-2670(94)85112-3.

Cantle, J.E., 1982. Atomic Absorption Spectrometry. p. 465.

Chan, L., 1981. The determination of tetraalkyl lead compounds in petrol using combined gas chromatography-atomic absorption spectrometry. Forensic Sci. Int. 18, 57–62. https://doi.org/10.1016/0379-0738(81)90139-0.

Cheremisinoff, N.P., 1996. Elemental and structural characterization tests. In: Polymer Characterization Laboratory Techniques and Analysis. Noyes Publications, pp. 43–81.

Cheshire, K., Morgan, R.M., Holmes, J., 2017. The potential for geochemical discrimination of single- and mixed-source soil samples from close proximity urban parkland locations. Aust. J. Forensic Sci. 49, 161–174. https://doi.org/10.1080/00450618.2016.1144789.

Chow, P.Y.T., Chua, T.H., Tang, K.F., Ow, B.Y., 1995. Dilute acid digestion procedure for the determination of lead, copper and mercury in traditional Chinese medicines by atomic absorption spectrometry. Analyst 120, 1221. https://doi.org/10.1039/an9952001221.

Fierro, I., Deban, L., Pardo, R., Tascón, M., Vázquez, D., 2006. Determination of mercury and arsenic in ecstasy tablets by electrochemical methods. Forensic Toxicol. 24, 70–74. https://doi.org/10.1007/s11419-006-0017-z.

French, H.E., Went, M.J., Gibson, S.J., 2013. Graphite furnace atomic absorption elemental analysis of ecstasy tablets. Forensic Sci. Int. 231, 88–91. https://doi.org/10.1016/j.forsciint.2013.04.021.

Gagliano-Candela, R., Colucci, A.P., Napoli, S., 2008. Determination of firing distance. Lead analysis on the target by atomic absorption spectroscopy (AAS). J. Forensic Sci. 53, 321–324. https://doi.org/10.1111/j.1556-4029.2008.00668.x.

Gurler, A.S., Gundogmus, U.N., Buyuk, Y., Ince, H.C., Sezer, Y., 2015. Contribution of biventricular cardiac blood and pleural effusion strontium concentrations to the diagnosis of drowning. Rom. J. Leg. Med. 23, 101–105. https://doi.org/10.4323/rjlm.2015.101.

Heye, C.L., Thornton, J.I., 1994. Firearm cartridge case comparison by graphite furnace atomic absorption spectrochemical determination of nickel, iron and lead. Anal. Chim. Acta 288, 83–96. https://doi.org/10.1016/0003-2670(94)85118-2.

Hickman, D.A., 1983. Elemental analysis and the discrimination of sheet glass samples. Forensic Sci. Int. 23, 213–223.

Hickman, D.A., Harbottle, G., Sayre, E.V., 1983. The selection of the best elemental variables for the classification of glass samples. Forensic Sci. Int. 23, 189–212. https://doi.org/10.1016/0379-0738(83)90148-2.

Houck, M.M., 2009. The use of spectroscopy for textile fiber identification. In: Identification of Textile Fibers. Elsevier, pp. 158–164. https://doi.org/10.1533/9781845695651.2.158.

Howden, C.R., German, B., Smalldon, K.W., 1977. The determination of iron and magnesium in small glass fragments using flameless atomic absorption spectrophotometry. J. Forensic Sci. Soc. 17, 153–159.

Husáková, L., Šrámková, J., Staňková, J., Němec, P., Večeřa, M., Krejčová, A., Štancl, M., Akštein, Z., 2008. Characterization of industrial explosives based on the determination of metal oxides in the identification particles by microwave digestion and atomic absorption spectrometry method. Forensic Sci. Int. 178, 146–152. https://doi.org/10.1016/j.forsciint.2008.03.009.

Kelestemur, S., Özcan, M., 2015. Determination of Pb in glasses by direct solid sampling and high-resolution continuum source graphite furnace atomic absorption spectrometry: method development and analyses of glass samples. Microchem. J. 118, 55–61. https://doi.org/10.1016/j.microc.2014.08.005.

Klotz, K., Weistenhöfer, W., Drexler, H., 2013. Determination of cadmium in biological samples. In: Sigel, A., Sigel, H., Sigel, R.K. (Eds.), Cadmium: From Toxicity to Essentiality, Metal Ions in Life Sciences. Springer Netherlands, Dordrecht, pp. 85–98. https://doi.org/10.1007/978-94-007-5179-8_4.

Kłys, M., Lech, T., Zięba-Palus, J., Białka, J., 1999. A chemical and physicochemical study of an Egyptian mummy 'Iset Iri Hetes' from the Ptolemaic period III–I B.C. Forensic Sci. Int. 99, 217–228. https://doi.org/10.1016/S0379-0738(98)00192-3.

Lajunen, L., 1992. Spectrochemical Analysis by Atomic Absorption and Emission. The Royal Society of Chemistry, Cambridge.

Lee, J., Lee, C., Lee, K., Lee, Y., 2008. TOF-SIMS study of red sealing-inks on paper and its forensic applications. Appl. Surf. Sci. 255, 1523–1526. https://doi.org/10.1016/j.apsusc.2008.05.094.

Manutsewee, N., Aeungmaitrepirom, W., Varanusupakul, P., Imyim, A., 2007. Determination of Cd, Cu, and Zn in fish and mussel by AAS after ultrasound-assisted acid leaching extraction. Food Chem. 101, 817–824. https://doi.org/10.1016/j.foodchem.2005.12.033.

Mohammadzai, I.U., Hinze, W.L., 2004. Indirect atomic absorption determination of cocaine via formation and liquid-liquid extraction of the iron(III)-cocaine complex. J. Chin. Chem. Soc. 51, 1297–1302. https://doi.org/10.1002/jccs.200400188.

Molina, D.K., Martinez, M., Garcia, J., DiMaio, V.J.M., 2007. Gunshot residue testing in suicides: part II analysis by inductive coupled plasma–atomic emission spectrometry. Am J Forensic Med Pathol 28, 191–194. https://doi.org/10.1097/PAF.0b013e31813e62cf.

Moore, G.L., 1989a. Principles and history of atomic emission spectroscopy. In: Introduction to Inductively Coupled Plasma Atomic Emission Spectrometry. Elsevier, pp. 1–17. https://doi.org/10.1016/B978-0-444-43029-8.50006-2.

Moore, G.L., 1989b. Inductively coupled plasmas. In: Introduction to Inductively Coupled Plasma Atomic Emission Spectrometry. Elsevier, pp. 65–90. https://doi.org/10.1016/B978-0-444-43029-8.50009-8.

Pérez-Cárceles, M.D., Sibón, A., Gil del Castillo, M.L., Vizcaya, M.A., Osuna, E., Casas, T., Romero, J.L., Luna, A., 2008. Strontium levels in different causes of death: diagnostic efficacy in drowning. Biol. Trace Elem. Res. 126, 27–37. https://doi.org/10.1007/s12011-008-8180-1.

Pye, K., Blott, S.J., Wray, D.S., 2006. Elemental analysis of soil samples for forensic purposes by inductively coupled plasma spectrometry—precision considerations. Forensic Sci. Int. 160, 178–192. https://doi.org/10.1016/j.forsciint.2005.09.012.

Rahil-Khazen, R., Bolann, B.J., Myking, A., Ulvik, R.J., 2002. Multi-element analysis of trace element levels in human autopsy tissues by using inductively coupled atomic emission spectrometry technique (ICP-AES). J. Trace Elem. Med. Biol. 16, 15–25. https://doi.org/10.1016/S0946-672X(02)80004-9.

Rodriguez Pereiro, I., Łobiński, R., 1997. Fast species-selective screening for organolead compounds in gasoline by multicapillary gas chromatography with microwave-induced plasma atomic emission detection. J. Anal. At. Spectrom. 12, 1381–1385. https://doi.org/10.1039/A704050D.

Saracoglu, S., Divrikli, U., Soylak, M., Elci, L., Dogan, M., 2003. Determination of trace elements of some textiles by atomic absorption spectrometry. J. Trace Microprobe Tech. 21, 389–396. https://doi.org/10.1081/TMA-120020273.

Schrön, W., Liebmann, A., Nimmerfall, G., 2000. Direct solid sample analysis of sediments, soils, rocks and advanced ceramics by ETV-ICP-AES and GF-AAS. Fresenius J. Anal. Chem. 366, 79–88. https://doi.org/10.1007/s002160050015.

Simon, P.J., Giessen, B.C., Copeland, T.R., 1977. Categorization of papers by trace metal content using atomic absorption spectrometric and pattern recognition techniques. Anal. Chem. 49, 2285–2288. https://doi.org/10.1021/ac50022a045.

Sungur, Ş., Gülmez, F., 2015. Determination of metal contents of various fibers used in textile industry by MP-AES. J. Spectrosc. 2015, 1–5. https://doi.org/10.1155/2015/640271.

Tonetti, C., Innocenti, R., 2009. Determination of heavy metals in textile materials by atomic absorption spectrometry: verification of the test method. Autex Res. J. 9, 66–70.

Turillazzi, E., Di Peri, G.P., Nieddu, A., Bello, S., Monaci, F., Neri, M., Pomara, C., Rabozzi, R., Riezzo, I., Fineschi, V., 2013. Analytical and quantitative concentration of gunshot residues (Pb, Sb, Ba) to estimate entrance hole and shooting-distance using confocal laser microscopy and inductively coupled plasma atomic emission spectrometer analysis: an experimental study. Forensic Sci. Int. 231, 142–149. https://doi.org/10.1016/j.forsciint.2013.04.006.

Tuzen, M., Onal, A., Soylak, M., 2008. Determination of trace heavy metals in some textile products produced in Turkey. Bull. Chem. Soc. Ethiop. 22https://doi.org/10.4314/bcse.v22i3.61213.

Van Loon, J.C., 1980. Analytical Atomic Absorption Spectroscopy: Selected Methods. Academic Press, New York.

Welz, B., Sperling, M., 1999. Atomic Absorption Spectrometry, third revised ed. Wiley-VCH Verlag GmbH, Germany.

Yeung, V., Miller, D.D., Rutzke, M.A., 2017. Atomic absorption spectroscopy, atomic emission spectroscopy, and inductively coupled plasma-mass spectrometry. In: Nielsen, S.S. (Ed.), Food Analysis, Food Science Text Series. Springer International Publishing, Cham, pp. 129–150. https://doi.org/10.1007/978-3-319-45776-5_9.

Yourd, E.R., Tyson, J.F., Koons, R.D., 2001. On-line matrix removal of lead for the determination of trace elements in forensic bullet samples by flow injection inductively coupled plasma-mass spectrometry. Spectrochim. Acta B At. Spectrosc. 56, 1731–1745.

CHAPTER 6

NMR spectroscopy for forensic samples

1 Introduction

Forensic science is largely based on the analysis of evidence found at a crime scene. The analysis of such evidence usually involves determining its structure, composition, and behavior (Pandey et al., 2017; Rawtani et al., 2019). One such technique that provides direct structural determination without any sample preparation is nuclear magnetic resonance (NMR) spectroscopy. NMR is a technique in that the characteristic electromagnetic signal produced by the nuclei of a molecule under high magnetic fields is detected. NMR can completely determine the structure of a molecule or compound. It is a powerful tool capable of performing both quantitative and qualitative analysis and is therefore employed by industrial as well as the academic institutions across the world. NMR has several advantages that make it suitable for the analysis of forensic evidence. One major advantage is that this technique is capable of observing the molecular dynamics in the liquid or solid state without any sample preparation and keeps the samples preserved even after testing. Apart from this, the instrument does not require any major preparation prior to the sample testing, and it speeds up the data acquisition and analysis of the information obtained from the sample. Most NMR spectrometers are also equipped with a library of molecules that can be used to compare with the sample to be analyzed. Conventionally, NMR has been widely used in forensic chemistry, that deals with the application of chemical principles to identify the structure of a compound. While there are certain advantages to this technique, it is not free of limitations such as low sensitivity and resolution, high cost, and loss of information. However, despite the limitations, NMR is still the most efficient tool to analyze compounds without any damage to the sample. Therefore, in this chapter, the principles, instrumentations, and types of NMR spectroscopy and how they have been used in various forensic applications such as drug analysis, postmortem changes, xenobiotic examination, and the identification of body fluids, explosives, and counterfeit products will be discussed.

Handbook of Analytical Techniques for Forensic Samples
https://doi.org/10.1016/B978-0-12-822300-0.00006-9

2 Principle and instrumentation of NMR spectroscopy

^1H NMR or proton NMR is the most widely used type of NMR spectroscopy, in that the structure of the compounds is determined with respect to the hydrogen nucleus. Hydrogen is the simplest atom with an atomic number and atomic mass of 1. It is denoted as ^1H, where the 1 indicates the atomic mass, that is, the sum of the total number of protons and neutrons in the atom. The most crucial aspect of the hydrogen atom is its angular momentum, that depicts the spin of the nucleus. The spinning of the hydrogen nucleus is analogous to a sphere spinning on its axis. It is positively charged and generates a magnetic field with the magnetic moment. This magnetic moment has magnitude and direction as well and is characteristic to every element. NMR spectroscopy employs this principle to analyze the structure of different compounds (Lambert et al., 2019).

In the case of heavier elements, this spin is determined by combining the overall spin of the protons and neutrons. In cases of an even atomic number, the number of protons and neutrons is equal and the spins cancel each other, meaning the nucleus has no magnetic properties. However, in the case of an odd atomic number, the nucleus possesses magnetic properties due to the extra spin accounted by the odd number of protons and electrons. The magnitude of the magnetic moment generated is highly dependent on a factor known as the gyromagnetic or magnetogyric ratio. The magnetic moment is directly proportional to this ratio. In the presence of a high magnetic field with a direction, the magnetic moment of each element tends to align in the direction of the magnetic field. The frequency of the magnetic moment in this case is known as the Larmor Frequency, and it is directly proportional to the magnitude of the magnetic field. The magnetic moment also tends to move in a circular fashion in the presence of a magnetic field. This movement is also known as precession. When the frequency of the magnetic field matches the frequency of the magnetic moment, resonance occurs, due to that there is a net absorption in energy. This absorbance is detected by the NMR spectroscopy and is displayed in the form of a plot of frequency vs energy absorbed. The gyromagnetic ratio also affects the energies of the spin state and is directly proportional to the intensity of the signals produced. It is highly crucial that the magnetic field applied is high, as it affects the signal-to-noise ratio, magnitude, and coupling of the elements. With the increase in the magnetic field, the signal-to-noise ratio increases proportionally. Also, two major phenomena occur when the magnetic field is increased, namely the chemical shifts and J coupling. A chemical shift is an interesting phenomenon in that the electron cloud that is specific to the nucleus has its own magnetic moment, that alters the external magnetic field applied in the microenvironment. The total magnetic field of the element is thus dependent on the magnetic moment of the electron cloud and the magnetic moment of the nucleus. Depending on the shielding effect of the element, the overall resonant frequency gets affected and this variation in the resonant frequency is known as the chemical shift. These frequencies are dependent on the local environment and are specific to the structure of the molecule. The resonant frequency is also affected by the presence of other magnetic nuclei in the nearby environment. In NMR, if there are multiple proton groups neighboring each other, this may result in multiple or split peaks; it is known as spin-spin splitting or indirect coupling. The J coupling constant is another measure that depicts the extent of the influence of nuclear spins on their neighboring proton groups through the distance between the peaks, that is measured in terms of hertz (Hz). The magnitude of the external magnetic field is also crucial as it helps in the formation of a simple spectrum. This is possible due to the fact that the chemical shift is directly proportional to the external electric field applied. With an increase in the chemical shift, the resonances between two proton groups also increase significantly until they are relatively larger than the J coupling, thus resulting in a simple spectrum (Bai, 2016; Lambert et al., 2019).

2.1 Types of NMR

NMR is a versatile technique as it is capable of determining the molecular structure by determining the protons associated with hydrogen followed by the protons attached to carbon. The ^{1}H NMR takes into account the chemical shift, J-coupling constant, splitting, and the protons attached to ^{1}H in the molecule while another type of NMR, commonly known as ^{13}C NMR, takes into account the protons attached to ^{13}C carbon (Bales et al., 1984; Ellis, 1973; Haasnoot et al., 1980; Wagner et al., 1983). Here, ^{13}C carbon rather than ^{12}C is used as it is the only form of carbon that has one-half spin and thus possesses a magnetic moment. This form of carbon is very rare and covers only 1.1% of the main isotope. These techniques are also known as 1D NMR (Sarkar and Bax, 1985). Sometimes, 1D NMR can generate spectra that are too complex and therefore additional techniques are required that are capable of producing a simple spectrum. In such cases, 2D NMR is used, that consists of techniques such as correlational spectroscopy (COSY), total correlated spectroscopy (TOCSY), and nuclear overhauser effect spectroscopy (NOESY). Apart from these techniques, ^{31}P NMR or phosphorous NMR is another NMR technique in that compounds containing phosphorous can be analyzed. Phosphorous has a nonspherical symmetry due to that the chemical shift occurring in this case comes from the paramagnetic nature of this element, that is actually contributed by the electron cloud surrounding the element. The paramagnetic term also takes into account the energies of the excited states, radial expansion, and bond overlap. This technique is widely used for the analysis of phospholipid bilayers and cell membranes. Because most of the biological compounds have phosphorous as their key element, they are highly useful in analyzing evidence comprising body fluids. Here, ^{31}P is used as it is easily available with a relatively high gyromagnetic ratio (Gorenstein, 1984).

In COSY NMR, only the protons that are separated by two or three bonds give the signal. This is highly useful for protein structure interpretation as the torsion angle between the hydrogen attached to the amine group and the alpha hydrogen gives excellent NMR signals in this case (Dufour et al., 2015). In TOCSY, every proton in the spin state gives the NMR signal. This is advantageous because the signal emitted is characteristic to every molecule and can therefore be used to differentiate between different molecules such as amino acids, that have similar functional groups but varying side chains (Hansen et al., 2017). The NOESY experiment employs the nuclear Overhauser effect in that two nuclei that are in close proximity to each other tend to transfer their spin state to the nearby nonspinning nucleus. This technique is useful in cases of predicting the stereochemistry of proteins (Jacobsen, 2007; Aguilar et al., 2012).

2.2 Instrumentation of NMR spectroscopy

The instrumentation of NMR comprises certain common components such as a magnet to generate the magnetic field; a radio frequency (RF) generator; an apparatus to generate the magnetic pulse of the sample and receive the NMR signal; a sample holder for positioning the sample near the magnet; hardware and controllers for feedback, control, and optimization of the signal; and a computer as an end-user interface for the processing and interpretation of the signals.

Initially, electromagnets were used to provide a continuous RF field that would be constantly showered on the sample. Any signal from the sample would result in a change in the quality factor of the electromagnetic coil, that could be interpreted in terms of the absorption of energy by the sample.

The spectrometer employing this mode of detection is known as a continuous wave spectrometer. Nowadays, the pulse Fourier transform mode is used in that superconducting magnets are used. This technique was developed after a brief use of permanent magnets, that provided better stability than the electromagnets but were still not strong enough to provide a clean NMR spectrum. Superconducting magnets have several benefits such as high stability, low resistance, similar signal averaging for multiple analysis, and repeatability. Currently, a superconducting solenoid that is capable of producing 600 MHz of magnetic field is used. This apparatus is encased in a chamber that prevents any leakage of the magnetic field. There are separate RF generators for the hydrogen proton and any other nuclei. There is also a separate RF generator for the deuterium lock that is responsible for controlling the magnetic field. The deuterium lock controls the magnetic field with an inbuilt microprocessor and maintains the homogeneity of the field. Samples in NMR are usually held together with a probe that is usually kept in a cryogenically cooled unit. This is because the sample holder itself is comprised of transmitter coils that are responsible for providing the magnetic fields and receiver coils for receiving the NMR signals. The primary computer is another important component in the NMR spectrometer that is responsible for the acquisition of data as well as the maintenance of the frequency and homogeneity of the RF and the magnetic field. It also serves as an end-user interface and carries out the data processing and acquisition of the signals received from the instrument. The NMR spectrometer is composed of separate entities that allow it to be modified or customized according to use; it can also be used along with other techniques such as high-performance liquid chromatography (Bai, 2016; Lambert et al., 2019).

One of the most crucial aspects of analysis using an NMR spectrometer is the sample preparation. NMR initially was used for the analysis of liquid samples, but later modifications to the sample probe were made such that, when accompanied with appropriate sample preparation techniques, allowed excellent analysis of the samples. In the case of solid samples, small rotors are fixed in the sample probe and the sample is stuffed in those rotors. When the analysis starts, the rotors start to mix the sample such that their movement mimics the free tumbling of the molecules in liquids. When the particles move rapidly at a particular angle, the spectral lines formed are narrow and a high-resolution spectrum is produced. In the case of liquid samples, the sample is usually dispersed in a suitable solvent in clean tubes and placed in the sample holder for further analysis. Fig. 1 gives a brief overview of the NMR instrumentation.

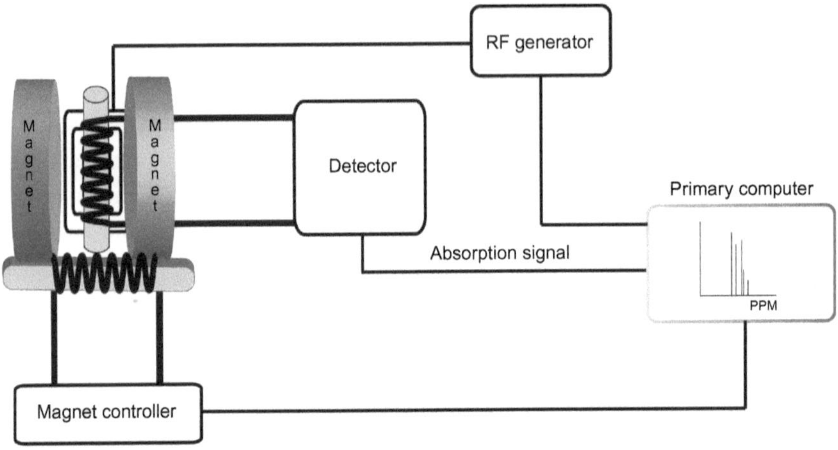

Fig. 1 Instrumentation of NMR.

3 NMR in the world of forensic science

This section discusses the applications of NMR spectroscopy in the world of forensic science. Forensic applications that employ NMR as an analytical tool, such as drug analysis, explosive detection, postmortem change analysis, examination of xenobiotics, identification of body fluids, and identification of counterfeit products, are discussed in detail (Fig. 2). The details of NMR analysis for the aforementioned applications are shown in Table 1.

3.1 Drug analysis

NMR spectroscopy has been used by scientists for drug analysis for many years. Various pharmaceutical industries use this spectroscopic technique to identify drugs and their metabolites. The applications of ^{1}H NMR and ^{13}C NMR spectroscopy for the analysis of different drugs have been very well reviewed in the past (Pieters and Vlietinck, 1989). Over the years, forensic scientists and investigators have started using NMR to identify drugs in different samples or evidence collected from the crime scene.

In the past, Wilson and Nicholson used NMR spectroscopy in combination with solid phase extraction chromatography (SPEC) to detect and identify metabolites of drugs in urine samples. The drugs used were aspirin, oxpentifylline, ibuprofen, and paracetamol (collected from a male donor) and naproxen (collected from a female donor). The combination of SPEC helped in removing the endogenous constituents from the urine sample that contained drugs and their metabolites before they could be detected and

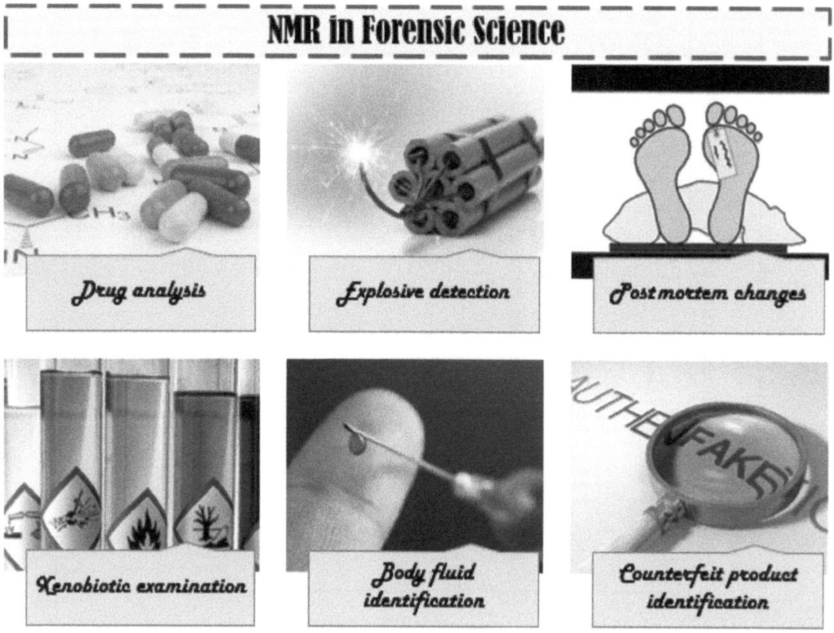

Fig. 2 Applications of NMR in forensic science.

identified through NMR (Wilson and Nicholson, 1988). Research work before this study also demonstrated the ability of ^{1}H NMR for the quantitative and qualitative analysis of drugs and their metabolites in the urine (Bales et al., 1984). In a comparatively recent study, proton NMR was utilized for the detection of paracetamol and its metabolites in urine (Galinski et al., 2014). In another study from the mid-1990s, chiral drug molecules, namely ephedrine, methamphetamine, pseudoephedrine, and metacathione, were identified and quantified using NMR in combination with gas chromatography. The solvating agent used [(R)-(+)-1,1′-bi-2-naphthol] was chiral in nature, which helped in the detection of the enantiomers of the aforementioned drugs. The forensic importance of some of these drugs such as ephedrine and pseudoephedrine makes this study very relevant to forensic investigators dealing with cases of drug abuse (LeBelle et al., 1995).

Studies from this decade as well have shown the potential of NMR for the detection and identification of drugs and their metabolites. A study from 2011 reported the use of low field NMR for the detection of micromolar concentrations of drugs such as nicotine, harmine, and morphine. The study utilized an advanced para-hydrogen induced polarization (PHIP) technique to polarize the drugs that selectively enhance the ^{1}H signals coming from

Table 1 NMR analysis of forensic samples.

S. no.	Application	Analyte/sample	Type of NMR used	Operating frequency	Temperature (in K)	Reference
1	Drug analysis	Nicotine, harmine, morphine	^{1}H	166 kHz	–	Glöggler et al. (2011)
		Cocaine	^{1}H	700 MHz	298	Pagano et al. (2013)
2	Detection of explosives	2,4,6-Trinitrophenol	^{1}H	400 MHz	300	Paul et al. (2019)
		Torpex	^{1}H, ^{13}C, ^{15}N	500 MHz	Room temperature	Nawała et al. (2020)
3	Analysis of postmortem changes	Human brain tissue	^{1}H	600 MHz	300	Graham et al. (2016)
		Cerebrospinal fluid	^{1}H	300 MHz	296	Kanawaku et al. (2017)
4	Examination of xenobiotics	Ethyl glucuronide	^{1}H	700 MHz	298	Nicholas et al. (2006)
		Ecstasy	^{1}H	500 MHz	–	Liu et al. (2010)
5	Identification of body fluids	Blood, semen, urine, and saliva	^{1}H	500 MHz	300	Scano et al. (2013)
6	Identification of counterfeit products	Counterfeit Viagra	^{1}H	4 kHz to 40 MHz	–	Wilczyńki et al. (2017)
		Counterfeit branded spirits (vodka, rum, and whiskey)	^{1}H	–	300	Kuballa et al. (2018)

them. This polarization helped in detecting trace levels of the aforementioned drugs (Glöggler et al., 2011). A group of researchers from Italy and Denmark used NMR for the chemical profiling of cocaine, a well-known illicit drug that was seized from different places and at different times from Naples, Italy. The study helped in identifying the trafficking path for the drug, which was detected from the chemical's fingerprint. It was concluded that the fingerprint of cocaine was dependent on its origin plant, method of cultivation, and process adopted for its extraction and refinement (Pagano et al., 2013).

3.2 Detection of explosives

Explosives have been widely used in the preparation of bombs and ammunition. Their traces can be found in terrorist activities or criminal activities involving firearms and ammunition. Sometimes, this evidence has such minute levels of explosive that it becomes difficult to detect. In such cases, NMR can be used for detection and can identify various primary and secondary explosives, even in trace quantities. An article compiling the application of NMR for explosive detection has been published (King and De Los Santos, 2004).

The first study regarding the investigation of explosive residue in environmental samples was carried out in 2004 by Godejohann et al. The samples were collected from the Susten-Stone glacier (Switzerland), which was a disposal site for ammunition. ^{1}H NMR in combination with high-performance liquid chromatography, solid phase extraction, and mass spectroscopy was used to detect traces of explosives such as trinitrotoluene, hexogen (RDX), trinitrobenzene, dinitrotoluene, octogen (HMX), dinitrophenol, dinitroaniline, centralite, acardite, dinitroanthranile, and bisphenol A. The concentrations of these residues detected through NMR were in the range of 0.1–48 ppm (Godejohann et al., 2009). Another study discussed the detection of liquid explosives, which are a major security threat in public places and airplanes, using an NMR-based detector. The spin relaxation times and self-diffusion coefficient of different liquids helped to differentiate among them (Gradišek and Apih, 2010). In the same year, pentaerythritol tetranitrate (PETN), a highly explosive compound from the family of nitrocellulose and nitroglycerine, was detected using NMR. The study also employed ^{14}N NQR (nuclear quadrupole resonance) spectroscopy along with ^{1}H NMR to detect PETN. The spin-lattice relaxation time was measured between the frequencies of 1.8 kHz to 40 MHz in the temperature range of 250–390 K (Smith et al., 2010). In a recent study, norharmane (NHM) was used to detect 2,4,6-trinitrophenol (TNP) in the presence of potential contaminants, which can usually be found in real explosive samples collected from a crime scene. Ratiometric spectroscopic responses were responsible for the interaction between NHM and TNP. NMR was used to identify the mechanism behind these responses, which was proton transfer phenomenon at the ground state (Paul et al., 2019). Another recent study employed ^{1}H and ^{13}C NMR to detect explosives that were excavated from the bed of the Baltic sea (Nawała et al., 2020).

3.3 Analysis of postmortem changes

Whenever any crime is committed and a dead body is found, the estimation of time since death becomes very important. The dead body starts undergoing many changes, such as pallor mortis, algor mortis, rigor mortis, and livor mortis, followed by putrefaction and decomposition. The issues that investigators face while determining the time since death is studied under the science of thanatochronology (Santos et al., 2018). Forensic scientists have used NMR for many years as an analytical tool to determine the postmortem time interval and study the postmortem changes a cadaver undergoes.

A study from the last decade used ^{1}H NMR to study the biochemical alterations that take place in the skeletal muscle of a rat after death. The biochemical changes in the perchloric acid extracted from the skeletal muscle at different postmortem time intervals were observed. The NMR study was performed at 298 K, with a frequency of 200 MHz. The NMR spectra of extracts collected at different time intervals were different, indicating the biochemical changes taking place inside the body after death (Fineschi et al., 1990). Another study from the same decade used ^{31}P NMR to analyze the muscle metabolism and changes after death. A rabbit was taken as the model animal. The NMR study was performed at 162 MHz, using a dual tune probe (^{1}H and ^{31}P). It was observed that after muscle death, the amount of energy-rich compounds and changes in pH increased. This could be correlated to the time since death (Miri et al., 1991). In another study, postmortem changes were correlated to time since death by monitoring the metabolic changes in the femoral muscles of the rat. The rats were sacrificed through an overdose of cocaine, suffocation, and induced respiratory failure. The NMR spectra of the metabolites differed for extracts collected at different time intervals as well as among the different modes of death. The frequency used for the NMR study was 270 MHz (Hirakawa et al., 2009). ^{1}H

NMR has also been used to study the metabolites produced in the brains of dead people who suffered from Huntington disease. The metabolites were studied in the frontal lobe and the striatum regions of the brain. Urea was the metabolite that underwent maximum changes in the striatum. All the NMR measurements were done at 300 K with a frequency of 600.13 MHz (Graham et al., 2016). In a recent study, the cerebrospinal fluids collected after the deaths of rats that had induced seizures and comas using drugs were also analyzed through NMR (Kanawaku et al., 2017).

3.4 Examination of xenobiotics

Any chemical substance that is foreign to the body can be considered xenobiotic. Broadly, this can be drugs, food additives, agrochemicals, organic compounds, carcinogens, etc. Natural compounds can also behave as xenobiotic compounds if they are taken up by any organism whose natural composition does not contain that compound. In forensic science, the examination of xenobiotics becomes important in cases related to poisoning. NMR has been used by different forensic researchers to detect various poisons in the body.

Imbenotte et al. used NMR for the qualitative and quantitative analysis of xenobiotic compounds in cases related to poisoning. Serum and urine samples were used in cases of lysine acetylsalicylate, valproic acid, paraquat, tetrahydrofuran, methanol, and ethylene glycol poisoning. The frequency of operation for NMR analysis was 300 MHz. The NMR spectra from the body fluids revealed the metabolites of the aforementioned xenobiotics that were responsible for the poisoning in patients (Imbenotte et al., 2003). In another study, glyphosate, a well-known herbicide, was detected and quantified using ^{31}P and ^{1}H NMR spectroscopy. The proton NMR spectra was obtained at 300.09 MHz while the phosphorus NMR spectra was obtained at 121.48 MHz. The study showed the ability of NMR spectroscopy for the identification and quantification of organophosphates, which form a major class of pesticides (Cartigny et al., 2004). Proton NMR has also shown its applicability in the detection of ethyl glucuronide present in the liver extract. Ethyl glucuronide is a metabolite of ethanol, whose presence in the biological fluid can be correlated with alcohol consumption. The samples were collected from the livers of rats who were allowed to consume alcohol at a concentration of 8.6 g/kg/day. The NMR spectra was obtained at 700 MHz (Nicholas et al., 2006). Ecstasy, a common synthetic drug widely abused at rave parties, has also been detected and quantified using NMR spectroscopy. The study used a urine sample as a source for the drug. The NMR spectra was obtained at 500 MHz frequency. A linear response was observed while performing the calibration curve for the concentrations ranging from 0.01 to 1 mg/mL. Spiked as well as real samples were analyzed for the presence and quantification of ecstasy (Liu et al., 2010).

3.5 Identification of body fluids

Body fluids such as urine, semen, saliva, sweat, blood, and vomit have served as sources for different types of analyzes for diagnostic and medical use. These body fluids are also key evidence for crime scene investigators because they contain useful information about the identities of victims and suspects. A number of spectroscopic (Zapata et al., 2015) and chromatographic (Deyl et al., 1986) techniques have been adopted by analysts to detect different analytes and metabolites in body fluids. These fluids can be identified via different biochemical tests, but these require preliminary and confirmatory examination through various chemicals. The utilization of analytical techniques such as NMR can identify these fluids in a more specific and sensitive manner.

However, until now, only one research work has been published by Scano et al. for identifying different types of body fluids. Different body fluids such as blood, saliva, semen, and urine were collected from donors and analyzed through ^{1}H NMR at 300 K with a frequency of 500 MHz. The analysis was done for single body fluid samples as well as binary mixtures of the aforementioned fluids. The body fluids were identified on the basis of the presence of different metabolites such as lactate, citrate, propionate, and various amino acids in them. In order to identify the fluids in the binary mixture, a fitting procedure was developed that was based on the spectra obtained for the individual fluids (Scano et al., 2013).

3.6 Identification of counterfeit products

Counterfeit products have created havoc in the global market. Counterfeiting is practiced for products related to jewelry, medicine, clothing, food, etc. Counterfeit drugs are consciously and deceitfully labeled regarding their origin and individuality. Counterfeit medications can be divided into five types (Holzgrabe and Malet-Martino, 2011): counterfeit drugs (a) with a low active pharmaceutical ingredient (API); (b) wrongly labeled; (c) wrong API; (d) no API; and (e) low quality. Counterfeit food products are also a matter of concern because they directly affect the public health. Various types of analytical techniques have been utilized for detecting the counterfeiting of medications and food products. NMR has also been utilized for this purpose (Malet-Martino and Martino, 2015).

In a study, ^{13}C NMR was used for the assessment of the APIs in counterfeit medicines such as aspirin and paracetamol. More than 15 samples for both medicines were collected from pharmacies in different countries and analyzed through NMR. Tablets and capsules containing around 1 g of API were crushed and API was extracted. The NMR spectra of the extracts containing API were recorded at 303 K with a frequency of 125.76 MHz. The spectra differed with the suppliers and batches of the drugs. The study can be useful to identify any counterfeiting or infringement of patents on pharmaceutical industries (Silvestre et al., 2009). In another study, ^{1}H spin-lattice NMR was used to identify and distinguish between counterfeit and original Viagra. The analysis was performed in the frequency range of 4 kHz to 40 MHz at 310 K. The NMR study revealed that the original Viagra had a single exponential relaxation process while for the counterfeit Viagra, the relaxation process was biexponential. This helped in differentiating between the original and the duplicate drug (Wilczyńki et al., 2017). In a recent study, counterfeiting in branded alcoholic beverages was detected using ^{1}H NMR. The counterfeiting of such beverages usually involves a mixture of methanol that poses many health hazards but is difficult to detect organoleptically by the consumers. The counterfeit samples (vodka, rum, and whiskey of different brands) were collected from Russia and Kenya and compared with authentic ones. The NMR spectra were obtained at 300 K. The coefficient of determination was different for the authentic and counterfeit samples, which helped in distinguishing between the two (Kuballa et al., 2018). Another recent study has utilized NMR with a recording frequency of 600 MHz for the analysis of counterfeiting and impurities in weight loss-related drugs (Lee et al., 2019).

4 Conclusion

NMR is a widely used technique to determine the structure of a molecule. This technique takes into account the magnetic moments of the nucleus and electron cloud of a particular element when exposed to a strong external magnetic field. This technique is highly useful in forensic science wherein the structure of evidence commonly encountered at a crime scene needs to be determined. NMR has several advantages such as high stability and negligible sample preparation while also keeping the samples intact even after testing. However, this technique is not devoid of limitations. NMR has less sensitivity, especially while analyzing bodily fluids that are majorly comprised of water and have an abundance of hydrogen. Therefore, the smaller molecules become difficult to differentiate. It is also an expensive technique as it requires strong magnetic fields that consume extensive energy and have high maintenance costs. However, with recent advances in superconducting magnets, these limitations may be easily overcome. Also, techniques such as ^{31}P NMR that target the phosphorous elements in the samples can be used in combination with H-NMR in order to improve the sensitivity of the analysis. Also, the entire setup of an NMR spectrometer is modular in nature, thereby allowing room for any customizations or modifications that can improve the analysis. NMR can also be used in conjunction with other analytical techniques such as high-performance liquid chromatography, thereby making this hyphenated technique highly useful in the toxicological analysis of forensic evidence.

In forensic science, this method has been used for the detection of illicit drugs that are abused at rave parties such as cocaine, morphine, nicotine, etc. Different types of primary as well as secondary explosives such as TNT, PETN, RDX, Torpex, etc., have also been analyzed through this spectroscopic technique. The biochemical changes that occur inside the body after death have also been analyzed through NMR. Such studies are helpful in determining the time since death because the NMR spectra were different for the tissue extracts collected at different postmortem time intervals. Xenobiotics such as ecstasy, alcohol, etc., have also been identified and quantified using NMR. In the case of body fluid identification, only one study has been published so far as per our knowledge. Counterfeit drugs such as Viagra and those related to weight loss as well as counterfeited branded spirits have also been investigated through this technique. Due to drawbacks associated with NMR such as long analysis time and costly instrumentation, forensic laboratories avoid the use of this instrument. However, partnerships of such laboratories with local universities and research institutes can assist the forensic analysts in utilizing these facilities and increasing the efficiency of the investigation many-fold.

References

Aguilar, J.A., Colbourne, A.A., Cassani, J., Nilsson, M., Morris, G.A., 2012. Decoupling two-dimensional NMR spectroscopy in both dimensions: pure shift NOESY and COSY. Angew. Chem. Int. Ed. 51, 6460–6463. https://doi.org/10.1002/anie.201108888.

Bai, S., 2016. Nuclear magnetic resonance instrumentation. In: Encyclopedia of Analytical Chemistry. American Cancer Society, pp. 1–26. https://doi.org/10.1002/9780470027318.a6108.pub3.

Bales, J.R., Higham, D.P., Howe, I., Nicholson, J.K., Sadler, P.J., 1984. Use of high-resolution proton nuclear magnetic resonance spectroscopy for rapid multi-component analysis of urine. Clin. Chem. 30, 426–432.

Cartigny, B., Azaroual, N., Imbenotte, M., Mathieu, D., Vermeersch, G., Goullé, J.P., Lhermitte, M., 2004. Determination of glyphosate in biological fluids by 1H and 31P NMR spectroscopy. Forensic Sci. Int. 143, 141–145. https://doi.org/10.1016/j.forsciint.2004.03.025.

Deyl, Z., Hyanek, J., Horakova, M., 1986. Profiling of amino acids in body fluids and tissues by means of liquid chromatography. J. Chromatogr. B Biomed. Sci. Appl. 379, 177–250. https://doi.org/10.1016/S0378-4347(00)80685-4.

Dufour, G., Evrard, B., de Tullio, P., 2015. 2D-Cosy NMR spectroscopy as a quantitative tool in biological matrix: application to cyclodextrins. AAPS J. 17, 1501–1510. https://doi.org/10.1208/s12248-015-9806-9.

Ellis, P.D., 1973. Carbon-13 NMR spectra—a collection of assigned, coded, and indexed spectra. J. Magn. Reson. 11, 110–111. https://doi.org/10.1016/0022-2364(73)90092-9.

Fineschi, V., Picchi, M.P., Tassini, M., Valensin, G., Vivi, A., 1990. 1H-NMR studies of postmortem biochemical changes in rat skeletal muscle. Forensic Sci. Int. 44, 225–236. https://doi.org/10.1016/0379-0738(90)90253-U.

Galinski, M., Racine, S.-X., Bossard, A.-E., Fleyfel, M., Hamza, L., Bouchemal, N., Adnet, F., Moyec, L.L., 2014. Detection and follow-up, after partial liver resection, of the urinary paracetamol metabolites by proton NMR spectroscopy. Pharmacology 93, 18–23. https://doi.org/10.1159/000357095.

Glöggler, S., Emondts, M., Colell, J., Müller, R., Blümich, B., Appelt, S., 2011. Selective drug trace detection with low-field NMR. Analyst 136, 1566. https://doi.org/10.1039/c0an01048k.

Godejohann, M., Heintz, L., Daolio, C., Berset, J.-D., Muff, D., 2009. Comprehensive non-targeted analysis of contaminated groundwater of a former ammunition destruction site using 1H-NMR and HPLC-SPE-NMR/TOF-MS. Environ. Sci. Technol. 43, 7055–7061. https://doi.org/10.1021/es901068d.

Gorenstein, D.G. (Ed.), 1984. Phosphorus-31 NMR: Principles and Applications. Academic Press, Orlando, FL.

Gradišek, A., Apih, T., 2010. NMR-based liquid explosives detector. Appl. Magn. Reson. 38, 485–493. https://doi.org/10.1007/s00723-010-0145-9.

Graham, S.F., Kumar, P.K., Bjorndahl, T., Han, B., Yilmaz, A., Sherman, E., Bahado-Singh, R.O., Wishart, D., Mann, D., Green, B.D., 2016. Metabolic signatures of Huntington's disease (HD): 1H NMR analysis of the polar metabolome in post-mortem human brain. Biochim. Biophys. Acta (BBA) - Mol. Basis Dis. 1862, 1675–1684. https://doi.org/10.1016/j.bbadis.2016.06.007.

Haasnoot, C.A.G., de Leeuw, F., Altona, C., 1980. The relationship between proton-proton NMR coupling constants and substituent electronegativities—I. Tetrahedron 36, 2783–2792. https://doi.org/10.1016/0040-4020(80)80155-4.

Hansen, A.L., Li, D., Wang, C., Brüschweiler, R., 2017. Absolute minimal sampling of homonuclear 2D NMR TOCSY spectra for high-throughput applications of complex mixtures. Angew. Chem. Int. Ed. 56, 8149–8152. https://doi.org/10.1002/anie.201703587.

Hirakawa, K., Koike, K., Uekusa, K., Nihira, M., Yuta, K., Ohno, Y., 2009. Experimental estimation of postmortem interval using multivariate analysis of proton NMR metabolomic data. Legal Med. 11, S282–S285. https://doi.org/10.1016/j.legalmed.2009.02.007 Selected Papers from the 7th International Symposium Advances in Legal Medicine (ISALM).

Holzgrabe, U., Malet-Martino, M., 2011. Analytical challenges in drug counterfeiting and falsification—the NMR approach. J. Pharm. Biomed. Anal. 55, 679–687. https://doi.org/10.1016/j.jpba.2010.12.017.

Imbenotte, M., Azaroual, N., Cartigny, B., Vermeersch, G., Lhermitte, M., 2003. Identification and quantitation of xenobiotics by NMR spectroscopy in poisoning cases. Forensic Sci. Int. 133, 132–135. https://doi.org/10.1016/S0379-0738(03)00059-8.

Jacobsen, N.E., 2007. Advanced NMR theory: NOESY and DQF-COSY. In: NMR Spectroscopy Explained. John Wiley & Sons, Ltd, pp. 408–488. https://doi.org/10.1002/9780470173350.ch10.

Kanawaku, Y., Hirakawa, K., Koike, K., Kanetake, J., Ohno, Y., 2017. Pattern recognition analysis of proton nuclear magnetic resonance spectra of postmortem cerebrospinal fluid from rats with drug-induced seizure or coma. Legal Med. 25, 52–58. https://doi.org/10.1016/j.legalmed.2017.01.004.

King, J.D., De Los Santos, A., 2004. Development and evaluation of magnetic resonance technologies, particularly NMR, for detection of explosives. Appl. Magn. Reson. 25, 535–565. https://doi.org/10.1007/BF03166547.

Kuballa, T., Hausler, T., Okaru, A.O., Neufeld, M., Abuga, K.O., Kibwage, I.O., Rehm, J., Luy, B., Walch, S.G., Lachenmeier, D.W., 2018. Detection of counterfeit brand spirits using 1H NMR fingerprints in comparison to sensory analysis. Food Chem. 245, 112–118. https://doi.org/10.1016/j.foodchem.2017.10.065.

Lambert, J., Mazzola, E.P., Ridge, C., 2019. Nuclear Magnetic Resonance Spectroscopy: An Introduction to Principles, Applications, and Experimental Methods, second ed. John Wiley & Sons, Ltd., NJ, USA

LeBelle, M.J., Savard, C., Dawson, B.A., Black, D.B., Katyal, L.K., Zrcek, F., By, A.W., 1995. Chiral identification and determination of ephedrine, pseudoephedrine, methamphetamine and metecathinone by gas chromatography and nuclear magnetic resonance. Forensic Sci. Int. 71, 215–223. https://doi.org/10.1016/0379-0738(94)01669-0.

Lee, J.H., Park, H.N., Kim, N.S., Park, S., Bogonda, G., Oh, K., Kang, H., 2019. Application of screening methods for weight-loss compounds and identification of new impurities in counterfeit drugs. Forensic Sci. Int. 303, 109932. https://doi.org/10.1016/j.forsciint.2019.109932.

Liu, J., Decatur, J., Proni, G., Champeil, E., 2010. Identification and quantitation of 3,4-methylenedioxy-N-methylamphetamine (MDMA, ecstasy) in human urine by 1H NMR spectroscopy. Application to five cases of intoxication. Forensic Sci. Int. 194, 103–107. https://doi.org/10.1016/j.forsciint.2009.10.022.

Malet-Martino, M., Martino, R., 2015. Analysis of counterfeit medicines and adulterated dietary supplements by NMR. In: Harris, R.K., Wasylishen, R.L. (Eds.), EMagRes. John Wiley & Sons, Ltd, Chichester, UK, pp. 159–170. https://doi.org/10.1002/9780470034590.emrstm1423.

Miri, A., Foucat, L., Renou, J.P., Rodet, L., Talmant, A., Monin, G., Kozak-Reiss, G., Gascard, J.P., Bérenger, G., 1991. Use of perfused isolated muscle, as studied by 31P NMR, to investigate metabolism and post-mortem changes. Meat Sci. 30, 327–336. https://doi.org/10.1016/0309-1740(91)90041-N.

Nawała, J., Szala, M., Dziedzic, D., Gordon, D., Dawidziuk, B., Fabisiak, J., Popiel, S., 2020. Analysis of samples of explosives excavated from the Baltic Sea floor. Sci. Total Environ. 708, 135198. https://doi.org/10.1016/j.scitotenv.2019.135198.

Nicholas, P.C., Kim, D., Crews, F.T., Macdonald, J.M., 2006. Proton nuclear magnetic resonance spectroscopic determination of ethanol-induced formation of ethyl glucuronide in liver. Anal. Biochem. 358, 185–191. https://doi.org/10.1016/j.ab.2006.08.033.

Pagano, B., Lauri, I., De Tito, S., Persico, G., Chini, M.G., Malmendal, A., Novellino, E., Randazzo, A., 2013. Use of NMR in profiling of cocaine seizures. Forensic Sci. Int. 231, 120–124. https://doi.org/10.1016/j.forsciint.2013.04.028.

Pandey, G., Rawtani, D., Agrawal, Y.K., 2017. Future aspects of halloysite nanotubes in forensic investigations. J. Nanomed. Res. 6. https://doi.org/10.15406/jnmr.2017.06.00153.

Paul, S., Karar, M., Paul, P., Mallick, A., Majumdar, T., 2019. Dual mode nitro explosive detection under crowded condition: conceptual development of a sensing device. J. Photochem. Photobiol. A Chem. 379, 123–129. https://doi.org/10.1016/j.jphotochem.2019.04.038.

Pieters, L.A.C., Vlietinck, A.J., 1989. Applications of quantitative 1H- and 13C-NMR spectroscopy in drug analysis. J. Pharm. Biomed. Anal. 7, 1405–1417. https://doi.org/10.1016/0731-7085(89)80145-1.

Rawtani, D., Tharmavaram, M., Pandey, G., Hussain, C.M., 2019. Functionalized nanomaterial for forensic sample analysis. TrAC Trends Anal. Chem. 120, 115661. https://doi.org/10.1016/j.trac.2019.115661.

Santos, A.D.C., Dutra, L.M., Menezes, L.R.A., Santos, M.F.C., Barison, A., 2018. Forensic NMR spectroscopy: just a beginning of a promising partnership. TrAC Trends Anal. Chem. 107, 31–42. https://doi.org/10.1016/j.trac.2018.07.015.

Sarkar, S.K., Bax, A., 1985. A Simple and Sensitive One-Dimensional NMR Technique for Correlation of Proton and Carbon Chemical Shifts.

Scano, P., Locci, E., Noto, A., Navarra, G., Murgia, F., Lussu, M., Barberini, L., Atzori, L., De Giorgio, F., Rosa, M.F., d'Aloja, E., 2013. ^{1}H NMR metabolite fingerprinting as a new tool for body fluid identification in forensic science. Magn. Reson. Chem. 51, 454–462. https://doi.org/10.1002/mrc.3966.

Silvestre, V., Mboula, V.M., Jouitteau, C., Akoka, S., Robins, R.J., Remaud, G.S., 2009. Isotopic 13C NMR spectrometry to assess counterfeiting of active pharmaceutical ingredients: site-specific 13C content of aspirin and paracetamol. J. Pharm. Biomed. Anal. 50, 336–341. https://doi.org/10.1016/j.jpba.2009.04.030.

Smith, J.A.S., Rayner, T.J., Rowe, M.D., Barras, J., Peirson, N.F., Stevens, A.D., Althoefer, K., 2010. Magnetic field-cycling NMR and 14N, 17O quadrupole resonance in the explosive pentaerythritol tetranitrate (PETN). J. Magn. Reson. 204, 139–144. https://doi.org/10.1016/j.jmr.2010.02.019.

Wagner, G., Pardi, A., Wuethrich, K., 1983. Hydrogen bond length and proton NMR chemical shifts in proteins. J. Am. Chem. Soc. 105, 5948–5949. https://doi.org/10.1021/ja00356a056.

Wilczyńki, S., Petelenz, M., Florek-Wojciechowska, M., Kulesza, S., Brym, S., Błońska-Fajfrowska, B., Kruk, D., 2017. Verification of the authenticity of drugs by means of NMR relaxometry—Viagra® as an example. J. Pharm. Biomed. Anal. 135, 199–205. https://doi.org/10.1016/j.jpba.2016.12.018.

Wilson, I.D., Nicholson, J.K., 1988. Solid phase extraction chromatography and NMR spectroscopy (SPEC-NMR) for the rapid identification of drug metabolites in urine. J. Pharm. Biomed. Anal. 6, 151–165. https://doi.org/10.1016/0731-7085(88)80041-4.

Zapata, F., Fernández de la Ossa, M.Á., García-Ruiz, C., 2015. Emerging spectrometric techniques for the forensic analysis of body fluids. TrAC Trends Anal. Chem. 64, 53–63. https://doi.org/10.1016/j.trac.2014.08.011.

CHAPTER 7

Raman spectroscopy in forensic science

1 Introduction

Spectroscopic techniques use electromagnetic radiation with variable energies to analyze the characteristics of a molecule. The electromagnetic radiation is composed of electric and magnetic vectors that oscillate at right angles with each other. Normally, when the radiation hits a molecule, the electrons on different levels absorb energy, get excited, and move to a higher level where they emit energy until they fall back to a ground state. Most spectroscopic techniques are based on this principle and the instrumentation and type of molecule that can be detected vary with the type of energy emitted by the molecules upon excitation by the radiation.

Over the years, many spectroscopic techniques have been developed to analyze the composition and characteristics of a molecule. Raman spectroscopic techniques are the most widely used techniques to analyze the composition of a molecule as they are easy to use, require minimal sample preparation, and the spectra is also easy to interpret. Raman spectroscopy (RS) works on the basis of Raman scattering, which was discovered by Krishna and Raman in 1928. The theory behind Raman scattering will be discussed in brief in this chapter.

Raman radiation lies in the range of 10^3–10^5 nm with a frequency between 10^{13} and 10^{14} Hz. Early Raman spectrometers were a bit more complicated and needed the expertise of experienced spectroscopists. However, recently several modifications have been done to the spectrometer and now they are fully automated and even portable. With appropriate statistical and sample preparation techniques, it is possible to develop spectra of samples with high reliability. This technique can also positively identify the elements and molecules as well. This technique can also with utmost clarity differentiate between the pure and impure forms of germanium, sulfur, carbon, and silicon. Amorphous and crystalline carbon can also be differentiated on the basis of the sharpness of the band. In this chapter, however, the main focus shall be on the use of RS in the analysis of various forensic evidence.

Handbook of Analytical Techniques for Forensic Samples
https://doi.org/10.1016/B978-0-12-822300-0.00007-0

In forensic science, evidence such as biological fluids, drugs, and explosives is often encountered and is of immense significance. Also, evidence such as paints, pigments, dyes, gunshot residue, and even fibers and soil are commonly encountered. The use of RS in such cases allows the facile, rapid, and accurate determination of the composition of the evidence (Hofmann, 2010; Pandey et al., 2017; Rawtani et al., 2019; Smith and Dent, 2005).

2 Principle of Raman spectroscopy

Light, that is, electromagnetic radiation, is made up of small packets of energy known as photons. When these photons hit a molecule, a transference of energy takes place. If the energy of the incident photon matches the energy gap between the ground state and the excited state of the molecule, then the energy by the photon is absorbed and the molecule moves to the higher state. However, sometimes the energy absorbed by the photon may also result in its scattering, which may be detected. This scattering is characteristic to every molecule and is emitted with a strong intensity as long as the scattering is not absorbed by nearby electronic transitions. The scattering that occurs is also of two types, namely elastic and inelastic scattering. In elastic scattering, the kinetic energy of the photon is conserved, albeit with a change in its direction. In inelastic scattering, the kinetic energy is not conserved and the energy is transferred to its interacting matter. Raman scattering is a type of inelastic scattering in which the unconserved kinetic energy is taken as the difference between two vibrational states of the molecule, due to which the molecule gets promoted to a higher energy level. This difference is also known as the Stokes shift. Sometimes, the scattered light tends to have more energy than the incident light, due to which there is energy transition to lower wavelengths. This phenomenon is known as the anti-Stokes shift (Fig. 1). The entirety of the Raman spectrum comprises these shift lines. While scattering due to the Stokes shift will be stronger and is often the preferred shift during analysis, the anti-Stokes shift, which is significantly weaker than the Stokes shift, is preferred when there is additional interference such as fluorescence that occurs at an energy lower than the excitation energy (John and George, 2017; McCreery, 2005; Slater et al., 2001).

A phenomenon associated with Raman scattering is surface enhanced Raman scattering, or SERS. In this technique, through the interaction of the light with rough metal surfaces or nanomaterials, there is an enhancement in the electromagnetic field due to which even low concentrations

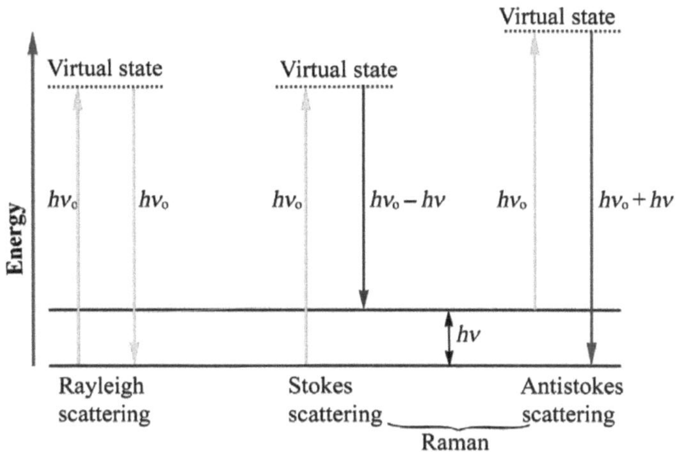

Fig. 1 Overview of Raman scattering (John and George, 2017).

can be easily detected. It is an excellent technique to enhance the structurally rich signal of Raman scattering. Enhancement in the electromagnetic radiation and chemical enhancement are the two main theories behind this phenomenon. Electromagnetic enhancement happens when the light is amplified due to the excitement of localized surface plasmon resonances (LSPR). In chemical enhancement, there is a series of charge transfer mechanisms in which the excitation wavelength corresponds with the charge transfer electronic states between the metal and the molecules. SERS allows the enhancement of the signal by a factor of 10^{10}–10^{11}, thus offering a detection accuracy of even a single molecule. Factors affecting the SERS phenomena are the substrates to be analyzed, the excitation source, and a detector capable of detecting Raman spectra across a wide spectral region (Bruzas et al., 2018; Schlücker, 2014; Sharma et al., 2012).

2.1 Instrumentation of Raman spectrometer

The Raman spectrometer consists of three main components: the light source, the spectrometer, and the detector. These three main components are also supported by additional faculties such as monochromators (Fig. 2). In RS, in order to achieve adequate scattering, a source with very high energy needs to be used. Initially, mercury arc lamps were widely used before transitioning to the use of a laser. The laser seemed to a be a suitable choice for this analysis because of its high frequency. The spectrometer consists of monochromators that are responsible for minimizing any stray light

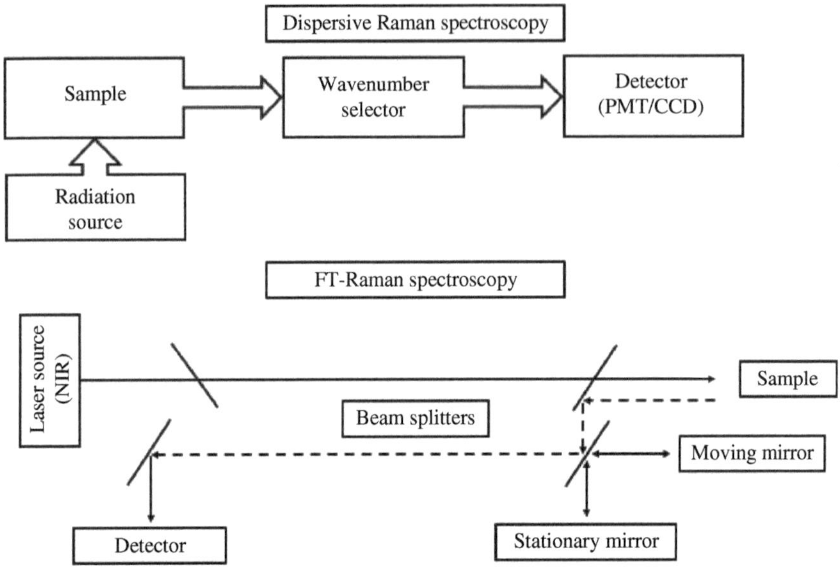

Fig. 2 Instrumentation of Raman spectroscopy (Biswas et al., 2010).

due to the scattering, thus allowing the formation of a clean spectra. The detectors used depend on the type of sample to be analyzed. They can be a dispersive monochromator with a charge coupled device, or a Fourier transform (FT) interferometer that employs indium gallium arsenide or liquid nitrogen cold germanium as the detector. Dispersive RS employs a laser of a shorter wavelength, as this will result in the enhancement of the Raman signals that occur at shorter wavelengths. In this type of spectroscopy, graters or monochromators also assist in the removal of any residual fluorescence. The charge coupled devices consist of light-sensitive receptors that capture every single dispersed wavelength. This type of detector has certain disadvantages as well, as it provides weak signals for high wavenumber rays such as in the case of NIR. Apart from this, FTRS is also used that uses a 1 μm excitation laser combined with an interferometer and an NIR detector. The interferometer receives the scattering signals and converts them into a uniform signal, thus producing an interferogram. This technique offers high resolution, an umbrella analysis of wavelengths, an increased signal-to-noise ratio, and excellent wavelength calibration. The type of detector chosen depends on the sample. Charge coupled devices are more suitable for the microscopic analysis of samples while FTRS is more suitable for the analysis of bulk substances.

Recently, several advancements have been made in RS. For instance, flexible optics have been the go-to option as a light source in this technique. Flexible optics have several advantages such as the detection of molecules in a liquid sample. Raman scattering as such is weak, but when used in combination with liquid chromatography, it can result in variable results. However, with flexible optics, it is possible to view the samples in the liquid with higher control and accuracy, thus enabling better results. Optical tweezers are an example of flexible optics in which the intensity of the light is focused on a particle. Once the light hits the particle from the top, replusive forces are created at the surface and at the base of the particle. Due to this there is a "caging" or "trapping" effect due to which the motion of the particle can be carefully controlled. By capturing the scattered light obtained from the particles, it is possible to obtain a very clean and accurate Raman spectra. The laser source used until now has been a continuous wave type in which the light emitted is of the same frequency. However, due to recent developments, tuneable or dye lasers are also used. Dye lasers involve the excitation of dyes that are kept in a continuous flow. The emissions obtained cover a wide range of frequency, thus allowing the analysis to be carried out over a wide spectral range. Double frequency lasers such as an Nd/YAG system are also used. These lasers offer a higher frequency range but at a lower power (Smith and Dent, 2005).

A great advantage with RS is that it requires minimal or no sample preparation. Samples such as fabrics and plastic sheets do not require sample preparation. However, samples that are in a lesser amount or are irregularly shaped require sample holders. Sample holders are also required when the laser beam needs to be focused on the sample at a particular angle in order to cancel out excessive irregular scattering such as in the case of crystalline samples. Samples can also be dispersed in a matrix that does not have a Raman spectrum such as potassium bromide or potassium chloride. Sometimes, samples cannot be dispersed in the mentioned solvents and need to be analyzed directly from the container. In such cases, it is important to consider the Raman scattering of the surrounding matrix as well. Samples that are less in amount or are microscopic may also be visualized and analyzed using a microscope linked with a Raman spectrometer (John and George, 2017).

3 Raman spectroscopy for forensic sample analysis

RS has remained an analytical tool for forensic scientists for many years (Fikiet et al., 2018; Izake, 2010). This section discusses the major types of forensic samples that can be examined using this characterization technique. These samples include biological samples such as body fluids, body remains, and fingerprints; recreational drugs; paints, dyes, and pigments; explosives and gunshot residue (GSRs); fibers; soil; and inks in questionable documents.

3.1 Samples of biological origin

Biological samples are among the most common evidence found at the crime scene. These samples primarily include body fluids, fingerprints, and decaying cadavers. A number of analytical techniques can be employed for the chemical and morphological profiling of these samples. RS has also served as a helping hand in the chemical profiling of the biological samples. Some research on biological samples including this technique are discussed here.

A group of researchers from Australia, France, and the United Kingdom using RS and a Raman microscope to examine the human remains of the Rwandan genocide that were excavated and put on display in the Murambi Technical School, Rwanda. The analysis at 1064 and 785 nm using a Raman microscope gave details about the deterioration of the remains due to environmental factors. Because the remains were preserved with lime, RS confirmed that the lime was converted to calcium carbonate. The presence of organic compounds was also confirmed through RS (Schotsmans et al., 2020). In addition to this, RS has also been used to study latent fingerprints. In a study, RS was used to study the interaction of cyanoacrylate with the fingerprints and find the anomalies in this mode of fingerprint detection. The cyanoacrylate polymerizes under suitable environmental conditions and develops a white film for fingerprint visualization. However, the RS spectra of the film indicated the presence of unreacted cyanoacrylate that could interact with the other substances such as drugs present in the fingerprint, thereby hindering their detection through fingerprints (Edwards and Day, 2006). SERS has also been used for detecting latent fingerprints. In a study, the presence of amino acid residues was taken as the base for detecting latent fingerprints. The sensitivity of the process was enhanced by using metal nanoparticles (vapors of silver deposited on polydimethylsiloxane). The study was the first one to carry out SERS-based imaging of fingerprints (Connatser et al., 2010). In a similar recent study, a SERS-based core shell nanoprobe functionalized with aptamer was developed for the Raman imaging of latent fingerprints found on the surface of curved or large immovable

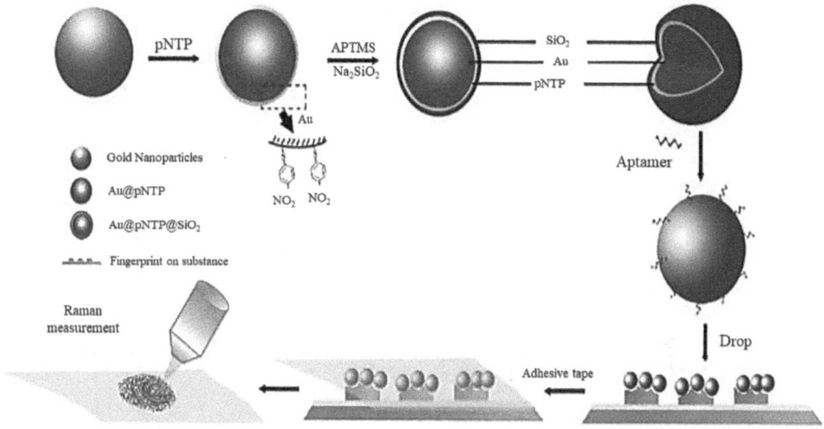

Fig. 3 SERS-based nanoprobe functionalized with an aptamer for the detection of latent fingerprints (Zhou et al., 2019).

objects (Fig. 3). The characteristic peaks obtained for the fingerprints were at 1572, 1343, 1081, 855, and 723 cm^{-1} (Zhou et al., 2019).

Because body fluids are very common evidence collected from the crime scene, therefore many forensic research works have been done on these body fluids by employing RS. A compilation of research works related to the age estimation of bloodstains through RS was published a few years back (Zadora and Menżyk, 2018). Lednev and coresearchers have shown the potential of RS for the identification of different types of body fluids such as semen (Virkler and Lednev, 2009), saliva (Virkler and Lednev, 2010), vaginal fluid (Sikirzhytskaya et al., 2012), and sweat (Sikirzhytski et al., 2012). Lednev also carried out a research work with his team to detect and identify stains of very common human body fluids such as blood, saliva, and semen. Fourteen blood, 50 semen, and 15 saliva samples were collected, dried, and further used for RS analysis. The study employed statistical analysis with RS for producing the signature spectra of these body fluids, and these spectra can be referred to for the identification of unknown body fluid samples (Sikirzhytski et al., 2010). These body fluids are often found contaminated with dust, soil, and sand particles when collected. In this regard, a study demonstrated the ability of Raman microspectroscopy to identify blood in the presence of these contaminants (Sikirzhytskaya et al., 2013). Another group of researchers also used RS to differentiate between dried samples of saliva, sweat, semen, blood, and vaginal fluid (Fig. 4) (Muro et al., 2016). The detection of urine on polyester and cotton fabric materials has also been

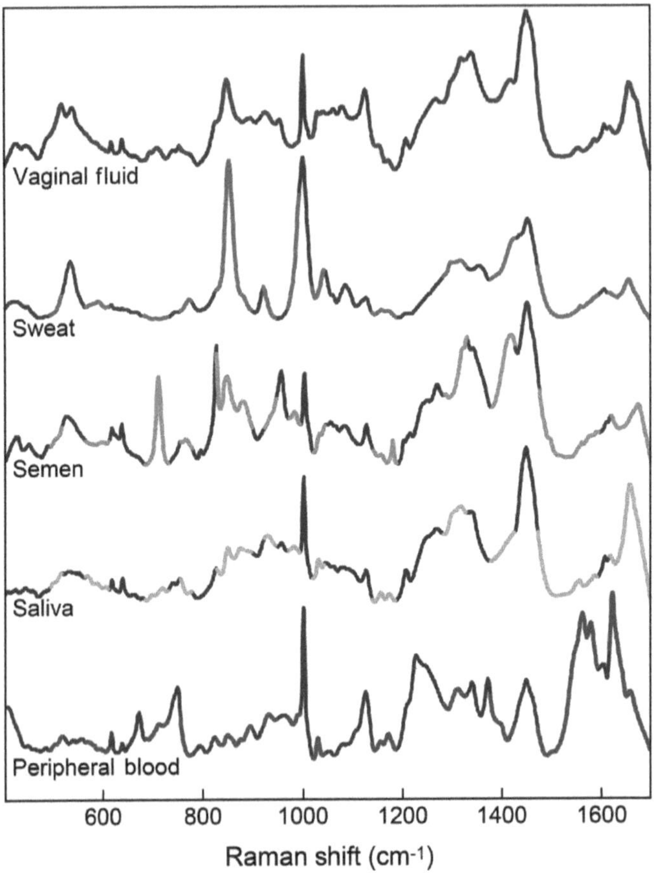

Fig. 4 Average Raman spectra of commonly found human body fluids (Muro et al., 2016).

made possible by researchers through a portable Raman spectrometer. The Raman spectra showed a single characteristic peak of the N—C—N bond found in urea at 1000 cm^{-1}. The study concluded that the combination of RS with partial least squares discriminant analysis can help in the detection of urine in these fabrics with high accuracy (Hager et al., 2018). Another recent study demonstrated the potential of RS to differentiate between human and animal blood (Doty and Lednev, 2018).

3.2 Recreational drugs

A recreational drug can be defined as a drug taken for its psychoactive nature, with users thinking that their sporadic consumption cannot be addictive.

These drugs can be synthetic or naturally available. Common examples of such drugs include cocaine, morphine, marijuana, etc. The utilization of RS for the analysis of drugs has been compiled in articles published a few years back (de Oliveira Penido et al., 2016; Khandasammy et al., 2018). Forensic scientists have utilized RS very efficiently to analyze these recreational drugs as well as their analogs.

In a study, RS was used to examine the cocaine hydrochloride collected from the nails of a human. The nail clippings were donated by the researchers themselves. The RS spectra of cocaine collected from the nail was also compared to that of pure drug and nail varnish in order to eliminate any extra peaks. A near-infrared laser (785 nm) was used for the RS analysis, which helped in reducing the fluorescence that is generated with visible range lasers (Ali et al., 2008). In another study, RS was used to analyze β-ketophenethylamine (β-KP), a class of legal high recreational drugs from the cathinone family. The analysis was performed with reference as well as seized samples. Calcium carbonate was used as the excipient in the seized β-KP drugs, which was identified through RS (Stewart et al., 2012). In a recent study, UV resonance RS was used to detect the presence of cocaine in oral fluid. The reference cocaine sample was dissolved in acetonitrile and water, and the spectra of both these solutions were compared with the cocaine's spectra in the oral fluid. Upon eliminating the peaks for pure oral fluid, the characteristic peaks of cocaine obtained on the RS spectra were at 1714, 1607, 1281, 1178, and 908 cm^{-1} (Fig. 5) (D'Elia et al., 2018). Phencyclidine, also called angel dust, and its analogs such as rolicyclidine, tenocyclidine, 3-methoxycyclidine, and 4-methoxycyclidine, have also been detected using Raman microspectroscopy in a recent study (Quinn et al., 2020).

3.3 Paints, dyes, and pigments

Paint analysis is very important in the automobile industries. However, for forensic investigators, the analysis of paints is important mostly in hit-and-run cases. If Locard's principle of exchange is taken into consideration, paint scrapings can be found in such accidents on the vehicle, person, or object that was hit. In addition to this, dyes and pigments that can be used in different fibers and inks also need to be examined by forensic scientists. Research works employing RS for such investigations are discussed here.

In a study, automotive paints were analyzed through RS. Paint flakes as well as a cross-section of paint layers were used to obtain the RS spectra. The

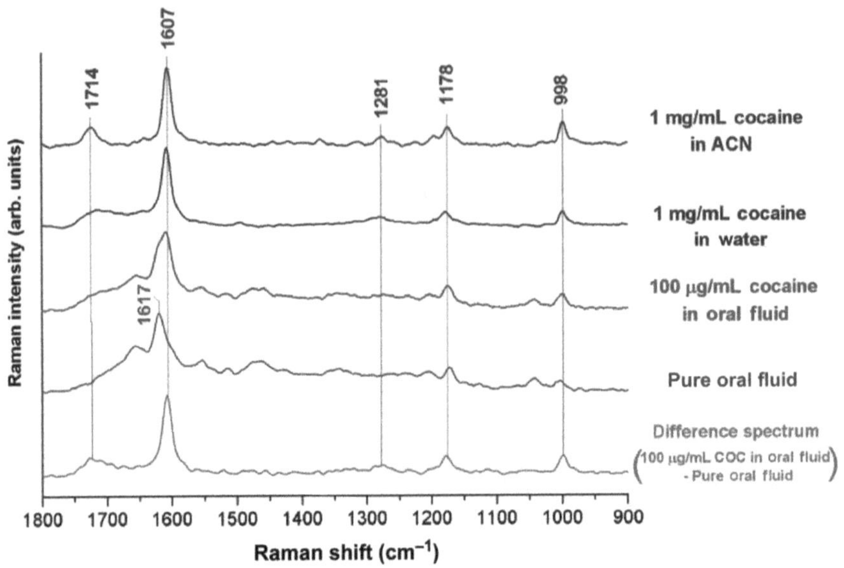

Fig. 5 Resonance Raman spectra of cocaine at different concentrations in different solvents and oral fluid (D'Elia et al., 2018).

study not only focused on the identification of pigments, but also on the identification of extenders and binders that are present in different layers of paint. Regarding the RS analysis, it was concluded that paint samples can be easily distinguished upon comparing the RS spectra of base coats (De Gelder et al., 2005). In another study, multiple layers of paint coats were examined through RS. The analysis was also supported by other techniques such as infrared and X-ray fluorescence spectroscopy. The RS analysis with an excitation laser in the visible range (633 nm) gave information about the pigments used in the paints. However, it did not give any details about the polymers used (Zięba-Palus and Borusiewicz, 2006). In addition to paints, dyes have also been analyzed through this technique. In a recent study, SERS was used for the analysis of trace quantities of dye samples with $<5\%$ weight of dyes. The constraint of synthesizing a SERS substrate for the SERS analysis was overcome in this study, which demonstrated a less time-consuming process for synthesizing silver nanoparticles that acted as SERS substrates. Eleven different dyes (basic, acidic, mordant, and direct) such as orange 1, yellow 2, red 3, green 4, blue 9, yellow 9, violet 19, blue 27, red 28, orange 52, and blue 74 were analyzed through SERS and the RS technique. The excitation wavelength for the laser was in the visible range (473 nm). In order to perform the SERS analysis, dye solutions were mixed with silver nanoparticles while for Raman analysis, dye solutions were directly placed on a glass slide covered with aluminum foil (Lux et al., 2019).

3.4 Explosives and gunshot residue

Explosive analysis is a very important part of forensic investigations, especially in cases related to terrorist attacks, bombings, and firearm incidents. RS has been utilized by researchers for the examination and chemical profiling of explosives and their derivatives. Gulia et al. used time gated RS for the analysis of trace quantities of explosives as well as their derivatives. The improvised explosive devices have traces of explosives on their surface up to the ppm level that need to be detected with high sensitivity and selectivity. This study helped in this regard through RS analysis. The laser's pulse energy, number of pulses getting accumulated, and the gain of the intensifier were the parameters that were optimized for increasing the sensitivity of detection through RS. The RS spectra of trinitrobenzene (TNB), trinitroperhydro triazine (RDX), and para-nitrobenzoic acid were obtained. The study showed the potential of RS, especially the time gated one for the detection of traces of explosives even on contaminated surfaces (Gulia et al., 2016). Almeida et al. conjugated an independent component analysis with Raman hyperspectral imaging for the detection of explosives. The study was focused on the detection of the ANFO (ammonium nitrate fuel oil) explosive on currency notes after an explosion experiment in ATM. The RS spectra were obtained through a near-infrared laser with an excitation wavelength of 785 nm. The characteristic peak for nitrate was obtained at 1044 cm^{-1}. Fig. 6 shows the Raman spectra of ANFO and compares it with that of other nitrate compounds (Almeida et al., 2017). In another study, the Stokes RS and infrared spectra of explosive-related materials were comparatively examined. The explosive-related compounds were urea nitrate, ammonium nitrate, nitroguanidine, and ammonium perchlorate. It was concluded that both analytical techniques were completing each other because the vibrations that were weak in one had strong bands in the other (Elbasuney and El-Sherif, 2017). Different SERS substrates have also been used for the detection of explosives in recent studies (Liszewska et al., 2019; Wu et al., 2020).

In addition to explosives, GSRs have also been detected using RS. In a study, macroscopic GSRs were examined through RS for the assessment of the memory effect of the weapon. Twenty shots were fired at the paper at a close distance of approximately 30 m, using the same weapon but two

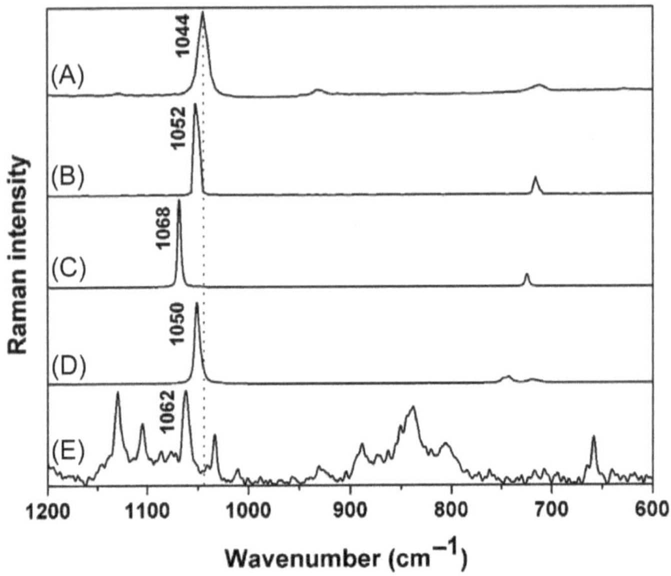

Fig. 6 Raman spectra of nitrate containing compounds: (A) ANFO; (B) potassium nitrate; (C) sodium nitrate; (D) calcium nitrate; (E) miconazole nitrate (Almeida et al., 2017).

different types of ammunition. The 1st, 3rd, 9th, and 20th shots were with the 1st ammunition while the rest were with the 2nd ammunition. The RS spectra of the GSRs were different at 1342 cm^{-1}, which corresponded to the diphenylamine and its derivatives that were present in one of the ammunitions (López-López et al., 2013). In another study that demonstrated the applicability of luminescent markers in GSR as indicators, RS was used to study the chemical profile of the modified GSR. The RS spectra helped in concluding that the marker merges with the GSR and travels along with it once shot from the weapon (Lucena et al., 2017). In a recent study, RS was used to analyze the GSR of two different ammunitions which had organic composition. The GSR spectra were compared with that of stabilizers and nitrocellulose. The study showed the potential of RS to differentiate between rifle calibers having propellants with ethyl centralite and diphenylamine as stabilizers (Karahacane et al., 2019).

3.5 Fibers

Fiber analysis in forensic science is important because it gives an idea about the clothing and fiber-containing belongings associated with the victim and the suspect. Microscopic techniques such as optical, bright field, fluorescence, and polarization are widely used for forensic fiber analysis (De Wael and Lepot, 2017). RS has been used by forensic scientists for fiber analysis mainly for three purposes: to identify the dyes present in the fibers; to compare the spectra of dyes in fibers with the reference dyes; and to determine the class and subclass to which the fiber belongs. Fig. 7 shows the Raman spectra of different types of vegetable, animal, and synthetic fibers using a visible range laser (633 nm) (De Wael and Lepot, 2017).

In a study, black or gray and blue cotton fibers were analyzed through RS, and the effect of changing the laser's wavelength was also investigated. Most of the dye components were identified through RS. The variable excitation wavelengths of the laser were 830, 785, 632.8, and 514.5 nm. It was observed that the lasers with 830 and 785 nm excitation wavelengths gave the most promising results. It was also concluded that for preparing the dye sample for RS analysis, the best way is to prepare their aqueous solutions and dry them on an aluminum slide (Thomas et al., 2005). In another study, the ability of RS to compliment other techniques such as microspectrophotometry (MSP) and optical microscopy in forensic fiber analysis was shown. Fibers were collected from five different criminal cases through the tape-lifting technique and mounted on glass slides or cover slips using mounting resin. The RS spectra were obtained using a laser with an excitation wavelength of 514.5 nm. The study showed the potential of RS-based fiber analysis in criminal cases such as murder, burglary, and car accidents (Lepot et al., 2008).

3.6 Soil

Soil analysis in forensic science can be helpful in finding the mineral composition of soil, which can be further correlated to the location of the soil. Soil samples collected from vehicles as well as the clothing of victims and suspects can be analyzed to identify the origin of the soil and the location where the crime was committed. RS has been used in research works that focused on the forensic examination of soil. However, a limitation of using RS for soil analysis is the fluorescence emitted by the soil sample, which interferes with the analysis.

In a study, RS was used to identify arsenates in the soil. The study was helpful for forensic and environmental scientists. The arsenic compounds that leach into the soil combine with metals and form metal arsenates. RS was used to analyze one such compound called vivianite arsenate. The RS spectra was obtained in the range of 4000–100 cm^{-1} using a laser with

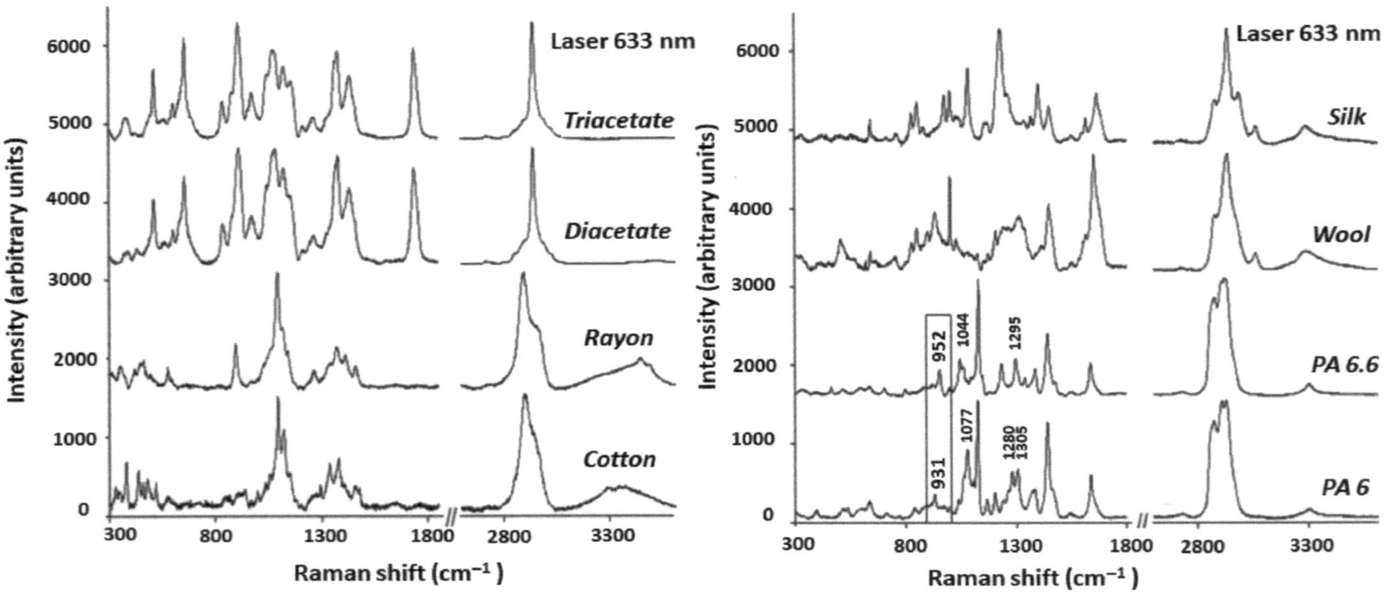

Fig. 7 Raman spectra of different types of fibers (De Wael and Lepot, 2017).

an excitation wavelength of 633 nm (Frost et al., 2003). Another study focused on the use of near-infrared RS for forensic soil analysis. The water soluble and mineralogical organic fractions were examined using RS. Ten soil samples were collected each from Bradford (urban location) and Lincolnshire (rural location). The RS spectra were obtained through the laser that operated at a 785 nm excitation wavelength. The soil samples were prepared through oxidation using hydrogen peroxide for removing aromatic and aliphatic contaminants from soil. Oxidation also helped in reducing the fluorescence from soil samples, which had been a major issue with the RS analysis of soil. The inorganic and organic constituents of soil appeared between the vibrational bands of 2000–1000 cm^{-1} (Edwards et al., 2012). Farmland soils have also been examined for their organic matter using RS. A total of 194 soil samples were collected from farmlands of China, and the RS spectra were obtained in the spectral range of 180–3200 cm^{-1}. There were 49 fluvo-aquic soil samples, 50 paddy soil samples, 46 black soil samples, and 49 red soil samples. The analysis was carried out using a near-infrared laser with an excitation wavelength of 785 nm. The study proposed the combination of Fourier transform infrared photo acoustic spectroscopy (FTIR-PAS) along with RS for better analysis of the organic matter in soil (Xing et al., 2016). In another study, a novel variant of RS called morphologically directed RS (MDRS) was used for the forensic examination of soil. The MDRS technique combines the RS with automated particle imaging in order to determine the particle shape and size. The analysis was performed using a semiconductor laser with an excitation wavelength of 785 nm. The difference in mineral content of different soil samples was examined through RS, which could assist criminalists in soil-based investigations (Kammrath et al., 2018).

3.7 Inks in questionable documents

The analysis of questionable documents becomes important for forensic investigators in cases related to forgery, counterfeiting, or murder. Documents are examined for line crossings and ink composition using different analytical tools. The inks in stamps can be used for hand signatures as a proof of the person's identity. RS has also shown its potential for analyzing the inks present on such questionable documents. The inks can be from a printer or a pen. In addition, line crossings can also be analyzed through RS (Braz et al., 2013).

Geiman et al. applied RS and SERS for the chemical profiling and examination of synthetic dyes that are present in inks of ballpoint pens. Ten dye samples were studied, and the spectra were recorded using a laser with excitation wavelengths of 633, 785, and 1064 nm. After performing the thin layer chromatography of the dyes, samples for RS and SERS were directly taken for the chromatography plate. Raman spectra often face hindrance due to the fluorescence offered by the dyes. However, SERS was found to be helpful in characterizing the dye samples (Geiman et al., 2009). In another study that focused on the forensic analysis of documents printed from black and colored inkjet printers, RS was used to distinguish the color printed documents. Ten inkjet ink cartridges were taken (seven from HP, one each from Canon, Epson, and Lexmark) for the analysis. The Raman spectra for different color printed documents were recorded with lasers having different excitation wavelengths (458, 532, 633, and 785 nm). The study also suggested the use of SERS for eliminating any chances of fluorescence from inks during RS (Heudt et al., 2012). Another study focused on the analysis of stamp inks present on questionable documents through RS. Nine different samples of stamp pad inks were collected from across India and analyzed through RS. All the Raman spectra were obtained with a laser having 785 nm as the excitation wavelength. The Raman scattering tool helped in analyzing intersection strokes as well by comparing the spectrum of individual inks with that of the intersection point (Raza and Saha, 2013).

4 Conclusion

RS is a versatile, nondestructive technique capable of analyzing various molecules with minimal or no sample preparation. RS works on the principle of Raman scattering that occurs when a high energy incident photon strikes a molecule, resulting in the absorption of the photon by the molecule and their shift to a higher or lower wavelength. This shifting to a higher or lower wavelength is known as the Stokes or anti-Stokes shift, respectively. The vibrations caused due to this are specific to each molecule and are therefore suitable for the characterization of every molecule. Any type of liquid or solid sample can be directly mounted on the spectrometer for analysis. A variation of the RS is surface enhanced Raman spectroscopy, or SERS. In this technique, when the incident light hits a rough metal surface or any nanomaterial, it results in an increase in the amplitude of the weak Raman signal by many-fold due to electromagnetic or chemical enhancement. The instrumentation of RS consists of three main components: the light source, the spectrometer, and the detector. The laser is the most used light source and the type of detector selected depends on the sample to be analyzed.

In this chapter, a brief overview of the principle, theory, and instrumentation was given. Also, the applications of RS in forensic science was also discussed. The application includes the use of this technique in the identification of samples of biological origin such as human remains, fingerprints, body fluids, and other samples such as recreational drugs, dyes, paints and pigments, and even explosives and gunshot residue. Bulk samples such as fabric and soil can also be analyzed using RS. Portable Raman spectrometers are also widely used. In the future, further miniaturizations can be done such that a lab-on-chip device with this technique can be used.

References

Ali, E.M.A., Edwards, H.G.M., Hargreaves, M.D., Scowen, I.J., 2008. Raman spectroscopic investigation of cocaine hydrochloride on human nail in a forensic context. Anal. Bioanal. Chem. 390, 1159–1166. https://doi.org/10.1007/s00216-007-1776-z.

Almeida, M.R., Logrado, L.P.L., Zacca, J.J., Correa, D.N., Poppi, R.J., 2017. Raman hyperspectral imaging in conjunction with independent component analysis as a forensic tool for explosive analysis: the case of an ATM explosion. Talanta 174, 628–632. https://doi.org/10.1016/j.talanta.2017.06.064.

Biswas, A., Wang, T., Biris, A.S., 2010. Single metal nanoparticle spectroscopy: optical characterization of individual nanosystems for biomedical applications. Nanoscale 2, 1560–1572. https://doi.org/10.1039/C0NR00133C.

Braz, A., López-López, M., García-Ruiz, C., 2013. Raman spectroscopy for forensic analysis of inks in questioned documents. Forensic Sci. Int. 232, 206–212. https://doi.org/10.1016/j.forsciint.2013.07.017.

Bruzas, I., Lum, W., Gorunmez, Z., Sagle, L., 2018. Advances in surface-enhanced Raman spectroscopy (SERS) substrates for lipid and protein characterization: sensing and beyond. Analyst 143, 3990–4008. https://doi.org/10.1039/C8AN00606G.

Connatser, R.M., Prokes, S.M., Glembocki, O.J., Schuler, R.L., Gardner, C.W., Lewis, S.-A., Lewis, L.A., 2010. Toward surface-enhanced Raman imaging of latent fingerprints*: surface-enhanced Raman imaging of latent fingerprints. J. Forensic Sci. 55, 1462–1470. https://doi.org/10.1111/j.1556-4029.2010.01484.x.

D'Elia, V., Montalvo, G., Ruiz, C.G., Ermolenkov, V.V., Ahmed, Y., Lednev, I.K., 2018. Ultraviolet resonance Raman spectroscopy for the detection of cocaine in oral fluid. Spectrochim. Acta A Mol. Biomol. Spectrosc. 188, 338–340. https://doi.org/10.1016/j.saa.2017.07.010.

De Gelder, J., Vandenabeele, P., Govaert, F., Moens, L., 2005. Forensic analysis of automotive paints by Raman spectroscopy. J. Raman Spectrosc. 36, 1059–1067. https://doi.org/10.1002/jrs.1408.

de Oliveira Penido, C.A.F., Pacheco, M.T.T., Lednev, I.K., Silveira, L., 2016. Raman spectroscopy in forensic analysis: identification of cocaine and other illegal drugs of abuse: Raman spectroscopy in forensic toxicology. J. Raman Spectrosc. 47, 28–38. https://doi.org/10.1002/jrs.4864.

De Wael, K., Lepot, L., 2017. Forensic science, applications of Raman spectroscopy to fiber analysis. In: Encyclopedia of Spectroscopy and Spectrometry. Elsevier, pp. 712–719, https://doi.org/10.1016/B978-0-12-409547-2.12168-7.

Doty, K.C., Lednev, I.K., 2018. Differentiation of human blood from animal blood using Raman spectroscopy: a survey of forensically relevant species. Forensic Sci. Int. 282, 204–210. https://doi.org/10.1016/j.forsciint.2017.11.033.

Edwards, H.G.M., Day, J.S., 2006. Anomalies in polycyanoacrylate formation studied by Raman spectroscopy: implications for the forensic enhancement of latent fingerprints for spectral analysis. Vib. Spectrosc. 41, 155–159. https://doi.org/10.1016/j.vibspec.2005.12.001.

Edwards, H., Munshi, T., Scowen, I., Surtees, A., Swindles, G.T., 2012. Development of oxidative sample preparation for the analysis of forensic soil samples with near-IR Raman spectroscopy. J. Raman Spectrosc. 43, 323–325. https://doi.org/10.1002/jrs.3031.

Elbasuney, S., El-Sherif, A.F., 2017. Instant detection and identification of concealed explosive-related compounds: induced stokes Raman versus infrared. Forensic Sci. Int. 270, 83–90. https://doi.org/10.1016/j.forsciint.2016.11.036.

Fikiet, M.A., Khandasammy, S.R., Mistek, E., Ahmed, Y., Halámková, L., Bueno, J., Lednev, I.K., 2018. Surface enhanced Raman spectroscopy: a review of recent applications in forensic science. Spectrochim. Acta A Mol. Biomol. Spectrosc. 197, 255–260. https://doi.org/10.1016/j.saa.2018.02.046.

Frost, R.L., Kloprogge, T., Weier, M.L., Martens, W.N., Ding, Z., Edwards, H.G.H., 2003. Raman spectroscopy of selected arsenates—implications for soil remediation. Spectrochim. Acta A Mol. Biomol. Spectrosc. 59, 2241–2246. https://doi.org/10.1016/S1386-1425(03)00068-4.

Geiman, I., Leona, M., Lombardi, J.R., 2009. Application of Raman spectroscopy and surface-enhanced Raman scattering to the analysis of synthetic dyes found in ballpoint pen inks*. J. Forensic Sci. 54, 947–952. https://doi.org/10.1111/j.1556-4029.2009.01058.x.

Gulia, S., Gulati, K.K., Gambhir, V., Sharma, R., Reddy, M.N., 2016. Trace detection of explosive and their derivatives in stand-off mode using time gated Raman spectroscopy. Vib. Spectrosc. 87, 207–214. https://doi.org/10.1016/j.vibspec.2016.10.009.

Hager, E., Farber, C., Kurouski, D., 2018. Forensic identification of urine on cotton and polyester fabric with a hand-held Raman spectrometer. Forensic Chem. 9, 44–49. https://doi.org/10.1016/j.forc.2018.05.001.

Heudt, L., Debois, D., Zimmerman, T.A., Köhler, L., Bano, F., Partouche, F., Duwez, A.-S., Gilbert, B., De Pauw, E., 2012. Raman spectroscopy and laser desorption mass spectrometry for minimal destructive forensic analysis of black and color inkjet printed documents. Forensic Sci. Int. 219, 64–75. https://doi.org/10.1016/j.forsciint.2011.12.001.

Hofmann, A., 2010. Spectroscopic techniques: I spectrophotometric techniques. In: Wilson, K., Walker, J. (Eds.), Principles and Techniques of Biochemistry and Molecular Biology. Cambridge University Press, Cambridge, pp. 477–521, https://doi.org/10.1017/CBO9780511841477.013.

Izake, E.L., 2010. Forensic and homeland security applications of modern portable Raman spectroscopy. Forensic Sci. Int. 202, 1–8. https://doi.org/10.1016/j.forsciint.2010.03.020.

John, N., George, S., 2017. Raman Spectroscopy. Elsevier, https://doi.org/10.1016/B978-0-323-46140-5.00005-4.

Kammrath, B.W., Koutrakos, A., Castillo, J., Langley, C., Huck-Jones, D., 2018. Morphologically-directed Raman spectroscopy for forensic soil analysis. Forensic Sci. Int. 285, e25–e33. https://doi.org/10.1016/j.forsciint.2017.12.034.

Karahacane, D.S., Dahmani, A., Khimeche, K., 2019. Raman spectroscopy analysis and chemometric study of organic gunshot residues originating from two types of ammunition. Forensic Sci. Int. 301, 129–136. https://doi.org/10.1016/j.forsciint.2019.05.022.

Khandasammy, S.R., Fikiet, M.A., Mistek, E., Ahmed, Y., Halámková, L., Bueno, J., Lednev, I.K., 2018. Bloodstains, paintings, and drugs: Raman spectroscopy applications in forensic science. Forensic Chem. 8, 111–133. https://doi.org/10.1016/j.forc.2018.02.002.

Lepot, L., De Wael, K., Gason, F., Gilbert, B., 2008. Application of Raman spectroscopy to forensic fibre cases. Sci. Justice 48, 109–117. https://doi.org/10.1016/j.scijus.2007.09.013.

Liszewska, M., Bartosewicz, B., Budner, B., Nasiłowska, B., Szala, M., Weyher, J.L., Dzięcielewski, I., Mierczyk, Z., Jankiewicz, B.J., 2019. Evaluation of selected SERS substrates for trace detection of explosive materials using portable Raman systems. Vib. Spectrosc. 100, 79–85. https://doi.org/10.1016/j.vibspec.2018.11.002.

López-López, M., Delgado, J.J., García-Ruiz, C., 2013. Analysis of macroscopic gunshot residues by Raman spectroscopy to assess the weapon memory effect. Forensic Sci. Int. 231, 1–5. https://doi.org/10.1016/j.forsciint.2013.03.049.

Lucena, M.A.M., Ordoñez, C., Weber, I.T., Torre, M., García-Ruiz, C., López-López, M., 2017. Investigation of the use of luminescent markers as gunshot residue indicators. Forensic Sci. Int. 280, 95–102. https://doi.org/10.1016/j.forsciint.2017.09.013.

Lux, C., Lubio, A., Ruediger, A., Robert, S., Muehlethaler, C., 2019. Optimizing the analysis of dyes by surface-enhanced Raman spectroscopy (SERS) using a conventional-microwave silver nanoparticles synthesis. Forensic Chem. 16, 100186. https://doi.org/10.1016/j.forc.2019.100186.

McCreery, R.L., 2005. Raman Spectroscopy for Chemical Analysis. John Wiley & Sons, Inc., Hoboken, NJ, USA, https://doi.org/10.1002/0471721646.fmatter.

Muro, C.K., Doty, K.C., de Souza Fernandes, L., Lednev, I.K., 2016. Forensic body fluid identification and differentiation by Raman spectroscopy. Forensic Chem. 1, 31–38. https://doi.org/10.1016/j.forc.2016.06.003.

Pandey, G., Rawtani, D., Agrawal, Y.K., 2017. Future aspects of halloysite nanotubes in forensic investigations. J. Nanomed. Res. 6. https://doi.org/10.15406/jnmr.2017.06.00153.

Quinn, M., Brettell, T., Joshi, M., Bonetti, J., Quarino, L., 2020. Identifying PCP and four PCP analogs using the gold chloride microcrystalline test followed by Raman microspectroscopy and chemometrics. Forensic Sci. Int. 307, 110135. https://doi.org/10.1016/j.forsciint.2019.110135.

Rawtani, D., Tharmavaram, M., Pandey, G., Hussain, C.M., 2019. Functionalized nanomaterial for forensic sample analysis. TrAC Trends Anal. Chem. 120, 115661. https://doi.org/10.1016/j.trac.2019.115661.

Raza, A., Saha, B., 2013. Application of Raman spectroscopy in forensic investigation of questioned documents involving stamp inks. Sci. Justice 53, 332–338. https://doi.org/10.1016/j.scijus.2012.11.001.

Schlücker, S., 2014. Surface-enhanced Raman spectroscopy: concepts and chemical applications. Angew. Chem. Int. Ed. 53, 4756–4795. https://doi.org/10.1002/anie.201205748.

Schotsmans, E.M.J., Wessling, R., McClue, W.A., Wilson, A.S., Edwards, H.G.M., Denton, J., 2020. Histology and Raman spectroscopy of limed human remains from the Rwandan Genocide. J. Forensic Legal Med. 70, 101895. https://doi.org/10.1016/j.jflm.2020.101895.

Sharma, B., Frontiera, R.R., Henry, A.-I., Ringe, E., Van Duyne, R.P., 2012. SERS: materials, applications, and the future. Mater. Today 15, 16–25. https://doi.org/10.1016/S1369-7021(12)70017-2.

Sikirzhytskaya, A., Sikirzhytski, V., Lednev, I.K., 2012. Raman spectroscopic signature of vaginal fluid and its potential application in forensic body fluid identification. Forensic Sci. Int. 216, 44–48. https://doi.org/10.1016/j.forsciint.2011.08.015.

Sikirzhytskaya, A., Sikirzhytski, V., McLaughlin, G., Lednev, I.K., 2013. Forensic identification of blood in the presence of contaminations using Raman microspectroscopy coupled with advanced statistics: effect of sand, dust, and soil. J. Forensic Sci. 58, 1141–1148. https://doi.org/10.1111/1556-4029.12248.

Sikirzhytski, V., Virkler, K., Lednev, I.K., 2010. Discriminant analysis of Raman spectra for body fluid identification for forensic purposes. Sensors 10, 2869–2884. https://doi.org/10.3390/s100402869.

Sikirzhytski, V., Sikirzhytskaya, A., Lednev, I.K., 2012. Multidimensional Raman spectroscopic signature of sweat and its potential application to forensic body fluid identification. Anal. Chim. Acta 718, 78–83. https://doi.org/10.1016/j.aca.2011.12.059.

Slater, J.B., Tedesco, J.M., Fairchild, R.C., Lewis, I.R., 2001. Raman Spectrometry and Its Adaptation to the Industrial Environment., p. 104.

Smith, E., Dent, G., 2005. Modern Raman Spectroscopy—A Practical Approach. John Wiley & Sons Ltd, England.

Stewart, S.P., Bell, S.E.J., Fletcher, N.C., Bouazzaoui, S., Ho, Y.C., Speers, S.J., Peters, K.L., 2012. Raman spectroscopy for forensic examination of β-ketophenethylamine "legal highs": reference and seized samples of cathinone derivatives. Anal. Chim. Acta 711, 1–6. https://doi.org/10.1016/j.aca.2011.10.018.

Thomas, J., Buzzini, P., Massonnet, G., Reedy, B., Roux, C., 2005. Raman spectroscopy and the forensic analysis of black/grey and blue cotton fibres. Forensic Sci. Int. 152, 189–197. https://doi.org/10.1016/j.forsciint.2004.08.009.

Virkler, K., Lednev, I.K., 2009. Raman spectroscopic signature of semen and its potential application to forensic body fluid identification. Forensic Sci. Int. 193, 56–62. https://doi.org/10.1016/j.forsciint.2009.09.005.

Virkler, K., Lednev, I.K., 2010. Forensic body fluid identification: the Raman spectroscopic signature of saliva. Analyst 135, 512–517. https://doi.org/10.1039/B919393F.

Wu, J., Zhang, L., Huang, F., Ji, X., Dai, H., Wu, W., 2020. Surface enhanced Raman scattering substrate for the detection of explosives: construction strategy and dimensional effect. J. Hazard. Mater. 387, 121714. https://doi.org/10.1016/j.jhazmat.2019.121714.

Xing, Z., Du, C., Tian, K., Ma, F., Shen, Y., Zhou, J., 2016. Application of FTIR-PAS and Raman spectroscopies for the determination of organic matter in farmland soils. Talanta 158, 262–269. https://doi.org/10.1016/j.talanta.2016.05.076.

Zadora, G., Menżyk, A., 2018. In the pursuit of the holy grail of forensic science—spectroscopic studies on the estimation of time since deposition of bloodstains. TrAC Trends Anal. Chem. 105, 137–165. https://doi.org/10.1016/j.trac.2018.04.009.

Zhou, Y.-Y., Du, Y.-M., Bian, X.-J., Yan, J., 2019. Preparation of aptamer-functionalized Au@pNTP@SiO_2 core-shell surface-enhanced Raman scattering probes for Raman imaging study of adhesive tape transferred-latent fingerprints. Chin. J. Anal. Chem. 47, 998–1005. https://doi.org/10.1016/S1872-2040(19)61171-0.

Zięba-Palus, J., Borusiewicz, R., 2006. Examination of multilayer paint coats by the use of infrared, Raman and XRF spectroscopy for forensic purposes. J. Mol. Struct. 792–793, 286–292. 8th Polish Molecular Spectroscopy Conference https://doi.org/10.1016/j.molstruc.2006.03.072.

CHAPTER 8

HPLC for the toxicological analysis of forensic samples

1 Introduction

Liquid chromatography is one of the most commonly employed analytical techniques for the separation, identification, and quantification of different constituents present in a mixture. The solvents, commonly called the mobile phase, are moved through the stationary phase, where the separation takes place. Different kinds of drugs, explosives, pesticides, toxins, etc., are analyzed using liquid chromatography (Oberacher and Arnhard, 2016). High-performance liquid chromatography (HPLC) (previously called high-pressure liquid chromatography) is the advanced version of traditional chromatographic techniques such as paper chromatography, thin layer chromatography, etc. It employs high pressure to push the mobile phase through the stationary phase made up of small particles that are densely packed inside a column (Gumustas et al., 2013). The merits of HPLC over other techniques include its amenability for myriad samples such as biomolecules, ions, and labile organic compounds; great sensitivity, precision, and resolution; autosampling (through an autosampler); and sample recovery postanalysis. However, there are some drawbacks also associated with this technique such as its costly instrumentation and operation; extensive sample processing; time consuming; the requirement of strict experimental conditions such as a dust-free environment and constant temperature; and the requirement of a skilled operator (Pragst, 2008). With advancements in different components of the HPLC instrument, most of the aforementioned drawbacks have been overcome. HPLC with new technological support in its pump, detector, autosampler, etc., has taken over the market, and this new HPLC is termed ultrahigh-performance liquid chromatography (UPLC) (Jerkovich et al., 2003).

This analytical technique has gained the attention of forensic investigators, and it has been used, especially in forensic toxicology, for the analysis of different toxicological evidence. Forensic toxicology is the division of forensic science that helps in the legal or medical examination of drug abuse,

Handbook of Analytical Techniques for Forensic Samples
https://doi.org/10.1016/B978-0-12-822300-0.00008-2

poisoning, or death due to a drug overdose or the consumption of poison or any pesticide (Pandey et al., 2017; Rawtani et al., 2019). The substance being ingested may get metabolized in the body, and can be found in body fluids and vomiting. The amount of the abused substance may also get reduced in concentration in the evidence collected from the crime scene (such as blood, sweat, urine, saliva, etc.), and given to the toxicologist for analysis. These kinds of analyses of illicit drugs, toxins, pesticides, etc., and their metabolites, even at very low concentrations, have become possible because of HPLC (El Sherbiny and Wahba, 2019). The technique has also been combined with other techniques such as mass spectroscopy for more specific and accurate identification of analytes and their metabolites.

This chapter discusses the utilization of HPLC in forensic science investigations with respect to toxicological analysis. Initially, the chapter sheds light on the principle, classification, and instrumentation of HPLC. Further, the application of HPLC in the detection of toxicological analytes such as illicit drugs, plant toxins, and pesticides is elaborated as well.

2 High performance liquid chromatography: An insight

2.1 Principle

The major principle behind any liquid chromatography (LC) is the interaction between the sample present in the mobile phase and the small particles present in the stationary phase. Similarly, in HPLC, whenever any mixture with myriad constituents is passed through the column with a stationary phase, different types of physical as well as chemical interactions occur between the analyte and the column packing material. Principally, LC and HPLC work in a similar manner except that HPLC is more efficient, sensitive, and easy to operate compared to LC. In HPLC, different components of the mixture have different affinities toward the column packing material (stationary phase). The components with higher affinity move slower while the components with lesser affinity travel faster in the column. Because of the different affinities of each component, they get separated and elute from the column at different times. The time duration for which the component is present inside the column is termed the retention time for that component. At the exit, the amount of these components is measured through a detector that gives a graphical output, known as a chromatogram (Lozano-Sánchez et al., 2018; Weston and Brown, 1997a).

2.2 Classification

HPLC techniques can be classified on the basis of separation mode, the principle of separation, the operating scale, the type of elution, and the analysis type. These classifications on the basis of the aforementioned parameters are as follows:

2.2.1 On the basis of separation mode

- **Normal phase**: In this case, the mobile phase is nonpolar in nature while the stationary phase is its opposite, that is, polar.
- **Reverse phase**: Here, the stationary phase is nonpolar while the mobile phase is polar. This technique is more common for drug analysis because most of the drugs are hydrophilic in nature.

2.2.2 On the basis of principle of separation

- **Ion exchange**: In this type, polar compounds and ions get separated on the basis of their charge. Resin is used in the stationary phase, which has anions and cations attached on it. The oppositely charged ions in the analyte interact electrostatically with the resin and get separated.
- **Gel permeation**: The analytes get separated on the basis of their size. There is no interaction between the analyte and the stationary phase.
- **Affinity**: The interaction between the stationary phase and the analyte is highly specific and selective. The technique is used to selectively separate any desirable compound from a mixture. Usually, proteins are separated using this method by using its antibody in the stationary phase.
- **Adsorption**: Here, the analyte molecules directly interact with the surface of the stationary phase. Depending on their affinity with the stationary phase, they elute at different times and thereby get separated.
- **Chiral**: The technique is use to separate stereoisomers. Enantiomers, which are stereoisomers with nonsuperimposable mirror images, cannot be separated using normal chromatography. To separate them, either the stationary phase or the mobile phase needs to be chiral in nature.

2.2.3 On the basis of scale of operation

- **Preparative**: Discrete constituents of a compound can be recovered.
- **Analytical**: Different constituents of the compound cannot be recovered.

2.2.4 On the basis of type of elution

- **Isocratic**: The composition of the mobile phase remains the same throughout the process. The technique is suitable only for simple separations. Upon increasing the complexity, the peaks become broad and flat.
- **Gradient**: During the separation process, the composition of the mobile phase keeps on changing. The technique helps in reducing the retention time for the components that get eluted last, thereby forming sharp and narrow peaks and thus assisting in complex separations.

2.2.5 On the basis of type of analysis

- **Qualitative**: The technique is useful only for the separation of compounds and analyzing the nature of the chemical constituents of a mixture.
- **Quantitative**: The technique in addition to the separation of compounds also gives data about their quantity and proportion in the mixture.

2.3 Instrumentation

The basic instrumentation of any HPLC includes a solvent delivery system, pumps, a sample injector, a guard column, an analytical column, a detector, and a data processing unit (computer) (LaCourse and LaCourse, 2017; Weston and Brown, 1997b). These components of the HPLC instrument are shown in Fig. 1 and elaborated below in detail.

2.3.1 Solvent delivery system

The delivery system of a solvent or the mobile phase mainly consists of reservoirs and degassers. In the case of isocratic elution, the full mobile phase is present in the reservoir while in case of gradient elution, the reservoirs contain different solvents that constitute the mobile phase. The mixing of these components takes place in a mixing chamber, which is usually present before pumps. The characteristics of solvents that can be used as the mobile phase in HPLC are that they are noncorrosive; pure, cheap, and UV transparent; less viscous; able to solubilize different analytes; nonflammable; and less toxic. An online vacuum degasser is used for degassing in the case of gradient elution in order to achieve a proper blending of the components of the mobile phase. If an online degasser is not available, a sonicator is used for degassing.

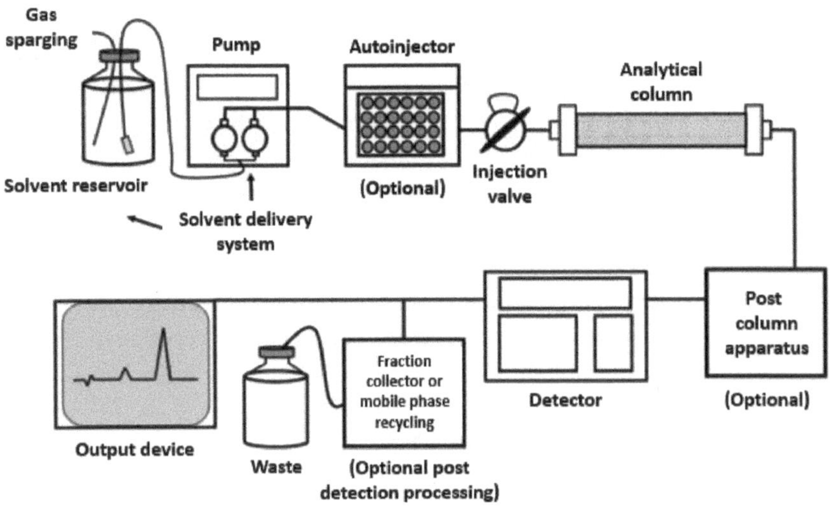

Fig. 1 Basic instrumentation of HPLC (LaCourse and LaCourse, 2017).

2.3.2 Pumps

Pumps are used to push the mobile phase through the column at the desired pressure and flow rate, which is usually expressed in mL/min. Typically, the flow rate during operation is kept at 1 or 2 mL/min, and the pressure is kept in the range of 6000–9000 psi. The type of elution (isocratic or gradient) is decided by the delivery of the mobile phase through the pump. The different types of pumps used in an HPLC instrument include:

- **Syringe pump**: This pump delivers a fixed volume (mostly between 250 and 500 mL) of the mobile phase and needs to be refilled again and again, which makes it suitable only for columns with a small bore size. The pump is driven by a mechanical lead screw that helps in delivering the mobile phase at a perpetual rate.
- **Reciprocating piston pump**: The piston on this pump is connected to a motor and moves back and forth inside a hydraulic compartment of around 30–400 μL. The back stroke pulls the mobile phase from the reservoir while the separation valve is shut. The following forward stroke pushes the solvent out from the column. Alterations in the stroke frequency or stroke volume vary the rate of flow for the mobile phase.
- **Constant pressure pump**: The flow rate remains in continuous phase in this pump, which drives the mobile phase through the column by using the pressure from a gas cylinder. The gas source is normally at low pressure in order to achieve high pressure in the solvent.

2.3.3 Sample injector

The sample injector helps to introduce the sample into the flow stream of the solvent (mobile phase). The sample is injected just before the column with the help of a Teflon and stainless steel valve, which is fitted with a loop that has two positions: load and inject. In the load position, the sample is introduced into the sample loop using a syringe, during which the injector is not in the flow path of the mobile phase. In the inject position, the injector comes in the flow path of the mobile phase and delivers the sample just ahead of the column. Fig. 2 depicts these two positions. The typical volume of a sample injection (manually) is 10–20 μL. The advanced versions of HPLC come equipped with autosamplers that can continuously inject the sample with a volume ranging from 1 μL to 1 mL.

2.3.4 Guard column

It is a small column, fitted just before the analytical column. The purpose of using this column is to protect the analytical column from any kind of contamination from foreign particles. It is also packed with the same material as the analytical column. These columns should not degrade the analytical performance of the instrument and should not cause any significant variation in the pressure of the mobile phase.

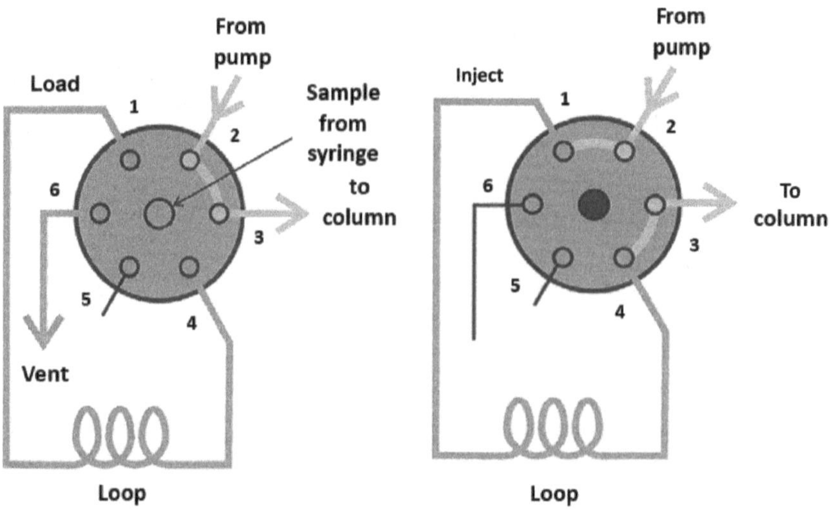

Fig. 2 Load and Inject position in a sample injector of HPLC (Dong, 2005).

2.3.5 *Analytical column*

The analytical column is considered the heart of the HPLC instrument. Different types of physical and chemical parameters govern the separation of components through the stationary phase present inside the column. Mostly, columns are made up of stainless steel, which allows them to withstand high pressure of the mobile phase created by pumps. Other materials include glass and polyether ether ketone (PEEK). The particles used inside the column are referred to as packing material. Depending on the type of HPLC, the bonding groups in the packing materials change. For normal HPLC, polar groups such as silica and aminopropyl are used; for reverse phase HPLC, nonpolar groups such as C18 (octadecyl), C8 (octyl), etc., are used. Among these bonding groups, silica is the most common. However, a drawback of silica-based packing material is that the column is suitable for separations only in the pH range of 2–8. Below pH 2, the Si—O bonds gets hydrolyzed, and above pH 8, the structure of silica becomes susceptible to dissolution. Depending upon the internal diameter of the column, it can be of different types such as preparative (>25 mm), semiprep (10 mm), minibore (3.0 mm), narrow bore (2.0 mm), microbore (1 mm), micro-LC (<0.5 mm), and nano-LC (<0.1 mm). However, these diameters are not universal, and may fluctuate with different manufacturers.

2.3.6 *Detector*

The detector is used to identify and quantify the analytes that get eluted from the column. The detector collects the data, converts it into an electrical signal, and sends it as an output signal to the recorder or the computer. The typical characteristics of an HPLC detector include high sensitivity, stability, and reproducibility; less response time; nondestructive; and a similar response toward different analytes. Various types of detectors used in HPLC are discussed below:

- **Absorbance or UV-visible detector**: It has a typical Z shape, and measures the absorbance of the eluents from the column. It is of three types: Fixed wavelength, variable wavelength, and diode array. In a fixed wavelength detector, the wavelength is around 254 nm, and the source is a mercury vapor lamp. In a variable wavelength detector, a complete spectrum of the UV-visible range is available for analysis (190–900 nm). In a diode array detector (DAD), the absorbance of the analyte is simultaneously measured at various wavelengths.
- **Fluorescence detector**: The detector has the same design as a typical fluorometer. Molecules with a fluorescence property can be very easily

detected using this detector. The sensitivity of these detectors is dependent on the fluorescent property of the eluent.
- **Refractive index detector**: The ability of the eluent to bend or refract light is measured through this detector. The degree of deflection can be correlated to the concentration of the analyte present in the mobile phase. A deflection refractometer and a Fresnel refractometer are the two main kinds of refractive index detectors.
- **Electrochemical detector**: The current produced due to the redox reaction going on at the electrode because of the analyte is measured through this detector. The intensity of the current produced is in direct proportionality with the concentration of the analyte.
- **Conductivity detector**: The ability of the analyte to conduct or resist current when it is placed between two electrodes to form a Wheatstone bridge is measured through this detector. The resistance offered is in proportionality with the concentration of the analyte.
- **Evaporative light-scattering detector (ELSD)**: The scattering of light by the particles present in the mobile phase postevaporation is measured in this detector. The mobile phase gets evaporated upon passing through a nebulizer, and the remaining analyte is thrown with a laser beam. The scattering of light is measured at a right angle.
- **Chiral detector**: Depending upon the technique, it is one of two types: circular dichroism (CD) and optical rotary dispersion (ORD). CD detectors differentiate between enantiomers by quantifying the difference between the absorption of circularly polarized light. ORD detectors calculate the difference in the refractive index of the chiral compounds.

2.3.7 Data processing unit

The data are obtained from the detector and utilized to decide the retention time and amount of analyte for the qualitative and quantitative analysis, respectively, by the data processing unit. The concentration of each component or analyte is measured by calculating the area of the peak corresponding to that analyte.

3 Applications of high-performance liquid chromatography in forensic toxicology

HPLC finds extensive application in forensic toxicology, especially in the detection of drugs, toxins, and pesticides. It can be used to analyze compounds that are nonvolatile, have a higher molecular weight, and are thermally sensitive. A validated HPLC method is capable of analyzing a compound with high sensitivity, reproducibility, automation, and robustness. It can also be applied to a wide range of samples in different matrices. Every class of analyte has a specific technique of sample preparation; analyzing a compound is highly dependent on the type of sample preparation, the mobile phase, the column, and the detector. Based on the sample, the HPLC conditions can be varied and modified accordingly. In this section, the use of HPLC for the analysis of various drugs, pesticides, and plant toxins will be discussed with special mention to sample preparation. Table 1 outlines the various HPLC parameters for the analysis of opiates, pesticides, and plant toxins.

3.1 HPLC analysis of drugs

In toxicology, drugs are the most widely used compounds that contribute to a person's death. Forensic drug analysis using HPLC involves the identification and quantification of illicit drugs. In this section, the HPLC analysis of cannabis and opiates will be discussed in brief.

Cannabis has long been used as a medicinal and recreational drug and for various industrial purposes such as the production of fiber and oil. The active constituent of cannabis is Δ^9-tetrahydrocannabinol or THC, which is known to produce euphoria and relaxation as well as chronic dependence and addiction. Apart from THC, cannabidiol is also an active constituent of this drug. Cannabis can also be detected through gas chromatography (GC) and mass spectrometry techniques, albeit with certain limitations. A major drawback with using GC for the detection of THC and cannabidiol is that upon heating during injection at the thermal port, these compounds often get oxidized from their acidic form to neutral forms, thus resulting in their decarboxylation. Therefore, they can be detected using HPLC. In a study, THC, cannabidiol, Δ^9-tetrahydrocannabinolic acid (THCA), and cannabidiolic acid (CBDA), which are some of the common constituents found in cannabis, were detected through HPLC-DAD. In the study, sample preparation was carried out by drying the plant and grinding it evenly, followed by the extraction of the components using a mixture of methanol and hexane. The column used was TCC-3000RS with a gradient elution of water and acetonitrile adjusted with 0.1% formic acid. The compounds showed peaks at retention times of 6–12 at a wavelength of 210 nm (Hädener et al., 2019). In yet another study, a more complex sample

Table 1 Parameters for toxicological analysis of analytes using HPLC.

S. no.	Analyte	Class	Stationary phase	Mobile phase	Retention time (min)	Flow rate (mL/min)	Detector	References
1.	THC, THCA, CBD, CBDA, CBG, CBGA, CBN, Δ^8-THC	Cannabis	MS C18 analytical column (5 mm, 250 mm × 2.1 mm)	MeOH/water with 50 nM ammonium formate	THC-23.52, THCA-18.55, CBD-16.52, CBDA-10.77, CBG-17.19, CBGA-12.47, CBN-31.0, Δ^8-THC-24.32	0.3	HPLC-DAD	De Backer et al. (2009)
2.	Morphine, Codeine	Opiate	C18 analytical column, 150 mm × 3.9 mm	Sodium heptane sulfonate/ MeOH with glacial acetic acid and trimethylamine	0–10	0.8	HPLC-UV	Freiermuth and Plasse (1997)
3.	Heroine	Opiate	Eclipse XDB-C8 column (4.6 × 150 mm)	Acetonitrile: 20 mM ammonium formate–100 mM potassium chloride buffer	0–5	1.5	HPLC-DAD	Elbardisy et al. (2019)

4.	Fluopyram, Spinosadm, Cyprodinil	Pesticide	UHPLC BEH C18 column (100 mm × 2.1 mm)	Ammonium formate/ MeOH/formic acid	Fluopyram-8.45, Spinosad-8.9, Cyprodinil-8.3	0.35	UHPLC-MS/MS	Valera-Tarifa et al. (2020)
5.	Thiabendazole, Fuberidazole, Carbofuran, Carbaryl and 1-naphtol	Pesticide	Zorbax Eclipse XDB-C18, 75 mm × 4.6 mm, 3.5 μm	Acetonitrile/ water	1–3.5	0.80	HPLC-UV-DAD	Montemurro et al. (2016)
6.	Carbamates	Pesticide	Centurysil C18 column (250 mm × 4.6 mm i.d., 5.0 m)	Acetonitrile/ water	6–25	1.0	HPLC-DAD	Wang et al. (2019)
7.	Norditerpenoid	Alkaloid	5 μm C18 Betasil HPLC column (100 × 2.1 mm)	Ammonium acetate and methanol	2–10	0.3	HPLC-ESI-MS/MS HPLC-ESI-MS/MS	Gardner and Pfister (2009)
8.	Taxine B and Iso-taxine B	Alkaloid	LiChrospher 60 RP-select B (250 × 4 mm, 5 μm)	Ammonium acetate, solvent B of methanol	32–34	1	HPLC-MS	Beike et al. (2003)
9.	Digoxigenin, Diginatin	Cardenolides	LiChrospher 100 RP-18 (5 pm)	Acetonitrile/ water	4.51, 24.92 resp.	1.2	HPLC-UV	Wiegrebe and Wichtl (1993)

THC, Δ^9-tetrahydrocannabinol; *THCA*, THC acid; *CBD*, Cannabidiol; *CBDA*, CBD acid; *CBG*, cannabigerol; *CBGA*, CBG acid; *CBN*, cannabinol; Δ^8-*THC*, Δ^8-tetrahydrocannabinol; *DAD*, diode array detector.

preparation was used in which dynamic maceration, ultrasound assisted extraction, microwave extraction, and supercritical fluid extraction were employed. For analysis, a C18 reverse phase column was used with a gradient elution of water and acetonitrile adjusted with 0.1% formic acid. The detection was done through HPLC-UV/DAD and further detection was done using electrospray ionization-mass spectrometry. The dynamic maceration technique that was done in ethanol proved to be the most efficient sample preparation technique. Overall, the mobile phase used, the sample preparation, and the detector were commonalities in several studies (Brighenti et al., 2017; Burnier et al., 2019). However, in a study, an innovative methodology was developed using HPLC-DAD for the analysis of the cannabis constituents as well as for the determination of the plant phenotype, the psychoactive potency of the plant, and its quality. Here, several multilinear models were developed that had the ability to predict the most optimum chromatographic conditions (De Backer et al., 2009).

Apart from cannabis, opiates have also been widely analyzed using HPLC. Opiates are derived from the plant *Papaver somniferum*, or the poppy plant, and are commonly referred to as opium. Opium contains morphine, codeine, and their derivatives such as heroin, oxymorphone, methadone, and fentanyl. These drugs are known to induce a state of stupor and euphoria, and can also cause hallucinations (Ling and Wesson, 1990). Similar to cannabis, opiates have been analyzed using an HPLC-DAD detector. In a study, cocaine, cocaethylene, and methadone were dissolved in methanol and analyzed using an X-Terra RP-8 column with a mobile phase of acetonitrile and a phosphate buffer. The analysis was also performed in real-life urine samples (Fernández et al., 2005). In another study, morphine was detected in an HPLC-PDA (photodiode array detector). Here, the morphine standard was ground and dissolved in methanol, followed by its analysis in HPLC using a Kinetex Su PFP 100A column with the mobile phase acetonitrile and ammonium formate (Chan, 2017). While these studies used similar sample preparation techniques, a paper reported the use of HPLC-UV for the detection of morphine and codeine through solid phase extraction. For the extraction, C18 bond Elut cartridges were used that were conditioned with methanol. The HPLC column used was a C18 column with a mobile phase consisting of sodium heptane sulfonate, glacial acetic acid, and triethylamine. This procedure offered single-step extraction and analysis of the drugs (Freiermuth and Plasse, 1997). Interestingly, HPLC was also used to analyze opiates present in hair. In this case, hair was usually dissolved in NaOH, followed by further analysis using HCl. The final extraction was done by isoamyl alcohol for the extraction of morphine. A similar extraction technique was also conducted for the analysis of codeine and other opiates. 6-monoacetylmorphine was considered the main byproduct of heroin that is widely found in hair, which was hydrolyzed in strong acidic and basic conditions as mentioned above (Staub, 1995).

3.2 HPLC analysis of pesticides

Pesticides have high forensic relevance as agricultural workers often get accidently exposed, resulting in acute and chronic poisoning. Quite often, the lack of appropriate regulations, surveillance systems, and enforcement can lead to misuse of pesticides and contribute to higher incidences of pesticide poisoning (Kulkarni, 2011; Rawtani et al., 2018).

Just like any other analyte for the analysis of HPLC, pesticides are also extracted from the matrix using a myriad of techniques. Some of the extraction techniques include solid-phase extraction (SPE) such as QuEChERS, which is an acronym for quick, easy, cheap, effective, and rugged; stir bar sorptive extraction; liquid-liquid extraction; and coupling techniques in which two different types of extraction techniques are combined (Nasiri et al., 2020). SPE is a technique in which the extraction takes place with the help of a solid matrix. This technique has several advantages such as ease of use and handling as well as automation. Over the years, adsorbents such as metal organic frameworks, graphene oxide sheets, and TiO_2 nanotubes have been widely used for this method of extraction (Huang and Lee, 2015; Li et al., 2016; Zhou and Fang, 2015). There are various forms of SPE such as dispersive SPE in which the adsorbent is directly added to the pesticide solution for extraction. Magnetic SPE is another technique in which the adsorbent itself is magnetic. Once the extraction process is complete, the adsorbent is magnetically separated. These techniques often pose difficulties when coupling with GC or HPLC. Therefore, an alternative method known as solid phase microextraction (SPME) was developed that performed extraction with increased speed, less solvent usage, and, most importantly, offered easy pairing with GC and HPLC. In this technique, the sorbent is coated on fibers and then exposed to the pesticide solution. Of recent, the most widely used SPE technique is QuEChERs, in which the sample-containing pesticide such as fruits or vegetables is homogenized in a suitable analyte and is later centrifuged before analysis through HPLC (Chan, 2017). Apart from solid phase extraction, liquid-liquid extraction is also used. However, the use is limited as it is expensive, requires large amounts of liquid sorbent. and is time consuming without any proper resolution of the analytes. Apart from liquid-liquid extraction, liquid-phase microextraction is also used in which the aqueous solution of the pesticide is mixed with a few microliters of a solvent that is not miscible with water. This technique can also be easily coupled with HPLC (Nasiri et al., 2020).

The various sample preparation techniques have been employed by researchers over the years for the extraction and detection of pesticides from various matrices. In a study, 250 pesticides in processed fruits were detected through ultraperformance liquid chromatography that was coupled with MS/MS. The extraction method used in this case was the QuEChERS method. The entire method of analysis took only 16 min and it achieved excellent linearity, trueness, and limit of quantification. This technique was later applied to real fruit samples and pesticides such as fluopyram, spinosad, and cyprodinil, which are organochlorine pesticides, were detected (Valera-Tarifa et al., 2020). A group of researchers reported the use of HPLC-DAD for the detection of carbamate pesticides in milk, wine, and juice. In this study, solvent solid phase extraction technique was used to extract the pesticide from the matrix. It was later analyzed using HPLC-DAD with a C18 column and a mobile phase of acetonitrile and water. Apart from HPLC-MS/MS, researchers have also used fluorescence emission for the detection of pesticides in the microgram range using HPLC-DAD. The researchers modeled the data with three commonly known algorithms and coupled the data with the fluorescence emission data. This technique allowed the detection of pesticides as low as 0.02 μg/L (Montemurro et al., 2016). Interestingly, HPLC has also been used for the analysis of pesticides in cannabis plants. Because many cannabis users are medically fragile, they may be at an increased threat of pesticides. Also, during processing, the oils present in cannabis may get pyrolyzed, resulting in the development of several harmful byproducts. The extraction methods for the pesticides from cannabis are similar to the extraction techniques discussed before. For their detection, HPLC-UV and HPLC-DAD have been used (Atapattu and Johnson, 2020).

3.3 HPLC in analysis of plant toxins and alkaloids

The accidental ingestion of plants can often result in toxicity due to plant alkaloids, which are the largest group of plant toxins. The alkaloids consist of a nitrogen containing heterocyclic ring. Examples include pyrrolizidine, piperidine, pyridine, indole, and ergot alkaloids. Apart from alkaloids, glycosides are also commonly occurring toxins. Examples include digitalin and toxins from *Nerium oleander* (Aguete et al., 2001; Beyer et al., 2009; Elbardisy et al., 2019; Jones and Dargan, 2016; Poppenga, 2010; Romanucci et al., 2016; Wiegrebe and Wichtl, 1993). Just like any other analyte, plant alkaloids also must be subjected to a series of extraction and preparation procedures. The most commonly used method of extraction is through sonication in a suitable solvent. The pH of the solvent is highly important to allow extraction with high efficiency. Apart from sonication, acid-base extraction is also used in which a crude extract is developed from the plant, followed by a liquid-liquid extraction using an acid that is often diluted, which is followed by neutralization of the extract with a suitable solvent. Extraction techniques such as SPME and liquid phase microextraction are also used for the extraction of plant alkaloids. The HPLC analysis of alkaloids is usually done in a stationary phase composed of alkyl silica while the mobile phases are usually composed of methanol, acetonitrile, and water. A major factor in the analysis of alkaloids using HPLC is the pH of the mobile phase. At low pH, the silanol group of the stationary phase does not undergo ionization and therefore allows better retention of the analyte, thus improving the peak shape. The reverse happens when the pH is high (de Oliveira, 2014). For the analysis of opium alkaloids, HPLC with reverse phased ion chromatography is excellent because they have a bulky nature and tend to form strong ion-pair bonds. [Gaillard and Pepin (1999)] analyzed a group of 40 plant toxins using HPLC-MS. In order to perform their analysis, the researchers employed a C18 column along with a gradient elution of acetonitrile-methanol with a formate buffer as the solvent system. Apart from this, pyrrolizidine alkaloids have also been detected in liqueurs, elixirs, and herbal juices. For sample preparation, a small amount of the sample was added in sulfuric acid and for further analysis, HPLC-ESI-MS/MS was used. The amount of pyrrolizidine detected was as low as 9.5 μg/kg. Also, the alcohol content used for the extraction determined the amount of alkaloid that could be extracted. Extraction using alcohol greater than 75% showed alkaloid content greater than the limit of detection (Chmit et al., 2019). Apart from pyrrolizidine alkaloids, norditerpenoid alkaloids have also been analyzed using HPLC-MS. In this case, two plant species, *Delphinium nuttallianum* and *D. andersonii*, were selected. They found four major alkaloids in these plants along with three additional alkaloids found in lower concentrations. The HPLC method used was reverse phase with a C18 column and a gradient elution with ammonium acetate and methanol was performed. Concentrations of the alkaloids detected using this technique ranged from 0.72 to 7.02 mg/g (Gardner and Pfister, 2009).

4 Conclusion

HPLC has remained as a favored analytical technique of forensic toxicologists for the separation, identification, and quantification of different analytes such as drugs, toxins, explosives, pesticides, etc. The technique has many advantages over conventional chromatographic techniques such as high sensitivity, selectivity, reproducibility, fast analysis, etc. HPLC can be classified on the basis of separation mode, principle of separation, scale of operation, type of elution, and type of analysis. The basic components of a typical HPLC instrument include a solvent delivery system, pumps, a sample injector, a guard column, an analytical column, a detector, and a data processing unit. The solvent delivery system consists of reservoirs (for storing the mobile phase) and degassers. Pumps help in pushing the mobile phase through the column at a desirable pressure and flow rate. The guard column prevents any kind of contamination from entering into the analytical column. The heart of the HPLC instrument, called the analytical column, is responsible for interacting with the analytes in the mobile phase. Due to differences in retention time for different components, they elute at different times from the column, which is identified and quantified by the detector. The data are finally expressed as peaks by the data processing unit in the form of a liquid chromatogram.

HPLC has a crucial role in the analysis of drugs, pesticides, and plant toxins in forensic science. A very important factor in the analysis of any analyte is its sample preparation. Over the years, various sample preparation techniques have been developed that are capable of extracting the main constituents from a sample. Based on the sample, the type of solvent used is also determined. If the sample has a high pH in nature, then the constituent is usually treated with an organic solvent of low pH. For the HPLC analysis of any sample, the mobile phase, the pH of the mobile phase, the analytical column, and the detector play significant roles in the analysis of any compound. In this chapter, we presented a brief discussion on the sample preparation of opiates, cannabis, pesticides, and plant toxins such as alkaloids and glycosides as well as their analysis using various HPLC detectors. HPLC is also a suitable alternative over GC in samples that are nonvolatile or that can get damaged upon injection at the thermal port of the GC. Recently, several new adsorbents have been made that can be incorporated in the analytical column for the analysis. With developments such as these, the time required and the cost required for HPLC analysis may become less, making it a prime candidate for facile toxicological analysis.

References

Aguete, E.C., Gago-Martínez, A., Rodríguez-Vázquez, J.A., O'Connell, S., Moroney, C., James, K.J., 2001. Application of HPLC and HPCE to the analysis of cyanobacterial toxins. Chromatographia 53, S254–S259. https://doi.org/10.1007/BF02490338.

Atapattu, S.N., Johnson, K.R.D., 2020. Pesticide analysis in cannabis products. J. Chromatogr. A 1612, 460656https://doi.org/10.1016/j.chroma.2019.460656.

Beike, J., Karger, B., Meiners, T., Brinkmann, B., Köhler, H., 2003. LC-MS determination of Taxus alkaloids in biological specimens. Int. J. Legal Med. 117, 335–339. https://doi.org/10.1007/s00414-003-0399-y.

Beyer, J., Drummer, O.H., Maurer, H.H., 2009. Analysis of toxic alkaloids in body samples. Forensic Sci. Int. 185, 1–9. https://doi.org/10.1016/j.forsciint.2008.12.006.

Brighenti, V., Pellati, F., Steinbach, M., Maran, D., Benvenuti, S., 2017. Development of a new extraction technique and HPLC method for the analysis of non-psychoactive cannabinoids in fibre-type *Cannabis sativa* L. (hemp). J. Pharm. Biomed. Anal. 143, 228–236. https://doi.org/10.1016/j.jpba.2017.05.049.

Burnier, C., Esseiva, P., Roussel, C., 2019. Quantification of THC in Cannabis plants by fast-HPLC-DAD: a promising method for routine analyses. Talanta 192, 135–141. https://doi.org/10.1016/j.talanta.2018.09.012.

Chan, K.-W., 2017. Validation of a straightforward high performance liquid chromatographic method for morphine quantitation. Egypt. J. Forensic Sci. 7, 1. https://doi.org/10.1186/s41935-017-0003-0.

Chmit, M.S., Wahrig, B., Beuerle, T., 2019. Quantitative and qualitative analysis of pyrrolizidine alkaloids in liqueurs, elixirs and herbal juices. Fitoterapia 136, 104172. https://doi.org/10.1016/j.fitote.2019.104172.

De Backer, B., Debrus, B., Lebrun, P., Theunis, L., Dubois, N., Decock, L., Verstraete, A., Hubert, P., Charlier, C., 2009. Innovative development and validation of an HPLC/DAD method for the qualitative and quantitative determination of major cannabinoids in cannabis plant material. J. Chromatogr. B 877, 4115–4124. https://doi.org/10.1016/j.jchromb.2009.11.004.

de Oliveira, B.H., 2014. High pressure liquid chromatography analysis of plant alkaloids. In: Meyers, R.A. (Ed.), Encyclopedia of Analytical Chemistry. John Wiley & Sons, Ltd, Chichester, UK, pp. 1–20. https://doi.org/10.1002/9780470027318.a9921.

Dong, M.W., 2005. HPLC instrumentation in pharmaceutical analysis: status, advances, and trends. In: Ahuja, S., Dong, M.W. (Eds.), Separation Science and Technology, Handbook of Pharmaceutical Analysis by HPLC. Academic Press, pp. 47–75. https://doi.org/10.1016/S0149-6395(05)80047-9.

El Sherbiny, D., Wahba, M.E.K., 2019. Micellar liquid chromatographic method for the simultaneous determination of citalopram hydrobromide with its two demethylated metabolites. Utility as a diagnostic tool in forensic toxicology. J. Pharm. Biomed. Anal. 164, 173–180. https://doi.org/10.1016/j.jpba.2018.10.032.

Elbardisy, H.M., Foster, C.W., Cumba, L., Antonides, L.H., Gilbert, N., Schofield, C.J., Belal, T.S., Talaat, W., Sutcliffe, O.B., Daabees, H.G., Banks, C.E., 2019. Analytical determination of heroin, fentanyl and fentalogues using high-performance liquid chromatography with diode array and amperometric detection. Anal. Methods 11, 1053–1063. https://doi.org/10.1039/C9AY00009G.

Fernández, P., Vázquez, C., Morales, L., Bermejo, A.M., 2005. Analysis of opiates, cocaine and metabolites in urine by high-performance liquid chromatography with diode array detection (HPLC-DAD). J. Appl. Toxicol. 25, 200–204. https://doi.org/10.1002/jat.1052.

Freiermuth, M., Plasse, J.-C., 1997. Determination of morphine and codeine in plasma by HPLC following solid phase extraction. J. Pharm. Biomed. Anal. 15, 759–764. https://doi.org/10.1016/S0731-7085(96)01911-5.

Gaillard, Y., Pepin, G., 1999. Poisoning by plant material: review of human cases and analytical determination of main toxins by high-performance liquid chromatography—(tandem) mass spectrometry. J. Chromatogr. B Biomed. Sci. Appl. 733, 181–229. https://doi.org/10.1016/S0378-4347(99)00181-4.

Gardner, D.R., Pfister, J.A., 2009. HPLC-MS analysis of toxic norditerpenoid alkaloids: refinement of toxicity assessment of low larkspurs (*Delphinium* spp.). Phytochem. Anal. 20, 104–113. https://doi.org/10.1002/pca.1104.

Gumustas, M., Kurbanoglu, S., Uslu, B., Ozkan, S.A., 2013. UPLC versus HPLC on drug analysis: advantageous, applications and their validation parameters. Chromatographia 76, 1365–1427. https://doi.org/10.1007/s10337-013-2477-8.

Hädener, M., König, S., Weinmann, W., 2019. Quantitative determination of CBD and THC and their acid precursors in confiscated cannabis samples by HPLC-DAD. Forensic Sci. Int. 299, 142–150. https://doi.org/10.1016/j.forsciint.2019.03.046.

Huang, Z., Lee, H.K., 2015. Micro-solid-phase extraction of organochlorine pesticides using porous metal-organic framework MIL-101 as sorbent. J. Chromatogr. A 1401, 9–16. https://doi.org/10.1016/j.chroma.2015.04.052.

Jerkovich, A.D., Mellors, J.S., Jorgenson, J.W., 2003. The use of micrometer-sized particles in ultrahigh pressure liquid chromatography. LCGC N. Am. 21, 600–610.

Jones, A.L., Dargan, P.I., 2016. Poisoning: overview and statistics. In: Encyclopedia of Forensic and Legal Medicine. Elsevier, pp. 1–9. https://doi.org/10.1016/B978-0-12-800034-2.00311-6.

Kulkarni, K.V., 2011. Identification of Pesticide and Drugs of Forensic Interest by Using Modern Analytical Method. Shri Chhatrapati Shivaji College, Omerga, India.

LaCourse, M.E., LaCourse, W.R., 2017. General instrumentation in HPLC*. In: Fanali, S., Haddad, P.R., Poole, C.F., Riekkola, M.-L. (Eds.), Liquid Chromatography. second ed. Elsevier, pp. 417–429. https://doi.org/10.1016/B978-0-12-805393-5.00017-8.

Li, M., Wang, J., Jiao, C., Wang, C., Wu, Q., Wang, Z., 2016. Graphene oxide framework: an adsorbent for solid phase extraction of phenylurea herbicides from water and celery samples. J. Chromatogr. A 1469, 17–24. https://doi.org/10.1016/j.chroma.2016.09.056.

Ling, W., Wesson, D.R., 1990. Drugs of abuse—opiates. West. J. Med. 152, 565–572.

Lozano-Sánchez, J., Borrás-Linares, I., Sass-Kiss, A., Segura-Carretero, A., 2018. Chromatographic technique: high-performance liquid chromatography (HPLC). In: - Sun, D.-W. (Ed.), Modern Techniques for Food Authentication. second ed. Academic Press, pp. 459–526. https://doi.org/10.1016/B978-0-12-814264-6.00013-X.

Montemurro, M., Pinto, L., Véras, G., de Araújo Gomes, A., Culzoni, M.J., Ugulino de Araújo, M.C., Goicoechea, H.C., 2016. Highly sensitive quantitation of pesticides in fruit juice samples by modeling four-way data gathered with high-performance liquid chromatography with fluorescence excitation-emission detection. Talanta 154, 208–218. https://doi.org/10.1016/j.talanta.2016.03.078.

Nasiri, M., Ahmadzadeh, H., Amiri, A., 2020. Sample preparation and extraction methods for pesticides in aquatic environments: a review. TrAC Trends Anal. Chem. 123, 115772. https://doi.org/10.1016/j.trac.2019.115772.

Oberacher, H., Arnhard, K., 2016. Current status of non-targeted liquid chromatography-tandem mass spectrometry in forensic toxicology. TrAC Trends Anal. Chem. 84, 94–105. https://doi.org/10.1016/j.trac.2015.12.019.

Pandey, G., Rawtani, D., Agrawal, Y.K., 2017. Future aspects of halloysite nanotubes in forensic investigations. J. Nanomed. Res. 6. https://doi.org/10.15406/jnmr.2017.06.00153.

Poppenga, R., 2010. Poisonous plants. In: Molecular, Clinical and Environmental Toxicology. Springer Nature, Birkhäuser Basel.

Pragst, F., 2008. High performance liquid chromatography in forensic toxicological analysis. In: Bogusz, M.J. (Ed.), Handbook of Analytical Separations. Elsevier, pp. 447–489. https://doi.org/10.1016/S1567-7192(06)06013-X.

Rawtani, D., Khatri, N., Tyagi, S., Pandey, G., 2018. Nanotechnology-based recent approaches for sensing and remediation of pesticides. J. Environ. Manag. 206, 749–762. https://doi.org/10.1016/j.jenvman.2017.11.037.

Rawtani, D., Tharmavaram, M., Pandey, G., Hussain, C.M., 2019. Functionalized nanomaterial for forensic sample analysis. TrAC Trends Anal. Chem. 120, 115661. https://doi.org/10.1016/j.trac.2019.115661.

Romanucci, V., Pisanti, A., Di Fabio, G., Davinelli, S., Scapagnini, G., Guaragna, A., Zarrelli, A., 2016. Toxin levels in different variety of potatoes: alarming contents of α-chaconine. Phytochem. Lett. 16, 103–107. https://doi.org/10.1016/j.phytol.2016.03.013.

Staub, C., 1995. Analytical procedures for determination of opiates in hair: a review. Forensic Sci. Int. 70, 111–123. https://doi.org/10.1016/0379-0738(94)01610-H.

Valera-Tarifa, N.M., Santiago-Valverde, R., Hernández-Torres, E., Martínez-Vidal, J.L., Garrido-Frenich, A., 2020. Development and full validation of a multiresidue method for the analysis of a wide range of pesticides in processed fruit by UHPLC-MS/MS. Food Chem. 315, 126304. https://doi.org/10.1016/j.foodchem.2020.126304.

Wang, X., Meng, X., Wu, Q., Wang, C., Wang, Z., 2019. Solid phase extraction of carbamate pesticides with porous organic polymer as adsorbent followed by high performance liquid chromatography-diode array detection. J. Chromatogr. A 1600, 9–16. https://doi.org/10.1016/j.chroma.2019.04.031.

Weston, A., Brown, P.R., 1997a. High-performance liquid chromatography. In: Weston, A., Brown, P.R. (Eds.), HPLC and CE. Academic Press, San Diego, pp. 1–23. https://doi.org/10.1016/B978-012136640-7/50001-5.

Weston, A., Brown, P.R., 1997b. Instrumentation for high-performance liquid chromatography. In: Weston, A., Brown, P.R. (Eds.), HPLC and CE. Academic Press, San Diego, pp. 71–133. https://doi.org/10.1016/B978-012136640-7/50003-9.

Wiegrebe, H., Wichtl, M., 1993. High-performance liquid chromatographic determination of cardenolides in Digitalis leaves after solid-phase extraction. J. Chromatogr. A 630, 402–407. https://doi.org/10.1016/0021-9673(93)80478-Q.

Zhou, Q., Fang, Z., 2015. Graphene-modified TiO_2 nanotube arrays as an adsorbent in micro-solid phase extraction for determination of carbamate pesticides in water samples. Anal. Chim. Acta 869, 43–49. https://doi.org/10.1016/j.aca.2015.02.019.

CHAPTER 9

Gas chromatography in forensic science

1 Introduction

Gas chromatography (GC) is a powerful analytical tool used for the analysis of various forensic evidence that can be made volatile. It offers rapid analysis and can efficiently separate the analytes present in the sample. This technique can also be efficiently coupled with mass spectrometry and thus can be used to accurately identify the molecular composition of the sample.

Chromatography itself was first created by Ramsey and Tswett in 1905 and 1906, respectively, for which they received the Nobel prize. The first GC was developed by James and Martin in 1951 and 1952. While the first ones created were very fragile and poorly constructed, they did the job of separating the analytes in the sample. Over the years, this instrument has undergone a revolution, and now almost every forensic lab contains a GC equipment.

In forensic science, evidence such as drugs, toxins, and explosives is commonly encountered (Pandey et al., 2017; Rawtani et al., 2019). Also, any forensic evidence that can be easily solubilized in a solvent and volatilized can be analyzed using this technique. Some equipment also has an inbuilt library of molecules through which the analysis and identification of the evidence can be easily done. In this chapter, a brief overview of the principle and theory of GC, its instrumentation, and its applications in the analysis of various forensic evidence such as alcohol, organic compounds, drugs, explosives, ignitable materials, fingerprints, inks, paints, and toxins has been discussed.

2 Principle and theory of gas chromatography

Gas chromatography, just like every other chromatography technique, separates analytes through adsorption by a stationary phase, followed by exposure to the mobile phase that in this case is a carrier gas. More information on the type of stationary phases and the carrier gases used is given in the next

Handbook of Analytical Techniques for Forensic Samples
https://doi.org/10.1016/B978-0-12-822300-0.00009-4

section. When the sample is introduced to the stationary phase, it gets adsorbed on it. Based on the affinity of the analytes present in the sample toward the stationary phase or the mobile phase, they get separated into separate components. These analytes are further detected by a suitable detector. The affinity of the sample toward the stationary phase is represented by the partition coefficient and provides a number for the total amount of analytes adsorbed on the stationary phase. In GC, the flow of the carrier gas is kept constant and the flow rate itself is measured in terms of mL/min. This flow rate is highly dependent on the column volume and varies with different volumes; it is known as volumetric flow. Another parameter that represents the flow of the gas is the average linear velocity, which is a representative of the velocity of the gas within the column according to the pressure inside it. This parameter is more useful than the volumetric flow rate because it is independent of the volume of the column and once calculated for a sample, it can be applied to any instrument in any laboratory as it is not affected by the column size.

Other highly crucial parameters for analysis using GC are the retention time and factor. In GC, the mobile phase carries the samples that get adsorbed on the stationary phase, thus making them travel at separate times. The retention time is therefore the amount of time the sample spends on the stationary phase and thereby in the column. This retention time is an excellent parameter that indicates the time spent by the sample in the column and the velocity of the mobile phase. Every compound therefore has its characteristic retention time and is represented as a peak at that particular retention time. Apart from retention time, the retention factor is yet another parameter that indicates the retention capacity of the sample. It is a ratio of the time spent by the analyte in the stationary phase to the time spent in the mobile phase, and it is calculated from the retention time. Because this technique too allows the separation of the sample through two different phases, therefore there may be an overlapping of two peaks. This overlapping can be prevented by increasing the resolution in which either the mobile phase or the stationary phase can be modified. Appropriate sample extraction techniques can also increase the sample concentration. Also, by focusing on factors such as the distance between two peaks and the area or width of each peak, it is possible to optimize the resolution of a particular sample. While resolution deals with how efficiently a chromatographic method can be developed to separate two compounds and thus prevent the overlapping of their peaks, another factor known as selectivity gives an idea of how the selectively distinguishes the two compounds and is represented through the alpha value.

This value is directly proportional to the separation of the two compounds, that is, the greater the alpha value, the greater the separation of the compounds.

Theories such as plate theory and rate flow theory as well as the Van Demeter equation are applied in every chromatographic technique to hypothesize the different forms of chromatographic separation. In plate theory, it is assumed that the column is made up of a series of plates stacked closely to each other vertically. The column efficiency is determined by the number and height equivalent of the plates, which is also known as the height equivalent to the theoretical plate (HETP) and is usually calculated from the retention time and the peak area. Because HETP is an indicator of column efficiency, it is dependent on the length of the column and the number of plates stacked while being inversely proportional to the column efficiency. The rate theory is another hypothesis in which emphasis is given to the dynamics inside the column. Factors such as the flow rate, the type of flow and diffusion, and the temperature and pressure that contribute to band broadening are indicators of the column efficiency. The main phenomenon considered in this theory is the flow of the sample in the column, either through longitudinal diffusion or the flow paths. In longitudinal diffusion, the sample diffusion occurs along the length of the column and occurs when the linear velocity is very slow, which gives the samples more time to stay in the column. However, this does not mean that the linear velocity can be increased many-fold. Too much linear velocity can prevent the samples from actually getting adsorbed by the stationary phase. Also, every sample has its own path of movement in the sample and is made uniform according to the flow of the mobile phase, also known as eddy diffusion. Here, the faster the flow of the mobile phase, the greater will be the deviation of the sample flow path, which may further influence the band broadening. All these factors had have been summarized efficiently in an equation known as the Van Deemter equation, which is as follows:

$$\mathrm{HETP} = A + B/\mu + C \times \mu$$

where A represent the eddy diffusion (cm), B represents the diffusion coefficient (cm^2/s), and C is the resistance to mass transfer between the stationary phase and the mobile phase. μ is the parameter for the linear velocity (cm/s). Fig. 1 gives a brief overview of the principle of gas chromatography.

By considering these parameters, the HETP can be easily found and the column efficiency can also be determined (Blumberg, 2012; McNair and Miller, 1933).

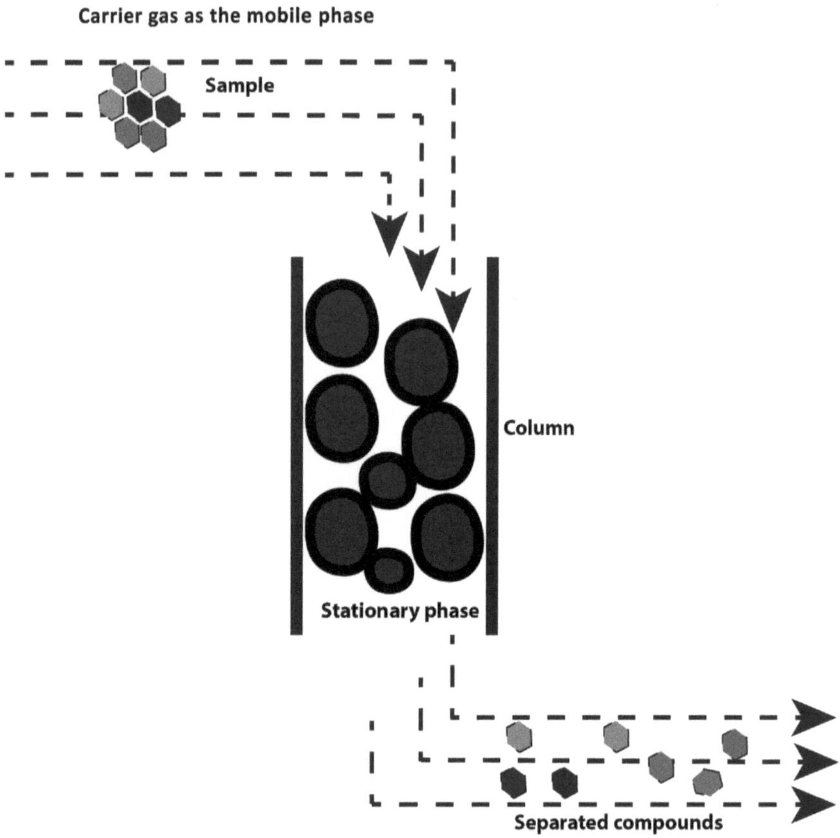

Fig. 1 Principle scheme of gas chromatography.

3 Instrumentation of gas chromatography

GC typically consists of a sample injector, carrier gas, flow controller, column, detector, and data acquisition system. The sample injector is responsible for the inlet of samples into the stream of the carrier gas. It is usually built to handle a myriad of samples such as solids, liquids, and gases and their conversion to a volatile form so that they can be introduced into the carrier gas. The sample inlets vary according to the columns used. For instance, a flash vaporizer is used in case of a packed column and split or splitless inlets are used in case of a capillary column. More information on the different types of columns used is discussed further. Gas samples are often obtained as mixtures of gases and liquids. These samples are further converted to liquids before injection into the carrier gas. Typically, syringes and valves that are specific to gas sampling are used. Meanwhile, liquid sampling is also done using syringes that are capable of injecting very small amounts of samples. Extreme small amounts of liquid samples are inserted because upon vaporization, they tend to expand and may therefore pose problems in the columns. Solid samples are the most challenging types of samples because of the challenges in volatilizing them. One of the most common ways of preparing a solid sample is by solubilizing it in a suitable solvent. Once solubilized, it is treated just like any liquid sample. Here too, syringes are used to inject the samples. However, these days, autosamplers are used in which the amount of sample drawn from the prefitted vial is automatically done according to the volume given in the software of the system.

The carrier gas as mentioned before serves as the mobile phase and carries the samples from the sample inlet to the column and the detector. The type of carrier gas used depends on the type of detector used because this gas also serves as a matrix for the detector. For instance, a flame ionization detector requires helium or nitrogen gas. Factors such as the purity, flow control, measurement, and compressibility of the carrier gas are highly important. The carrier gas used must be free of any impurities because it can react with the stationary phase in the column, possibly hindering and corrupting the sampling. Impurities here mean any foreign substance apart from the gas. Therefore, even oxygen and water are considered impurities in such cases. These impurities also tend to produce their own peaks that may affect the movement of samples as well as their resolution and width. The flow rate of the carrier gas is also important as it can affect the column efficiency and the band broadening. The flow of the carrier gas is controlled by using specific controls that take care of the pressure, temperature, and flow rate. The measurement is also typically done either by a soap bubble flowmeter or an electronic device.

The columns are the most crucial part of GC and are considered the heart of the instrument. There are basically two types of columns: the packed column and the capillary column. The packed column consists of small particles typically of diatomaceous earths that are usually modified with various organosilanes to enhance the adsorption capacity of the stationary phase; they come in variable heights, with the longest at 12 ft. Apart from this silica gel, alumina, carbon adsorbents, and zeolites have also been used as stationary phases. Meanwhile, the capillary column is a simple column with extreme diameters, and it does not consist of any packing. These columns are more efficient than the packed columns as they have a high separation efficiency, as long columns can be easily used. These columns are also coated by the stationary phases as mentioned above.

Another equally crucial part of the instrument is the detector. A good detector has excellent sensitivity, signal-to-noise ratio, detectability, specificity, linearity, and response time. Detectors commonly used for GC are thermal conductivity detectors (TCD), flame ionization detectors (FID), electron capture detectors (ECD), helium ionization detectors (HeD), alkali flame ionization detectors (AFID), and flame photometric detectors (FPD).

In TCD, two filaments are used that are heated sufficiently and the power to heat them is kept constant. Carrier gases such as hydrogen and helium are passed on these filaments. One filament that is the reference filament receives only the carrier gas while the other filament receives the carrier gas with the sample. Due to this difference, there is a also a difference in the amount of power required to maintain the temperature of the filament. This difference is considered and processed further. Meanwhile, FID is highly suitable for the detection of low concentrations of analytes by detecting the ions that are formed when an analyte is burned in a hydrogen flame. The greater the amount of organic groups in the analyte, the greater the amount of ions generated. ECD, as the name suggests, works by capturing electron absorbing compounds such as halogens present in the carrier gas. The greater the electronegativity, the greater the amount of electrons captured, thus causing an increase in the current detected. HeD uses metastable helium ions that produce ions of the analytes that hit the metastable ions. The generated ions are captured, and an electric current is generated that serves as the signal and is further processed. AFID is a modification of the flame ionization detector in that the hydrogen flame enters the FID perpendicularly to the FID jet. Due to this, there is a constant supply of alkalis, thus limiting the detector fatigue time. Lastly, FPD is commonly used to detect sulfur and phosphorous compounds by measuring the fluorescence emissions of the molecules in organic samples. Fig. 2 highlights the overall instrumentation of GC (Conte and Barry, 1993; Hartmann, 1971; Kitson et al., 1996; McNair and Miller, 1933; Šev̂cík, 1976; Sparkman et al., 2011).

4 Forensic sample analysis via gas chromatography

GC has been used widely in different laboratories and industries for the detection and quantification of myriad analytes. In forensic science, GC in combination with other techniques has helped in the investigation of analytes such as alcohol and other volatile organic compounds (VOCs), drugs,

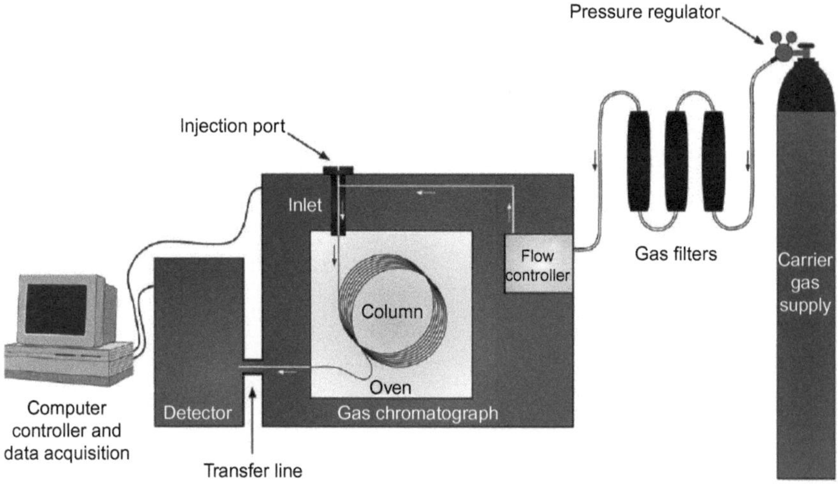

Fig. 2 Instrumentation of gas chromatography (Stauffer et al., 2008).

explosives and ignitable materials, fingerprints, inks, paints, and toxins (Sampat et al., 2016a). The research on the use of GC for such samples is elaborated in this section.

4.1 Alcohol and other volatile organic compounds

Alcohol and VOCs are routinely analyzed in forensic laboratories in cases of abuse, poisoning, leakage, etc. GC has assisted in this regard in many investigations, and various research works have also been conducted.

In a study by Davis and Cortivo, biological specimens such as the blood, brain, liver, and kidney were analyzed to check for the presence of isopropanol using headspace GC (HS-GC). Acetone, being the metabolic product of isopropanol, was also detected. MS was used for the identification of such products. The concentration range (mg/100 g) of isopropanol detected in these specimens was 1–29 (for blood), 7–59 (for liver), 2–12 (for brain), and 6–26 (for kidney) (Davis et al., 1984). Ethyl glucuronide, a metabolic product of ethanol, has also been determined using GC with tandem MS. Solid phase microextraction (SPME) was used to extract the metabolite from hair. Helium was used as the carrier gas with a flow rate of 1 mL/min. A retention time of 6.6 min was observed for ethyl glucuronide (Fig. 3). The limit of detection (LOD) was 0.6 pg/mg of hair (Agius et al., 2010). Brasseur et al. used 2D GC-MS to study the VOCs released by cadavers inside the soil by burying pig carcasses that were already decaying. The flow rate for the carrier gas (helium) was maintained at 1 mL/min. VOCs such as ketone,

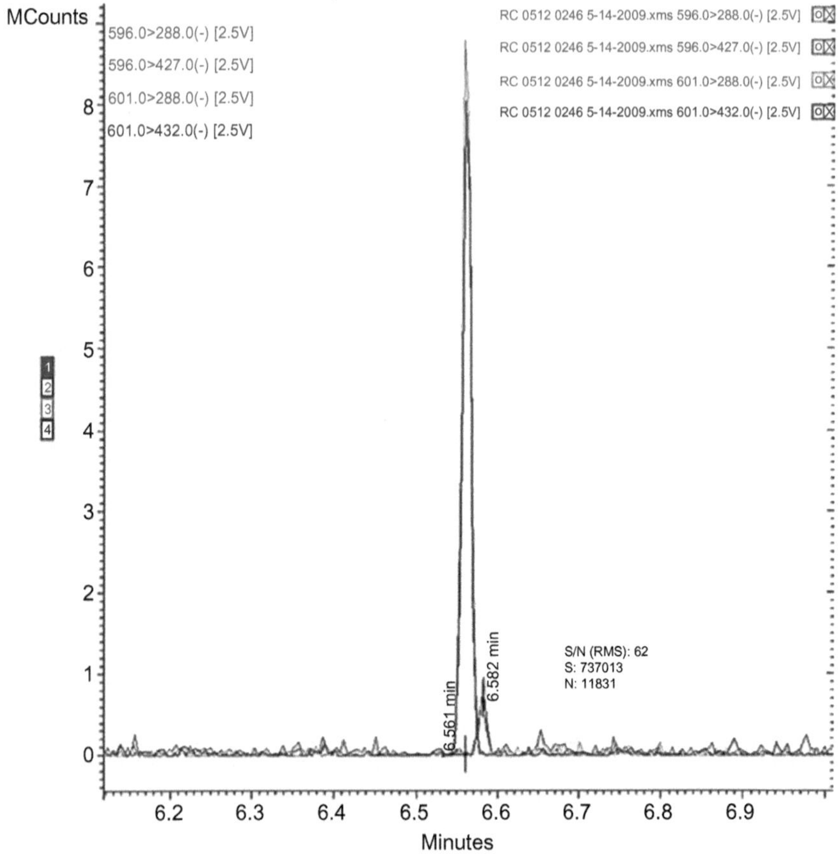

Fig. 3 GC chromatogram of ethyl glucuronide in a real hair sample (Agius et al., 2010).

aldehyde, alcohol, alkanes, etc., were found in the samples. An increasing trend in the concentration of alkanes was observed upon moving up from the buried cadaver toward the soil's surface (Brasseur et al., 2012). Another study focused on the analysis of VOCs such as sevoflurane, acetaldehyde, *n*-propanol, ethanol, methanol, acetone, and desflurane present in biological fluids such as blood and urine using GC coupled with a flame ionization detection (FID) system. In blood, the lowest LOD was recorded for acetone (0.7 mg/dL) while in urine, the lowest LOD recorded was for sevoflurane (1.3 mg/dL). Fig. 4 shows the chromatogram with retention time for VOCs discussed in this study (Kovatsi et al., 2011). Another study focusing on the exposure of neonates to ethanol and other VOCs such as acetic acid, acetaldehyde, methanol, and acetone used HS-GC along with MS for the investigation. Helium at a flow rate of 1 mL/min was used as the

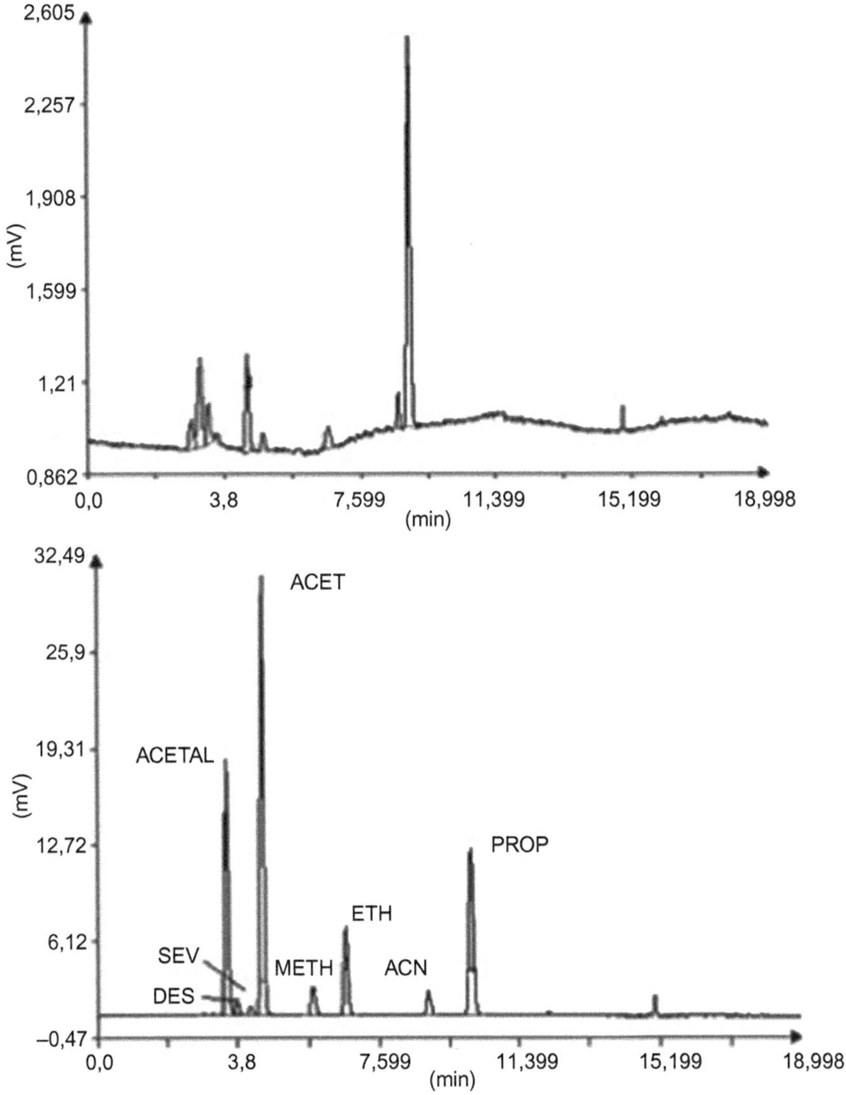

Fig. 4 GC chromatogram of blank (acetonitrile) and VOCs present in blood (Kovatsi et al., 2011).

carrier gas. The LOD values ranged from 0.1 to 0.2 mg/L. The consumption of ethanolic medications by neonates was the reason for the presence of such VOCs in the blood (Cordell et al., 2013). VOCs coming from fabric specimens were evaluated through GC-MS in a research work. The VOCs were extracted from fabrics such as polyester, acrylic, and cotton via SPME (headspace). Helium as the carrier gas flowed through the stationary phase at the rate of 1 mL/min. The LOD for different VOCs varied from 0.44 to 45.45 nM (Gherghel et al., 2018).

4.2 Drugs

Illicit drugs have been globally consumed, mainly by youngsters for recreational purposes. Rave parties are one of the most common locations for the possibility of such evidence. GC has been used by a number of research teams for the detection of these drugs.

In a study from the late 1990s, cocainics and opiates extracted from human hair though solid phase extraction (SPE) were quantified on GC-MS. A flow rate of 1.3 mL/min was selected for helium gas, which was the carrier for the analytes. For both the drugs and their metabolites, the LOD value ranged from 0.12 to 0.28 ng/mg of hair. The method developed can be used for the extraction and quantification of such drugs from human hair evidence (Gaillard and Pépin, 1997). Hadidi et al. extracted tramadol, a commonly abused analgesic, from hair via SPE and analyzed it through GC coupled with MS. The flow for the helium gas that was acting as the carrier was 1.2 mL/min. The LOD value for the drug in the hair sample was 0.5 ng/mg of hair. It was concluded that the method developed could detect tramadol in hair for patients prescribed with multiple drugs as well (Hadidi et al., 2003). 2D GC with MS was used by Song et al. in order to screen and confirm the presence of 78 drugs (underivatized) of forensic interest. The 2D analysis in GC was employed to enhance the separation of the drugs on the chromatographic column. Two different columns were used for first and second dimensional separation. The flow rate was 2 mL/min for helium gas (carrier). The retention time of all the analyzed drugs was found to be in the range of 250–1750 s (Song et al., 2004). Cannabinoids, particularly tetrahydrocannabinolic acid, was also separated using GC, and further analyzed by MS. The flow rate of helium gas was maintained at 1 mL/min. The study concluded that due to external contamination of hair with cannabinoids, majorly because of the smoke, the cannabinoids enter the matrix of the hair and cannot be removed even after multiple washes (Auwärter et al., 2010). Benzodiazepines and their metabolites were also detected in blood samples using GC coupled with MS with an electron impact ionization source. The carrier gas (helium) was maintained at a flow rate of 1 mL/min. The lowest retention time (14.43 min) was for medazepam while the highest (23.24 min) was for triazolam (Fig. 5). The developed protocol could be used for the analysis of many benzodiazepine drugs simultaneously from blood (Papoutsis et al., 2010). GC

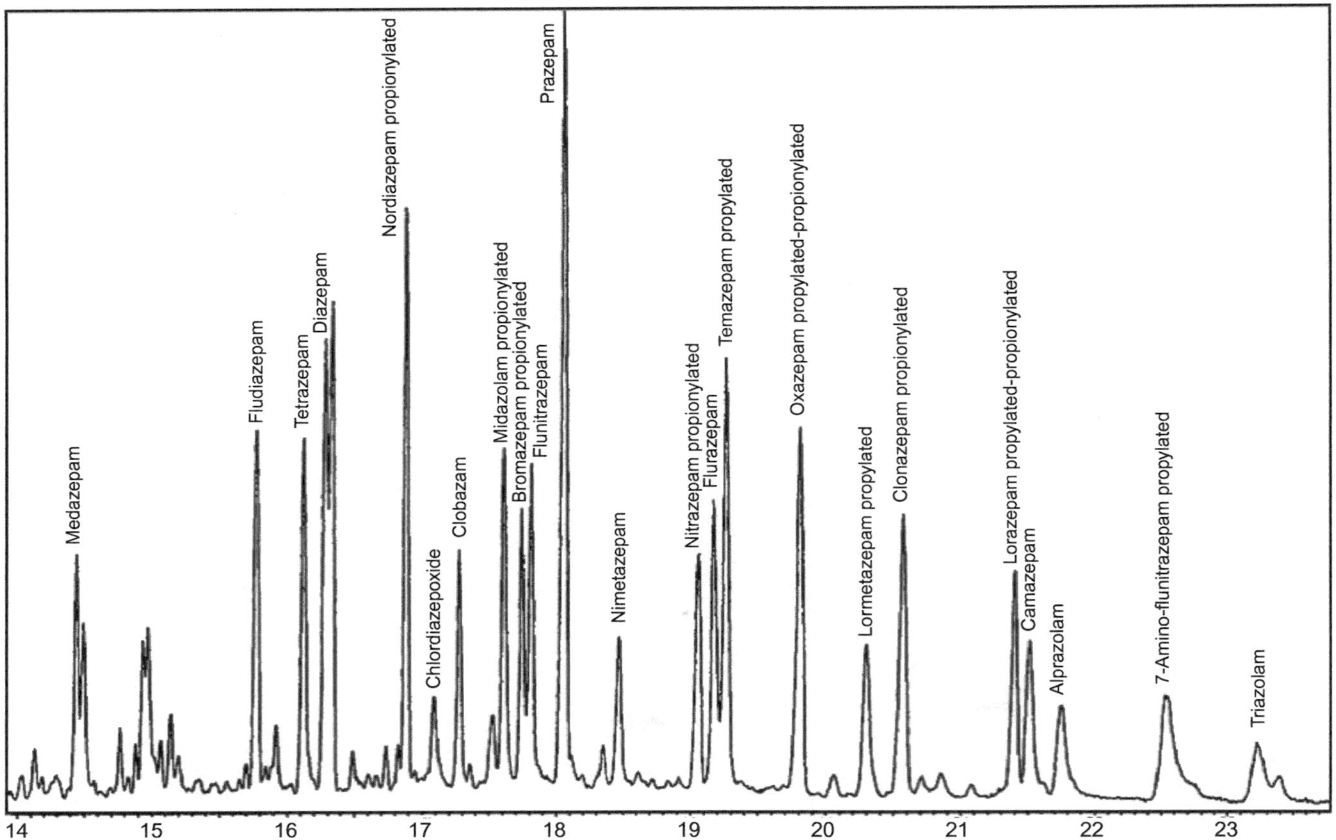

Fig. 5 GC chromatogram of benzodiazepine drugs in a spiked blood sample (Papoutsis et al., 2010).

along with MS has also been used by Philipp et al. for monitoring kratom, a herbal drug commonly abused in Thailand. O-demethyltramadol (ODT) has also been used to produce another herbal blend called Krypton using kratom. The drugs were detected in the urine samples. The flow rate of the carrier gas was 1.2 mL/min. The LOD value for ODT was 50 ng/mL while for parent alkaloids, it was around 100 ng/mL. The procedure could be suitable for kratom and Krypton detection in humans having similar kinds of metabolisms (Philipp et al., 2011). Psychoactive drugs such as antidepressants, anticonvulsants, analgesics, benzodiazepines, sedatives, anesthetics, hypnotics, and other drugs of abuse have been detected in serum samples using GC-MS (Grapp et al., 2016). GC in combination with tandem MS has also been used by Truta et al. to detect as well as quantify antidepressants present in blood samples (Truta et al., 2016).

4.3 Explosives and ignitable materials

Explosives and ignitable materials are often collected as evidence from bomb blast sites and in cases of fire and arson. These materials along with their residue are readily analyzed through chromatographic techniques. GC has also assisted many researchers for this purpose, which is discussed in this section.

In a study, components of kerosene as well as gasoline were detected in blood via GC coupled with MS. The LOD value for gasoline was 0.01 μg while for kerosene, it was 50 pg. The samples exposed to kerosene vapor showed the presence of aliphatic hydrocarbons with more than nine carbons and aromatic hydrocarbons with nine carbons such as trimethylbenzene, cumene, pseudocumene, etc. Gasoline-exposed samples showed the presence of aliphatic hydrocarbons with 5–8 carbons, and aromatic hydrocarbons such as benzene, xylene, and toluene (Kimura et al., 1988). In another research work, petroleum-based ignitable liquids were also analyzed using 2D GC coupled with MS. The flow rate was 2 mL/min fixed for helium as the carrier gas. The study focused on attaining the near theoretical maximum for the gain capacity in peaks during the investigation of flammable liquids via 2D GC-MS (Nizio et al., 2016). Sampat et al. also employed 2D GC for the chemical profiling of flammable liquid, mainly white spirits. The GC analysis was combined with FID as well as MS separately for the analysis. The carrier gas (helium) had a 1 mL/min flow rate. It was observed that for aromatic compounds, the maximum retention time (44.87 min) was for tetramethylbenzene while for aliphatic compounds, it was for pentylcyclohexane (44.20 min) (Sampat et al., 2016b). In the case of explosives,

Calderara et al. used GC with an electron capture detector for the analysis of explosives after a blast, extracted via SPME. The flow rate for helium was fixed at 8 mL/min. It was observed that the retention time for the standards of organic explosives was in the range of 4–17 min (Calderara et al., 2003). HS-GC has also been used by Stamboli et al. for the analysis of triacetone triperoxide (TATP) residue in the debris collected after an explosion. The method developed in this study was able to detect TATP levels as low as 0.1 ng. The use of HS in this analysis helped in removing the tedious process of sample treatment with organic solvents (Stambouli et al., 2004).

4.4 Fingerprints

Fingerprint analysis is often done using microscopic techniques. However, in order to chemically profile the fingerprints, which can give an idea about the chemicals or compounds the victim or the suspect was subjected to, chromatographic and spectroscopic techniques come into the picture. This section deals with the research that employed GC for the chemical analysis of fingerprints.

In a study by Croxton et al., GC coupled with MS was used for the investigation of components present in a latent fingerprint. The fingerprint samples were collected from nonporous substrates. Helium was flowing at a rate of 1 mL/min through the stationary column. Through the analysis, 12 amino acids as well as 10 fatty acids were identified in the fingerprints (Croxton et al., 2006). Weyemann et al. also used GC in combination with MS for the analysis of fingerprint composition, which was present at the time of deposition of the fingerprint. Helium carried the analytes through the stationary phase at a constant flow rate, which was 1 mL/min. Cholesterol and squalene were the target compounds for the analysis in fingerprints collected from porous and nonporous surfaces. It was observed that with time, squalene was depleted from the nonporous surface as compared to cholesterol (Weyermann et al., 2011).

4.5 Inks

Ink chemical profiling and solvent analysis often become important in the forensic examination of questionable documents. GC has been utilized by researchers over the years for such purposes, which are elaborated here.

In a study by Brazeau and Gaudreau, GC coupled with MS was used to quantify the ink solvents extracted via SPME. For chromatographic separation, helium was maintained at a flow rate of 1.2 mL/min. The study was the first of its kind to report the analysis of volatiles in inks via headspace after extraction by SPME (Brazeau and Gaudreau, 2007). Weyermann analyzed the ink from a ballpoint pen that was dried on paper via GC coupled with MS in order to estimate the age of the ink. The flow rate was 3 mL/min for helium. It was found that the concentration of ink solvents kept decreasing in the ink on paper with time, which could help in its age estimation (Weyermann et al., 2007). In another study, black ink from a gel pen was analyzed for its volatile components under variable conditions via GC with an FID system. Nitrogen at a flow rate of 2 mL/min was used as the carrier gas. Fig. 6 shows the retention time of different solvents on the chromatogram (Li et al., 2014). Aging parameters have also been analyzed for ballpoint pen inks with the help of GC coupled with MS (Koenig et al., 2015).

4.6 Paints

Paint analysis in forensic laboratories is mostly done by a combination of chromatographic and spectroscopic techniques. Pyrolysis GC (Py-GC) has also assisted in this regard. The research works that used GC for paint analysis are discussed here.

Burns and Doolan used Py-GC coupled with MS for the characterization of paints used in vehicles. The injector temperature in GC was maintained at 250°C. It was observed that such an analysis could differentiate paint samples from different car producers on the basis of solid paint, base, and clear coat. The components found in solid paint samples from an Audi car, such as pentanoic acid methyl methyl ester (PAMME), styrene, and diisocyanatohexane, are shown in Fig. 7 (Burns and Doolan, 2005). Another study used

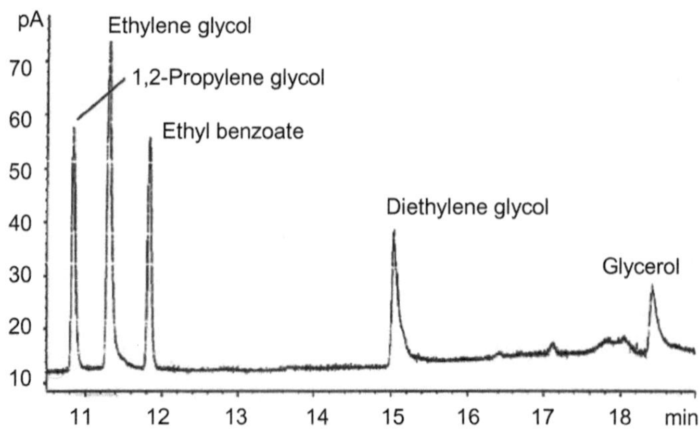

Fig. 6 GC chromatogram of ink solvents (Li et al., 2014).

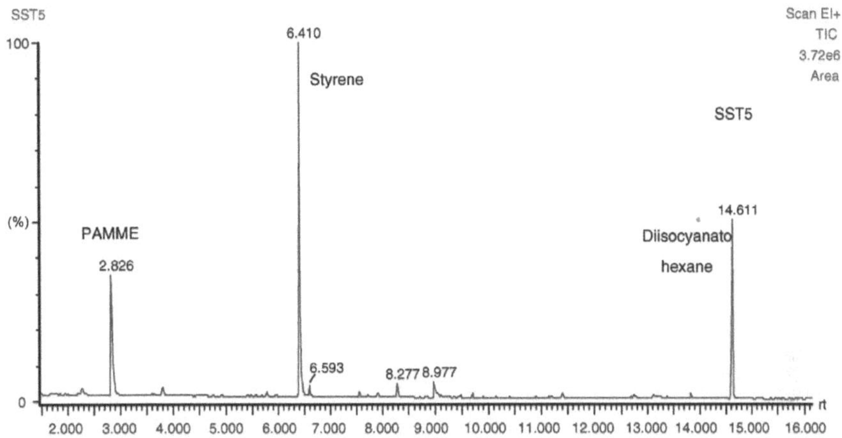

Fig. 7 Py-GC analysis of solid paint from an Audi car (Burns and Doolan, 2005).

Py-GC with MS for the analysis of polymer binders present in the clear coat of paint traces. Helium was passed through the column at a pressure of 70 kPa. The study was able to discriminate between different paint samples, but not all were discriminated. This could be due to the focus on only one component of paint, which was the clear coat (Zięba-Palus et al., 2008). Paint components that consist of polystyrene were also analyzed using Py-GC-MS by Yang et al. for their quantification (Yang et al., 2012). In addition to vehicle paints, spray paints have also been analyzed through this characterization tool (Milczarek and Zięba-Palus, 2009).

4.7 Toxins

Excessive sniffing of organic solvents or pesticide intake often becomes toxic and fatal in many cases. Forensic analysis of such cases requires the use of GC to identify the cause of poisoning.

A research work from Boatto et al. used GC coupled with MS for the analysis of a lethal intoxication case of *o*-cresol and phenol. The concentrations of these compounds were examined in the stomach content, blood, and urine. Helium was used as the carrier gas. The concentration of phenol varied from 3.3 to 115 ppm in different specimens while for *o*-cresol, it was in the range of 1.9 to 20.5 ppm (Boatto et al., 2004). Another research group used GC coupled with MS for the analysis of pesticides present in blood, which inhibit the activity of cholinesterase. The flow rate of helium was maintained at a rate of 1.1 mL/min. The LOD value ranged between 1 and 10 μg/L for different pesticides. The developed method could be helpful in solving suicide or accidental cases related to pesticide intake (Papoutsis et al., 2012). A similar kind of work has also been done on organophosphate pesticides, which are also a potent cholinesterase inhibitor. The pesticides were extracted by SPE from blood. For the analysis, the quantification limit was 500 ng/mL (Valente et al., 2015). A case report by Boumba et al. used GC-MS for a suicidal case due to the intake of mirtazapine (antidepressant) and pyrethroids (deltamethrin and α-cypemethrin). Toxic levels of these compounds were detected in the blood and urine of the subject. In addition to these, GC-MS analysis revealed the presence of a high ethanol concentration in the blood, which could have become the combined cause of fatality (Boumba et al., 2017).

5 Conclusion

Gas chromatography is the most widely used analytical technique for the analysis of various forensic evidence. It is widely used because it offers rapid and efficient analysis with high resolution. It is also a very sensitive technique due to which it can easily detect extreme low concentrations of analytes. Another major advantage with this technique is that it requires only a microliter amount of samples and is a one-time investment. However, despite these advantages, it has certain disadvantages such as it can be used to analyze only the samples that can be volatilized without damaging their overall composition. Therefore, thermally labile compounds cannot be easily analyzed. However, this limitation can be overcome through appropriate extraction techniques or by exposing the samples to lower temperatures to ensure their proper elution. In forensic science, it has been mainly used for the analysis of volatile compounds such as alcohol and other organic compounds, drugs, explosives and flammable materials, fingerprints, inks, paints, and toxins. Such myriad applicability of GC in the analysis of forensic evidence makes it a ubiquitous tool in forensic science laboratories across the world. In this chapter, a brief overview of the theory behind GC and its applications in the analysis of various forensic evidence has been given.

References

Agius, R., Nadulski, T., Kahl, H.-G., Schräder, J., Dufaux, B., Yegles, M., Pragst, F., 2010. Validation of a headspace solid-phase microextraction–GC–MS/MS for the determination of ethyl glucuronide in hair according to forensic guidelines. Forensic Sci. Int. 196, 3–9. https://doi.org/10.1016/j.forsciint.2009.07.023.

Auwärter, V., Wohlfarth, A., Traber, J., Thieme, D., Weinmann, W., 2010. Hair analysis for Δ9-tetrahydrocannabinolic acid A—new insights into the mechanism of drug incorporation of cannabinoids into hair. Forensic Sci. Int. 196, 10–13. https://doi.org/10.1016/j.forsciint.2009.12.023.

Blumberg, L.M., 2012. Theory of gas chromatography. In: Gas Chromatography. Elsevier, pp. 19–78, https://doi.org/10.1016/B978-0-12-385540-4.00002-X.

Boatto, G., Nieddu, M., Carta, A., Pau, A., Lorenzoni, S., Manconi, P., Serra, D., 2004. Determination of phenol and o-cresol by GC/MS in a fatal poisoning case. Forensic Sci. Int. 139, 191–194. https://doi.org/10.1016/j.forsciint.2003.10.023.

Boumba, V.A., Rallis, G.N., Vougiouklakis, T., 2017. Poisoning suicide with ingestion of the pyrethroids alpha-cypermethrin and deltamethrin and the antidepressant mirtazapine: a case report. Forensic Sci. Int. 274, 75–78. https://doi.org/10.1016/j.forsciint.2016.11.023.

Brasseur, C., Dekeirsschieter, J., Schotsmans, E.M.J., de Koning, S., Wilson, A.S., Haubruge, E., Focant, J.-F., 2012. Comprehensive two-dimensional gas chromatography–time-of-flight mass spectrometry for the forensic study of cadaveric volatile organic compounds released in soil by buried decaying pig carcasses. J. Chromatogr. A 1255, 163–170. https://doi.org/10.1016/j.chroma.2012.03.048.

Brazeau, L., Gaudreau, M., 2007. Ballpoint pen inks: the quantitative analysis of ink solvents on paper by solid-phase microextraction. J. Forensic Sci. 52, 209–215. https://doi.org/10.1111/j.1556-4029.2006.00299.x.

Burns, D.T., Doolan, K.P., 2005. A comparison of pyrolysis–gas chromatography–mass spectrometry and fourier transform infrared spectroscopy for the characterisation of automative paint samples. Anal. Chim. Acta 539, 145–155. https://doi.org/10.1016/j.aca.2005.02.048.

Calderara, S., Gardebas, D., Martinez, F., 2003. Solid phase micro extraction coupled with on-column GC/ECD for the post-blast analysis of organic explosives. Forensic Sci. Int. 137, 6–12. https://doi.org/10.1016/S0379-0738(03)00256-1.

Conte, E.D., Barry, E.F., 1993. Alkali flame ionization detector for gas chromatography using an alkali salt aerosol as the enhancement source. J. Chromatogr. A 644, 349–355. https://doi.org/10.1016/0021-9673(93)80719-O.

Cordell, R.L., Pandya, H., Hubbard, M., Turner, M.A., Monks, P.S., 2013. GC-MS analysis of ethanol and other volatile compounds in micro-volume blood samples—quantifying neonatal exposure. Anal. Bioanal. Chem. 405, 4139–4147. https://doi.org/10.1007/s00216-013-6809-1.

Croxton, R.S., Baron, M.G., Butler, D., Kent, T., Sears, V.G., 2006. Development of a GC-MS method for the simultaneous analysis of latent fingerprint components. J. Forensic Sci. 51, 1329–1333. https://doi.org/10.1111/j.1556-4029.2006.00203.x.

Davis, P.L., Dal Cortivo, L.A., Maturo, J., 1984. Endogenous isopropanol: forensic and biochemical implications. J. Anal. Toxicol. 8, 209–212. https://doi.org/10.1093/jat/8.5.209.

Gaillard, Y., Pépin, G., 1997. Simultaneous solid-phase extraction on C18 cartridges of opiates and cocainics for an improved quantitation in human hair by GC-MS: one year of forensic applications. Forensic Sci. Int. 86, 49–59. https://doi.org/10.1016/S0379-0738(97)02125-7.

Gherghel, S., Morgan, R.M., Arrebola-Liébanas, J., Romero-González, R., Blackman, C.S., Garrido-Frenich, A., Parkin, I.P., 2018. Development of a HS-SPME/GC–MS method for the analysis of volatile organic compounds from fabrics for forensic reconstruction applications. Forensic Sci. Int. 290, 207–218. https://doi.org/10.1016/j.forsciint.2018.07.015.

Grapp, M., Maurer, H.H., Desel, H., 2016. Systematic forensic toxicological analysis by GC-MS in serum using automated mass spectral deconvolution and identification system: systematic forensic toxicological screening by GC-MS in serum using AMDIS. Drug Test. Anal. 8, 816–825. https://doi.org/10.1002/dta.1848.

Hadidi, K.A., Almasad, J.K., Al-Nsour, T., Abu-Ragheib, S., 2003. Determination of tramadol in hair using solid phase extraction and GC–MS. Forensic Sci. Int. 135, 129–136. https://doi.org/10.1016/S0379-0738(03)00196-8.

Hartmann, C.H., 1971. Gas chromatography detectors. Anal. Chem. 43, 113–125.

Kimura, K., Nagata, T., Hara, K., Kageura, M., 1988. Gasoline and kerosene components in blood—a forensic analysis. Hum. Toxicol. 7, 299–305. https://doi.org/10.1177/096032718800700401.

Kitson, F.G., Larsen, B.S., McEwen, C.N., 1996. Gas Chromatography and Mass Spectrometry—A Practical Guide. Academic Press, USA.

Koenig, A., Magnolon, S., Weyermann, C., 2015. A comparative study of ballpoint ink ageing parameters using GC/MS. Forensic Sci. Int. 252, 93–106. https://doi.org/10.1016/j.forsciint.2015.03.027.

Kovatsi, L., Giannakis, D., Arzoglou, V., Samanidou, V., 2011. Development and validation of a direct headspace GC-FID method for the determination of sevoflurane, desflurane and other volatile compounds of forensic interest in biological fluids: application on clinical and post-mortem samples. J. Sep. Sci. 34, 1004–1010. https://doi.org/10.1002/jssc.201000921.

Li, B., Xie, P., Guo, Y., Fei, Q., 2014. GC analysis of black gel pen ink stored under different conditions. J. Forensic Sci. 59, 543–549. https://doi.org/10.1111/1556-4029.12313.

McNair, H.M., Miller, J.M., 1933. Basic Gas Chromatography-Techniques in Analytical Chemistry Series. John Wiley & Sons, Inc.

Milczarek, J.M., Zięba-Palus, J., 2009. Examination of spray paints on plasters by the use of pyrolysis-gas chromatography/mass spectrometry for forensic purposes. J. Anal. Appl. Pyrolysis 86, 252–259. https://doi.org/10.1016/j.jaap.2009.06.009.

Nizio, K., Cochran, J., Forbes, S., 2016. Achieving a near-theoretical maximum in peak capacity gain for the forensic analysis of ignitable liquids using GC × GC-TOFMS. Separations 3, 26. https://doi.org/10.3390/separations3030026.

Pandey, G., Rawtani, D., Agrawal, Y.K., 2017. Future aspects of halloysite nanotubes in forensic investigations. J. Nanomed. Res. 6, 00153.

Papoutsis, I.I., Athanaselis, S.A., Nikolaou, P.D., Pistos, C.M., Spiliopoulou, C.A., Maravelias, C.P., 2010. Development and validation of an EI–GC–MS method for the determination of benzodiazepine drugs and their metabolites in blood: applications in clinical and forensic toxicology. J. Pharm. Biomed. Anal. 52, 609–614. https://doi.org/10.1016/j.jpba.2010.01.027.

Papoutsis, I., Mendonis, M., Nikolaou, P., Athanaselis, S., Pistos, C., Maravelias, C., Spiliopoulou, C., 2012. Development and validation of a simple GC-MS method for the simultaneous determination of 11 anticholinesterase pesticides in blood-clinical and forensic toxicology applications. J. Forensic Sci. 57, 806–812. https://doi.org/10.1111/j.1556-4029.2011.02031.x.

Philipp, A.A., Meyer, M.R., Wissenbach, D.K., Weber, A.A., Zoerntlein, S.W., Zweipfenning, P.G.M., Maurer, H.H., 2011. Monitoring of kratom or krypton intake in urine using GC-MS in clinical and forensic toxicology. Anal. Bioanal. Chem. 400, 127–135. https://doi.org/10.1007/s00216-010-4464-3.

Rawtani, D., Tharmavaram, M., Pandey, G., Hussain, C.M., 2019. Functionalized nanomaterial for forensic sample analysis. TrAC Trends Anal. Chem. 120, 115661. https://doi.org/10.1016/j.trac.2019.115661.

Sampat, A., Lopatka, M., Sjerps, M., Vivo-Truyols, G., Schoenmakers, P., van Asten, A., 2016a. Forensic potential of comprehensive two-dimensional gas chromatography. TrAC Trends Anal. Chem. 80, 345–363. https://doi.org/10.1016/j.trac.2015.10.011.

Sampat, A.A.S., Lopatka, M., Vivó-Truyols, G., Schoenmakers, P.J., van Asten, A.C., 2016b. Towards chemical profiling of ignitable liquids with comprehensive two-dimensional gas chromatography: exploring forensic application to neat white spirits. Forensic Sci. Int. 267, 183–195. https://doi.org/10.1016/j.forsciint.2016.08.006.

Ševĉík, J., 1976. 9. The flame photometric detector (FPD). J. Chromatogr. Libr., 145–164. https://doi.org/10.1016/S0301-4770(08)60437-6. Detectors In Gas Chromatography. Elsevier.

Song, S.M., Marriott, P., Kotsos, A., Drummer, O.H., Wynne, P., 2004. Comprehensive two-dimensional gas chromatography with time-of-flight mass spectrometry (GC × GC-TOFMS) for drug screening and confirmation. Forensic Sci. Int. 143, 87–101. https://doi.org/10.1016/j.forsciint.2004.02.042.

Sparkman, O.D., Penton, Z.E., Kitson, F.G., 2011. Gas chromatography. In: Gas Chromatography and Mass Spectrometry: A Practical Guide. Elsevier, pp. 15–83, https://doi.org/10.1016/B978-0-12-373628-4.00002-2.

Stambouli, A., El Bouri, A., Bouayoun, T., Bellimam, M.A., 2004. Headspace-GC/MS detection of TATP traces in post-explosion debris. Forensic Sci. Int. 146, S191–S194. https://doi.org/10.1016/j.forsciint.2004.09.060.

Stauffer, E., Dolan, J.A., Newman, R., 2008. Gas chromatography and gas chromatography—mass spectrometry. In: Stauffer, E., Dolan, J.A., Newman, R. (Eds.), Fire Debris Analysis. Academic Press, Burlington, pp. 235–293 (Chapter 8). https://doi.org/10.1016/B978-012663971-1.50012-9.

Truta, L., Castro, A.L., Tarelho, S., Costa, P., Sales, M.G.F., Teixeira, H.M., 2016. Antidepressants detection and quantification in whole blood samples by GC–MS/MS, for forensic purposes. J. Pharm. Biomed. Anal. 128, 496–503. https://doi.org/10.1016/j.jpba.2016.06.027.

Valente, N.I.P., Tarelho, S., Castro, A.L., Silvestre, A., Teixeira, H.M., 2015. Analysis of organophosphorus pesticides in whole blood by GC-MS-μECD with forensic purposes. J. Forensic Legal Med. 33, 28–34. https://doi.org/10.1016/j.jflm.2015.03.006.

Weyermann, C., Kirsch, D., Vera, C.C., Spengler, B., 2007. A GC/MS study of the drying of ballpoint pen ink on paper. Forensic Sci. Int. 168, 119–127. https://doi.org/10.1016/j.forsciint.2006.06.076.

Weyermann, C., Roux, C., Champod, C., 2011. Initial results on the composition of fingerprints and its evolution as a function of time by GC/MS analysis. J. Forensic Sci. 56, 102–108. https://doi.org/10.1111/j.1556-4029.2010.01523.x.

Yang, S.-H., Shen, J.Y., Chang, M.S., Wu, G.J., 2012. Quantification of vehicle paint components containing polystyrene using pyrolysis-gas chromatography/mass spectrometry. Anal. Methods 4, 1989. https://doi.org/10.1039/c2ay05809j.

Zięba-Palus, J., Zadora, G., Milczarek, J.M., Kościelniak, P., 2008. Pyrolysis-gas chromatography/mass spectrometry analysis as a useful tool in forensic examination of automotive paint traces. J. Chromatogr. A 1179, 41–46. https://doi.org/10.1016/j.chroma.2007.09.044.

CHAPTER 10
HPTLC in forensic science

1 Introduction

High-performance thin-layer chromatography (HPTLC) is a type of chromatography technique that is used for the analysis of mixtures based on the migration of chemicals along the solid phase under the influence of a suitable mobile phase. This is a type of planar chromatography that offers direct visual comparison of the separation of mixtures whose results can be developed into high-quality color photos and can be easily produced as evidence and testimony in court. HPTLC offers excellent automation, scanning, and minimum sample preparation that can also be easily hyphenated with other techniques such as Fourier transform infrared spectroscopy (FTIR). It is also a highly cost-effective method that can offer high sample throughput as well.

In forensic science, HPTLC has been used for the detection of textile dyes and drugs and even in the detection of adulterants (Pandey et al., 2017; Rawtani et al., 2019). In this chapter, the principle, theory, and instrumentation of HPTLC as well as its applications in the analysis of various forensic evidence are discussed in brief.

2 Principle and theory of HPTLC

As mentioned briefly in the introduction section, HPTLC is a type of planar chromatography that separates the samples over a stationary phase under a suitable mobile phase and is developed after modification in thin-layer chromatography. The differences between TLC and HPTLC are highlighted in Fig. 1.

There are a few theoretical considerations associated with this type of chromatography, namely separation efficiency, partition coefficient, retention factor, spot factor, capacity, resolution, and selectivity. During chromatography, if a compound has an increased affinity toward the stationary phase rather than the mobile phase, then it tends to move slowly over the stationary phase and separates at a different rate. This separation efficiency is affected by factors such as the polarity, pH, and composition of the mobile phase. Conventionally, the stationary phases in this technique are made of

Handbook of Analytical Techniques for Forensic Samples
https://doi.org/10.1016/B978-0-12-822300-0.00010-0

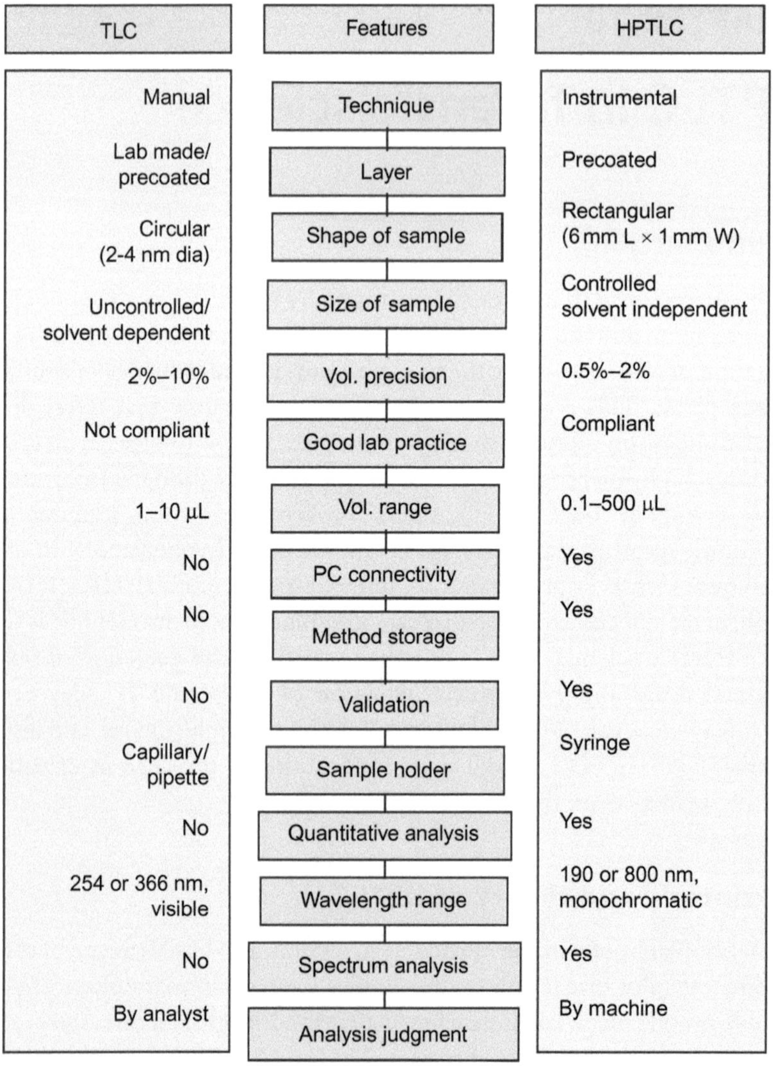

Fig. 1 Differences between TLC and HPTLC (Srivastava, 2011).

silica gel that offers an extremely low contact angle for all mobile phases. However, this water contact angle increases with the increase in the number of alkyl chains in the stationary phases. Another factor that effects the separation efficiency of the compounds is the particulate nature of the stationary phase. The mobile phase tends to travel at a higher velocity and to a long distance in cases of stationary phases with fine particle coating. However, this rate significantly slows in the case of a coarse particle layer. In HPTLC, there is a significantly shorter migration distance. Due to this, the number of theoretical plates, which is basically a hypothetical zone, is considered as the equilibrium stage between the liquid and the stationary phases. The greater the number of theoretical plates, the greater the efficiency of separation. The separation of the samples on the stationary phase also depends on the partition coefficient that serves as an indicator of the migration rate of the samples. During migration, a certain part of the compound gets absorbed by the stationary phase while a certain part remains in the mobile phase, due to which there is a variation in the migration times of the compound. This is presented in terms of the partition coefficient that is further represented as a ratio of the concentration of the sample in the stationary (C_s) and mobile phase (C_m) at equilibrium, and is written as K_a.

$$K_a = C_s / C_m$$

This coefficient is affected by the amount of sample loaded, as an excessive sample causes it to flow with the mobile phase irrespective of the affinity of the sample to the mobile phase. This further causes an increase in the partition coefficient.

The most crucial aspect for the interpretation of any chromatographic result is the retention time or retention factor (R_f), which is basically an indicator of the spot size of the sample on the stationary phase. It can be written as:

$$R_f = \text{analyte's distance travelled}\,(Z_s) / \text{solvent front's distance travelled}\,(Z_f)$$

This factor is dependent on factors such as the partition coefficient, the stationary phase, the thickness of the stationary phase, the composition of the mobile phase, the migration distance, the amount of the sample, the humidity, and the temperature. Apart from R_f values, R_x values are also considered, which is a ratio of the solute's travelled distance to that by the standard sample. While the R_f values usually range from 1 to 0, the R_x values can go greater than 1. Ideally, in order to compare the R_f values of different samples, they are usually run against a standard sample whose R_f value is known. A value of 0.5 is considered optimal for achieving excellent resolution. Apart from the retention factor, the capacity factor is also there, which is the ratio of the sample's retention time in the stationary and mobile phases. This factor allows the determination of any shifts in retention time due to either the stationary phase or the mobile phase. Apart from capacity factor, spot capacity also plays a crucial role in this technique as it determines the maximum number of clearly resolved substances between a retention time of 0 to 1 under isocratic conditions. It is also known as the separation number and is a simple parameter to measure with easy handling and sufficient accuracy. In HPTLC, the number of the plate height corresponds to the number of theoretical plates. It is useful to understand the broadening of the chromatographic spot and is dependent on the diffusion processes affecting the broadening of the spot. The Van Demeter equation is a highly useful chromatographic equation that takes into account the information regarding the flow rate of the mobile base and other constants representing the eddy diffusion, mass transfer, and mass diffusion that determine the broadening capacity of the spot. Therefore, with optimal plate height, it is possible to achieve a high resolution of the mixture separations. The resolution in HPTLC is represented as R_s and is defined as the measurable distance between two spots. The distance is measured from the center of the two spots. When the resolution is equal to 1 or more than 1, then the separation is said to be excellent. Here too, the loading of the sample plays a key role in determining the resolution, as too much of the sample can distort the distance between the two spots (Srivastava, 2011).

3 HPTLC instrumentation

Typically, in HPTLC, a series of steps is followed to develop the method. It first starts with the preparation of the standard and the sample. After that, the stationary phase and the mobile phase are prepared and conditioned. Once they are prepared, the sample and the standard are applied on the stationary phase and kept in a chamber saturated with the fumes of the mobile phases. This phase is known as chromatographic development and is highly crucial to allow the separation of the analytes in the sample and the standard. Once the chromatographic development is done, the separated spots are later detected and scanned. Additionally, a camera is also fixed to ensure the photographic development of the sample and standard separation.

The HPTLC instrumentation is therefore designed to support this methodology. The stationary phase is made up of equal layers of pristine or modified silica gel. The HPTLC techniques are usually divided into normal-phase and reverse-phase techniques based on the type of stationary phase used. For instance, if silica gel is used without any modification, then it is considered to be a normal phase technique where a nonpolar mobile phase such as chloroform and methanol are used. However, in reverse-phase HPLC, the silica gel is modified with compounds that are lipophilic such as C18 or C-12 phenyl compounds. In this case, the mobile phase used will be polar. Apart from silica gel, aluminum oxide, magnesium silicate, and amino or thiol modified silica gels are also used as adsorbents for the stationary phase. It is important that the plates are carefully stored and protected from direct sunlight or dust to prevent contamination. The handling of the plates is also extremely important as it can cause unnecessary contamination. Prior to analysis, layer prewashing with a suitable solvent is done to remove any impurity and make the stationary phase more reactive to adsorption.

Preparation of the sample and the standard is highly important in this technique, as too high a concentration of the sample or the presence of any additional analytes in the sample may affect the migration of the analytes on the plate. Sample preparation processes such as grinding, sonication, extraction, and centrifugation are the basic sample preparation techniques. Another crucial aspect is the solvent in which the sample will be dissolved. Typically, solvents such as methanol, ethanol, or chloroform are used that are nonpolar and volatile. In TLC, the sample is loaded manually but in HPTLC, special automated applicators exist that load the correct optimal amount of the sample in the form of narrow bands. The applicator consists of a syringe connected to a motor and the sample is loaded in the syringe. Due to such an applicator, problems such as overloading of the samples are minimized. Once the sample is developed, the plate must be chromatographically developed by placing it in a twin trough chamber. In this chamber, the mobile phase is first poured, and the chamber is tightly closed in order to ensure that the chamber is saturated with the vapors of the mobile phase. Once saturated, the plate is kept inside and developed for 10–25 min, during which the migration of the analytes occurs. During the migration of the analytes, specific zones are created that vary in color and other properties such as fluorescence or UV-absorbing. Once the chromatographic development is done, the zones are detected through special viewing chambers that allow visualization within a specific wavelength. UV, fluorescence, colorimetric zones, and bioluminescence can be easily detected through this method. Currently, the detection is done with the help of a densitometer or a scanner in which the light source is fixed and emitted from a narrow rectangular slit. The scanner then processes the zones and detects the absorbance or fluorescence of the zones. Once the scanning is done, photographic documentation of the developed film is done, which is usually done under UV light of 254 and 366 nm wavelengths and white light (Patel et al., 2011).

While the discussed steps are routinely used in any lab, quite often the postchromatographic derivatization of the plate is also done in which the analytes can be further derivatized to obtain additional information. Here again, the plate can be prepared by putting it in the chamber and factors such as the mobile phase and the positioning of the plate (vertically or horizontally) are considered according to the analysis of the samples. Also, similar to HPLC, the analysis here can be done using a gradient elution in which more than one mobile phase is used (Figs. 2 and 3) (Patel et al., 2011).

Because HPTLC is a planar chromatography and each step in the methodology development can be performed independently, the hyphenation of this technique with other techniques such as HPLC, mass spectrometry (MS), and FTIR can be easily done. Some of the advantages of using HPTLC for hyphenation are easy sample preparation, no loss of samples

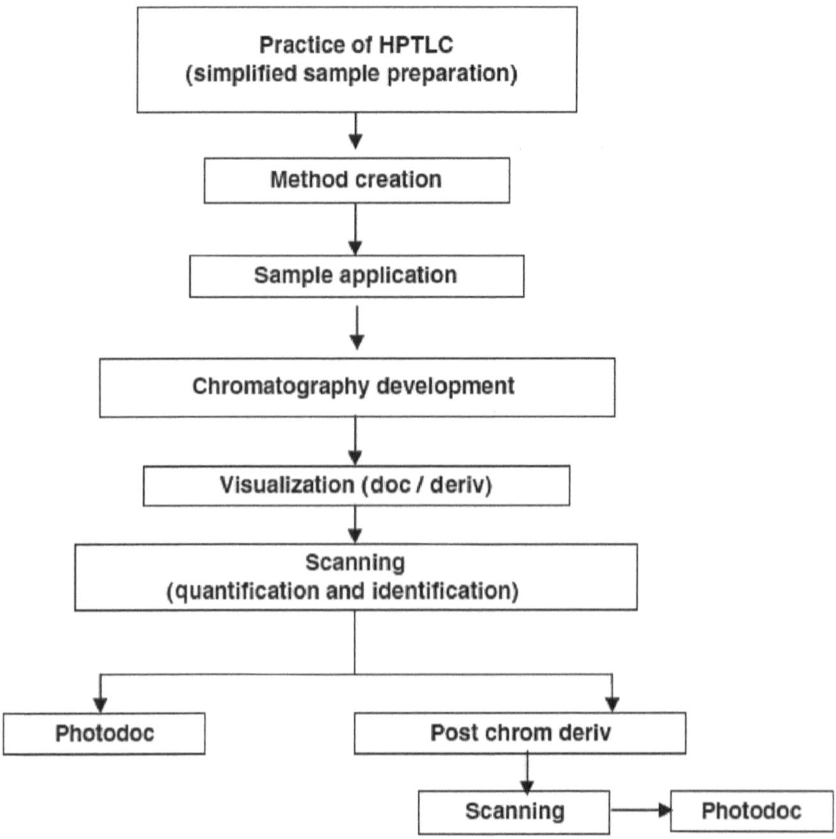

Fig. 2 HPTLC methodology for the separation of analytes (Charegaonkar, 2011).

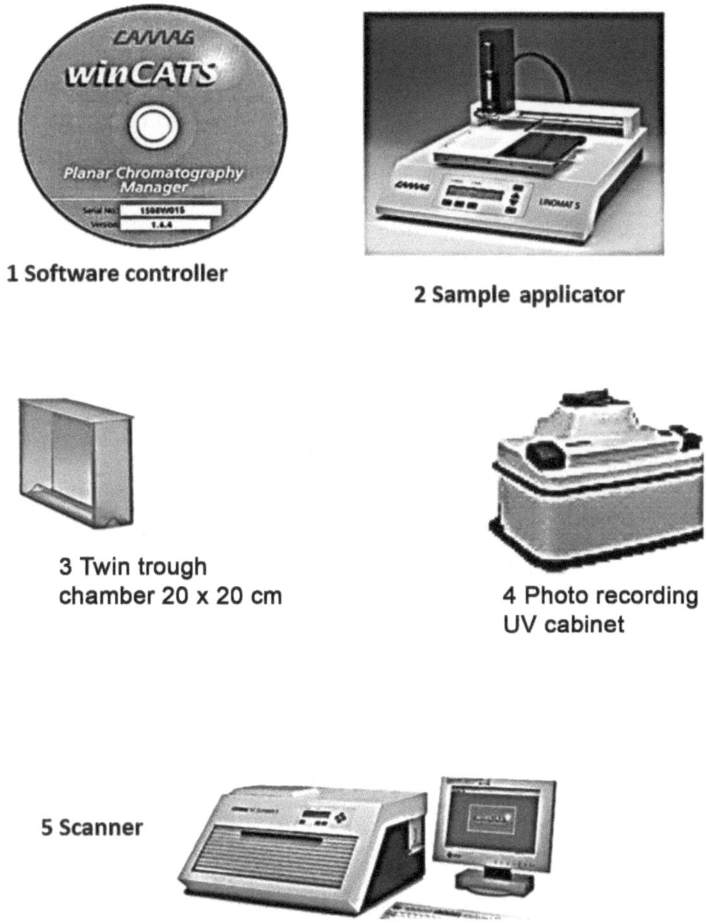

Fig. 3 Basic instrumentation of an automated HPTLC system (Charegaonkar, 2011).

once they are adsorbed on the plate, parallel chromatography in which 70 or more samples can be analyzed at a time, photographic development of the plate with the analytes, and a flexible system for easy hyphenation. If hyphenated with FTIR, then the scanning is done with an IR source using the diffuse reflectance infrared Fourier transform (DRIFT) technique in which the IR light hits the sample and comes back in diffusely remitted radiation (Cimpoiu, 2011). This technique has also been paired with MS through different ionization sources that are capable of working under ambient conditions and atmospheric pressure. Techniques such as matrix laser adsorption and desorption (MALDI) and other elution or desorption-based techniques have been widely used for the analysis of samples from HPTLC. For instance, HPTLC-UV-Vis/ESI-MS and HPTLC-UV-Vis/MALDI-TOF-MS have been used for the detection of food dyes and phospholipids, respectively (Morlock and Schwack, 2010). In the case of food dyes, upto 36 samples could be applied on the same plate. This is followed by scanning with a UV-visible detector and then analyzing the purity and composition of the food dyes with ESI-MS. Similarly, phospholipids, which are a major component of biological samples, can be easily separated with HPTLC, and then scanned by a UV-visible detector. This is followed by ionization using MALDI and mass analysis by time of flight (TOF) (Morlock and Schwack, 2010).

4 HPTLC analysis of forensic samples

HPTLC has been widely used in industries such as pharmaceuticals, food, textiles, etc., for the separation of various kinds of analytes on chromatographic plates. Forensic experts also require separation, identification, and quantification of analytes such as drugs, explosives, inks, dyes in fibers, and toxins. HPTLC has been utilized by forensic experts and researchers for such purposes. This section sheds light on the research works that employed HPTLC for the analysis of the aforementioned forensic samples (Table 1).

4.1 Drugs

Chromatographic techniques have been utilized by forensic experts for the separation and quantification of different drugs of forensic importance. HPTLC has also shown its potential in analyzing forensically significant drugs (mainly illicit drugs) in different research works, which are discussed in this section.

In a study by Varshney, HPTLC was utilized for heroin, a widely used illicit drug, in order to check its stability in methanol. The sample was pure heroin (reference standard), and the study was carried out at room temperature for a duration of 1 year. A small aliquot (10 μL) of heroin solution was applied as a 6 mm band on an HPTLC plate using a sample applicator. Through the study, it was observed that heroin starts degrading after its solution preparation in methanol on the second day, and was completely degraded in 38 weeks (Varshney, 2002). Dongre and Kamble also used HPTLC to detect and identify heroin in samples of forensic importance. The study developed a spray reagent with ferric chloride and dipyridyl

Table 1 Analysis of forensic samples using HPTLC.

S. no.	Type of forensic sample	Example of forensic sample	Stationary phase	Mobile phase	Sample volume (μL)	HPTLC plate dimension (cm*cm)	Detection	Reference
1.	Drug	Heroin	Silica gel 254	Ethanol and chloroform (1:9)	5	10*10	Spray reagent (ferric chloride and dipyridyl)	Dongre and Kamble (2003)
		Diazepam	Silica gel $60F_{254}$	Methanol and chloroform (1:9)	4, 8, 10	20*20	Chromogenic reagent (dinitrobenzene in dimethyl sulfoxide)	Daundkar et al. (2008)
		Ecstasy	–	TB {toluene (15): cyclehexane (75): diethylamine (5)} and TAEA {toluene (45): acetone (45): ethanol (7): ammonia (3)}	2	20*10	Densitometer	Wirasuta (2012)
2.	Explosive	PETN and TNT	Silica gel $60F_{254}$	Acetone and petroleum ether (1:2)	–	20*20	UV detector	Nejad-Darzi et al. (2009)
		GN and NQ	Silica gel $60F_{254}$	Tetrahydrofuran and dioxane (1:1)	GN: 2, 6, 8, 10, 12, 16 NQ: 1, 2, 4, 6, 8	20*20	UV detector	Chaloosi et al. (2007)
		Tetryl, TNT, NC, NG, RDX, and PETN	–	Chloroform and acetone (9:1); toluene; acetone and trichloroethylene (2:8)	10	–	Gries reagent	Krishnamurthy et al. (2006)
3.	Ink	Blue ballpoint pen ink	Silica gel 60	Butanol, ethanol, and water (5:0.05:1.5)	–	20*20	TLC scanner	Senior et al. (2012)
		Blue pen and ballpoint pen ink	Silica gel	Butanol, isopropanol, water, and acetic acid (20:10:10:1)	5	20*10	UV light and TLC visualizer device	Hosu et al. (2015)

Continued

Table 1 Analysis of forensic samples using HPTLC—cont'd

S. no.	Type of forensic sample	Example of forensic sample	Stationary phase	Mobile phase	Sample volume (μL)	HPTLC plate dimension (cm*cm)	Detection	Reference
4.	Dye	Fast Green FCF, Sudan IV, Rhodamine B, Bismarck Brown Y	Glass HPTLC plates from Sorbtech, EMD Millipore, Analtech, Machery-Nagel	Butanol, ammonia, acetone, and water (5:2:5:1)	–	10*10	–	Groves et al. (2018a)
5.	Toxin	Carbosulfan	Silica gel $60F_{254}$	Hexane and acetone (4:1)	5	10*10	Spray reagent (potassium ferrocyanide)	Kulkarni et al. (2010)
		Tolclofos-methyl, edifenfos, and ditalimfos	Silica gel $60F_{254}$	Hexane and acetone (6:4, 6.5:3.5, 7:3, 7.5:2.5, 8:2, 8.5:1.5, 9:1, 9.5:0.5, 10:0)	–	10*10	Densitometer	Nagaraju et al. (2011)
		Oleandrin	–	Hexane and ethyl acetate (4:6)	1 (for stan dard) and 10 for extrac ted sample	–	UV detector after spray with cerium sulfate reagent	Turkmen et al. (2013)

PETN, pentaerythritol tetranitrate; *TNT*, 2,4,6-trinitrotoluene; *UV*, ultraviolet; *GN*, guanidine nitrate; *NQ*, nitroguanidine; *NC*, nitrocellulose; *NG*, nitrogylcerine; *RDX*, cyclonite; *TLC*, thin layer chromatography.

for heroin detection, which was responsible for the red-colored spot of heroin on the HPTLC plate. After spraying of the reagent on the plate, it was subjected to heat at high temperature. An automated applicator was used to apply sample spots of 5 μL volume on the plate. The reagent also helped in developing spots for codeine, morphine, and thebaine, but not for narcotine and papaverine (Dongre and Kamble, 2003). Benzodiazepine drugs, which are abused worldwide by youngsters, have also been analyzed through HPTLC. A diode array variant of HPTLC was used for forensic analysis with the help of the R_f value and UV data acquired by the spectrometric analysis of HPTLC plates. This variant of HPTLC helped in detecting trace quantities of the drug up to 20 ng. 10 μL of drug samples were applied as 7 mm bands on HPTLC plates using an applicator. Three different mobile phases were used, and it was observed that drugs had different R_f values in different mobile phases (Spangenberg et al., 2005). Among benzodiazepenes, diazepam was selected as the target by a group of researchers to develop a chromogenic reagent specific to diazepam. HPTLC was utilized for the detection of diazepam. The drug was extracted from tissues, vomit, blood, urine, and tablets. Chloroform methanol was used as the mobile phase. For diazepam, violet-colored bands were observed while for other benzodiazepines, there was no band, showing the specificity of the reagent (Daundkar et al., 2008). A study from Jakarta focused on the chemical profiling of ecstasy that was recovered from places around Jakarta using densitometry HPTLC. 2 μL of samples were spotted at 19 positions on HPTLC plates, and after analysis, the plates were also analyzed at 210 nm. The R_f values were used to identify the drugs. The study proved helpful in the identification of the street drug (Wirasuta, 2012). The adulteration of regular slimming products with sibutramine has also been identified with the help of HPTLC densitometry. The mobile phase used was a mixture of hexane, acetone, and ammonia. The stationary phase on HPTLC plates was silica gel. A calibration curve was plotted for the drug in the range of 250–2000 ng/spot. The developed method was precise, accurate, linear, and robust in the detection of illegally added sibutramine (Ariburnu et al., 2012). NBOMe, which is an N2-methoxybenzyl substituted phenylethylamine hallucinogen, has also been detected in blotters using HPTLC. 2 μL of standards and 1 μL of samples were applied in the form of 3 mm bands on the HPTLC plates using an automatic sampler. The study concluded that NBOMes have replaced LSD (lysergic acid diethylamide) as the major hallucinogen impregnated in the blotters in Chile (Duffau et al., 2016).

4.2 Explosives

Explosives are commonly encountered in terrorist attacks and cases related to firearms and ammunition. They are detected using various kinds of chromatographic and spectroscopic techniques. Forensic experts have utilized these techniques to separate, identify, and quantify the explosives and their residue collected from a crime scene. This section discusses the research works on explosives using HPTLC.

In a study by Sharma and Lahiri, explosives and their residue were characterized and identified using HPTLC. The explosives taken for the study were nitroglycerine (NG), tetryl, cyclonite (RDX), trinitrotoluene (TNT), pentaerythritoltetranitrate (PETN), and ammonium nitrate fuel oil. 3–27 μL of each sample were spotted on the HPTLC plate using an autosampler. A benzene acetone mixture was used as the mobile phase. It was observed that upon increasing the volume of the sample, the area of the developed spots kept increasing (Sharma and Lahiri, 2005). Krishnamurthy et al. performed HPTLC analysis on explosives in a study focusing on the successive bomb blasts that took place in Mumbai. The explosives selected for the study were TNT, RDX, tetryl, NG, nitrocellulose (NC), and PETN. The samples of explosive debris were prepared in acetone. Three different mobile phases were used for the analysis: chloroform and acetone; toluene; and trichloroethylene and acetone. It was observed after spraying the Gries reagent that all the samples contained RDX (Krishnamurthy et al., 2006). Chaloosi et al. used HPTLC to separate and identify guanidine nitrate and nitroguanidine. The analysis was carried out in online and offline modes. Silica gel was used as the stationary phase while a mixture of tetrahydrofuran and dioxane was used as the mobile phase. A 10 μL Hamilton syringe was used to spot the samples in the form of bands of 2 mm size. Both explosives were differentiated after the plate development using their UV scanning (Chaloosi et al., 2007). Chaloosi et al. also used HPTLC for separating and quantifying PETN and TNT. The stationary phase was the same as the previous study (silica gel). However, two mobile phases were used in this study: petroleum ether and acetone; and acetone and trichloroethylene. The study was able to detect very low concentrations of the analyzed explosives. The detection limit for PETN was 0.180 μg/spot while for TNT, it was 0.006 μg/spot. In order to obtain the chromatograms of these explosives, UV analysis was carried out at 215 nm (for PETN) and 250 nm (for TNT) (Chaloosi et al., 2009). Another study used HPTLC in combination with the partial least squares regression method for the real-time quantification of PETN and TNT. A mixture of acetone and petroleum ether was used as the mobile phase for silica gel (stationary phase). A Hamilton syringe was used to apply the sample, and UV analysis of the spots was carried out in the range of 200–400 nm. The detection limit for PETN was 2.06 μg/spot while for TNT, it was 0.078 μg/spot. Both limits were higher than that reported in the previous study by Chaloosi et al. (2009) and Nejad-Darzi et al. (2009). Diethylaniline (DEA) and nitro explosives have also been analyzed through HPTLC densitometry along with spectroscopy by Kuila and Lahiri. The nitro explosives taken for the study were TNT, dinitrobenzene (DNB), and dinitrotoluene (DNT). Silica gel with a fluorescence indicator was used as the stationary phase. 3 μL of samples were spotted on the HPTLC plates using an autosampler. It was observed that the complexes of nitro explosives with DEA did not move, but spots having only nitro explosives showed movement on the HPTLC plate (Kumar Kuila and Chandra Lahiri, 2012).

4.3 Inks

Inks have been analyzed for many decades in forensic laboratories during the examination of questionable documents. The major aims behind these examinations is to know the source as well as chemical profile of the inks used in writings or prints on the questionable documents. Many types of techniques based on spectroscopy and chromatography have assisted forensic experts in this regard (Sherma, 2016). Many of the research works have also utilized HPTLC to investigate the inks, which are discussed in this section.

In 2009, Neumann and Margot presented their research work on ink analysis using HPTLC in three sets of publications. In the first set, HPTLC was used to develop a process for quality assurance during the forensic analysis of inks. For the study, 13 types of inks with variable dye profiles were selected from 40 pens collected from the market. These 13 samples were analyzed on 10 different HPTLC plates. Standard ladders of dyes were also analyzed for calibration purposes. The samples were compared with the standards, and the data were stored for future examiners (Neumann and Margot, 2009a). The second set focused on the development and analysis of mathematical algorithms, which can be used for comparing ink specimens analyzed by HPTLC. The sample volume for the samples was 10 μL, and the band size varied from 5 to 20 mm. The study helped in the automatic comparison of inks through the calibration and acquisition reports from the previous study (Neumann and Margot, 2009b). The third set focused on the operative applications and evaluation of the methodology proposed in previous research works in the context of forensic science. The study showed the ability of HPTLC, which is often considered to give nonreproducible results, to be used to develop digital libraries with information related to ink analysis for practitioners and researchers working in the field of forensic ink examination (Neumann and Margot, 2009c). Senior et al. published a technical note on questionable documents in which blue ballpoint pen inks were characterized and dated using HPTLC in combination with spectroscopic techniques. The inks were classified into three different groups on the basis of their R_f values and the tones of colored bands obtained on HPTLC plates. Silica gel was used as the stationary phase, and the samples were applied on plates using an autosampler. It was concluded that HPTLC lacked efficiency in analysis as compared to the spectroscopic techniques (Senior et al., 2012). Lee et al. also characterized blue ballpoint pen inks using HPTLC. The ink samples were collected after methanolic extraction from the paper. A mixture of ethyl acetate, water, and methanol was used as the mobile phase. The samples were manually spotted on the plates with the help of a capillary tube. The inks were classified into four groups based on the HPTLC analysis, and crystal violet was found to be the major constituent of all the inks, except the ones in the first group (Lee et al., 2014). Hosu et al. also used HPTLC for the analysis of pigments present in the ink samples. The mobile phase used was a mixture of butanol, isopropanol, acetic acid, and water while silica gel was the stationary phase. 5 μL of the sample was applied on the plates. From the results, multiple bands were observed for each ink sample, revealing the presence of many components in the inks collected in the ethanolic extracts from the paper (Hosu et al., 2015). Sharma and Kumar also carried out their research work on blue ballpoint pen inks using HPTLC (Sharma and Kumar, 2017).

4.4 Dyes in fibers

Fiber analysis is often comprised of structural as well as chemical profiling. Structural details are studied using microscopic techniques while spectroscopic and chromatographic techniques assist in the chemical profiling. This section discusses the use of HPTLC for the forensic analysis of dyes present in fibers.

Groves et al. in his study proved the reproducibility of HPTLC in the investigation of dyes used in textile fibers. The study stated that the R_f value is not enough to identify any particular dye. Therefore, the effect of variables affecting the variability of HPTLC analysis of a dye mixture with six components was studied. The variables were the saturation of the tank, pre-elution, the selection of the plate, the distance of development, and the stability of the eluent. The reference dye mixture that was used to compare other dye samples consisted of four dyes: Fast Green FCF, Sudan IV, Rhodamine B, and Bismarck Brown Y. It was concluded that HPTLC results can be reproduced on different plates and laboratories if the aforementioned variables are controlled and defined properly (Groves et al., 2018a). In a review article published by Groves et al. in the same year, HPTLC was included in the general approach toward the forensic identification of dyes in fibers. The potential of HPTLC to separate a dye mixture into its components was highlighted as an important step in dye identification (Groves et al., 2018b).

4.5 Toxins

Forensic toxicology deals with the study of toxins such as pesticides, plant toxins, heavy metals, etc. A number of techniques based on spectroscopy and chromatography are used by forensic toxicologists for the qualitative and quantitative estimation of toxins. This section throws light on the forensic research works that used HPTLC for the analysis of pesticides and plant toxins.

Carbosulfan, a pesticide from the class of carbamates, has been sensed and identified using a chromogenic spray reagent by Kulkarni et al. via HPTLC. The analysis was carried out using standards of carbosulfan and other pesticides of different classes as well as carbosulfan's residue extracted from biological tissues. The volume selected for spotting the samples was 5 μL, and a mixture of hexane and acetone was used as the mobile phase. The spray reagent consisted of potassium ferrocyanide, which was responsible for the specific detection of carbosulfan up to 0.5 μg (Kulkarni et al., 2010). A similar group of scientists also detected endosulfan, an organochlorine pesticide, using HPTLC. The samples applied on HPTLC plates were endosulfan, endrin, dichloro diphenyl trichloroethane (DDT), and extracts from viscera with endosulfan. It was observed that upon varying the mobile phase composition, the R_f value of endosulfan also varied from 0.78 to 0.97 (Kulkarni et al., 2016). Pyrethroids, a widely used class of insecticides, have also been detected by this research group using HPTLC (Kulkarni et al., 2017). Nagaraju et al. also used HPTLC for the separation of fungicides of the organophosphate class for forensic purposes. The selected fungicides were tolclofos-methyl, edifenfos, and ditalimfos. A mixture of acetone with hexane was taken as the mobile phase, which was run on silica gel (stationary phase). It was observed that the R_f value kept increasing upon increasing the content of acetone in the mobile phase (Nagaraju et al., 2011). Nagaraju et al. also evaluated the parameters responsible for the separation of fungicides. However, in this study, reverse-phase HPTLC was used for the analysis with a water and methanol mixture as the mobile phase. Bands of 6 mm size were applied using an autosampler. The study concluded that the use of this mobile phase in reverse-phase HPTLC efficiently separated the selected pesticides (Nagaraju et al., 2012). Oleandrin, a plant toxin coming from *Nerium oleander*, was detected using HPTLC by Turkman et al. The injection volume for standards and purified oleandrin was 1 μL while for toxin extracted from plant and biological samples, it was 10 μL. Hexane and ethyl acetate were mixed in order to prepare the mobile phase. The R_f value for the toxin was found to be around 0.24, with a limit of detection of about 2.2 ng/spot. The study was the first of its kind that reported the use of HPTLC to detect oleander poisoning (Turkmen et al., 2013). Normal- and reverse-phase HPTLC has also been used for the detection of organophosphate pesticides such as chlorpyrifos, triazophos, and quinalphos. Normal-phase HPTLC had a hexane and acetone mixture as the mobile phase in which the R_f value increased with increasing the acetone content. Reverse-phase HPTLC had a water and acetonitrile mixture as the mobile phase in which the R_f value increased by decreasing the water content (Sanganalmath et al., 2017). Normal- and reverse-phase HPTLC has also been used by Nareshkumar et al. for the detection of insecticides such as imidacloprid, flubendiamide, and buprofezin (Nareshkumar et al., 2018). More et al. developed a spray reagent with a chromogenic property for the detection of indoxacarb, an oxidiazine pesticide, via HPTLC (More et al., 2019). These research works show the potential of HPTLC for sensing different toxins at trace levels in normal- as well as reverse-phase mode.

5 Conclusion

HPTLC is a type of planar chromatography technique in which the samples are migrated across the plate coated with the adsorbent. The plate is also known as the stationary phase. It is a sophisticated technique capable of sample preparation, chromatographic development, scanning, detection, photographic development, and derivatization of the sample. It is a completely automated technique in that each step in the development method is completely independent. There are certain factors that determine the results of the HPTLC such as the stationary phase, layer thickness, mobile phase, purity of the solvent, size of the developing chamber, volume and concentration of the sample, size of the sample spot, flow rate of solvent, and mode of development. Once the sample has been adsorbed on the stationary phase, the plate can be reused for further analysis. Due to this feature, this technique can also be easily hyphenated with techniques such as FTIR and MS. In forensic science, this tool is especially good because it allows photographic development of the sample, due to which it can be easily produced in a court as scientific testimony. Apart from this, the cost of HPTLC is also much lower and the entire technique can be easily performed in any lab. Due to such reasons, HPTLC is already being used in the analysis of various forensic samples in various forensic science labs across the world. With minimal extraction of samples and other sample preparation techniques, it is relatively easy to perform HPTLC analysis for the analysis of forensic samples. In this chapter, the principle and theory of HPTLC, and its use in the analysis of forensic evidence such as drugs, explosives, inks in questionable documents, dyes in textile fibers, and toxins such as pesticides and plant toxins has been discussed. In the future, modifications in the sample application, chromatographic development, and scanning of the plate may increase the sensitivity and selectivity of this technique.

References

Ariburnu, E., Uludag, M.F., Yalcinkaya, H., Yesilada, E., 2012. Comparative determination of sibutramine as an adulterant in natural slimming products by HPLC and HPTLC densitometry. J. Pharm. Biomed. Anal. 64–65, 77–81. https://doi.org/10.1016/j.jpba.2012.02.004.

Chaloosi, M., Amoli-Diva, M., Gholamian, F., Mozaffari, S., 2007. Separation and determination of nitroguanidine and guanidine nitrate by HPTLC. Chromatographia 66, 295–296. https://doi.org/10.1365/s10337-007-0322-7.

Chaloosi, M., Nejad-Darzi, S.K.H., Ghoulipour, V., 2009. Separation and determination of PETN and TNT by HPTLC. Propellants Explos. Pyrotech. 34, 50–52. https://doi.org/10.1002/prep.200800012.

Charegaonkar, D., 2011. High-performance thin-layer chromatography: excellent automation. In: High-Performance Thin-Layer Chromatography (HPTLC). Springer-Verlag, Berlin Heidelberg, pp. 55–68.

Cimpoiu, C., 2011. HPTLC hyphenated with FTIR: principles, instrumentation and qualitative analysis and quantitation. In: Srivastava, M. (Ed.), High-Performance Thin-Layer Chromatography (HPTLC). Springer Berlin Heidelberg, Berlin, Heidelberg, pp. 385–394, https://doi.org/10.1007/978-3-642-14025-9_18.

Daundkar, B., Malve, M., Krishnamurthy, R., 2008. A specific chromogenic reagent for detection of diazepam among other benzodiazepines from biological and nonbiological samples after HPTLC. J. Planar. Chromatogr. - Mod. TLC 21, 249–250. https://doi.org/10.1556/JPC.21.2008.4.4.

Dongre, V., Kamble, V., 2003. HPTLC detection and identification of heroin (diacetylmorphine) in forensic samples. Part III. J. Planar. Chromatogr. - Mod. TLC 16, 458–460. https://doi.org/10.1556/JPC.16.2003.6.10.

Duffau, B., Camargo, C., Kogan, M., Fuentes, E., Cassels, B.K., 2016. Analysis of 25 C NBOMe in seized blotters by HPTLC and GC–MS. J. Chromatogr. Sci. 54, 1153–1158. https://doi.org/10.1093/chromsci/bmw095.

Groves, E., Palenik, S., Palenik, C.S., 2018a. Reproducibility of high-performance thin-layer chromatography (HPTLC) in textile dye analysis. Forensic Chem. 8, 104–110. https://doi.org/10.1016/j.forc.2018.03.004.

Groves, E., Palenik, S., Palenik, C.S., 2018b. A generalized approach to forensic dye identification: development and utility of reference libraries. J. AOAC Int. 101, 1385–1396. https://doi.org/10.5740/jaoacint.18-0052.

Hosu, A., Pop, B., Cimpoiu, C., 2015. The forensic analysis of pigments from some inks by HPTLC. J. Liq. Chromatogr. Relat. Technol. 38, 1109–1112. https://doi.org/10.1080/10826076.2015.1028289.

Krishnamurthy, R., Daundkar, B.B., Malve, M.K., 2006. Successive explosions in Mumbai the economic center of India. J. Clin. Forensic Med. 13, 311–315. https://doi.org/10.1016/j.jcfm.2006.06.020.

Kulkarni, K.V., Shinde, D.B., Mane, D.V., Toche, R.B., Garad, M.V., 2010. New chromogenic spray reagent for detection and identification of carbosulfan. J. Planar. Chromatogr. - Mod. TLC 23, 373–375.

Kulkarni, U.K., Kulkarni, K.V., Pardeshi, R.K., Mane, D.V., 2016. High-performance thin-layer chromatographic detection of endosulfan from biological samples. J. Planar. Chromatogr. - Mod. TLC 29, 391–393. https://doi.org/10.1556/1006.2016.29.5.11.

Kulkarni, U.K., Kulkarni, K.V., Pardeshi, R.K., Mane, D.V., 2017. HPTLC detection of pyrethroids in autopsy tissues. Chem. Res. J. 2, 153–155.

Kumar Kuila, D., Chandra Lahiri, S., 2012. Determination of the association constants of charge-transfer complexes between N,N-diethylaniline and nitro explosives using high performance thin layer chromatography with scanning video densitometry and spectrophotometry. J. Solut. Chem. 41, 36–52. https://doi.org/10.1007/s10953-011-9779-y.

Lee, L.C., Nunurung, S.M., Ishak, A.A., 2014. Forensic analysis of blue ballpoint pen inks on questioned documents by high performance thin layer chromatography technique (HPTLC). Malays. J. Anal. Sci. 18, 9.

More, P.V., Chandegaonkar, V.R., Gosavi, N.R., More, B.P., Kulkarni, K.V., 2019. A new chromogenic spray reagent for detection and identification of indoxacarb in visceral material by HPTLC. Int. J. Appl. Biol. Pharm. Technol. 10, 3–7.

Morlock, G., Schwack, W., 2010. Hyphenations in planar chromatography. J. Chromatogr. A 1217, 6600–6609.

Nagaraju, P., Sanganalmath, P., Kemparaju, K., Mohan, B., 2011. Separation of organophosphorus fungicides by high-performance thin-layer chromatography. A new approach in forensic analysis. J. Planar. Chromatogr. - Mod. TLC 24, 108–112. https://doi.org/10.1556/JPC.24.2011.2.4.

Nagaraju, P.M., Sanganalmath, P.U., Kemparaju, K., Mohan, B.M., 2012. Evaluation of separation parameters for selected organophosphorus fungicides of forensic importance by RP-HPTLC. Acta Chromatogr. 24, 253–262. https://doi.org/10.1556/AChrom.24.2012.2.8.

Nareshkumar, B., Sangnalmath, P.U., Gayatridevi, S., Sreeramulu, K., 2018. A simple method for the separation and detection of trace levels of buprofezin, flubendiamide and imidacloprid by NP-HPTLC and RP-HPTLC. Curr. Sci. 115, 895. https://doi.org/10.18520/cs/v115/i5/895-903.

Nejad-Darzi, S., Chaloosi, M., Gholamian, F., 2009. Simultaneous determination of pentaerythritol tetranitrate and 2,4,6-trinitrotoluene by high performance thin layer chromatography and partial least squares regression (PLSR) method. Propellants Explos. Pyrotech. 35, 66–71. https://doi.org/10.1002/prep.200800100.

Neumann, C., Margot, P., 2009a. New perspectives in the use of ink evidence in forensic science: part I. Development of a quality assurance process for forensic ink analysis by

HPTLC. Forensic Sci. Int. 185, 29–37. https://doi.org/10.1016/j.forsciint.2008.11.016.

Neumann, C., Margot, P., 2009b. New perspectives in the use of ink evidence in forensic science part II. Development and testing of mathematical algorithms for the automatic comparison of ink samples analysed by HPTLC. Forensic Sci. Int. 185, 38–50. https://doi.org/10.1016/j.forsciint.2008.12.008.

Neumann, C., Margot, P., 2009c. New perspectives in the use of ink evidence in forensic science part III: operational applications and evaluation. Forensic Sci. Int. 192, 29–42. https://doi.org/10.1016/j.forsciint.2009.07.013.

Pandey, G., Rawtani, D., Agrawal, Y.K., 2017. Future aspects of halloysite nanotubes in forensic investigations. J. Nanomed. Res. 6. https://doi.org/10.15406/jnmr.2017.06.00153.

Patel, R.B., Patel, M.R., Batel, B., 2011. Experimental Aspects and Implementation of HPTLC. Springer-Verlag Berlin Heidelberg, pp. 41–54.

Rawtani, D., Tharmavaram, M., Pandey, G., Hussain, C.M., 2019. Functionalized nanomaterial for forensic sample analysis. TrAC Trends Anal. Chem. 120, 115661. https://doi.org/10.1016/j.trac.2019.115661.

Sanganalmath, P.U., Bharath, N., Sreeramulu, K., 2017. Normal- and reversed-phase thin-layer chromatography of three structurally related organophosphorus pesticides of forensic importance. J. Planar. Chromatogr. - Mod. TLC 30, 154–163. https://doi.org/10.1556/1006.2017.30.3.1.

Senior, S., Hamed, E., Masoud, M., Shehata, E., 2012. Characterization and dating of blue ballpoint pen inks using principal component analysis of UV-vis absorption spectra, IR spectroscopy, and HPTLC. J. Forensic Sci. 57, 1087–1093. https://doi.org/10.1111/j.1556-4029.2012.02091.x.

Sharma, V., Kumar, R., 2017. Fourier transform infrared spectroscopy and high performance thin layer chromatography for characterization and multivariate discrimination of blue ballpoint pen ink for forensic applications. Vib. Spectrosc. 92, 96–104. https://doi.org/10.1016/j.vibspec.2017.05.006.

Sharma, S.P., Lahiri, S.C., 2005. Characterization and identification of explosives and explosive residues using GC-MS, an FTIR microscope, and HPTLC. J. Energ. Mater. 23, 239–264. https://doi.org/10.1080/07370650591006795.

Sherma, J., 2016. Advances in the thin-layer chromatographic forensic analysis of inks. J. Liq. Chromatogr. Relat. Technol. 39, 549–557. https://doi.org/10.1080/10826076.2016.1183130.

Spangenberg, B., Seigel, A., Kempf, J., Weinmann, W., 2005. Forensic drug analysis by means of diode-array HPTLC using R_F and UV library search. J. Planar. Chromatogr. - Mod. TLC 18, 336–343. https://doi.org/10.1556/JPC.18.2005.5.1.

Srivastava, M. (Ed.), 2011. High-Performance Thin-Layer Chromatography (HPTLC). Springer Berlin Heidelberg, Berlin, Heidelberg, https://doi.org/10.1007/978-3-642-14025-9.

Turkmen, Z., Mercan, S., Cengiz, S., 2013. An HPTLC method for the determination of oleandrin in *Nerium* plant extracts and its application to forensic toxicology. J. Planar. Chromatogr. - Mod. TLC 26, 279–283. https://doi.org/10.1556/JPC.26.2013.3.13.

Varshney, K., 2002. HPTLC study of the stability of heroin in methanol. J. Planar. Chromatogr. - Mod. TLC 15, 46–49. https://doi.org/10.1556/JPC.15.2002.1.9.

Wirasuta, I.M.A.G., 2012. Chemical profiling of ecstasy recovered from around Jakarta by high performance thin layer chromatography (HPTLC)-densitometry. Egypt. J. Forensic Sci. 2, 97–104. https://doi.org/10.1016/j.ejfs.2012.05.002.

CHAPTER 11

Hyphenated techniques for forensic sample analysis

1 Introduction

There has been a surge in the development of analytical techniques over the past century. Various chromatography techniques such as high-performance liquid chromatography (HPLC), thin-layer or high-performance thin-layer chromatography (TLC/HPTLC), and gas chromatography (GC) have been used for myriad applications. These techniques provide excellent separation and detection of analytes with high sensitivity, selectivity, and robustness. These features are further enhanced by coupling them with other analytical techniques such as UV-visible or infrared spectroscopy, mass spectrometry, and even nuclear magnetic resonance (NMR). This coupling of one technique with another to obtain a rounded analysis of compounds is known as hyphenation and it provides the best of both worlds. The most commonly used hyphenation techniques are the ones combining chromatographic techniques with spectroscopic or spectrometry techniques. The spectroscopic or spectrometric techniques provide selective information to identify the compound through spectral analysis while the chromatographic techniques perform an efficient separation of the analytes. However, not all hyphenation techniques are necessarily of two different techniques. Tandem techniques such as MS-MS are also considered hyphenation techniques.

Recently, there has been an increase in attention for these techniques in forensic science as they can provide both qualitative and the quantitative analysis of evidence that can be produced in court. In forensic science, evidence such as illicit drugs, adulterants in food and its packaging, explosives, and inks are widely detected using such hyphenation techniques (Pandey et al., 2017; Rawtani et al., 2019). In this chapter, a brief overview of the various hyphenated techniques commonly used in the detection of various forensic evidence is discussed. Also, the applications of the hyphenated techniques for the analysis is also discussed.

Handbook of Analytical Techniques for Forensic Samples
https://doi.org/10.1016/B978-0-12-822300-0.00011-2

2 Hyphenated techniques

In this section, a brief overview of hyphenated techniques such GC-MS, LC-MS, LC-IR, LC-NMR, capillary-electrophoresis-MS, HPTLC-FTIR, and HPTLC-MS will be discussed in brief.

2.1 Gas chromatography-mass spectrometry

The main principle behind GC is the volatilization of the sample in a special heated chamber followed by the separation of the analytes present in the sample and later detection using a detector. Here, in order to separate the samples, a special carrier gas, which is usually an inert gas such as hydrogen or helium, serves as the mobile phase carrying the samples over a specialized column coated with the stationary phase. Just like any other chromatography technique, the affinity of the analytes toward the mobile phase or the stationary phase determines their separation rates. The instrumentation of GC typically consists of a headspace or an injector that is responsible for the sample injection, followed by the GC column that consists of nonpolar stationary phases such as dimethyl siloxanes. The stationary phases are also made of different thicknesses, with the thin stationary phases suitable for analytes requiring higher temperature for volatilization and the thick stationary phases for compounds requiring lower temperatures for volatilization. Once separated, the detectors are used to detect the analytes. There are several detectors such as flame ionization detectors (FID), thermal conductivity detectors (TCD), thermionic specific detectors, and electron capture detectors (ECD). These detectors have different modes of working and are used according to the type of sample to be analyzed. In GC-MS, a special device responsible for the transport of the samples from the GC to the MS is fitted. This device ensures the integrity of the sample and maintains its phase prior to ionization at the MS. Here too, an inert carrier gas is used to transport the samples from one instrument to the other. The devices used can be capillary columns, packed columns, or jet separators, which are basically a series of two vacuums present under high vacuum. Precautions are taken to ensure that the columns used are completely inert in order to avoid any interaction of the samples with them. Once the samples enter the mass spectrometer, they undergo ionization followed by their separation by the mass analyzer and their detection (Fig. 1) (Kitson et al., 1996).

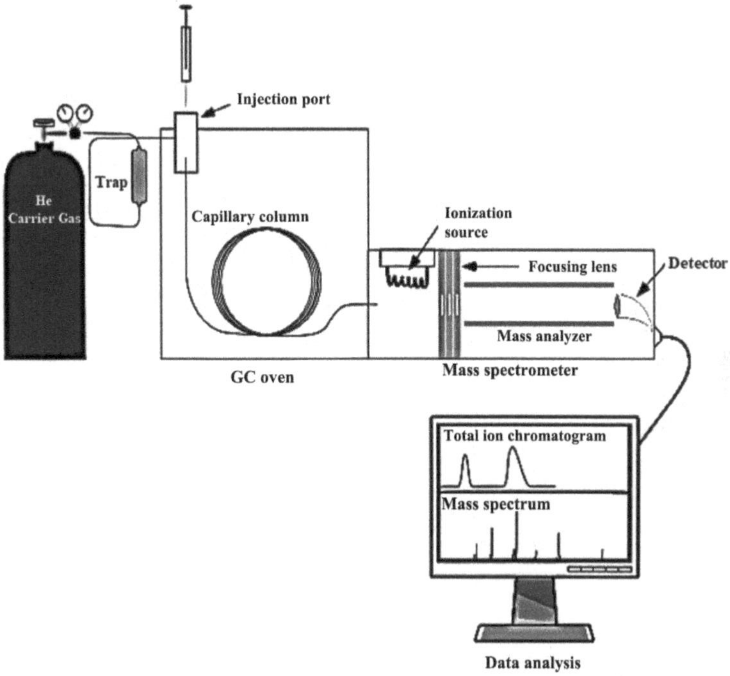

Fig. 1 Instrumental setup of GC-MS (Wu et al., 2012).

2.2 Hyphenated liquid chromatography techniques

2.2.1 LC-MS

The hyphenation of GC-MS is relatively easy as the sample is already in the vapor form, due to which the ionization at the MS becomes easier. Also, the interface of GC-MS is relatively simple due to this factor. However, with liquid chromatography, there are certain challenges to develop an interface as the sample is in a liquid phase. Over the years, there have been several developments to develop a suitable interface for these techniques and now LC-MS is one of the most widely used analytical techniques. Advantages such as accurate separation by the LC and its appropriate identification using the MS have influenced the development of such an interface. HPLC and ultra-HPLC (UHPLC) are the most commonly used LC techniques that have been hyphenated with MS. These instruments are commonly composed of a mobile phase reservoir and a pump to ensure the movement of the mobile phase from the reservoir to the column that consists of the stationary phase. A sample inlet injects the sample into the column and the analytes are later detected by the detector. Usually, in LC-MS, the

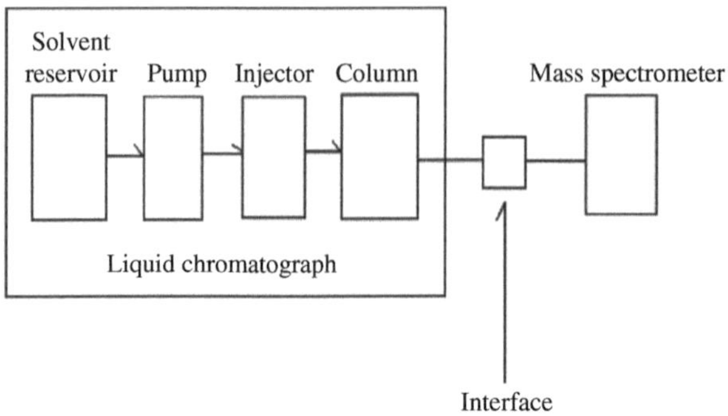

Fig. 2 Instrumental setup of LC-MS (Ligon, 2001).

ionization source is responsible for the ionization of the liquid sample. One of the earliest ionization sources was the moving belt interface in which the analytes and the mobile phase were present in a moving belt that moved continuously. The mobile phase was later removed through infrared heating followed by ionization using the fast atom bombardment (FAB) technique. However, these had certain disadvantages such as the nonuniformity of the sample. Later, soft ionization techniques such as electrospray ionization (ESI) and atmospheric pressure chemical ionization (APCI) were developed that are still most commonly used today. Once the samples are ionized, they undergo separation via the mass analyzers such as time of flight and quadrupole instruments (Fig. 2). Once analyzed, they undergo detection as usual in the MS (Ardrey, 2003; Bogusz, 1999).

2.2.2 LC-IR

IR or Fourier transform IR (FTIR) spectroscopy is a powerful vibrational spectroscopy technique capable of analyzing the composition of a sample. In this technique, an IR source is responsible for emitting IR radiation that hits the sample and excites the electrons present in the sample, due to which there are molecular vibrations. The vibrations are specific to a particular functional group, due to which they can be easily used to identify the nature and composition of different samples. LC itself is an efficient tool that separates the samples efficiently. However, developing an LC-FTIR interface is a challenging task due to the presence of the mobile phase in HPLC that may also contribute to its own vibrational information. Therefore, in order to overcome this challenge, specialized flow cells are created that allow the direct coupling of the LC with the FTIR. Typically, the most basic approach for the LC-IR interface is the use of a flow cell in which the sample from the

LC along with the sample are directly added. Once the sample is detected, the spectral data from the mobile phase are subtracted, thus allowing the information of only the sample to be retained. However, this method is vey much prone to errors, as any error during the subtraction may cause the development of erroneous spectra. Also, this technique is not suitable in cases where there is gradient elution of the sample in LC. Apart from the flow cell approach, a solvent removal method has also been developed in which the mobile phase is removed. The separated analytes are fixed on a substrate, after which the analytes are subjected to sample testing. This method is similar to allowing the flow of salt water on a heated belt, due to which the water evaporates and the salt is left behind. In this case, it is important that the mobile phase used is completely free of impurities in order to eliminate any contamination of the analytes present on the substrate. Once the analytes are immobilized, they can be easily analyzed using FTIR (Kuligowski et al., 2010; Somsen and Visser, 2006).

2.2.3 LC-NMR

NMR spectroscopy is a very sophisticated absorption technique responsible for the structural analysis of organic compounds by analyzing the magnetic field of atomic nuclei. There are many types of NMR such as CNMR, HNMR, and PNMR. These NMR types allow the detection of the respective elements and use that information to obtain their structural information. Parameters such as chemical shift, spin-spin coupling or J-constants, signal intensities, bond and spatial correlations are often considered during the complete analysis of the structure. There are many types of NMR based on the types of correlations as well. Two-dimensional correlation NMR such as correlation spectroscopy (COSY) and nuclear Overhauser effect spectroscopy (NOESY) look after the spatial orientation of the protons in a molecule to interpret the data. While LC separates the analytes efficiently, NMR elucidates the structure of the separated analytes. An analysis using LC-NMR starts with the usual analysis by LC in which the sample is injected and pumped in the column containing the stationary phase and then analyzed by a UV-visible or refractive index detector. Once the analytes in the sample are separated, they are sent to the NMR where a specialized flow cell is present and further NMR analysis occurs. The sample is sent to NMR through PEEK tubing. The NMR used in this case is liquid NMR, which requires its own set of solvents to analyze the analytes obtained from LC. Typically, deuterated solvents are used, preferably with reverse-phase HPLC as they are expensive in normal-phase HPLC. With the development of

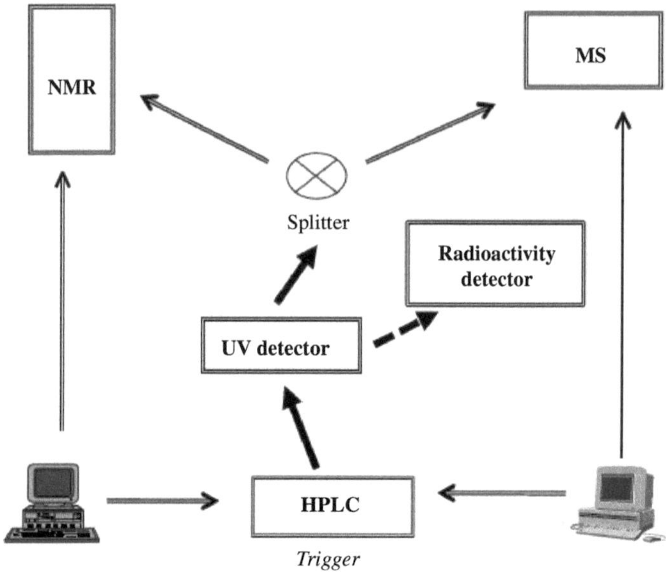

Fig. 3 Instrumental setup of LC-NMR (Victoria and Elipe, 2012).

suitable analytical techniques such as solid phase microextraction, it is now possible to significantly reduce the amount of deuterated solvents for the analysis using LC-NMR. Certain specifications such as the homogeneity and volume of the analyte are considered while injecting the analyte into the flow cell. Ensuring such proper conditions ensures that the analysis is highly accurate and sensitive. Fig. 3 shows the instrumental setup for LC-NMR.

2.3 IMS-MS

IMS is a technique that separates ions on the basis of the time the sample ions take to move from the source to the detector. Here, an inlet to inject the vaporized form of the sample is present; after that, the group of ions also known as a swarm enters a specific region known as the drift region where the movement of the ions is controlled by a voltage gradient. The ions move with a drift speed from the ionization source to the detector and the time taken for this movement is known as the drift time. Upon reaching the detector, they generate an electric current that is further amplified for analysis. The IMS results are interpreted through a mobility spectrum that is basically a plot between the intensity of the signals generated and the drift time. Typically, the parameters that are looked out for in the mobility spectrum are the mobility coefficient through which the properties or the characteristic ions can be determined, the peak shape through which the behavior of the ions in the drift region is known, and the secondary spectral details that offer information on the ion fragmentation and give an idea of the chemical class (Eiceman and Karpas, 2005).

In the case of IM-MS, two approaches are usually followed. One approach is where the sample ions are analyzed by the IMS, followed by their collection and detection by the MS. This approach focuses on the analysis and identification of compounds though their mass. In another approach, a typical tandem MS setup is present in which the drift cell is present instead of the collision cell. Both techniques are capable of analyzing the sample through their shape and mass. Fig. 4 shows a general setup of IMS-MS. The coupling of these techniques is relatively easy as the samples are used in a similar form.

2.4 HPTLC hyphenated techniques

2.4.1 HPTLC-FTIR

HPTLC is a highly sophisticated technique in which the separation of analytes can be directly seen and photographed. Typically in HPTLC, there is a stationary phase that is usually a plate coated with a layer of an adsorbent such as silica gel and the sample is spotted on a particular area of this plate. Later, this plate is placed in a twin trough chamber that is saturated with the mobile phase. Once developed, the analytes in the sample migrate over a distance and create zones that can be scanned or detected by a densitometer or any specific reagent. In forensic science, HPTLC has great potential to be hyphenated because the plate containing the migrated analytes can be easily reused for further analysis. HPLC FTIR is one such technique in which the molecular structures of the migrated analytes and any unknown compounds or impurities can be easily analyzed. In an HPTLC-FTIR interface, the scanner used is an FTIR instrument, typically diffuse reflectance infrared spectroscopy (DRIFT). In this technique, the use of multiple reflections and diffractions of the IR radiation is taken into account and converted into an interferogram. To this interferogram, Fourier transform analysis is applied to create a spectra of transmittance percentage or absorbance vs. the wavenumber. The DRIFT technique for the analysis of the HPTLC plate is useful because here, the signal-to-noise ratio is directly proportional to the square root of the concentration. Therefore, there is a very slight change in the signal-to-noise ratio with varying concentrations. While the DRIFT technique is very useful in HPTLC-FTIR analysis, challenges regarding the

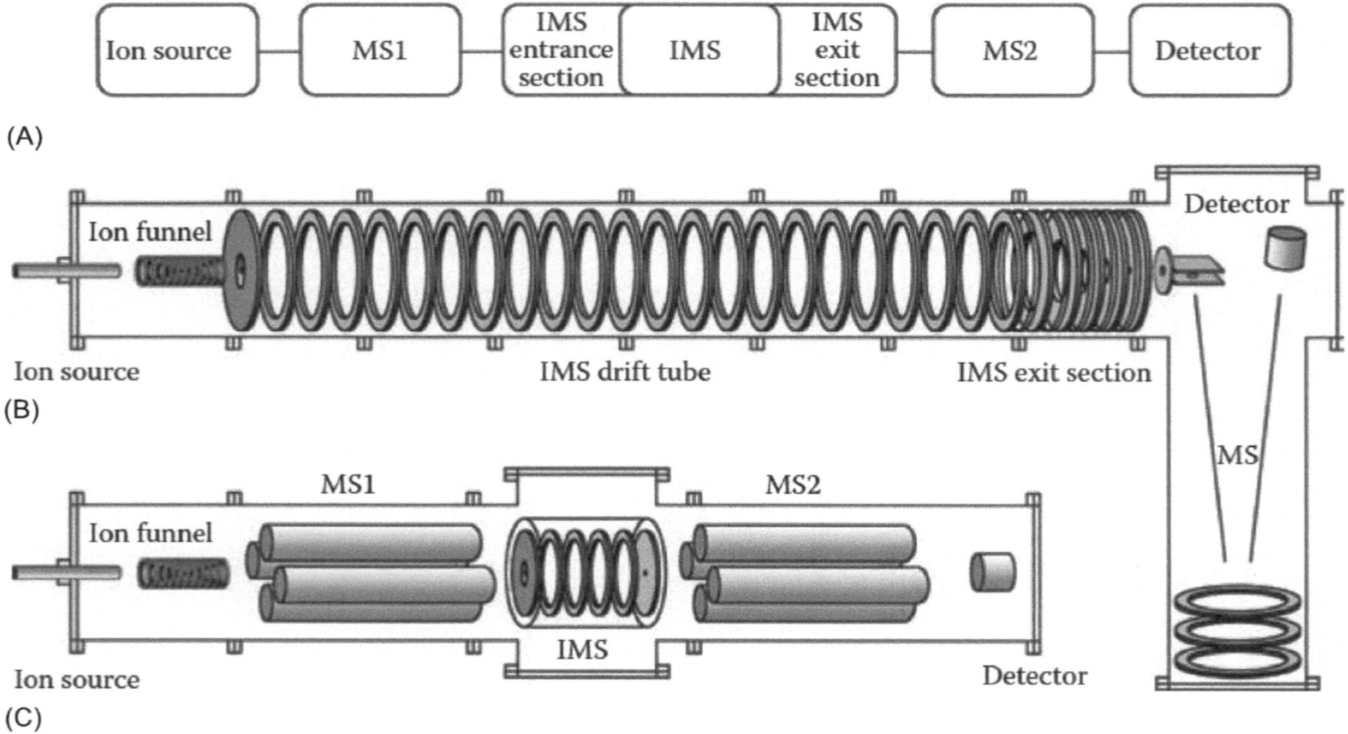

Fig. 4 (A) General schematic setup of IMS-MS, (B) IMS-TOF, and (C) IMS-quad-EMS-quad arrangement (Wilkins and Trimpin, 2011).

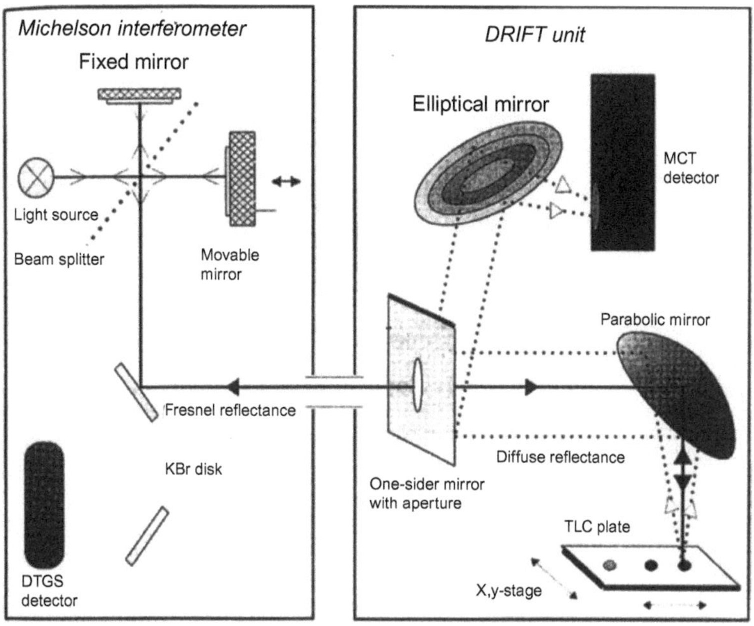

Fig. 5 Instrumental setup of HPTLC-FTIR (Cimpoiu, 2011).

samples exist. Silica gel is a very strong adsorbent of IR and therefore cannot be used directly for IR analysis. Therefore, stationary phases that contain a mixture of silica gel and magnesium tungstate significantly improve the signal-to-noise ratio. Fig. 5 highlights the instrumental setup of HPTLC-FTIR (Cimpoiu, 2011).

2.4.2 HPTLC-MS

HPTLC-MS is a unique interface because of its ability to elucidate the molecular structure very clearly (Fig. 6). The development of this technique had to happen with two basic needs, namely the conversion of the adsorbed samples on the TLC plate to the ionized part and the transfer of the ionized part to the mass analyzer, allowing its further detection. Two main ionization techniques for the ionization of the immobilized samples are electron spray ionization and MALDI. These techniques are based on the field desorption and particle beam desorption. Electron spray ionization is a technique in which the sample obtained from the plate is converted to a liquid form by dissolving it in a polar volatile solvent, and then it is nebulized in the presence of a strong electrostatic field. In MALDI, laser desorption takes place in which the laser is applied to the matrix due to which the samples get ionized. Once ionized, they get transported to the mass analyzer.

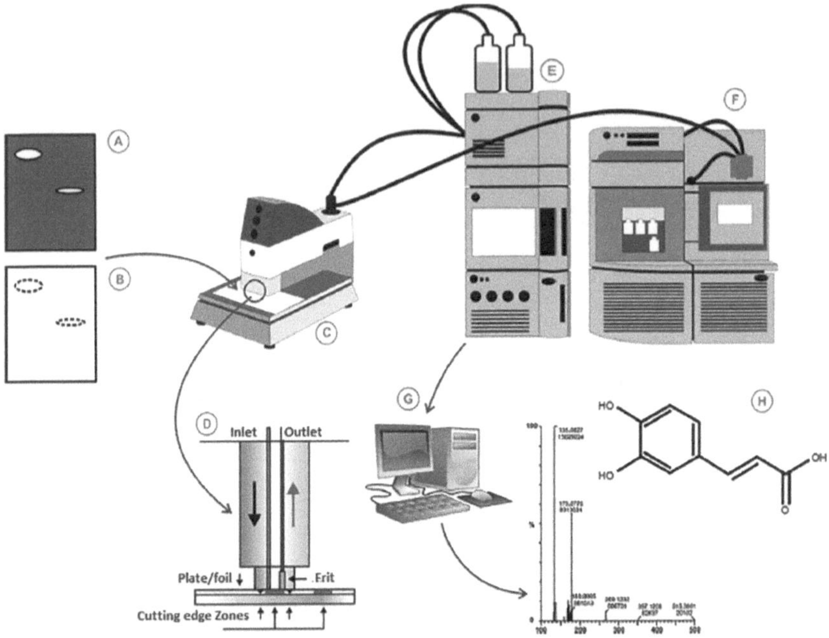

Fig. 6 Instrumental setup of HPTLC/MS. (A) Bioautography HPTLC plate, (B) reference HPTLC plate, (C) TLC interface, (D) flow-pump enabling the elution of target compounds from the plate, (E) UPLC pump, (F) time-of-flight mass spectrometer, (G) data analysis, and (H) mass spectral data allowing identification of compounds (Kasote et al., 2015).

TOF and quadrupole are the most commonly used analyzers in this case (Gupta and Gupta, 2011; Kasote et al., 2015).

3 Hyphenated technique-mediated analysis of forensic samples

A number of hyphenated techniques, as mentioned in Section 2, are present to assist analysts and researchers around the world in characterizing different kinds of samples. Forensic studies have also been done for many years using hyphenated techniques for the analysis of samples such as drugs, explosives, fibers, food products, inks, and toxins, which are discussed in this section.

3.1 Drugs

Drugs of forensic importance have been analyzed using chromatographic and spectroscopic techniques. However, the combined usage of two or more such techniques has proven to be more useful in clearly identifying and quantifying the drug present in the sample. The use of different techniques in combination for the forensic analysis of drugs is discussed here.

JWH-073, a manmade cannabinoid collected from urine and blood, was detected and quantified along with its metabolites using LC-MS/MS. A C-18 column was used for the LC part, and the sample injection volume was 5 μL. The limit of detection (LOD) for the drug was up to 0.08 ng/mL. The method proved to be robust and specific for the detection of JWH-073 and its metabolites (Ozturk et al., 2015). Vaino et al. detected propofol, a hypnotic drug, in blood and urine with the help of GC-MS and LC-MS. The study was able to detect low concentrations of the drug, which helped in concluding that the cause of death was acute propofol poisoning. The method developed in this study for LC-MS/MS proved to be more effective and reliable over the routinely used method for GC/MS because LC-MS/MS was faster in analysis, as evident in Fig. 7 (Vaiano et al., 2015). Al-Matrouk et al. also used GC-MS and LC-MS/MS to identify manmade cannabinoids, which were connected with deaths in Kuwait in 2018 (Al-Matrouk et al., 2019). Propofol and propofol glucuronide extracted simultaneously from hair evidence have also been detected using LC-MS/MS. The method validated in this study had an LOD of 3.6 pg/mg of hair for propofol and 7.8 pg/mg of

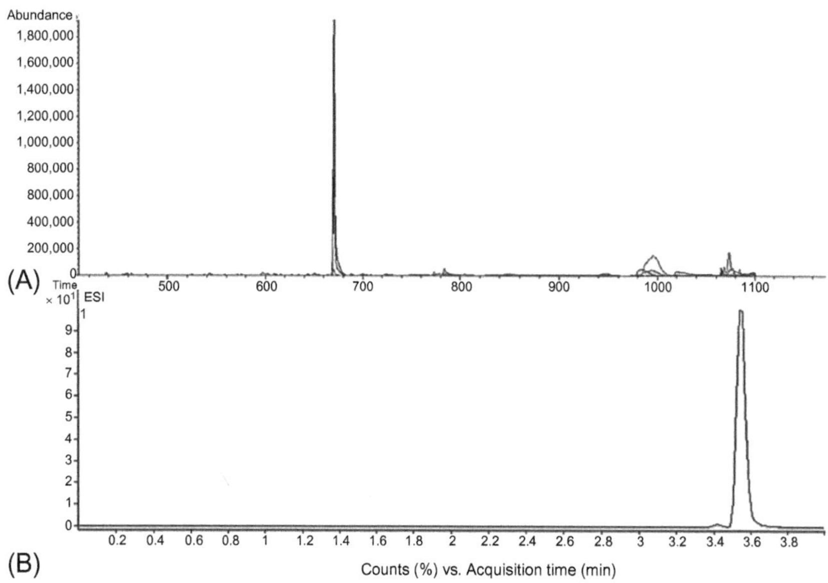

Fig. 7 (A) GC-MS and (B) LC-MS/MS chromatogram of propofol (Vaiano et al., 2015).

hair for propofol glucuronide (Maas et al., 2017). An LC-MS/MS mediated analysis of a drug abuse case with tramadol (an artificial opioid) extracted from hair evidence was carried out by Verri et al. Two mobile phases were used for the study: aqueous ammonium acetate and methanol:acetonitrile enriched with formic acid. The LOD for tramadol was 0.003 ng/mg of hair while for its metabolites, it was in the range of 0.005–0.009 ng/mg of hair (Verri et al., 2015). Urine samples have been examined for the presence of designer benzodiazepine drugs using LC-MS/MS. The urine samples were pretreated for hydrolyzing the glucuronide conjugates. Two mobile phases—acetonitrile and formic acid in water—were used for the analysis. The LOD ranged from 1 to 10 ng/mL for different drugs of the benzodiazepine class (Pettersson Bergstrand et al., 2016). A portable nano-LC-MS with electron ionization was used by Abonamah et al. for the detection (onsite) of fentanyl and associated derivatives. It took less than 10 min for the developed device, weighing around 37 kg (reported for the first time in this study) to analyze and identify heroin and manmade opioids such as fentanyl and its derivatives (Abonamah et al., 2019). Boldenone, a type of androgenic anabolic steroid used illegally to upsurge the body mass of animals, was detected in blood and urine using LC-MS/MS. The mobile phases used were mixtures of ammonium formate and formic acid in water and acetonitrile. It was observed that the drug's concentration was higher in blood than in urine (Park et al., 2019).

3.2 Explosives

Explosives and their traces are collected from crime scenes of firearm incidents and bomb blasts. This evidencs is detected and quantified using a number of techniques, and hyphenated techniques are among them. Research works focusing on explosive analysis using hyphenated techniques are discussed here.

Du et al. developed ion mobility spectroscopy (IMS)-MS (plug-type) for the onsite analysis of explosives and unlawful drugs. In order to characterize the ion chemistry in the gas phase, trinitrotoluene (TNT) was used for internal calibration. Nine different explosives were studied using this device in negative mode. It was observed that the gas phase ion chemistry of explosives was more complicated than that of drugs (Du et al., 2018). Triacetone triperoxide (TATP) and hexamethylene triperoxide diamine (HMTD) have been detected using LC-MS in Dunn et al.'s research work. The mobile phase used was a mixture of methanol and water. The LOD values for both explosives was 10 and 0.5 ng in the column, respectively. The developed method showed good chromatographic separation by retaining peroxide analytes inside the column in order to avoid peak overlap of solvents (Dunn et al., 2018). Homemade erythritol tetranitrate, a kind of nitrate ester explosive, was studied by Benzemer et al. using LC-MS. Acetonitrile with water was used as the mobile phase. The study was, however, not able to provide information regarding the precursor materials that were used to prepare the explosive (Bezemer et al., 2020). Headspace GC-IMS (HS-GC-IMS) was used for the detection of ignitable liquids that are present in fire debris. The carrier gas and drift gas taken for the study was nitrogen. The ignitable liquids studied included gasoline, diesel, paraffin, and ethanol. The ion mobility sum spectrum (IMSS) of substrates burned with these liquids is shown in Fig. 8, which helped in identifying the cause of the fire (Aliaño-González et al., 2019). LC-MS has also been used by Irlam et al. for the detection of organic explosives up to the femtogram level. Ammonium acetate added in a mixture of water and methanol in two different ratios was taken as the mobile phase. Among the different matrices chosen for the extraction of explosives, the approach showed most promising results for explosive detection in topsoil, oil, blood, and untreated wastewater (Irlam et al., 2019).

3.3 Fibers

Fiber analysis is an integral part of forensic investigations because it helps in giving information about the identity and clothing that a suspect or victim was wearing. Different types of hyphenated techniques are discussed in this section regarding forensic fiber analysis.

Causin et al. used pyrolysis GC-MS for the forensic investigation and differentiation of acrylic fibers. Phenylmethylpolysiloxane was the stationary phase while helium served as the mobile phase or the carrier gas. The pyrolysis products identified through the process helped in differentiating the structurally and visually similar fibers (Causin et al., 2006). In a study by Schotman et al., dyes present in textile fibers were initially separated by HPLC, followed by their detection using diode array detection (DAD) and MS. Ammonium acetate enriched mixtures of water:methanol and acetonitrile:methanol were utilized as the mobile phase. HPLC-DAD-MS helped in proving the presence of dye mixtures in textile fibers such as polyamide, polyester, and cotton. The study also helped in differentiating fibers of the same color on the basis of the dye concentration present

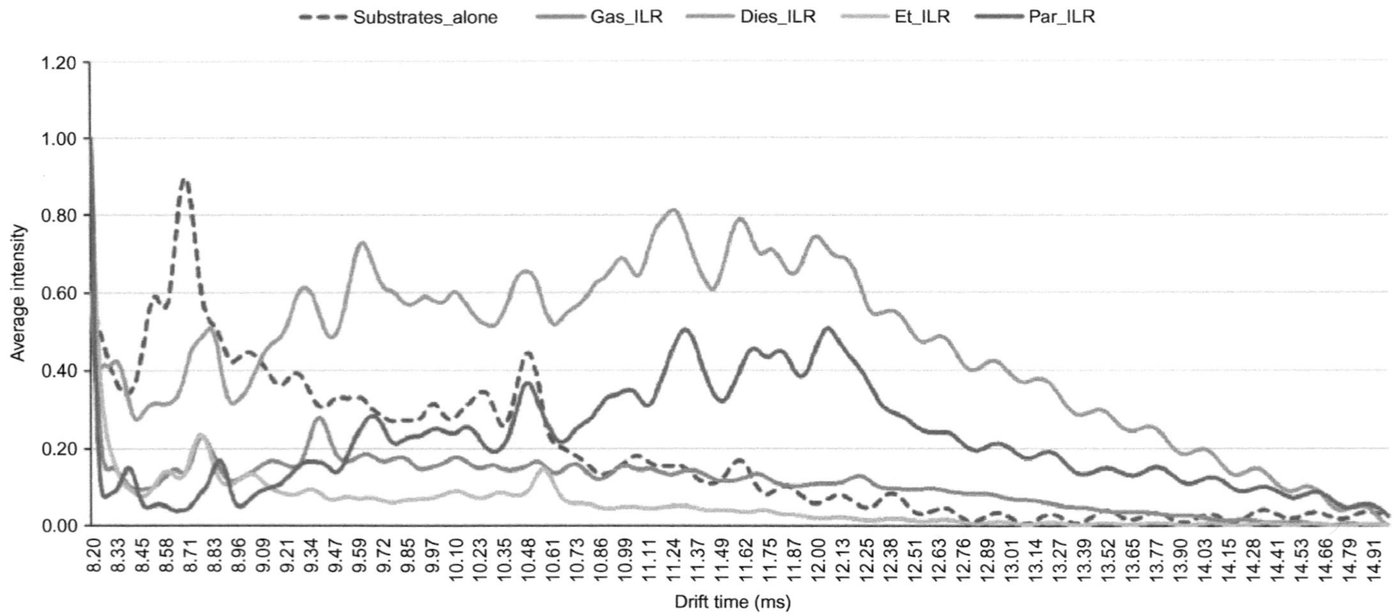

Fig. 8 IMSS of substrates burned alone and with different ignitable liquids (Aliaño-González et al., 2019).

(Schotman et al., 2017). Fiber dyes have also been analyzed by Hu et al. using HPLC-MS/MS. Different classes of dyes such as basic, acid, reactive, direct, and disperse were analyzed during the study. Mixtures of acetic acid with water and acetonitrile were used as the mobile phase. The LOD for dyes ranged from 0.01 to 16.7 ng/mL. The method proposed in this study was suitable and sensitive to detect dyes in even a few mm size of single fibers (Hu et al., 2018). Sultana et al. used a technologically advanced microfluidic device coupled with MS (MFD-MS) for the extraction and detection of textile dyes. The device required just 10 μL of solvent for the dye extraction from the fiber, and the analysis time was less than 12 min. The dyes such as anthraquinone and sulfonated azo dyes present in nylon fibers were tested in this work. Fibers of around 10 μm diameter could be analyzed by this technique to characterize the multiple dyes present in the fibers. The spectrum of acid red dye is shown in Fig. 9 (Sultana et al., 2018). Techniques such as TLC have been combined with a video spectral comparator (VSC) to enzymatically extract and differentiate dyes extracted from red cotton fibers (Góra and Wąs-Gubała, 2019).

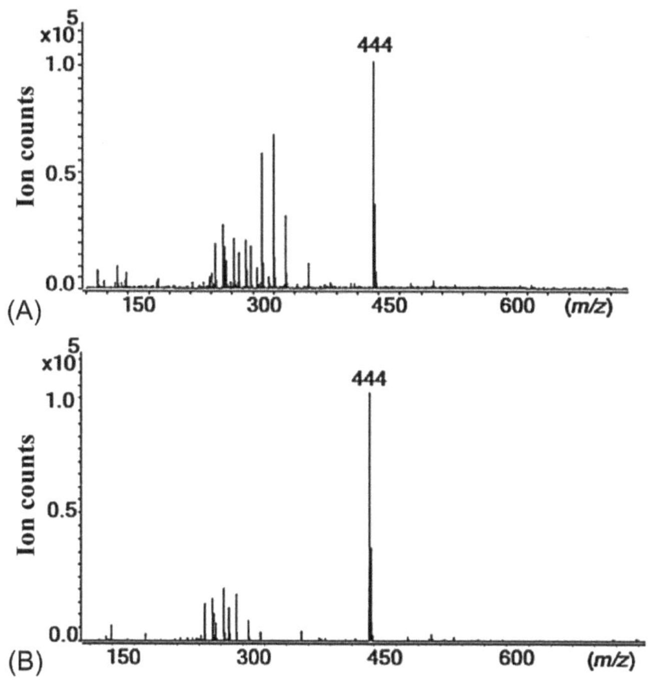

Fig. 9 Acid red 266 dye's MFD-MS spectrum (A) with background; (B) without background (Sultana et al., 2018).

3.4 Food products

Forensic food analysis is often associated with the investigation of illegally added or toxic substances present in food products. These harmful materials can be examined using different types of hyphenated techniques, among which LC-MS has been widely used by researchers (Gallart-Ayala et al., 2013; Gosetti et al., 2018).

In a study by Negreira et al., milk packed in cartons was analyzed by GC-MS in order to determine ink photo-initiators extracted from it via solid phase microextraction. Helium was utilized as the carrier gas while the electron impact was the source of ionization for the GC-MS analysis. The limit of quantification for the photo-initiators was in the range of 0.2–1 μg/L. The study was able to detect ink initiators in skimmed, semiskimmed, and whole milk samples (Negreira et al., 2010). LC-MS was also utilized by Sanchis et al. for the analysis of photo-initiators along with primary aromatic amines in food packaging materials such as pouches, bags, and juice tetrabricks (Sanchis et al., 2019). LC tandem MS has also been employed for the analysis of illegal dyes such as neutral and sulfonated azo dyes, basic dyes, indole dyes, xanthene dyes, quinolone dyes, and anthraquinone dyes present in food samples at trace levels (Hu et al., 2020). HS-GC-IMS has also been used to check the adulteration in honey samples. The study focused on the adulteration of honey with corn and sugarcane syrups. Pure nitrogen was used as the carrier as well as drift gas; the sample was eluted on a GC nonpolar column. The success rate of classifying pure and adulterated honey based on GC-IMS analysis was 97.44%. The study proved to be helpful in identifying and classifying honey based on the type of adulterant added (Arroyo-Manzanares et al., 2019).

3.5 Inks and questionable documents

Questionable documents are analyzed by forensic experts to determine their age, ink composition, line crossings in case of forgeries, etc. Hyphenated techniques have assisted researchers in the process of ink analysis, some of whose works are discussed here.

Koenig et al. used GC-MS to study the aging parameters of ballpoint pen inks. Helium was taken as the carrier gas, which was flowing on an HP5-MS column. The ionization source in MS was electron impact. It was found that factors such as ink composition, storage conditions, and writing pressure affect the aging of inks on documents (Koenig et al., 2015). Laser-induced breakdown spectroscopy (LIBS) was used with ICP-MS for the elemental analysis of inks used for printing documents such as inkjets, toners, offset, and intaglio. The CCD spectrometer in the LIBS and EMT detector in ICP-MS helped in identifying different elements present in the standard ink samples (Fig. 10) (Subedi et al., 2015). Redacted documents, which are the documents in which text is hidden with a marker or pen ink, have been analyzed by Huynh et al. using direct analyte probed nanoextraction coupled with MS with nanospray ionization (DAPNe-NSI-MS). Laser ablation was used to nondestructively extract the ink from the redacted document, and further, the analysis was carried out to study the thermal degradation of binding agents present in the inks (Huynh et al., 2016). LC-DAD-MS has been used by Sun et al. for the sensing and identification of blue roller ball pen and ballpoint pen inks. The degradation products of dyes such as Crystal Violet, Victoria blue, Acid blue, and Acid violet were identified in the ink samples (Sun et al., 2016). HPLC-MS has also been used for the analysis of red inks present in counterfeit Chinese currency notes, and the study was the first of its kind. A mixture of methanol, acetic acid, and water along with an ion pairing reagent were used in different ratios as the mobile phase. The method developed was able to discriminate between the real and counterfeit notes on the basis of data regarding the components of inks obtained by HPLC-MS and HPLC-MS/MS analysis (Xu et al., 2016).

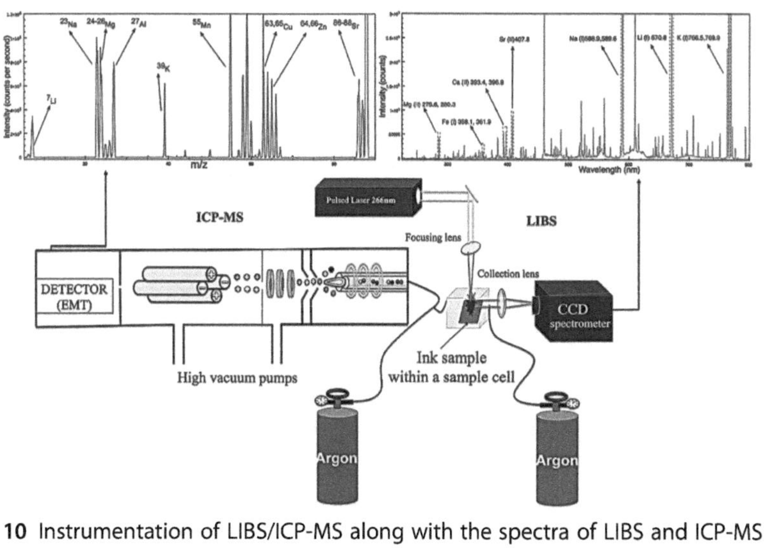

Fig. 10 Instrumentation of LIBS/ICP-MS along with the spectra of LIBS and ICP-MS for inks having different elements (Subedi et al., 2015).

3.6 Toxins

Toxins are readily studied in forensic toxicology laboratories by the use of hyphenated techniques for their separation, identification, and quantification. The research works related to toxins are discussed here. These works can be helpful for forensic practitioners to develop methods for detecting different toxins in illegally intoxicated people.

Coleman et al. used LC-MS/MS for the detection of tetrodotoxin, a potent paralytic toxin extracted from urine via solid phase extraction. Mobile phases used for the chromatographic separation were acetonitrile and ammonium formate enriched with formic acid. MS analysis was done in the positive mode. The study very precisely and accurately detected the toxin in individual and pooled urine samples (Coleman et al., 2016). In another study, poisoning due to tetradotoxin and its analogues coming from puffer fish was also studied in the body fluids of humans using LC-MS/MS. The toxin and its analogues were found in the urine, serum, and plasma, indicating the cause of intoxication for the incident from New Caledonia (Rambla-Alegre et al., 2018). Human plasma has also been analyzed to check the presence of mycotoxins using LC-MS/MS. Ammonium formate and a formic acid mixture in water as well as ammonium formate and formic acid in methanolic water were used as mobile phases. The LOD values for different toxins range from 0.04 to 2.7 ng/mL. The method developed in the study could be helpful to estimate the type of mycotoxin used to intoxicate the person (Arce-López et al., 2020). LC-MS/MS (high resolution) was used to identify biomarkers to detect the intoxication caused by the consumption of toxic *Ricinus communis* or mushrooms. Such materials can be deliberately added in food items for mass homicide, and thus a method for their analysis becomes an important task (Bambauer et al., 2020). Endocrine disruptors found in human tissues cause intoxication, and this has been detected using GC-MS. The endocrine disruptors selected for the study included bisphenol A (BPA), bisphenol Z (BPZ), bisphenol F (BPF), 4-tert-butylphenol (4-TBP), 4-tert-octylphenol (4-TOP), 4-pentyphenol (4-PP), 4-octylphenol (4-OP), 4-hexylphenol (4-HP), and 4-nonylphenol (4-NP). Fig. 11 shows the elution profile of these endocrine-disrupting compounds. The carrier gas used was helium, which was flowing through the capillary chromatographic column at a rate of 45 mL/min. The ionization source was electron impact while a quadrupole was the mass analyzer. In the case of alkylphenols, the limit of quantification was 0.05–4 ng/g of tissue while for bisphenols, it was in the range of

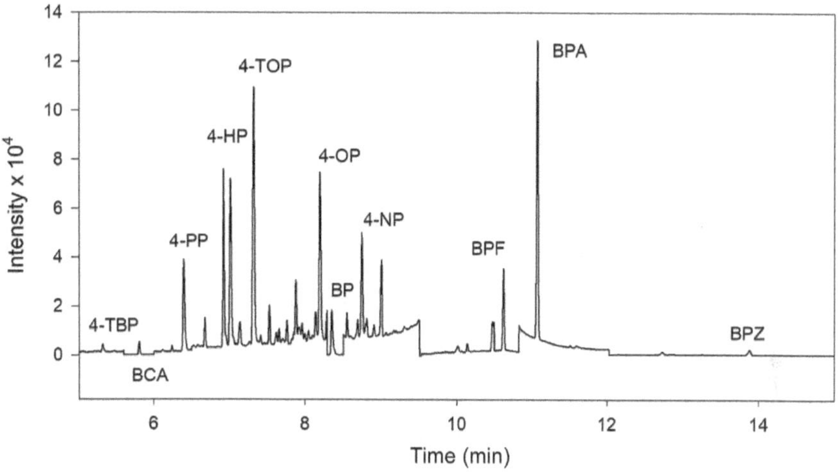

Fig. 11 Elution graph for different kinds of endocrine disruptors found in human tissues (Pastor-Belda et al., 2020).

0.26–2.6 ng/g of tissue. The study therefore developed a method to detect very minute quantities of such toxic compounds in human tissues.

These studies have developed methods for the separation, identification, and quantification of toxins coming from different origins, which will be helpful to forensic toxicologists in their investigations of criminal cases related to intoxication by such toxic substances.

4 Conclusion

Hyphenated techniques are those in which two different techniques are coupled in order to get the best of both worlds. In forensic science, these techniques are highly useful as the forensic evidence can be easily separated due to the chromatographic techniques and the structure of the molecule can also be easily elucidated because of techniques such as mass spectrometry. In this chapter, a brief overview of the different hyphenated techniques has been given and the applications of these techniques in the analysis of various forensic evidence has also been discussed.

References

Abonamah, J.V., Eckenrode, B.A., Moini, M., 2019. On-site detection of fentanyl and its derivatives by field portable nano-liquid chromatography-electron lonization-mass spectrometry (nLC-EI-MS). Forensic Chem. 16, 100180. https://doi.org/10.1016/j.forc.2019.100180.

Aliaño-González, M.J., Ferreiro-González, M., Barbero, G.F., Palma, M., 2019. Novel method based on ion mobility spectrometry sum spectrum for the characterization of ignitable liquids in fire debris. Talanta 199, 189–194. https://doi.org/10.1016/j.talanta.2019.02.063.

Al-Matrouk, A., Alqallaf, M., AlShemmeri, A., BoJbarah, H., 2019. Identification of synthetic cannabinoids that were seized, consumed, or associated with deaths in Kuwait in 2018 using GC-MS and LC-MS-MS analysis. Forensic Sci. Int. 303, 109960. https://doi.org/10.1016/j.forsciint.2019.109960.

Arce-López, B., Lizarraga, E., Flores-Flores, M., Irigoyen, Á., González-Peñas, E., 2020. Development and validation of a methodology based on Captiva EMR-lipid clean-up and LC-MS/MS analysis for the simultaneous determination of mycotoxins in human plasma. Talanta 206, 120193. https://doi.org/10.1016/j.talanta.2019.120193.

Ardrey, R.E., 2003. Liquid chromatography-mass spectrometry: an introduction. In: Analytical Techniques in the Sciences, first ed. Wiley, https://doi.org/10.1002/0470867299.

Arroyo-Manzanares, N., García-Nicolás, M., Castell, A., Campillo, N., Viñas, P., López-García, I., Hernández-Córdoba, M., 2019. Untargeted headspace gas chromatography—ion mobility spectrometry analysis for detection of adulterated honey. Talanta 205, 120123. https://doi.org/10.1016/j.talanta.2019.120123.

Bambauer, T.P., Wagmann, L., Maurer, H.H., Weber, A.A., Meyer, M.R., 2020. Development and application of a strategy for analyzing eight biomarkers in human urine to verify toxic mushroom or *Ricinus communis* ingestions by means of hydrophilic interaction LC coupled to HRMS/MS. Talanta 213, 120847. https://doi.org/10.1016/j.talanta.2020.120847.

Bezemer, K., McLennan, L., van Duin, L., Kuijpers, C.-J., Koeberg, M., van den Elshout, J., van der Heijden, A., Busby, T., Yevdokimov, A., Schoenmakers, P., Smith, J., Oxley, J., van Asten, A., 2020. Chemical attribution of the home-made explosive ETN. Part I. Liquid chromatography-mass spectrometry analysis of partially nitrated erythritol impurities. Forensic Sci. Int. 307, 110102. https://doi.org/10.1016/j.forsciint.2019.110102.

Bogusz, M.J., 1999. Hyphenated liquid chromatographic techniques in forensic toxicology. J. Chromatogr. B Biomed. Sci. Appl. 733, 65–91. https://doi.org/10.1016/S0378-4347(98)00520-9.

Causin, V., Marega, C., Schiavone, S., Guardia, V.D., Marigo, A., 2006. Forensic analysis of acrylic fibers by pyrolysis—gas chromatography/mass spectrometry. J. Anal. Appl. Pyrolysis 75, 43–48. https://doi.org/10.1016/j.jaap.2005.04.009.

Cimpoiu, C., 2011. HPTLC hyphenated with FTIR: principles, instrumentation and qualitative analysis and quantitation. In: Srivastava, M. (Ed.), High-Performance Thin-Layer Chromatography (HPTLC). Springer, Berlin Heidelberg, pp. 385–394, https://doi.org/10.1007/978-3-642-14025-9_18.

Coleman, R., Lemire, S.W., Bragg, W., Garrett, A., Ojeda-Torres, G., Hamelin, E., Johnson, R.C., Thomas, J., 2016. Development and validation of a high-throughput online solid phase extraction—liquid chromatography—tandem mass spectrometry method for the detection of tetrodotoxin in human urine. Toxicon 119, 64–71. https://doi.org/10.1016/j.toxicon.2016.05.009.

Du, Z., Sun, T., Zhao, J., Wang, D., Zhang, Z., Yu, W., 2018. Development of a plug-type IMS-MS instrument and its applications in resolving problems existing in in-situ detection of illicit drugs and explosives by IMS. Talanta 184, 65–72. https://doi.org/10.1016/j.talanta.2018.02.086.

Dunn, L., Al Obaidly, H.S.A., Khalil, S.E., 2018. Development and validation of fast liquid chromatography high-resolution mass spectrometric (LC-APCI-QToF-MS) methods for the analysis of hexamethylene triperoxide diamine (HMTD) and triacetone triperoxide (TATP). Forensic Chem. 10, 5–14. https://doi.org/10.1016/j.forc.2018.06.003.

Eiceman, G.A., Karpas, Z., 2005. Ion Mobility Spectrometry, second ed. CRC Press, London.

Gallart-Ayala, H., Núñez, O., Lucci, P., 2013. Recent advances in LC-MS analysis of food-packaging contaminants. TrAC Trends Anal. Chem. 42, 99–124. https://doi.org/10.1016/j.trac.2012.09.017.

Góra, P., Wąs-Gubała, J., 2019. Enzymatic extraction of dyes for differentiation of red cotton fibres by TLC coupled with VSC. Sci. Justice 59, 425–432. https://doi.org/10.1016/j.scijus.2019.02.007.

Gosetti, F., Bolfi, B., Mazzucco, E., Manfredi, M., Robotti, E., Marengo, E., 2018. LC-MS/MS approach for the identification of unknown degradation products of dyes in beverages. In: Grumezescu, A.M., Holban, A.M. (Eds.), Natural and Artificial Flavoring Agents and Food Dyes, Handbook of Food Bioengineering. Academic Press, pp. 229–260, https://doi.org/10.1016/B978-0-12-811518-3.00008-9.

Gupta, A.P., Gupta, S., 2011. HPTLC-MS coupling: new dimension of HPTLC. In: Srivastava, M. (Ed.), High-Performance Thin-Layer Chromatography (HPTLC). Springer, Berlin Heidelberg, pp. 311–333, https://doi.org/10.1007/978-3-642-14025-9_15.

Hu, C., Zhu, J., Mei, H., Shi, H., Guo, H., Zhang, G., Wang, P., Lu, L., Zheng, X., 2018. A sensitive HPLC-MS/MS method for the analysis of fiber dyes. Forensic Chem. 11, 1–6. https://doi.org/10.1016/j.forc.2018.08.001.

Hu, Z., Qi, P., Wang, N., Zhou, Q.-Q., Lin, Z.-H., Chen, Y.-Z., Mao, X.-W., Jiang, J.-J., Li, C., 2020. Simultaneous determination of multiclass illegal dyes with different acidic-basic properties in foodstuffs by LC-MS/MS via polarity switching mode. Food Chem. 309, 125745. https://doi.org/10.1016/j.foodchem.2019.125745.

Huynh, V., Sasiene, Z.J., Mach, P.M., Golden, T.D., Verbeck, G.F., 2016. Laser ablation coupled with DAPNe-NSI-MS applied to redacted documents. Sci. Justice 56, 329–340. https://doi.org/10.1016/j.scijus.2016.06.002.

Irlam, R.C., Parkin, M.C., Brabazon, D.P., Beardah, M.S., O'Donnell, M., Barron, L.P., 2019. Improved determination of femtogram-level organic explosives in multiple matrices using dual-sorbent solid phase extraction and liquid chromatography-high resolution accurate mass spectrometry. Talanta 203, 65–76. https://doi.org/10.1016/j.talanta.2019.05.047.

Kasote, D., Ahmad, A., Chen, W., Combrinck, S., Viljoen, A., 2015. HPTLC-MS as an efficient hyphenated technique for the rapid identification of antimicrobial compounds from propolis. Phytochem. Lett. 11, 326–331. https://doi.org/10.1016/j.phytol.2014.08.017.

Kitson, F.G., Larsen, B.S., McEwen, C.N., 1996. Gas Chromatography and Mass Spectrometry—A Practical Guide. Academic Press, USA.

Koenig, A., Magnolon, S., Weyermann, C., 2015. A comparative study of ballpoint ink ageing parameters using GC/MS. Forensic Sci. Int. 252, 93–106. https://doi.org/10.1016/j.forsciint.2015.03.027.

Kuligowski, J., Quintás, G., Garrigues, S., Lendl, B., de la Guardia, M., Lendl, B., 2010. Recent advances in on-line liquid chromatography-infrared spectrometry (LC-IR). TrAC Trends Anal. Chem. 29, 544–552. https://doi.org/10.1016/j.trac.2010.03.004.

Ligon, W.V., 2001. Liquid chromatography mass spectrometry. In: Encyclopedia of Materials: Science and Technology. Elsevier, pp. 1–5, https://doi.org/10.1016/B0-08-043152-6/00789-0.

Maas, A., Maier, C., Iwersen-Bergmann, S., Madea, B., Hess, C., 2017. Simultaneous extraction of propofol and propofol glucuronide from hair followed by validated LC-MS/MS analyses. J. Pharm. Biomed. Anal. 146, 236–243. https://doi.org/10.1016/j.jpba.2017.08.035.

Negreira, N., Rodríguez, I., Rubí, E., Cela, R., 2010. Solid-phase microextraction followed by gas chromatography-mass spectrometry for the determination of ink photo-initiators in packed milk. Talanta 82, 296–303. https://doi.org/10.1016/j.talanta.2010.04.037.

Ozturk, S., Ozturk, Y.E., Yeter, O., Alpertunga, B., 2015. Application of a validated LC-MS/MS method for JWH-073 and its metabolites in blood and urine in real forensic cases. Forensic Sci. Int. 257, 165–171. https://doi.org/10.1016/j.forsciint.2015.08.013.

Pandey, G., Rawtani, D., Agrawal, Y.K., 2017. Future aspects of Halloysite nanotubes in forensic investigations. J. Nanomed. Res. 6. https://doi.org/10.15406/jnmr.2017.06.00153.

Park, M., Sim, J., Jeon, Y., Yeon, S., Lee, J., In, S., 2019. Determination of boldenone in postmortem specimens including blood and urine samples using LC-MS/MS. J. Pharm. Biomed. Anal. 169, 111–115. https://doi.org/10.1016/j.jpba.2019.02.043.

Pastor-Belda, M., Drauschke, T., Campillo, N., Arroyo-Manzanares, N., Torres, C., Pérez-Cárceles, M.D., Hernández-Córdoba, M., Viñas, P., 2020. Dual stir bar sorptive extraction coupled to thermal desorption-gas chromatography-mass spectrometry for the determination of endocrine disruptors in human tissues. Talanta 207, 120331. https://doi.org/10.1016/j.talanta.2019.120331.

Pettersson Bergstrand, M., Helander, A., Beck, O., 2016. Development and application of a multi-component LC-MS/MS method for determination of designer benzodiazepines in urine. J. Chromatogr. B 1035, 104–110. https://doi.org/10.1016/j.jchromb.2016.08.047.

Rambla-Alegre, M., Leonardo, S., Barguil, Y., Flores, C., Caixach, J., Campbell, K., Elliott, C.T., Maillaud, C., Boundy, M.J., Harwood, D.T., Campàs, M., Diogène, J., 2018. Rapid screening and multi-toxin profile confirmation of tetrodotoxins and analogues in human body fluids derived from a puffer fish poisoning incident in New Caledonia. Food Chem. Toxicol. 112, 188–193. https://doi.org/10.1016/j.fct.2017.12.039.

Rawtani, D., Tharmavaram, M., Pandey, G., Hussain, C.M., 2019. Functionalized nanomaterial for forensic sample analysis. TrAC Trends Anal. Chem. 120, 115661. https://doi.org/10.1016/j.trac.2019.115661.

Sanchis, Y., Coscollà, C., Yusà, V., 2019. Comprehensive analysis of photoinitiators and primary aromatic amines in food contact materials using liquid chromatography high-resolution mass spectrometry. Talanta 191, 109–118. https://doi.org/10.1016/j.talanta.2018.08.047.

Schotman, T.G., Xu, X., Rodewijk, N., van der Weerd, J., 2017. Application of dye analysis in forensic fibre and textile examination: case examples. Forensic Sci. Int. 278, 338–350. https://doi.org/10.1016/j.forsciint.2017.07.026.

Somsen, G.W., Visser, T., 2006. Liquid chromatography/infrared spectroscopy. In: Meyers, R.A. (Ed.), Encyclopedia of Analytical Chemistry. John Wiley & Sons, Ltd, Chichester, UK, p. a5608, https://doi.org/10.1002/9780470027318.a5608.

Subedi, K., Trejos, T., Almirall, J., 2015. Forensic analysis of printing inks using tandem laser induced breakdown spectroscopy and laser ablation inductively coupled plasma mass spectrometry. Spectrochim. Acta B Atom. Spectrosc. 103–104, 76–83. https://doi.org/10.1016/j.sab.2014.11.011.

Sultana, N., Gunning, S., Furst, S.J., Garrard, K.P., Dow, T.A., Vinueza, N.R., 2018. Direct analysis of textile dyes from trace fibers by automated microfluidics extraction system coupled with Q-TOF mass spectrometer for forensic applications. Forensic Sci. Int. 289, 67–74. https://doi.org/10.1016/j.forsciint.2018.05.020.

Sun, Q., Luo, Y., Yang, X., Xiang, P., Shen, M., 2016. Detection and identification of dyes in blue writing inks by LC-DAD-orbitrap MS. Forensic Sci. Int. 261, 71–81. https://doi.org/10.1016/j.forsciint.2016.01.038.

Vaiano, F., Serpelloni, G., Focardi, M., Fioravanti, A., Mari, F., Bertol, E., 2015. LC-MS/MS and GC- -MS methods in propofol detection: evaluation of the two analytical procedures. Forensic Sci. Int. 256, 1–6. https://doi.org/10.1016/j.forsciint.2015.07.013.

Verri, P., Rustichelli, C., Palazzoli, F., Vandelli, D., Marchesi, F., Ferrari, A., Licata, M., 2015. Tramadol chronic abuse: an evidence from hair analysis by LC tandem MS. J. Pharm. Biomed. Anal. 102, 450–458. https://doi.org/10.1016/j.jpba.2014.10.002.

Victoria, M., Elipe, S., 2012. Basic technical aspects and operation of LC-NMR and LC-MS-NMR. In: LC-NMR and Other Hyphenated NMR Techniques Overview and Applications. Wiley & Sons Ltd., Canada, pp. 59–93.

Wilkins, C.L., Trimpin, S., 2011. Ion Mobility Spectrometry-Mass Spectrometry. CRC Press, p. 379.

Wu, S., Lv, G., Lou, R., 2012. Applications of chromatography hyphenated techniques in the field of lignin pyrolysis. In: Davarnejad, R. (Ed.), Applications of Gas Chromatography. InTech, https://doi.org/10.5772/32446.

Xu, Y., Zhou, X., Shi, X., 2016. HPLC and HPLC/MS analysis of red ink on counterfeit 100-yuan notes. Forensic Sci. Int. 259, 47–52. https://doi.org/10.1016/j.forsciint.2015.11.018.

CHAPTER 12

Optical microscopy for forensic samples

1 Introduction

Optical or light microscopes are highly crucial instruments that use light as a probe in order to determine the structure and morphology of a particular compound. Light as such is an electromagnetic radiation in which the two components that are primarily present—the electric field and the magnetic field—oscillate at right angles to each other. Light itself is composed of photons that contribute the particulate and the wave-like nature to light. Due to such properties, parameters such as amplitude that determines the height of the oscillation, wavelength, frequency, and the energy of light are manipulated in order to obtain a higher-resolution image of the sample. There are mainly two types of microscopes: simple and compound. While the simple microscope is composed of a single lens and is simply used to visually observe the samples, the compound microscope itself is made up of two or more lenses and a light source. Based on the type of light source and the way a sample is viewed, the compound microscopes are divided into various types.

In forensic science, light microscopes are often one of the earliest tools used for the preliminary and confirmatory analysis of evidence (Pandey et al., 2017; Rawtani et al., 2019). There is a myriad of evidence found and each sample has its own method of preparation and viewing. In this chapter, we present a brief overview on the different types of optical microscopes, namely the phase contrast, bright field, confocal, fluorescence, polarized, and stereomicroscope. Also, the applications of such microscopes in the analysis of biological samples such as bone, hair, skin, insects, fibers, explosives, glass, paints, questionable documents, soil, and minerals is also discussed.

Handbook of Analytical Techniques for Forensic Samples
https://doi.org/10.1016/B978-0-12-822300-0.00012-4

2 Fundamentals of optical microscopy

As mentioned in the introduction, light is made up of photons, which are packets of energies that contribute to its dual nature. This duality is usually expressed in the form of the equation:

$$E = hc/\lambda$$

Here, λ is the wavelength of the light, c is the speed of light, h is Plank's constant, and E is the energy that itself takes into account the frequency of the light. Apart from this, there are also various types of lights that vary based on their wavelengths and uniformity. Some of the types of light include monochromatic, polarized, coherent, and collimated. Monochromatic light has a single wavelength or frequency while in polarized light, the electric vectors of light oscillate at different planes. Typically, light itself changes phases when it enters different mediums. However, coherent light does not change the phase despite a change in the medium of propagation. Finally, collimated light propagates through mediums by maintaining its coaxial paths.

Typically, in a compound microscope, a light source with visible light is used to image a sample through an objective lens that is responsible for the collection of the light diffracted from the sample. This forms a real image that is significantly magnified. This real image can be visualized through an eye piece mounted on top of the microscope. Apart from this, condenser lenses are also used that are responsible for focusing the light from the light source onto the sample mounted on the sample holder. Apart from this, several fittings are also used that aid the entire process of light scattering and magnification (Fig. 1) (Murphy, 2001).

3 Types of optical compound microscopes

In this section, the different types of compound microscopes such as phase contrast, bright field, confocal, fluorescence, polarized, and stereomicroscope are discussed.

3.1 Bright-field microscope

The bright-field microscope is one of the simplest compound microscopes. It consists of a condenser lens that focuses the light from the source to the sample, and an objective that converges the light to form a magnified virtual image of the sample. The sample itself is placed between the object focal plane and the lens. This technique is incredibly simple and can be visualized

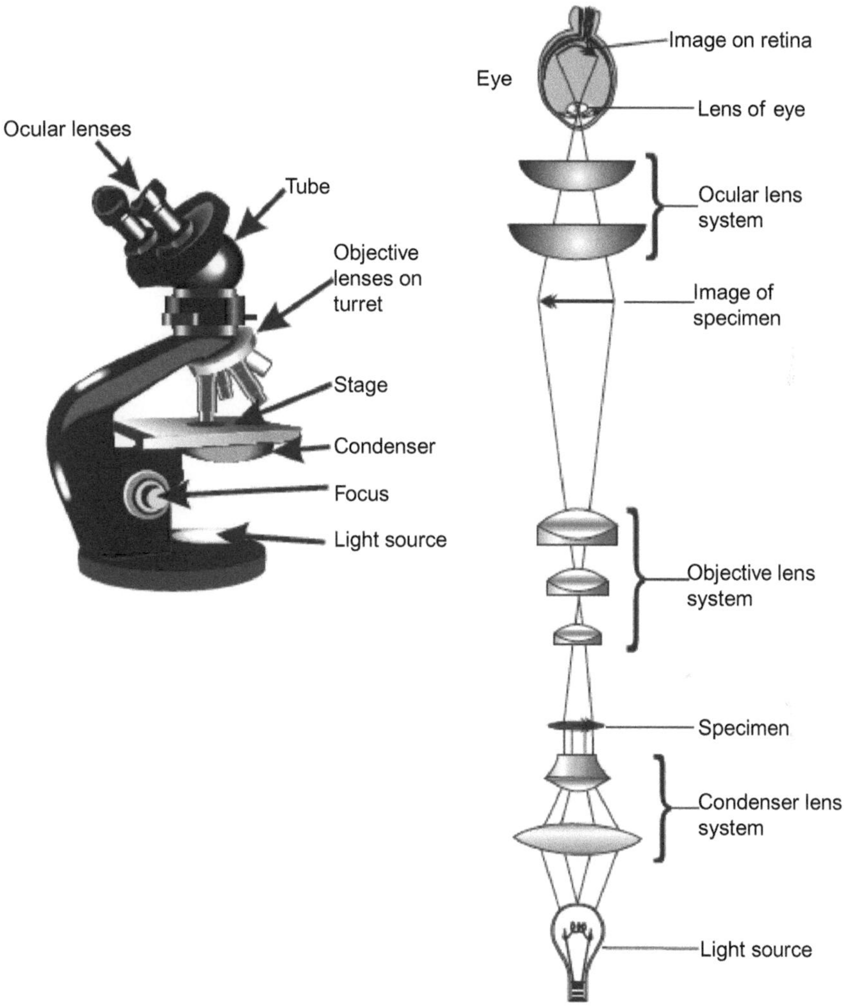

Fig. 1 Instrumental setup of a compound microscope (Roane and Pepper, 2015).

anywhere. Also, living cells such as any bacteria can also be easily visualized using this technique. It is also one of the most widely used microscopes for the preliminary analysis of forensic evidence. However, despite its use, it has certain limitations such as low contrast, limited magnification, and resolution. A crucial parameter of bright field microscopy and also of other techniques is the resolving power of the lens. The greater the resolving power of the lens, the higher the resolution of the image obtained. This parameter is dependent on another crucial factor known as the numerical aperture, which is basically the ability of the lens to clearly differentiate the positions of two closely situated points. The greater the numerical aperture, the greater the resolving power of the lens. Another factor related to the numerical aperture is the brightness. Here too, the greater the numerical aperture, the greater the brightness. The brightness is also inversely proportional to the transverse magnification that is basically a measure of the magnification. Therefore, it is important to keep in mind that too much brightness can be a hindrance to obtaining an image with high magnification and resolution. This limitation is why a bright-field microscope is not useful for myriad applications (Wayne, 2019a).

3.2 Phase contrast microscope

A phase contrast microscope typically functions on the contrast obtained from a sample. Contrast is mainly defined as the difference in the intensity of two points as observed from a single point of view. Typically, light travels at different velocities in different mediums and is represented according to the refractive index of the medium. As the refractive indices change, there is also a phase shift, due to which there is a difference in the contrast as well. The phase shift also occurs when there is diffraction, due to which there is a phase shift of $\lambda/4$. Based on the changes in the phases, the phase shift is represented in terms of changes in contrast, due to which the fine details of a particular sample can be easily observed. The instrumentation of this microscope is similar to any compound microscope and has a light source, condenser lens, objective lens, and an eye piece. The light source such as a halogen bulb or LED provides homogenous illumination. This light passes through a particular phase region known as the conjugate region. Apart from this, when the light hits the sample, it undergoes diffraction and is collected by the objective lens, after which it passes through a region known as the complementary region. The waves obtained from the different phase shifts interfere with each other, due to which it provides a brighter or darker image of the sample. If a sample has multiple regions of different media, then those regions can be visualized efficiently through this technique (Fig. 2) (Johnson, 2013).

3.3 Fluorescence microscope

Visualization of an object occurs when there is a collision of the photons with the sample and the scattered photons enter the eye. In the case of fluorescence, when the photons hit a sample, there is a transfer of energy from the photon to the electrons present in the sample. This energy transfer facilitates the movement of electron from the ground state (the preferred form of

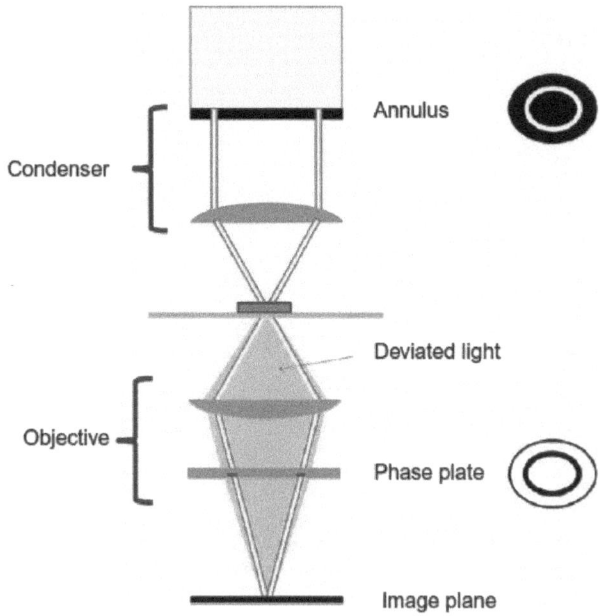

Fig. 2 Instrumentation of a phase contrast microscope (Johnson, 2013).

equilibrium for electron) to the excited state. However, it does not stay for a long time at the excited state, soon losing energy and falling back to the ground state. Many types of energies are released when the electron falls from its excited state to the ground state. This is dependent on the positioning of the electron in its shell. Fluorescence is one such energy that is emitted when the electron falls back. This lasts for a very small time, almost 10^{-8} s. Also, when it falls back to the ground state, the wavelength of the energy emitted is longer than the incident light. This phenomenon is known as the Stoke's shift and serves as the principle for fluorescence microscopy. A fluorescence microscope consists of a light source that is capable of emitting light in the UV range; currently, Kohler illumination is being used. The light from the light source then passes through the excitation filter in which only the wavelength of light that is capable of exciting the electrons passes through. This excitation light then passes through the condenser lens that focuses the light on the sample. The sample then emits fluorescence that is a mixture of both the emission and excitation wavelengths. This is further passed through an emission filter that focuses only on the emission wavelengths and is focused by the objective lens before it reaches the eye piece (Fig. 3) (Wayne, 2019b; Wolf, 2013).

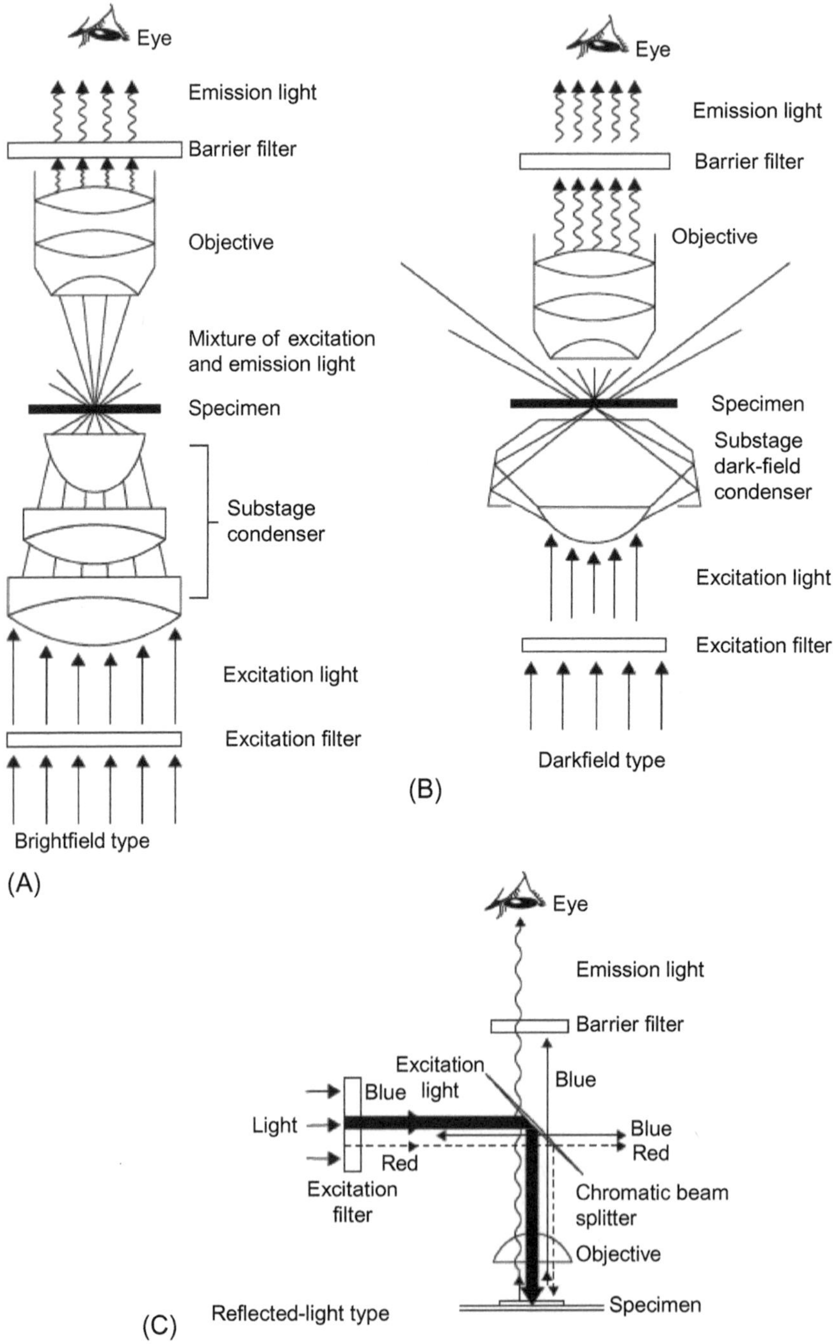

Fig. 3 Schematic setup of bright-field: (A) dark field, (B) reflected light, and (C) fluorescence microscope (Wayne, 2019b) .

3.4 Confocal microscope

Confocal microscopy is another type of optical microscope that increases the resolution and contrast of an image by focusing on it through a pinhole. Through this technique, it is possible to see a three-dimensional (3D) view of a sample by superimposing multiple two-dimensional (2D) images of the sample. This microscope is different from other techniques because it has the ability to scan the entire sample, whereas in other techniques, the entire sample is illuminated at once. Due to this, the small area of the sample that is illuminated can be focused more easily, thus increasing the resolution of the image. This focusing as mentioned before is done through a pinhole as it filters out the remaining areas.

Typically, the confocal microscope consists of a scanning stage that is capable of moving in the x-y direction. On this stage, the sample can also be kept. Due to this setup, the sample can be moved for scanning while the rest of the optical system remains fixed. However, the movement of the sample itself is very slow, especially when compared to a speed of at least 25 frames a second to achieve the optimal speed. Therefore, in order to do this, the optical beam is moved with the help of acousto-optical device-controlled mirrors. However, this technique has certain disadvantages such as the development of poor-quality beams. Therefore, later the laser itself was split into a thousand parts and each individual beam was controlled and focused on a particular area. A special pinhole disk called a Nipkow disk is used to generate the thousand beams and allow their efficient control. Fig. 4 shows the instrumentation of a confocal microscope (Lu, 2005; Yao and Wang, 2005).

3.5 Polarizing microscope

Typically, light is made up of the sinusoidal waves of the electric and magnetic components propagating at right angles to each other; such a light is commonly referred to as nonpolarized light. However, when light is polarized, there is only one direction or plane of vibration. Light that is completely polarized propagates in a sinusoidal uniform helical motion and can be visualized as an ellipse at the end; this is known as elliptically polarized light. Apart from this, linearly and plane polarized light also exist. The polarizing microscope uses the linearly polarized light to visualize various samples and is used to image samples with higher contrast, resolution, and magnification.

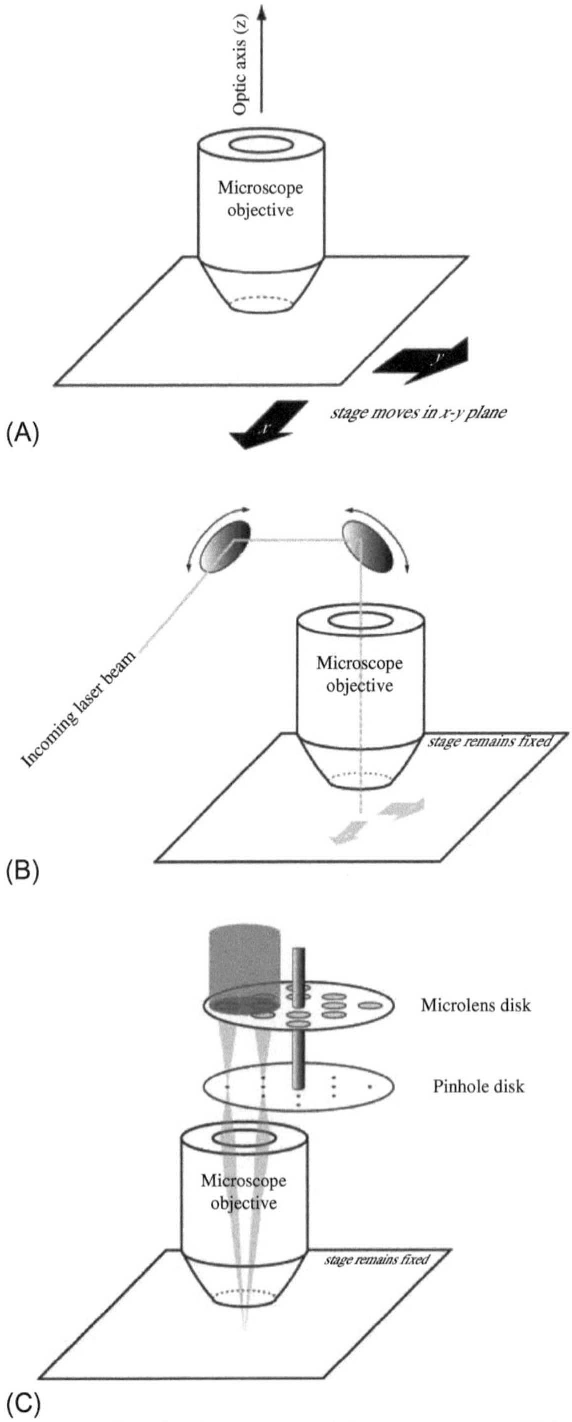

Fig. 4 Instrumentation of confocal microscope (A) scanning stage, (B) beam scanning, and (C) Nipkow disk (Lu, 2005).

Initially, the source used in the polarizing microscope to generate polarized light was a crystal called a tourmaline. Later, these were switched to calcite. Currently, a Nicol prism is the most commonly used polarized light source. From the source, a condenser is kept that is responsible for carefully diverting the polarized light onto the sample, which is set on a rotating stage. The light obtained from the specimen is further passed through the objective lens with low numerical apertures. It also passes through a compensator and analyzer that are highly useful in determining the birefringence. Birefringence is a phenomenon that typically occurs in anisotropic materials and whose refractive index changes based on the polarization and the direction of light propagation. Birefringence is also responsible for double refraction, due to which the polarized light may take two different paths, thus resulting in a distorted image. The image analysis is also done in which background subtraction and any errors are removed (Fig. 5) (Wayne, 2019c).

3.6 Stereomicroscope

Stereomicroscopes are typically used to image a sample with a lower magnification. Also, another crucial difference is that this microscope uses reflected light rather than transmitted light, as used in other microscopes. By capturing the reflected light from the sample, it is possible to image opaque or thick samples. Factors such as working distance and depth of field are important in order to obtain an image with higher resolution. Upon decreasing the working distance, the resolution as well as depth of field increases, which are the measures of acceptable sharpness. The stereomicroscopes are also designed according to the CMO (common main objective) design in which the objective lens is exactly parallel to the sample and is kept at the base of both optical paths.

The stereomicroscope consists of two optical paths in order to cater the reflection from the samples to both eyes, thus offering a 3D image of the sample. The light source used is generally fiber optic as it offers uniform high intensity light and can also be easily placed in the microscope. The light hits the sample from where the reflected light passes through the objective lens that is kept at the base of the two optical paths. Apart from this design, the stereomicroscope can also be operated at fixed and zoom magnifications. In the case of fixed magnification, simply by changing the eyepiece, the magnification of the microscope can be controlled. In zoom magnification, the eyepiece can be easily changed to another eyepiece of a higher range (Fig. 6) (Wilson and Wheeler, 2009).

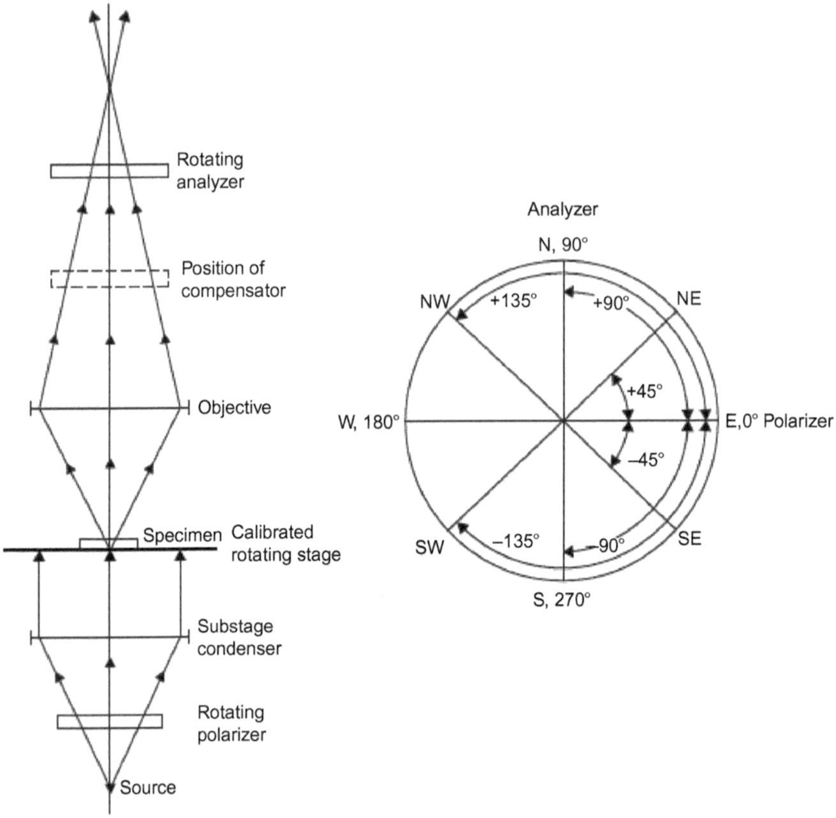

Fig. 5 Schematic of a polarized light microscope (Wayne, 2019c).

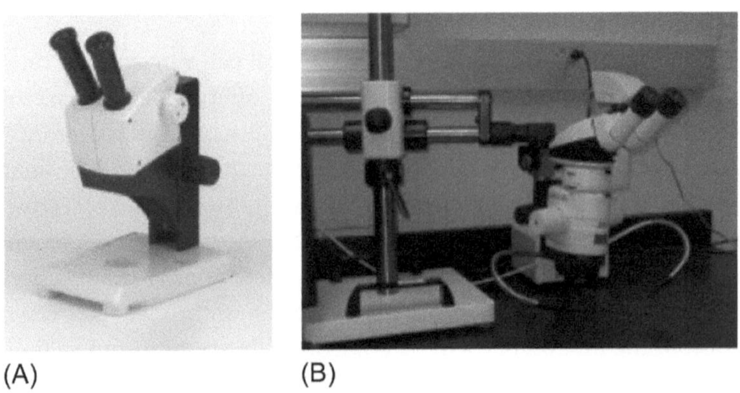

Fig. 6 Instrumentation of a stereomicroscope (Wilson and Wheeler, 2009).

4 Optical microscopic analysis of forensic samples

Optical microscopy has always assisted scientists in understanding the structural details of the microscopic world around us. Forensic scientists have used optical microscopy, especially polarized light microscopy (PLM) and stereomicroscopy, for the ultrastructural examination of forensic samples such as those of biological origin, fibers, explosives, glass, paint, questionable documents, soil, and minerals (Wilson and Wheeler, 2009). Some of the research works employing optical microscopy of forensic evidence are discussed here.

4.1 Biological samples

Different types of biological samples are encountered at a crime scene. This evidence can be of forensically important insects and their developmental stages as well as bones, hair, skin, etc. These samples are analyzed under an optical microscope for the in-depth investigation of their morphological features.

In a study by Sousa et al. that was focused on age determination from the incremental lines present in the cementum of teeth, light microscopy was used to take the images of cross-sections of almost all types of teeth, be they incisor, canine, molar, or premolar. The bar for taking the images was 100 μm. The incremental lines were clearly visible, and were counted to estimate the age of the individual. There was no significant difference observed in their estimated and real age (Sousa et al., 1999). A similar kind of study was also performed by Kasetty et al. using PLM at 10 × and stereomicroscopy at 5 ×. From the study, it was concluded that the annulations in cementum are less significant as compared to the thickness of cementum in the age estimation (Kasetty et al., 2010). Maat et al. used bright-field light microscopy for the analysis of bone, which could be helpful for forensic anthropologists. For bright-field microscopy, images were taken at 6.6 × for the parietal bone and a rib sample of 12 and 12 μm size. In addition, PLM was also used at 66 × for the detailed analysis of these bones. The method developed in this study for bone sample preparation could be used for inhumed, cremated, and preserved bone remains of forensic value (Maat et al., 2001). Cerutti et al. showed the ability of light microscopy for examining toolmarks on bone samples. Bones were cut by different types of tools such as metal cutting and wood cutting saws, symmetrical and asymmetrical blades, and box cutters. The injuries were made in two ways: a single swipe and repeated motion. The images were recorded at 40 × magnification.

The microscopic images clearly revealed the presence of fractures due to different tools that were surrounding as well as penetrating the osteons (Cerutti et al., 2016).

The morphology of flies of forensic importance mostly belongs to the order Diptera. Their morphological details along with the study of their developmental stages give an idea about the time since death. In a study by Sukontason et al., *Chrysomya villeneuvi* Patton, which belongs to the Calliphoridae family, was studied using optical microscopy for the morphological details of its second and third instars. The structural details observed under a light microscope such as the cephalopharyngeal skeleton, the anterior and posterior spiracles, the intersegmental spines, the tubercles, and the papillae were clearly visible for both the instars. The detailed structures for the second instar are shown in Fig. 7 (Sukontason et al., 2005). Another member of this family, *Lucilia porphyrina,* was also studied in its immature (instars and puparia) form using light microscopy. Magnification for the image procurement varied from 4 × to 40 ×. All the structural details were visible in the microscopic images (Klong-klaew et al., 2012). Hair has also been analyzed using light microscopy. In a study by Wilson and coresearchers, structural changes in hair shafts due to biodegradation by microbes was studied using light microscopy along with electron microscopy. 0.5 μm thick sections of hair were taken for high-resolution light microscopic analysis after treatment with toluidine blue. It was observed that a layer in the cuticle and melanin granules in the cortex was the most resistant while the intercellular cell membrane complex was the least resistant to biodegradation (Wilson et al., 2007). Yates et al. used light microscopy for the forensic identification of giraffe and elephant species using hair samples collected from their tails. The images were taken at 100 × magnification. As observed in the images, giraffe hair was different from elephant hair due to the radially arranged tubules in the periphery of the hair shaft (Yates et al., 2010).

A study from the late 1970s by Danielsen et al. showed the potential of light microscopy in the analysis of skin after thermal and electrical injuries. Pig skin was used for this purpose. The samples for microscopic analysis were stained by eosin and hematoxylin. The images revealed a clear distinction between the electrocuted and heated skin samples. It was observed that the changes in dermis and epidermis were distributed in a diffused manner in heated skin while for electrocuted skin, they were present in segments (Danielsen et al., 1978). Nakatome and coresearchers used light microscopy to detect apoptosis in cardiomyocytes, which could help to develop a

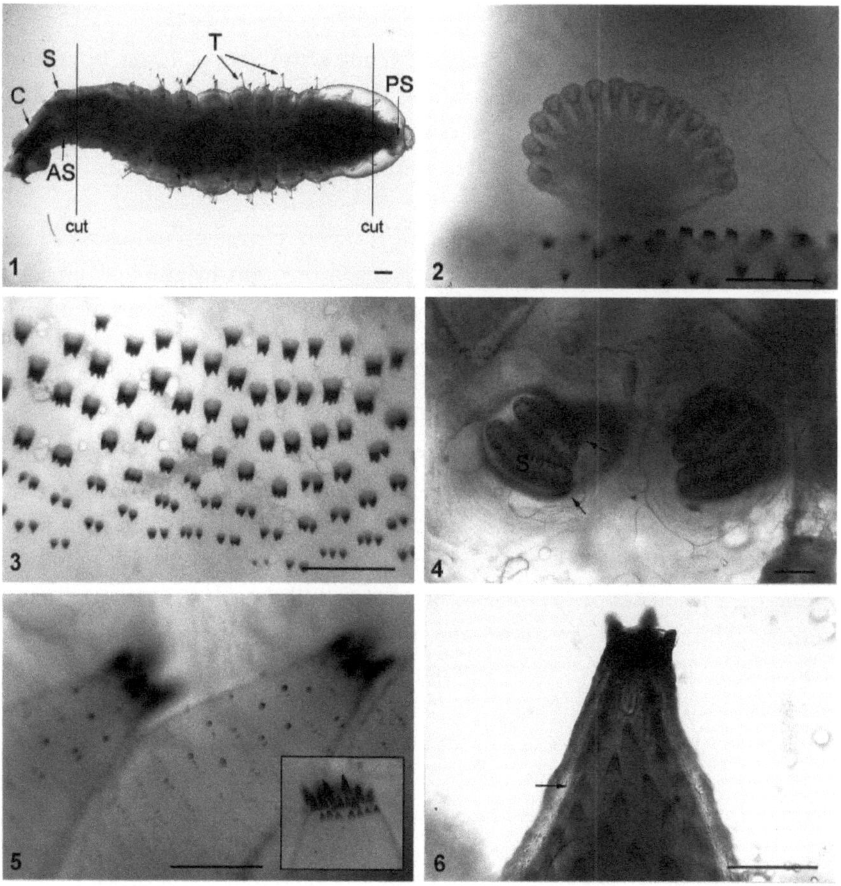

Fig. 7 Images from a light microscope of *Chrysomya villeneuvi* (second instar). (1) whole body sideways view: C, cephalopharyngeal skeleton; AS, anterior spiracle; S, spines in between segments; T, tubercles; PS, posterior spiracle. (2) Papillae in AS. (3) IS. (4) PS with spiracular slits. (5) T covered with dark spots. (6) Third instar cuticle (Sukontason et al., 2005).

biomarker for the diagnosis of acute myocardial infarction (AMI) during an autopsy. The myocardial cells were specifically stained and were visible under the microscope for samples coming from people who died of AMI (Nakatome et al., 2002). Wiltshire used light microscopy to check for the presence of fungal and psychotropic plant materials in people who had suspicious deaths. The images revealed the presence of fungal spores as well as pollens of different psychotropic plants in the gut contents as well as containers (Wiltshire et al., 2015).

4.2 Fibers

Fibers are very commonly encountered at any crime scene. These fibers can be of animal or plant origin or synthetic in nature. Optical microscopy has contributed much in the structural characterization of these fibers in forensic laboratories. Research works related to these structural characterizations are discussed here.

In a study by Grieve et al., acrylic fibers were characterized using PLM and bright-field microscopy. It was observed through bright-field images that fibers such as Orlon 78 and 21 as well as X-24 had an irregular mushroom-like morphology and were bicomponent in nature. The PLM analysis revealed that fibers such as Acrilan B94 and B57 as well as Dralon K exhibited birefringence property (Grieve et al., 1988). Another study by Flynn et al. used bright-field, fluorescence, and PL microscopy for the analysis of bicomponent fibers. The images were recorded at 100 × and 500 × magnification. It was observed that none of the fibers had a fluorescence property while the bicomponent nature of fibers was observed under a bright-field microscope and PLM, but not for all the fibers. Orlon 21 was seen to exhibit a side-by-side configuration of both components (Flynn et al., 2006). Kemp et al. used a stereomicroscope for the analysis of fabrics that were torn after a stabbing. The fabrics taken for the study were cotton single jersey knit and cotton bull drill (same as denim). The weapons taken were a kitchen knife, a hunting knife, and a screwdriver. It was observed through microscopic analysis that knives produced tapered and sharp severances on the fabric while screwdrivers produced rough and spherical severances (Kemp et al., 2009). De Wael and coresearchers have done a lot of work in this field. This research team did a forensic analysis of different fibers and published their research work in three parts. All the fibers were examined for their dichroism behavior in plane polarized light under an optical microscope. In the first part, polyester fibers colored with different disperse dyes were examined. Initially, the color of the fiber was noticed using bright-field microscopy, followed by analysis under a polarizer oriented in the north-south direction. It was observed that the dyed polyester fibers showed strong dichroic behavior by exhibiting hypochromic and hypsochromic or bathochromic effect (De Wael and Vanden Driessche, 2011a). The second part focused on the analysis of fiber sof polyamide, silk, and wool colored with different acid dyes. It was observed that these fibers had a weaker dichroic property than polyester fibers. However, these fibers showed inverse dichroism (hyperchromic effect) (De Wael and Vanden

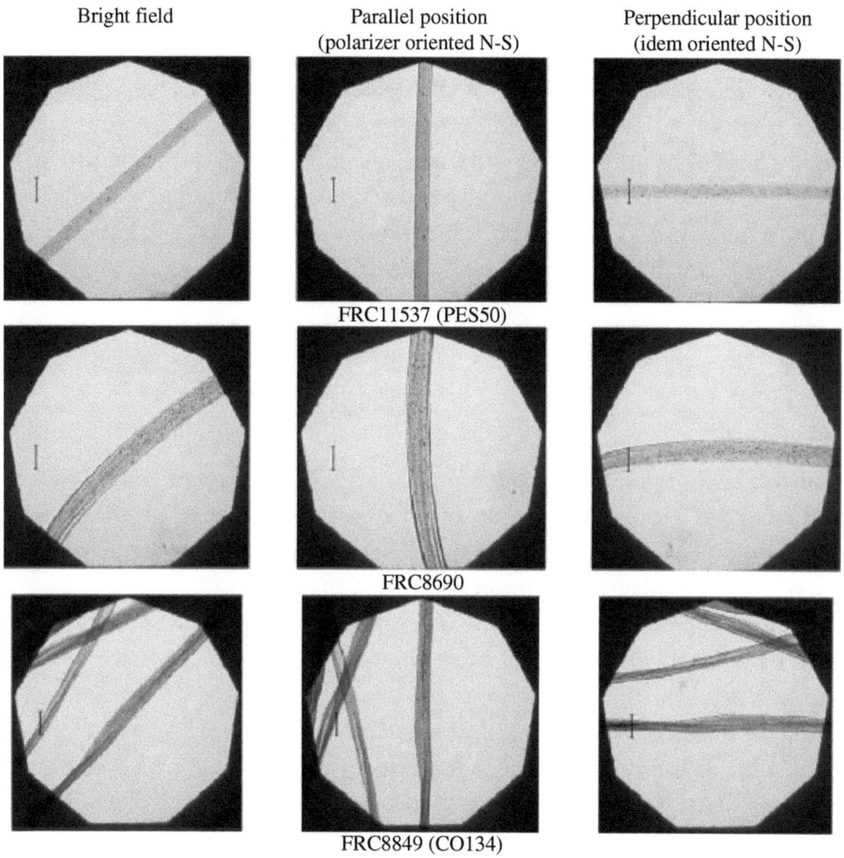

Fig. 8 Bright-field and polarized light microscopic images of three types of fibers (bar is 25 μm) (De Wael et al., 2012).

Driessche, 2011b). The third part was focused on the analysis of viscose and cotton fibers colored with various types of dyes. Hypochromic and hypsochromic effects were observed in this case as well (De Wael and Lepot, 2011). The PLM images of different fibers in comparison with bright-field light microscopy taken at 400 × are shown in Fig. 8 (De Wael et al., 2012).

4.3 Explosives

Explosives and their residues are found as fragments that penetrate the skins of people present at a crime scene. The analysis of such samples becomes important in order to identify the type of device utilized for the explosion.

In this regard, PLM and confocal microscopy were used by Turillazzi et al. for the structural analysis of skin samples of people killed in the explosion of a letter bomb (masked) for the detection of traces of explosives and their residues. A histological study for the samples was performed under a bright-field light microscope. The samples that exhibited stratified strange materials were taken for confocal microscopic analysis along with the use of polarized light. The presence of explosive residues in the skin samples was visible on the epidermal and dermal layer of the skin (Turillazzi et al., 2010).

4.4 Glass

Glass is a common trace evidence analyzed by forensic investigators in cases of hit and run, burglary, or assault. In these cases, fragments of glass are found on the body as well as clothing of the people involved in the crime. Optical microscopy has played an important role in these cases.

In a study by Welch et al., a phase contrast microscope was used for the analysis of band patterns in glass fragments. The fragments were ground and mixed with silicone oil for microscopic analysis. In the image, light and dark bands of around 5 and 50 μm size were observed, with the best resolution at the temperature when the refractive index of oil and glass matched. The examination of the dark and light bands in glass fragments could be correlated to its original source or the window from which it was broken (Welch et al., 1989). Petterd el al. published a survey from southeast Australia regarding the analysis of glass particles in the clothes of people. A stereomicroscope at magnification ranging from 6 to 40 × was used for the analysis. The study concluded that the possibility of finding glass fragments on the clothing of a suspect was very high because fragments were found on most of the garment samples examined (Petterd et al., 1999). A similar kind of study was conducted by Roux et al. for analyzing fragments of glass. However, the samples were collected from the footwear of people rather than clothing, as mentioned in the previous study. In this study, a stereomicroscope along with PLM was used in order to differentiate between isotropic and anisotropic wreckage. It was observed that the presence of glass fragments was more in the sole as compared to the upper region of the shoe (Roux et al., 2001). In another study focusing on the analysis of glass fragments present on the jackets of police officers and forensic laboratory workers, PLM was used to differentiate between the anisotropic (may be quartz or something similar) and isotropic (may be glass) (Daéid et al., 2009).

4.5 Paint

Paint analysis is usually performed in cases related to hit and run, or for the samples collected from tools involved in a burglary. Microscopic analysis reveals the layers of paint, by which it can be correlated to the vehicle or source from where it came. This analysis is commonly practiced in the forensic examination of vehicles.

Hamer published an article in the early 1980s regarding the use of light microscopy along with plane polar and crossed polar lights for the forensic analysis of pigments present in paints. The images were recorded in the magnification range of 150 × to 625 ×. Properties such as color, shape, size, crystallinity, and refractive index of paint samples were analyzed (Hamer, 1982). Allen also used light microscopy for the investigation of thin sections of household paint samples. The illumination type was both bright-field and dark-field. These microscopic techniques helped in the discrimination of paint samples on the basis of their color as well as particle distribution. In addition, PLM was used for analyzing the birefringence and fluorescence microscopy for the fluorescent property of the paints (Allen, 1992). In another study that focused on the investigation of architectural paints, a stereomicroscope was used for the analysis and discrimination of such paint samples on the basis of properties such as topcoat sheen and color as well as the layer color, texture, number, and thickness. The microscopic investigation helped in discriminating paints on the basis of their topcoat as well as the structure of layers they had (Wright et al., 2011).

4.6 Questionable documents

Questionable documents are analyzed in cases of counterfeiting or forgery. Chromatographic analysis usually helps in the chemical profiling of inks while microscopic analysis helps in the examinations of line crossings.

In this regard, Saini et al. used stereomicroscopy for the analysis of line crossings through gel pens and strokes of a laser printer. The magnification used for capturing the images was in the range of 2.5 × to 40 × while the illumination angles for the light source were 15 and 45 degrees. Features such as the sheen, gap, and skipping of the ink as well as specular reflection were visualized. Fig. 9 shows the sheen discontinuity, gel ink skipping and gaps, and specular reflection in a writing sample (Saini et al., 2009).

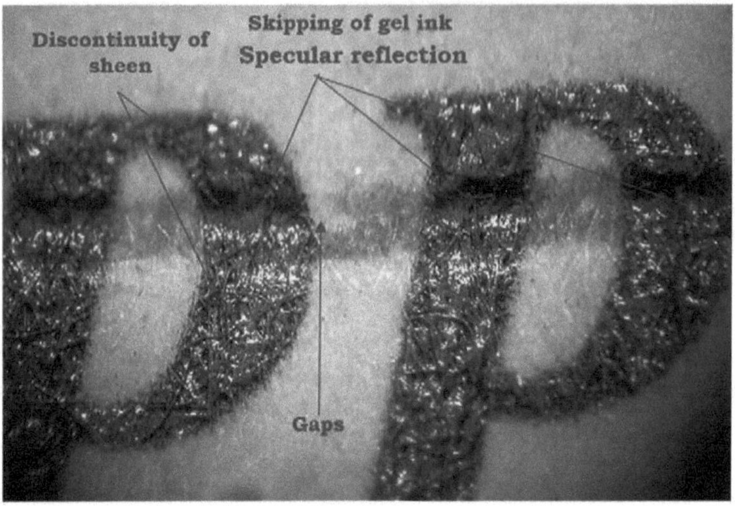

Fig. 9 Characteristics of writing through green gel pen ink observed via stereomicroscope at a 15-degree illumination angle (Saini et al., 2009).

4.7 Soil and minerals

The analysis of soil and mineral particles collected from the body, clothing, or shoes of a suspect or victim can be directly correlated to the crime scene or any other location related to the crime. The microscopic analysis of these particles is done in forensic laboratories using different optical microscopes.

In a study related to environmental forensics, a soil sample was analyzed with PLM to check the morphology of mineral crystals in the soil and also to detect the presence of perchlorate, a kind of xenobiotic substance, in those samples. The analysis was done at 200 ×. The study reported the presence of perchlorate in soil using PLM along with other characterization techniques. Fig. 10 shows the soil sample with perchlorate analyzed by PLM (Duncan et al., 2005). Bull et al. used a light microscope (binocular) to examine the sediment and soil particles collected from the cast footprint. Minerals such as calcite, feldspar, mica, iron, shale, quartz, and limestone along with fibers and hair were reported through microscopic analysis to be present in the soil samples (Bull et al., 2006). In a study related to the forensic characterization of soil from Portugal, light microscopy was used for the analysis of pollen grains present in the soil samples by capturing images at 400 × magnification (Guedes et al., 2011). A study from forensic geology showed the ability of PLM in the mineral analysis. The analysis helped in creating data about the mineral properties such as color, Becke line movement,

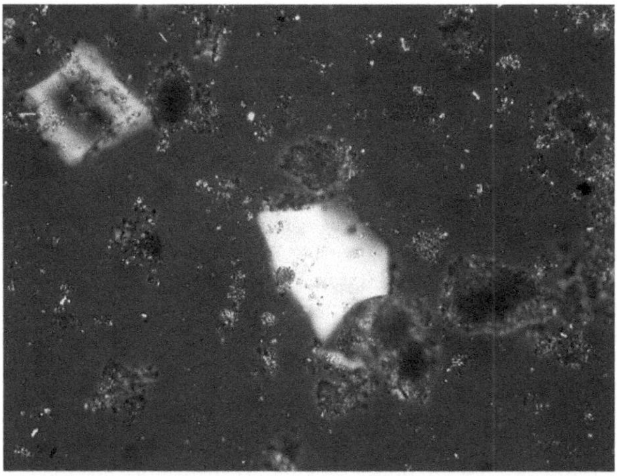

Fig. 10 PLM analysis of a soil sample with the presence of perchlorate (Duncan et al., 2005).

pleochroism, relief, shape, and texture (Shieh and Chen, 2013). In Giampaoli et al.'s study, which was related to the NGS (next-generation profiling) profiling of soil for forensic purposes, a stereomicroscope in the range of 10 × to 50 × along with PLM were used for the geological analysis of soil. The analysis revealed the presence of different minerals and fragments of rocks in the soil samples (Giampaoli et al., 2014).

5 Conclusion

Optical microscopes are among the most widely used tools for the preliminary and confirmatory analysis of different forensic evidence. There are two types of optical microscopes, namely the simple and the compound microscope. The magnifying glass is a simple microscope while the compound microscope consists of two or more lenses. The two most commonly used lenses are the objective and the condenser lens, which help in controlling the light paths from the source to the specimen and then to the user. Compound microscopes such as bright-field, phase-contrast, fluorescence, confocal, polarizing, and stereomicroscopes are commonly used in the analysis of various forensic evidence such as biological samples, fibers, explosives, glass, paint, questionable documents, soil, and minerals. These microscopes, especially polarizing and stereomicroscopes, are widely used for the analysis of breech marks in bullets as well as soils and minerals from different crime scenes. In this chapter, a brief overview of such microscopes along with their application in the analysis of different forensic evidence has been given.

References

Allen, T.J., 1992. The examination of thin sections of coloured paints by light microscopy. Forensic Sci. Int. 57, 5–16. https://doi.org/10.1016/0379-0738(92)90040-4.

Bull, P.A., Parker, A., Morgan, R.M., 2006. The forensic analysis of soils and sediment taken from the cast of a footprint. Forensic Sci. Int. 162, 6–12. https://doi.org/10.1016/j.forsciint.2006.06.075.

Cerutti, E., Spagnoli, L., Araujo, N., Gibelli, D., Mazzarelli, D., Cattaneo, C., 2016. Analysis of cutmarks on bone: can light microscopy be of any help? Am. J. Forensic Med. Pathol. 37, 248–254. https://doi.org/10.1097/PAF.0000000000000260.

Daéid, N.N., McColl, D., Ballany, J., 2009. The level of random background glass recovered from fleece jackets of individuals who worked in law enforcement or related professions. Forensic Sci. Int. 191, 19–23. https://doi.org/10.1016/j.forsciint.2009.06.003.

Danielsen, L., Thomsen, H.K., Nielsen, O., Aalund, O., Nielsen, K.G., Karlsmark, T., Genefke, I.K., 1978. Electrical and thermal injuries in pig skin—evaluated and compared by light microscopy. Forensic Sci. Int. 12, 211–225. https://doi.org/10.1016/0379-0738(78)90006-3.

De Wael, K., Lepot, L., 2011. Dichroism measurements in forensic fibre examination. Part 3. cotton and viscose fibres. Sci. Justice 51, 173–186. https://doi.org/10.1016/j.scijus.2011.03.001.

De Wael, K., Vanden Driessche, T., 2011a. Dichroism measurements in forensic fibre examination. Part 1. dyed polyester fibres. Sci. Justice 51, 57–67. https://doi.org/10.1016/j.scijus.2010.10.001.

De Wael, K., Vanden Driessche, T., 2011b. Dichroism measurements in forensic fibre examination. Part 2. Dyed polyamide, wool and silk fibres. Sci. Justice 51, 163–172. https://doi.org/10.1016/j.scijus.2010.10.005.

De Wael, K., Lepot, L., Lunstroot, K., 2012. The use of linear dichroism in forensic fibre examinations. Part 6. Validation and practical aspects of MSP-PPL. Sci. Justice 52, 249–258. https://doi.org/10.1016/j.scijus.2011.11.003.

Duncan, P.B., Morrison, R.D., Vavricka, E., 2005. Forensic identification of anthropogenic and naturally occurring sources of perchlorate. Environ. Forensic 6, 205–215. https://doi.org/10.1080/15275920590952883.

Flynn, K., O'Leary, R., Roux, C., Reedy, B.J., 2006. Forensic analysis of bicomponent fibers using infrared chemical imaging. J. Forensic Sci. 51, 586–596. https://doi.org/10.1111/j.1556-4029.2006.00116.x.

Giampaoli, S., Berti, A., Di Maggio, R.M., Pilli, E., Valentini, A., Valeriani, F., Gianfranceschi, G., Barni, F., Ripani, L., Romano Spica, V., 2014. The environmental biological signature: NGS profiling for forensic comparison of soils. Forensic Sci. Int. 240, 41–47. https://doi.org/10.1016/j.forsciint.2014.02.028.

Grieve, M.C., Dunlop, J., Kotowski, T.M., 1988. Bicomponent acrylic fibers—their characterization in the forensic science laboratory. J. Forensic Sci. Soc. 28, 25–33. https://doi.org/10.1016/S0015-7368(88)72800-5.

Guedes, A., Ribeiro, H., Valentim, B., Rodrigues, A., Sant'Ovaia, H., Abreu, I., Noronha, F., 2011. Characterization of soils from the Algarve region (Portugal): a multidisciplinary approach for forensic applications. Sci. Justice 51, 77–82. https://doi.org/10.1016/j.scijus.2010.10.006.

Hamer, P.S., 1982. Pigment analysis in the forensic examination of paints. III. A guide to motor vehicle paint examination by transmitted light microscopy. J. Forensic Sci. Soc. 22, 187–192.

Johnson, S.A., 2013. Phase contrast microscopy. In: Biomedical Optical Phase Microscopy and Nanoscopy. Elsevier, pp. 3–18, https://doi.org/10.1016/B978-0-12-415871-9.00001-6.

Kasetty, S., Rammanohar, M., Raju Ragavendra, T., 2010. Dental cementum in age estimation: a polarized light and stereomicroscopic study. J. Forensic Sci. 55, 779–783. https://doi.org/10.1111/j.1556-4029.2010.01363.x.

Kemp, S.E., Carr, D.J., Kieser, J., Niven, B.E., Taylor, M.C., 2009. Forensic evidence in apparel fabrics due to stab events. Forensic Sci. Int. 191, 86–96. https://doi.org/10.1016/j.forsciint.2009.06.013.

Klong-klaew, T., Sukontason, K., Sribanditmongkol, P., Moophayak, K., Sanit, S., Sukontason, K.L., 2012. Observations on morphology of immature *Lucilia porphyrina* (Diptera: Calliphoridae), a fly species of forensic importance. Parasitol. Res. 111, 1965–1975. https://doi.org/10.1007/s00436-012-3043-2.

Lu, P.J., 2005. Confocal scanning optical microscopy and nanotechnology. In: Yao, N., Wang, Z.L. (Eds.), Handbook of Microscopy for Nanotechnology. Springer US, Boston, MA, pp. 3–24, https://doi.org/10.1007/1-4020-8006-9_1.

Maat, G.J.R., Van Den Bos, R.P.M., Aarents, M.(J.)., 2001. Manual preparation of ground sections for the microscopy of natural bone tissue: update and modification of Frost's "rapid manual method". Int. J. Osteoarchaeol. 11, 366–374. https://doi.org/10.1002/oa.578.

Murphy, D.B., 2001. Fundamentals of Light Microscopy and Electronic Imaging. Wiley-Liss, New York.

Nakatome, M., Matoba, R., Ogura, Y., Tun, Z., Iwasa, M., Maeno, Y., Koyama, H., Nakamura, Y., Inoue, H., 2002. Detection of cardiomyocyte apoptosis in forensic autopsy cases. Int. J. Legal Med. 116, 17–21. https://doi.org/10.1007/s004140100249.

Pandey, G., Rawtani, D., Agrawal, Y.K., 2017. Future aspects of halloysite nanotubes in forensic investigations. J. Nanomed. Res. 6.

Petterd, C.I., Hamshere, J., Stewart, S., Brinch, K., Masi, T., Roux, C., 1999. Glass particles in the clothing of members of the public in South-Eastern Australia—a survey. Forensic Sci. Int. 103, 193–198. https://doi.org/10.1016/S0379-0738(99)00085-7.

Rawtani, D., Tharmavaram, M., Pandey, G., Hussain, C.M., 2019. Functionalized nanomaterial for forensic sample analysis. TrAC Trends Anal. Chem. 120, 115661. https://doi.org/10.1016/j.trac.2019.115661.

Roane, T.M., Pepper, I.L., 2015. Microscopic techniques. In: Environmental Microbiology. Elsevier, pp. 177–193, https://doi.org/10.1016/B978-0-12-394626-3.00009-0.

Roux, C., Kirk, R., Benson, S., Haren, T.V., Petterd, C.I., 2001. Glass particles in footwear of members of the public in South-Eastern Australia—a survey. Forensic Sci. Int. 116, 149–156.

Saini, K., Kaur, R., Sood, N.C., 2009. Determining the sequence of intersecting gel pen and laser printed strokes—a comparative study. Sci. Justice 49, 286–291. https://doi.org/10.1016/j.scijus.2009.07.003.

Shieh, C.E., Chen, Y., 2013. The application of polarized light microscopy to identify minerals—a preliminary study of forensic geology. Forensic Sci. J. 12, 15–30.

Sousa, E.M.D., Stott, G.G., Alves, J.B., 1999. Determination of age from Cemental incremental lines for forensic dentistry. Biotech. Histochem. 74, 185–193. https://doi.org/10.3109/10520299909047973.

Sukontason, K., Sukontason, K.L., Piangjai, S., Narongchai, P., Samai, W., Boonchu, N., Sripakdee, D., Ngern-klun, R., Siriwattanarungsee, S., 2005. Morphology of second and third instars of *Chrysomya villeneuvi* Patton (Diptera: Calliphoridae), a fly species of forensic importance. Forensic Sci. Int. 154, 195–199. https://doi.org/10.1016/j.forsciint.2004.10.009.

Turillazzi, E., Monaci, F., Neri, M., Pomara, C., Riezzo, I., Baroni, D., Fineschi, V., 2010. Collection of trace evidence of explosive residues from the skin in a death due to a disguised letter bomb. The synergy between confocal laser scanning microscope and

inductively coupled plasma atomic emission spectrometer analyses. Forensic Sci. Int. 197, e7–e12. https://doi.org/10.1016/j.forsciint.2009.12.012.
Wayne, R., 2019a. Bright-field microscopy. In: Light and Video Microscopy. Elsevier, pp. 95–116, https://doi.org/10.1016/B978-0-12-816501-0.00004-4.
Wayne, R., 2019b. Fluorescence microscopy. In: Light and Video Microscopy. Elsevier, pp. 151–170, https://doi.org/10.1016/B978-0-12-816501-0.00007-X.
Wayne, R., 2019c. Polarization microscopy. In: Light and Video Microscopy. Elsevier, pp. 189–242, https://doi.org/10.1016/B978-0-12-816501-0.00009-3.
Welch, A., Rickard, R., Underhill, M., 1989. The observation of banding in glass fragments and its forensic significance. J. Forensic Sci. Soc. 29, 5–13. https://doi.org/10.1016/S0015-7368(89)73208-4.
Wilson, L.J., Wheeler, B.P., 2009. Optical microscopy in forensic science. In: Meyers, R.A. (Ed.), Encyclopedia of Analytical Chemistry. John Wiley & Sons, Ltd, Chichester, UK, p. a9101, https://doi.org/10.1002/9780470027318.a9101.
Wilson, A.S., Dodson, H.I., Janaway, R.C., Pollard, A.M., Tobin, D.J., 2007. Selective biodegradation in hair shafts derived from archaeological, forensic and experimental contexts. Br. J. Dermatol. 157, 450–457. https://doi.org/10.1111/j.1365-2133.2007.07973.x.
Wiltshire, P.E.J., Hawksworth, D.L., Edwards, K.J., 2015. Light microscopy can reveal the consumption of a mixture of psychotropic plant and fungal material in suspicious death. J. Forensic Legal Med. 34, 73–80. https://doi.org/10.1016/j.jflm.2015.05.010.
Wolf, D.E., 2013. Fundamentals of fluorescence and fluorescence microscopy. In: Methods in Cell Biology. Elsevier, pp. 69–97, https://doi.org/10.1016/B978-0-12-407761-4.00004-X.
Wright, D.M., Bradley, M.J., Mehltretter, A.H., 2011. Analysis and discrimination of architectural paint samples via a population study. Forensic Sci. Int. 209, 86–95. https://doi.org/10.1016/j.forsciint.2011.01.001.
Yao, N., Wang, Z.L. (Eds.), 2005. Handbook of Microscopy for Nanotechnology. Kluwer Academic Publishers, Boston.
Yates, B.C., Espinoza, E.O., Baker, B.W., 2010. Forensic species identification of elephant (Elephantidae) and giraffe (Giraffidae) tail hair using light microscopy. Forensic Sci. Med. Pathol. 6, 165–171. https://doi.org/10.1007/s12024-010-9169-6.

CHAPTER 13

Electron microscopy for forensic samples

1 Introduction

Microscopes have been used over the years to magnify and image objects in much more detail. Light microscopes are most commonly used because of their cost efficiency, robustness, and noninvasiveness. There are also many types, and they can magnify images with high resolution. X-rays, which are another part of the electromagnetic spectrum, have also been employed in X-ray computed tomography to visualize materials with a resolution similar to that of a light microscope. However, this resolution does not exceed 0.1 μm and thus particles smaller than this cannot be visualized clearly. Therefore, electron microscopes have been developed that employ electron beams of shorter wavelengths that have been generated with accelerating voltages. Electron microscopes such as scanning electron microscopes (SEM), transmission electron microscopes (TEM), and scanning transmission electron microscopes (STEM) have been developed that are capable of visualizing an object in the range of pico- to nanometers as well. Such imaging offers new insights into the actual structural composition of the material. Electron microscopes have undergone a revolution in the past few years and are capable of visualizing the materials down to their atomic level. Even the surface quality and local chemistry of the material can be observed through these techniques. The resolution of the electron microscopes is similar to those of scanning probe microscopes such as atomic force microscopes (AFM) and scanning tunneling microscopes (STM) (Inkson, 2016).

In forensic science, evidence such as gunshot residue (GSR), bullets, paints, fibers, questionable documents, and counterfeit products are commonly encountered. This evidence is often characterized by analytical techniques such as spectroscopic techniques that are capable of determining the chemical composition and structural morphology of the material. The information obtained from these techniques can be paired with the data obtained from electron microscopes to ensure a thorough and positive identification

Handbook of Analytical Techniques for Forensic Samples
https://doi.org/10.1016/B978-0-12-822300-0.00013-6

of the materials. Interestingly, electron microscopes can also be combined with techniques such as energy dispersive X-ray spectroscopy (EDX) to analyze the elemental composition of the material. This technique is of importance in the identification of GSR where the composition of the primers used needs to be identified.

In this chapter, the principle, theory, working, and instrumentation of electron microscopes will be discussed. Also, emphasis will be given on sample preparation and the combination of electron microscopes with other techniques. The applications of electron microscopes in the analysis of various forensic evidence will also be discussed.

2 Principle and theory of electron microscopes

Electron microscopes use electron beams of a shorter wavelength that are developed with the help of accelerating voltages of 1–300 kV. Electrons behave in accordance with the wave particle duality in which a stream of electrons behaves like a wave, but upon impact with a material, each individual electron interacts with it. Due to this property of wave particle duality, the wavelength of the electrons depends on their momentum. Therefore, by using accelerating voltages, it is possible to generate electrons with very high energy and thereby shorter wavelengths.

The imaging in an electron microscope is done through the interaction of the material with two types of electrons, primary electrons that generate secondary electrons. Primary electrons are those that are targeted at the material while secondary electrons are those that escape from the material once the primary electron has hit the sample. The trajectories of these electrons are very much dependent upon the positioning of atoms and the types of elements in the material. Backscattered electrons are those electrons that reflect from the specimen and auger electrons are low-energy electrons. The scattered electrons are detected to develop a high-resolution image (Fig. 1).

The scattering of electrons mainly occurs in two different ways, elastic and inelastic scattering. In elastic scattering, the energy is conserved when the electron hits material while in inelastic scattering, there is a loss in energy upon impact. Uniform elastic scattering allows the imaging of crystalline structures with greater resolution as it is able to diffract better. The information obtained from inelastic scattering helps in getting an idea about the thickness, volume, and transparency of the material (Amelinckx et al., 1997; Kuo, 2014; Spence, 2009).

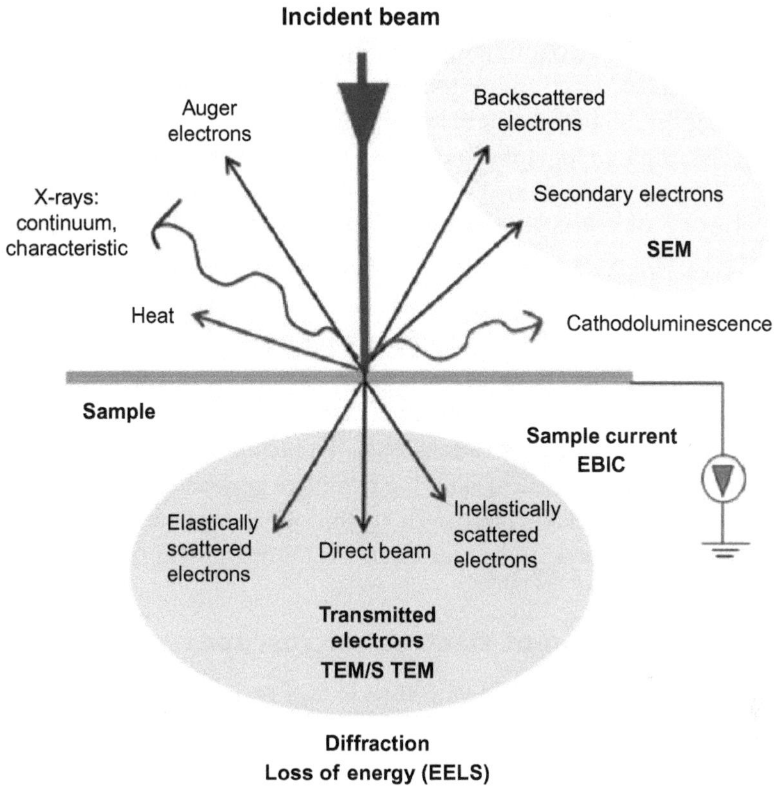

Fig. 1 Scattering of electrons upon impingement of electrons on a sample (Kwiecińska et al., 2019).

2.1 Theory of scanning electron microscopy

SEM is used to visualize the surface morphology of a material. It gives details such as topography, composition, and crystallographic information. However, it does not give any information about the underlying image. Quite often, cross-sections of a material are cut and visualized in order to get further insight into the structure and composition of the material. In SEM, electron beams are used that are emitted from a filament to which an accelerating voltage is applied. The sample that is conducting in nature is impinged by the electron beam. Any nonconducting materials are often coated with a coating material prior to analysis with SEM. In this technique, the primary electrons are released under vacuum through an electron beam with an accelerating voltage of 1–40 KeV. The electron beam is highly focused and monochromatic. It is sent through a series of magnetic lenses that keep the beam highly focused in order to increase the resolution. Here, the routinely used electrons are the backscattered and secondary electrons. The backscattered electrons give an idea about the multiple phases existing in the composition of the material while the secondary electrons give an idea about the sample topography and form (Joy, 2019; Sharma, 2018).

2.2 Theory of transmission electron microscopy

TEM is preferred when the internal microstructures of any material need to be visualized. It is highly useful for materials such as fibers, thin films, or even atoms. The principle of TEM is similar to SEM in that it uses electron beams that are highly focused. It also consists of a series of magnetic lens to focus the beams. However, the electrons detected in this case are the primary electrons that are able to pass through the material. The material therefore should be extremely thin and capable of allowing complete penetration of the electron beam. Scattering also takes place in this technique to produce elastically scattered electrons, X-rays, and auger electrons (Williams and Carter, 1996).

3 Instrumentation of electron microscopes

The instrumentation of TEM and SEM is similar in many ways; however, there are certain differences that need to be considered. For instance, both microscopes need to have an electron gun that is capable of accelerating electrons and generating electron beams. Also, the entire instrumental setup is done in a vacuum in order to prevent the loss of energy of the electrons upon collision with any gas molecules. Both instruments consist of a sample holder and detector to place the sample and detect the scattered and transmitted electrons.

The key differences between the instrumentation of SEM and TEM are in sample preparation, operating voltage, the number of apertures and condensers, and the positioning of the sample holder. In SEM, the sample holder is placed at the end of the instrument while in TEM, the sample holder is placed in the center of the instrument. Fig. 2 shows an overview of the SEM and TEM instrumentation.

Conventionally, the electron guns used employed thermal energy to emit electrons from a cathodic filament. The movement of electrons is assisted by the presence of an accelerating voltage. However, there were certain drawbacks such as the development of a wide area of the electron beam, due to which the narrowing and focusing are difficult. Later, field emission (FE) guns were developed that used tunneling or the Schottky effect to emit

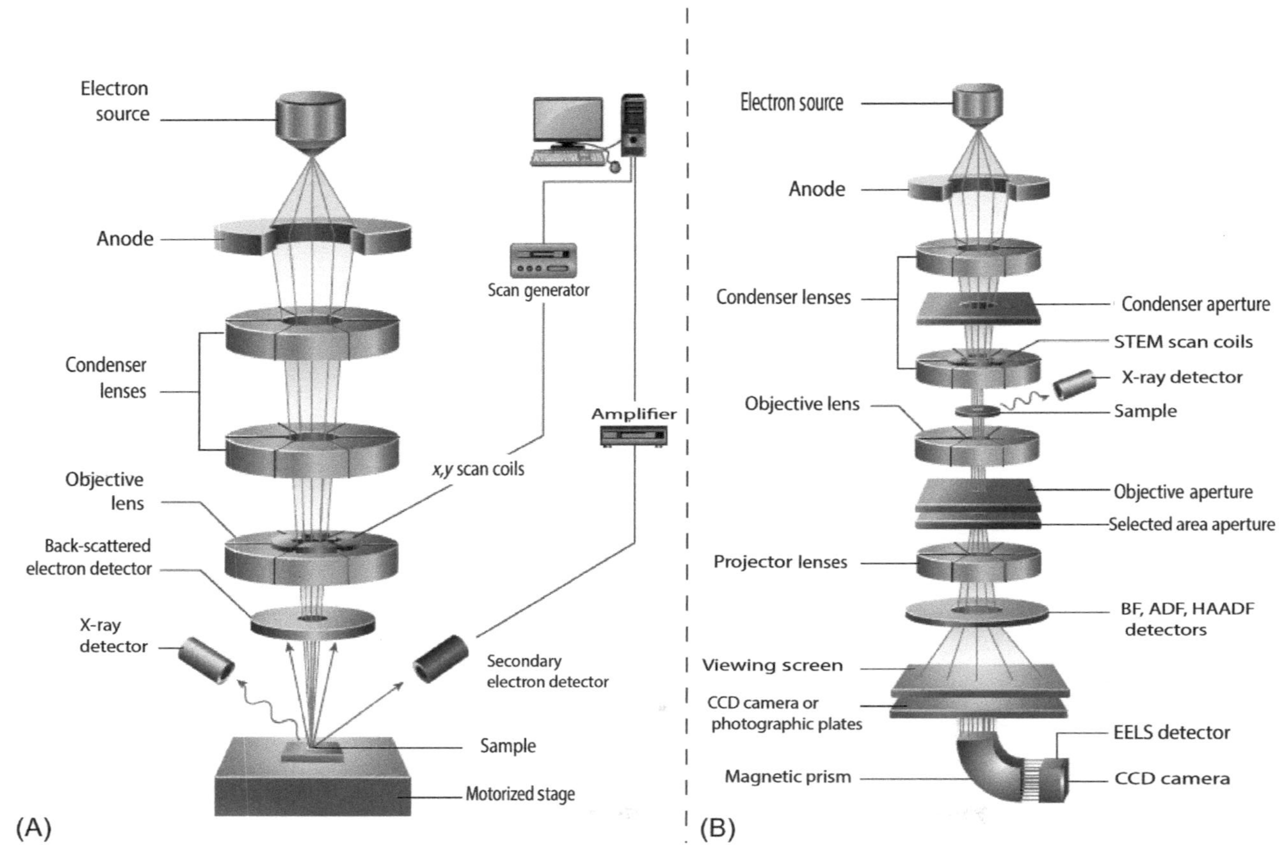

Fig. 2 Instrumentation of SEM and TEM (Inkson, 2016).

electrons. These are a remarkable development from the conventional cathode guns as they are capable of creating electron beams of 10 nm spread. However, these electron guns need to be operated in ultrahigh vacuum. SEM is usually operated at 0.1–10^{-4} Pa while TEM is operated at 10^{-7} to 10^{-9} Pa.

In SEM, the electron beam is controlled by a series of electromagnetic lenses and condensers that are designed to focus the beam and control the movement of the beam over the sample while scanning it. The lenses are also designed to remove or adjust any beam astigmatism. TEM uses electromagnetic lenses and condensers to control the beam, but it has more such lenses than SEM. Condenser, objective, and projector lenses are used sequentially before and after the sample holder to achieve the highest resolution. Also, apertures are built in to reduce any chromatic aberrations.

Another crucial aspect of electron microscopes are the detectors that are capable of capturing the scattered and transmitted electrons. In SEM, there are four types of detectors: scintillation detectors, semiconductor detectors, cathodoluminescence detectors, and microchannel plate detectors. In the scintillator detector, a scintillation material such as yttrium-aluminum-garnet (YAG) or yttriumaluminum-perovskite (YAP) converts the scattered electron signals to photons that are further amplified by a photomultiplier tube (PMT). Semiconductor detectors or solid-state detectors involve the generation of electron-hole pairs to generate current that is proportional to the number of electrons reaching the detectors. This type of detector is highly useful for detecting backscattered electrons. Microchannel plate detectors deal with the use of small PMTs present in a large number; they are kept between the sample holder and the objective lens. This detector is capable of working at both high and low accelerating voltages, and is also capable of efficiently detecting the scattered electrons. Quite often, when the high energy incident electrons hit the material, some amount of luminescence and visible light is also produced. The cathodoluminescence works by collecting the luminescence signal. The detector is kept at a very specific angle in order to collect the maximum amount of luminescence (Hawkes and Spence, 2019).

3.1 Sample preparation for electron microscope

Electron microscopes, while offering high resolution and information about the morphology and composition of the material, are still very complex and expensive tools. Over such complex instrumentation, it is therefore important to perform proper sample preparation procedures in order to achieve images of utmost clarity and detail. SEM and TEM both have very varied sample preparation procedures, but the one thing both techniques have in common is that the sample must be prepared under very clean and dry conditions, as any dirt may cause the presence of artifacts in the image. In SEM, it is necessary that the sample is conducting in nature. Therefore, in order to image nonconducting samples, especially biological samples, they are coated with conducting materials such as carbon, gold, or platinum. The coating is made ultrathin so that the sample is made conducting in nature while not bringing any change to the actual morphology of the sample. Also, prior to coating, the samples are usually fixed through slicing, acid etching, or abrasions. TEM, meanwhile, involves creating ultrathin slices of the sample such that the samples are at a maximum of 200 μm thick.

3.2 Modes of SEM and TEM

Various factors affect the imaging in SEM and TEM. In SEM, the factors affecting the imaging are the electron probe size, current, convergence angle, and accelerating voltage. Based on the variation of these factors, different modes of imaging by SEM have been developed. The electron probe size is the diameter of the beam spot when it hits the sample while the electron current is the one that, upon hitting the sample, generates the different signals required for imaging. The convergence angle is the angle of the cone that is developed when the electron beam hits the sample. In TEM, the geometry of the electron beam involved in the illumination of the sample, the path of the electron passing through the lenses attached below the specimen, the apertures, and the detector used are the main factors affecting imaging.

The four main modes of imaging in SEM are the resolution mode, the low-voltage mode, the high-current mode, and the depth-of-focus mode. Goldstein et al. (2003) very well summarized the different modes of imaging in SEM. In the resolution mode, the probe diameter is kept extremely small. Because the diameter of the probe is extremely small, the beam itself should contain enough current so that when it hits the sample, it can emit enough signals. This mode allows the imaging of the sample with greater detail and is highly useful to observe samples such as treated hair, where small striations are visible on it. The different imaging modes are often used in conjunction with each other. A smaller electron probe without adequate current generates a lot of noise and poor visibility of the image. Therefore, the high-current mode and the high-resolution mode are often used together. In the depth-of-focus mode, the convergence angle is manipulated such that the probe diameter is least affected. A small convergence angle allows greater focus while keeping the probe diameter small, thus allowing the focus to be on minute details of the sample. Manipulating the accelerating voltages controls the movement of the electron beam and its impingement on the sample. Electron beams generated with low voltages stay close to the surface of the sample and therefore give extremely high detail of the surface morphology. High-voltage electron beams have the ability to travel through the sample. While this is a great feature for TEM, it is not suitable in case of SEM as the sample prepared is thick and will not generate a clear image of the sample (Goldstein et al., 2003).

In TEM, bright-field imaging, electron diffraction, STEM, high-resolution TEM (HRTEM), and high-angle annular dark field (HAADF) are the most widely used. In bright-field imaging, the scattered electrons are blocked by an objective aperture that allows the transmittance of only unscattered electrons that form a bright field image. This mode is highly useful to analyze the thickness of the grain boundaries or any dislocations in the sample. Such features of the sample generate a darker contrast in the image as they do not allow the passing of electrons through them. Just like bright-field imaging, dark-field imaging can also be done in which the scattered electrons are detected. In electron diffraction mode, elastically scattered electrons are detected that can cause constructive and destructive interference. This phenomenon causes the impingement of the electrons on the sample at specific angles, thereby causing diffraction. The angles at which the electrons impinge are characteristic of the crystal structure and orientation of the sample. Therefore, they give an idea about the crystallinity of the sample. This mode can be used along with bright-field imaging and EDX. HRTEM, as the name suggests, is used when extremely high detailed imaging is required. In this case, it is necessary that the samples are ultrathin due to which the elastically scattered electrons are easily detected. STEM is an amalgamation of both SEM and TEM. In this technique, the focused electron beam is raster scanned on the sample. Unlike SEM and TEM, the electron in this case is sent from the bottom of the sample and is set upward. This technique also consists of a series of magnetic lens and condensers that focus the electron beam on the sample, and the elastically scattered electrons are detected (Peter, 2019). Also, STEM is used when the chemical composition of the entire sample has to be analyzed sequentially. HAADF collects the electrons that have been incoherently scattered at high angles. This mode is used when images with high contrast need to be generated and also when the chemistry at an atomic scale has to be analyzed (Inkson, 2016).

4 Electron microscopy-based forensic investigations

Electron microscopy has aided forensic investigations for many years. This technological advancement has helped forensic scientists to closely look at the morphological details of much evidence. Among different types of electron microscopic techniques, SEM has garnered the attention of most forensic scientists and is widely used for the morphological examination of forensic samples (Basu, 2006). The different evidence collected from the crime scene that needs to be analyzed through electron microscopy can be classified into three types: biological, anthropogenic, and geological (Palenik et al., 2018). This section discusses the use of electron microscopic techniques for the different aforementioned types of forensic samples.

4.1 Biological evidence

Crime investigations often encounter different kinds of biological evidence. Various types of biological evidence that can be found at a crime scene for microscopic analysis include hair, diatoms, wood, seeds, pollen, fungi, dental remains, skin, bones, etc. (Palenik et al., 2018). In this section, the electron microscopic analysis of biological forensic samples such as hair, skin, bones, and diatoms as well as flies and their developmental stages that are of forensic importance are discussed.

The morphological analysis of hair is an important task in forensic investigations. With changes in race and ethnicity, the morphology of hair changes. Drug consumption and cosmetic treatments also affect the structure of hair. The structure also varies between humans and animals. In a research work that focused on the analysis of cocaine incorporated in hair that was damaged by various cosmetic treatments, SEM was used to study the morphological changes in hair due to cosmetic treatments. The cosmetic hair treatments done were chemical (bleaching, dyeing) and thermal (straightening). The images were captured using an SEM microscope with 20 kV of acceleration potential and 60 μA of beam current. It was observed that thermal treatment produced localized damage while chemical treatments damaged the entire surface of the hair (Gerace et al., 2017). Another study also performed a similar kind of work in studying the effect of cosmetic treatments on hair. However, this study compared the effects produced by cosmetic treatments on hair from two different ethnic groups, Afro and Caucasian. The cosmetic treatments applied were thermal, straightening, and bleaching. It was observed that there was no significant morphological difference between the untreated hairs from the two ethnic groups. However, thermal treatment caused the removal or breakage of the cuticle layer (Fig. 3); bleaching caused the development of striations on the hair surface (Fig. 3); and straightening caused minor undulations and was responsible for cuticle detachment (dos Santos et al., 2019). Apart from hair, bone structure has also been examined by forensic scientists. In a case report published by Melki et al., SEM was used to investigate exhumed bones. The study

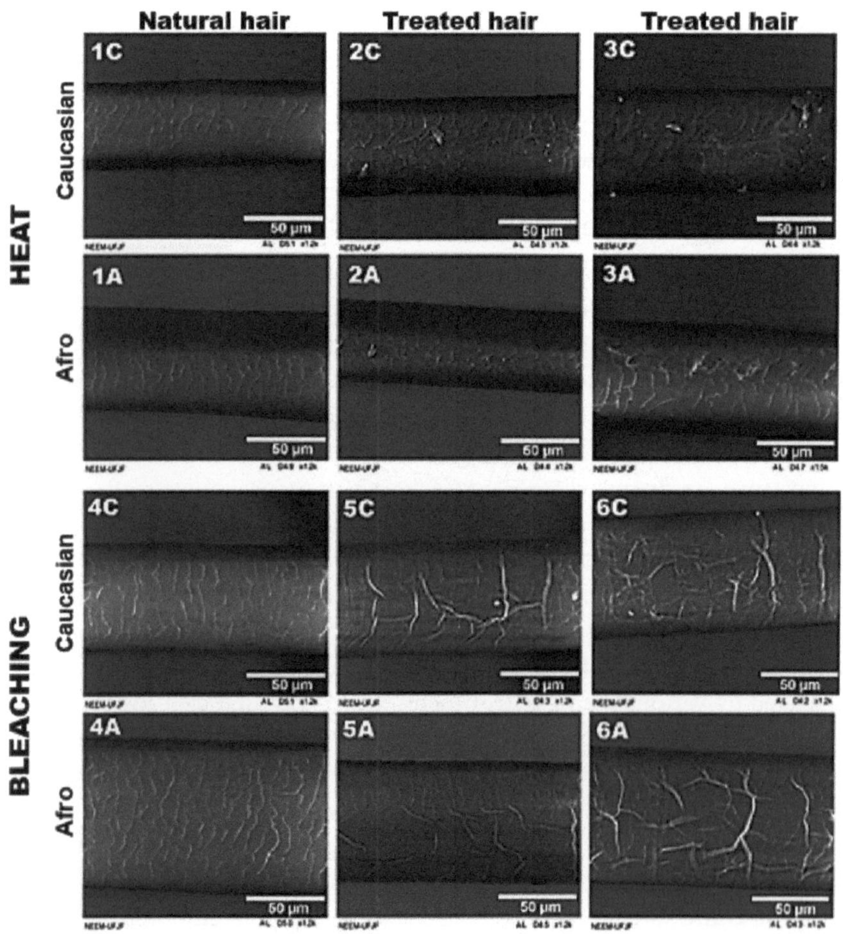

Fig. 3 SEM images of Caucasian and Afro hair after thermal and bleaching treatment (dos Santos et al., 2019).

focused on examining a fracture in the victim's skull to correlate it with a seized weapon that was probably used in the homicide. The first autopsy report discussed the use of a stabbing weapon for the homicide. The SEM results revealed the presence of red blood cells and fibrin in the bone tissue, which are not present in normal cases. It was concluded on the basis of the SEM images that the seized weapon was probably used for the homicide (Melki et al., 2011).

Skin samples are also analyzed to examine any kind of morphological changes in cases of burns, electrocution, etc. In a recent technical note, a preservation protocol for SEM analysis of skin samples of forensic importance has been developed. The skin samples with bone were used to perform SEM analysis just after sample collection and after storing the samples in 4% formalin for 7 days. The SEM images revealed that the preservation method was perfect because the samples had not lost any information, even after storage for a week (Montoriol et al., 2019). Skin samples of electrocuted people were also analyzed through SEM in a research work. The samples were collected from five different cases of electrocution, and all victims were male. The ultrastructural analysis of specimens showed that epidermal cells were elongated, the vacuolization of the horny layer became evident, and elastic fibers got coagulated (Fig. 4). The study suggested the use of SEM

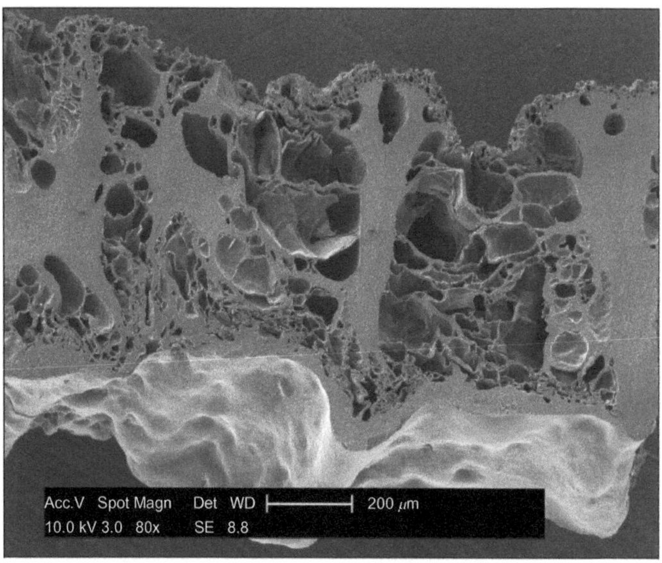

Fig. 4 SEM analysis of the skin of an electrocuted person (site of contact of skin with conductor was the index finger of the right hand) (Visonà et al., 2018).

along with EDS (energy dispersive X-ray spectroscopy) in cases of electrocution to identify metals that get embedded into the skin and to study the morphological changes in skin due to electrocution (Visonà et al., 2018). The detection of diatoms (unicellular algae with a transparent cell wall made up of silica) becomes essential in cases of drowning, where corpses are recovered from water in a mostly decomposed state. In a relevant study, SEM was used to identify different species of diatoms in cadavers found at the bank of a river and from the sea. Microwave digestion, followed by vacuum filtration and automated SEM, was used for this purpose. The identification of diatom species from SEM was compared with those identified through acid digestion (a conventional technique for a diatom test). Researchers were able to identify 25 species of diatoms through SEM, but only five with the conventional diatom test. Hence, SEM proved to be a better option for a diatom test over the conventional one (Zhao et al., 2015).

In addition to these biological samples, forensic entomology also utilizes SEM to study the ultrastructural details of different flies and their various developmental stages such as eggs, larvae, etc. Flies or insects are the pioneer species that start colonizing a dead body. They start depositing their eggs on the cadaver a few hours after death. Murder victims' bodies are often contaminated with eggs of various flies (Erzinclioglu, 1989). The correct examination of eggs and flies becomes important because they can be correlated to the postmortem interval, murder circumstances, movement or disturbance of the body after death, the presence of drugs, etc. (Benecke et al., 2004; Gagliano-Candela and Aventaggiato, 2001). SEM analysis has been widely used for this purpose. In a study, SEM was used to identify fly eggs. The eggs of dipteral species of forensic importance such as *Chrysomya putoria*, *Ophyra aenescens*, *Lucilia eximia*, *Chrysomya megacephala*, and *Lucilia cuprina* were identified through SEM analysis. The egg samples were prepared for SEM analysis by a thin gold coating of around 20–30 nm thick. On the basis of median area, anastomosis, and holes on islands of eggs, they were categorized into different species. These characteristics can be helpful to forensic entomologists for species identification, thereby aiding in forensic investigations (Mendonça et al., 2008). In another study, the ultrastructural details of forensically important flies belonging to the Muscidae family of the Diptera order were investigated. The species taken for the study were *Ophyra albuquerquei*, *Biopyrellia bipuncta*, *Ophyra aenescens*, and *Morellia humeralis*. The study was conducted to increase the database about the structural details of forensically important flies. It was observed that various groups of sensilla covered the katepisternals of *B. bipuncta* and *M. humeralis*. Sensilla were also densely populated on the ocellar region's surface of *O. chalcogaster* and *O. albuquerquei,* but absent on *O. aenescens.* The flagellum of *O. albuquerquei* also had a population of coeloconic sensilla. Fig. 5 reveals the ultrastructural details of *B. bipuncta* (Carriço et al., 2015). In another study, a fly from the Sarcophagidae family of the Diptera order, known as *Ravinia belforti* (commonly called the flesh fly), was ultrastructurally analyzed through SEM. Specimens of the developmental stages (larvae of the L1, L2, and L3 stages and puparium) of the flesh fly were analyzed in the study. The SEM images revealed the structural details of the larval stages, in which it was observed that the larvae had 12 segments with a pattern of muscoid vermiform. The posterior end of the larvae was thick while the anterior end was pointed. The cephalic collar's spines were flat and possessed double, triple, or quadruple points. The structural details of puparium were similar to those of the other species of this family (da Silva-Xavier and de Carvalho Queiroz, 2016). Another research work studied the antenna of the blow fly (*Hemipyrellia ligurriens*), which belongs to the Calliphoridae family of the Diptera order, through SEM. Both male and female blow flies were

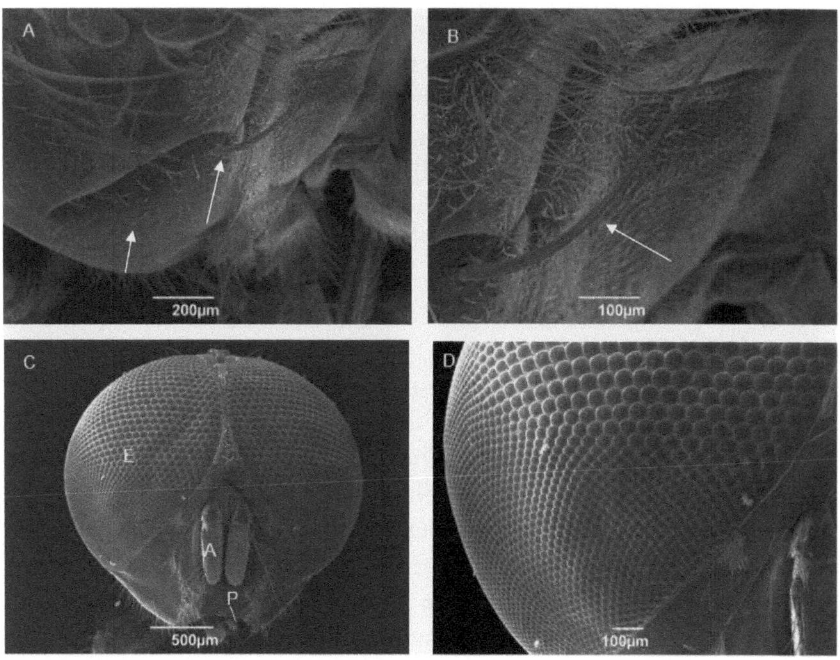

Fig. 5 SEM images of *Biopyrellia bipuncta*: (A) adult female's katepisternals; (B) Sensilla on katepisternal; (C) adult male's head with compound antenna (A), eye (E), and palp (P); and (D) compound eye's ommatidia (Carriço et al., 2015).

investigated for the ultrastructural details. The structural details of the antenna included the study of the flagella and sensilla because they play an important role in locating colonies and oviposition sites of these flies in cadavers. Through SEM images, it was observed that the width of flagellum was higher in females, but the length was almost the same in both sexes. In the length and width of coeloconic sensilla, females scored higher measurements than males (Hore et al., 2017). A similar group of researchers used SEM to examine the sensory organs of both sexes of *Ophyra capensis* belonging to the Muscidae family of the Diptera order. SEM images showed the existence of sexual dimorphism in the sensory organs of both sexes. The flagellum, ocellar region, arista, chaotic sensilla, and pedicel of females were larger in size compared to those of male flies. It was therefore concluded that larger sensory organs observed through SEM help with better vision and olfaction in female flies, thereby aiding in identifying sites for oviposition on cadavers (Hore et al., 2018).

4.2 Anthropogenic evidence

Anthropogenic evidence commonly encountered at a crime scene includes firearms and GSR; polymers such as paints, plastic, and fibers; metals and alloys; tool marks; pharmaceutical samples such as supplements and drugs; paper; glass; industrial samples such as chemicals, explosives, dyes, and pigments; and food (Palenik et al., 2018). In this section, the use of SEM for the ultrastructural analysis of anthropogenic forensic samples such as tool marks, line crossings in questionable documents, firearms, GSR, and fibers is discussed.

In a study published by Sehgal et al. in 1988, the potential of SEM for examining the ultrastructural details of tool marks was demonstrated. The cuts for the experiment were made using the complete cutting edge of the tool. SEM images showed the presence of focused and sharp striations across the sections cut for the experiment. The ultrafine structural details present between the striations were also visible in SEM. It was concluded that the SEM analysis can even give details about the tool marks produced from superfine tools (Sehgal et al., 1988). Questionable documents have also been analyzed through SEM in order to attest to the authenticity of the document. Line crossing examination is a common thing to do while examining the questionable documents. Brito et al. compiled an article in which the applications of SEM to investigate line crossings in questionable documents were described (e Brito et al., 2017). In a study by Kim et al., a focused ion beam was combined with SEM for the analysis of line crossings and to determine their sequence. Three kinds of samples were prepared for the analysis: (a) gel pen ink layered below red sealing ink; (b) red sealing ink layered below gel pen ink; and (c) red sealing ink layered below red sealing ink. Through SEM images, it was observed that the layers of the first ink got disturbed upon applying the second ink, and the particles of the second ink were found deposited over the paste of the first one. The study can be helpful in the examination of documents with forged signatures and overwriting (Fig. 6) (Kim et al., 2016).

During the analysis of firearm cartridges before and after shooting, the exact reproduction of the case head is difficult. In this regard, Valle et al. used polydimethylsiloxane (PDMS) to make a replica of the case head of the spent cartridge and identified the unique characteristics of the firearm through SEM. AFM and optical microscopy were also used for the comparison. A 20 keV SEM was used for the study and the samples were coated with a 10 nm thick layer of gold to reduce any electrostatic effects. The SEM analysis provided a 3D view of the characteristics of the cartridge through its replica, which in the 2D view in optical microscopy looked flat. Fig. 7 shows various marks present on the case head of the cartridge captured through SEM (Valle et al., 2012). GSR has also been analyzed thorough SEM. In a study, the distribution of GSR from shots fired using a silenced gun at close range was examined thorough SEM. Ammunition of 7.65 mm caliber was fired from the gun. The study suggested that there was a difference in the GSR pattern when a silencer was used with the gun (Brożek-Mucha, 2017). In another study, luminescent GSR that was used as a forensic marker was also investigated through SEM. The GSR was collected with tape from the hands of the shooter more than 50 times. On average, 393 luminescent GSRs were analyzed through SEM (Arouca et al., 2017). Fiber analysis is also an integral part of a crime investigation in order to relate an individual with the crime scene. In a study, SEM was used to study the morphological details of the PET (polyethylene terephthalate) fibers. The sample was sputtered with 10 nm thick gold/palladium and the analysis was carried out with an accelerating voltage of 30 kV. The SEM images of 14 PET fibers are shown in Fig. 8 (Farah et al., 2014).

4.3 Geological evidence

Geological evidence commonly encountered at any crime scene includes soil, sediments, minerals, and construction materials. The analysis of these samples can be directly correlated to the location or site where any incident related to the crime took place. This section discusses the utilization of SEM for the ultrastructural analysis of the aforementioned geological samples.

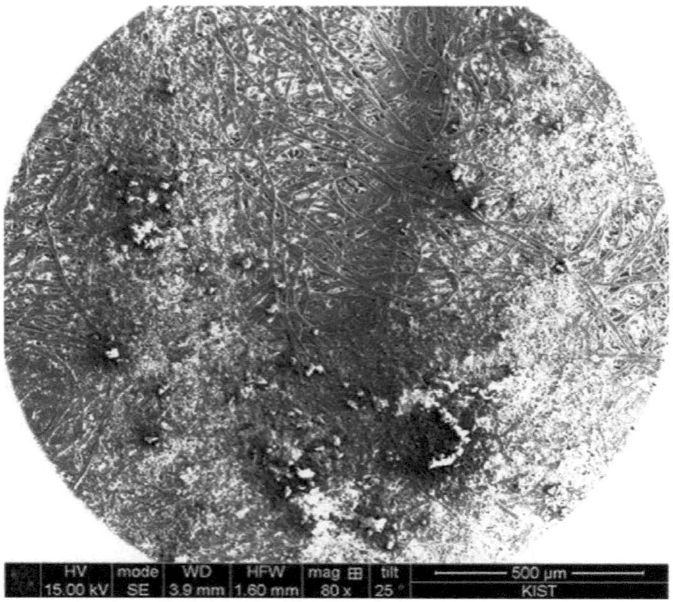

(A)

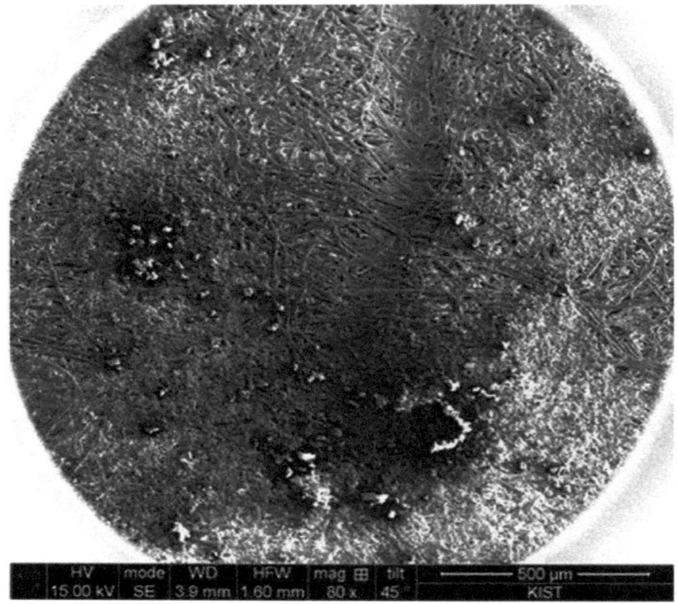

(B)

Fig. 6 SEM image of sample having red sealing layered under gel pen ink at 25 degrees (A) and 45 degrees (B) tilt (Kim et al., 2016).

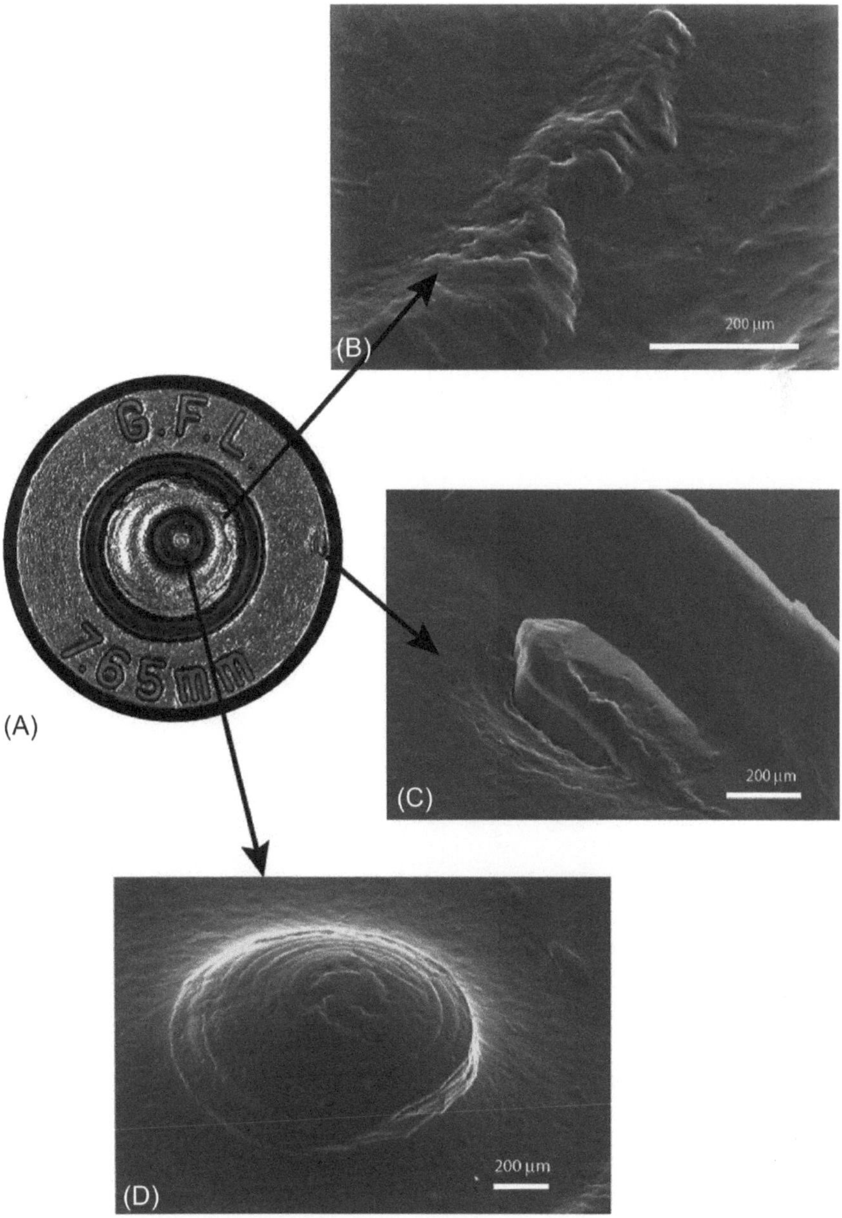

Fig. 7 SEM image of marks present on the case head of a cartridge that were visible on the PDMS replica (A) case head of cartridge; (B) breech face mark; (C) ejector mark; and (D) firing pin (Valle et al., 2012).

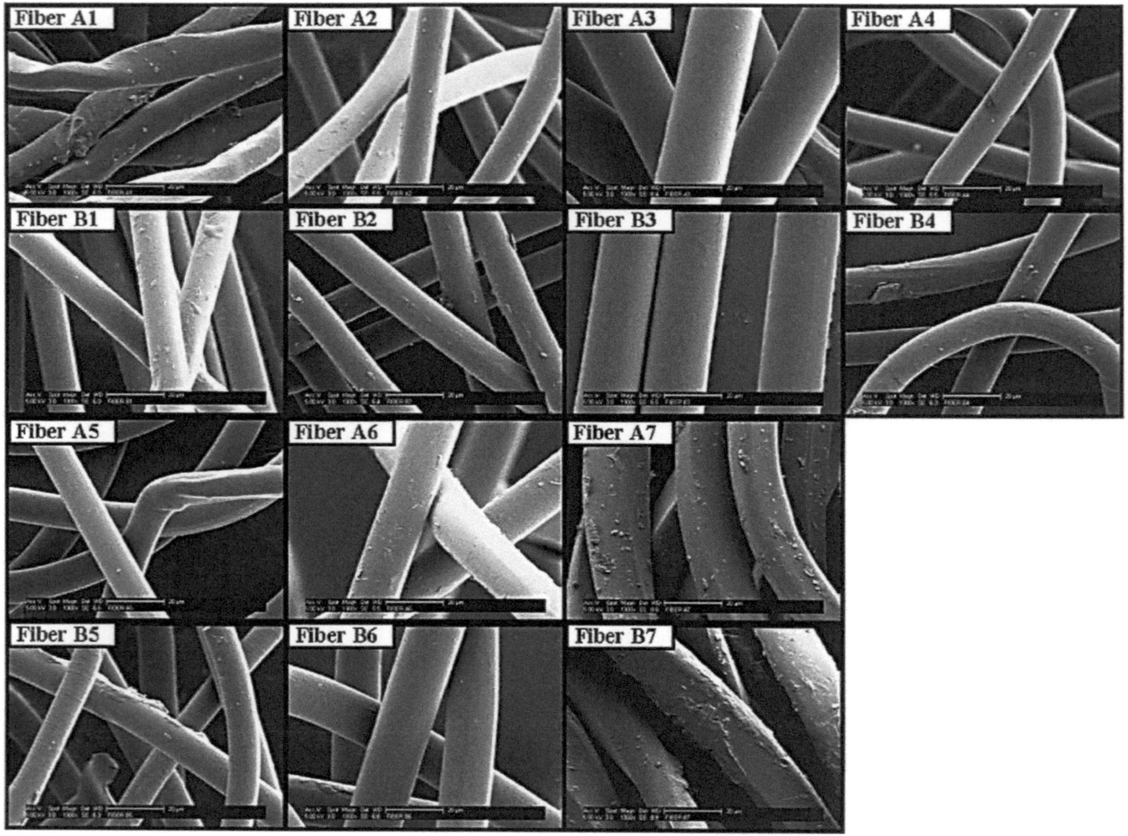

Fig. 8 SEM images of PET fibers (Farah et al., 2014).

A number of studies have been done regarding SEM-mediated forensic soil analysis. In a study from 2004, SEM along with EDS was used to analyze and discriminate the forensic soil. The soil samples were collected from various places in Istanbul. The aim of the study was to check the effect of pressure on the elemental composition and homogenization of soil samples. Through the SEM images, it was evident that the application of pressure breaks down the hard bits into a softer matrix, which further caused the soil samples to become homogenized (Cengiz et al., 2004). In a case study, soil and sediment samples collected from a footprint were examined through SEM. The samples were taken from an area in the English Midlands where a murder took place. In the SEM analysis, water lily pollens were found, which indicated that the perpetrator was standing in some nearby waterbody (Bull et al., 2006). A similar kind of soil and sediment analysis was carried out on experimental samples by Pye and Croft using SEM in combination with energy dispersive X-ray analysis (Pye and Croft, 2007). Woods et al. used SEM to discriminate between the forensic soil samples collected from Australia. The data obtained from SEM/EDS were efficient in discriminating soils from Canberra and other Australian locations (Woods et al., 2014). Another study used SEM/EDS to check the efficiency of soil recovery from clothes. Three different methods were used for recovering the soil from clothes: direct tag lifting, washing, and dry brushing. The major goal of the study was to study the mineralogy of the soil through EDS attached to SEM. The SEM images revealed that all the soil samples had textural similarities that were independent of the method of sample collection (Pirrie, 2018).

In addition to soil, construction materials (man-made) have also been analyzed through SEM. The analysis of such particles becomes important in investigations related to nuclear security. The SEM analysis along with EDS can aid in distinguishing man-made geological samples from soil minerals because each man-made construction material, be it cement, concrete, brick, render, construction block, or plaster, has a different texture and mineralogy (Pirrie et al., 2019). In a study related to an investigation in nuclear forensics, metastudtite (a mineral of uranium) and uranyl peroxide were analyzed through SEM at an accelerating voltage of 12 kV to study the effect of calcination on their decomposition. At different temperatures of calcination, a difference in circularity and particle area was observed (Schwerdt et al., 2018). In a recent study, coal particles (bituminous and anthracite) collected from an urban brownfield and an earlier rail yard were analyzed through SEM. The images obtained from SEM revealed the encrustations on the surface of coal particles that were resisting the sonication (Fig. 9) (Hagmann et al., 2020).

Fig. 9 SEM image of coal particles (Hagmann et al., 2020).

5 Conclusion

Electron microscopes are revolutionary instruments capable of imaging samples right down to their atomic positions. SEM and TEM are the most widely used electron microscopes and they are often used with different imaging modes. While there are several advantages to these techniques, there are certain limitations as well. For instance, the microscopes require the presence of an ultrahigh vacuum to prevent the collision of the electrons with any gas molecules, thereby making them lose their energy. This requirement makes the instruments very expensive and only specialized labs are able to acquire them. Also, any lack in sample preparation may cause the samples to be damaged by the high-intensity electron beams hitting them.

In forensic science, electron microscopes have been used for the analysis of various biological, anthropogenic, and geological samples. However, SEM has seen more popularity among forensic scientists, as evident in the third section of this chapter, mainly because of their requirement for surface analysis. However, TEM can also assist in crime investigations in cases where an effect on internal structures has to be analyzed, for example, in case of biological samples such as tissues, cells, etc. Over the years, several modifications have been made in the instrumentation of electron microscopes. With constant improvements, it is possible to make this tool a ubiquitous element in forensic labs across the world.

References

Amelinckx, S., Van Dyck, D., van Landuyt, J., van Tendeloo, G., 1997. Electron Microscopy: Principles and Fundamentals. Weinheim (Federal Republic of Germany), Germany, VCH Verlagsgescllschaft mbH.

Arouca, A.M., Lucena, M.A.M., Rossiter, R.J., Talhavini, M., Weber, I.T., 2017. Use of luminescent gunshot residues markers in forensic context. Part II. Forensic Sci. Int. 281, 161–170. https://doi.org/10.1016/j.forsciint.2017.09.022.

Basu, S., 2006. Scanning electron microscopy in forensic science. In: Encyclopedia of Analytical Chemistry. American Cancer Society, https://doi.org/10.1002/9780470027318.a1122.

Benecke, M., Josephi, E., Zweihoff, R., 2004. Neglect of the elderly: forensic entomology cases and considerations. Forensic Sci. Int. 146 (Suppl), S195–S199. https://doi.org/10.1016/j.forsciint.2004.09.061.

Brito, L.R.E., Martins, A.R., Braz, A., Chaves, A.B., Braga, J.W., Pimentel, M.F., 2017. Critical review and trends in forensic investigations of crossing ink lines. TrAC Trends Anal. Chem. 94, 54–69. https://doi.org/10.1016/j.trac.2017.07.005.

Brożek-Mucha, Z., 2017. A study of gunshot residue distribution for close-range shots with a silenced gun using optical and scanning electron microscopy, X-ray microanalysis and infrared spectroscopy. Sci. Justice 57, 87–94. https://doi.org/10.1016/j.scijus.2016.11.004.

Bull, P.A., Parker, A., Morgan, R.M., 2006. The forensic analysis of soils and sediment taken from the cast of a footprint. Forensic Sci. Int. 162, 6–12. https://doi.org/10.1016/j.forsciint.2006.06.075.

Carriço, C., Mendonça, P.M., Cortinhas, L.B., dos Santos-Mallet, J.R., Queiroz, M.M.D.C., 2015. Ultrastructural studies of some character of Diptera (Muscidae) of forensically importance. Acta Trop. 142, 96–102. https://doi.org/10.1016/j.actatropica.2014.11.005.

Cengiz, S., Cengiz Karaca, A., Çakır, İ., Bülent Üner, H., Sevindik, A., 2004. SEM-EDS analysis and discrimination of forensic soil. Forensic Sci. Int. 141, 33–37. https://doi.org/10.1016/j.forsciint.2003.12.006.

da Silva-Xavier, A., de Carvalho Queiroz, M.M., 2016. Ultrastructure analysis of the immature stages of *Ravinia belforti* (Diptera: Sarcophagidae), a species of medical-veterinary and forensic importance, by scanning electron microscopy. Acta Trop. 159, 192–199. https://doi.org/10.1016/j.actatropica.2016.03.039.

dos Santos, J.D., Edwards, H.G.M., de Oliveira, L.F.C., 2019. Raman spectroscopy and electronic microscopy structural studies of Caucasian and Afro human hair. Heliyon 5. https://doi.org/10.1016/j.heliyon.2019.e01582.

Erzinclioglu, Y.Z., 1989. The value of chorionic structure and size in the diagnosis of blowfly eggs. Med. Vet. Entomol. 3, 281–285. https://doi.org/10.1111/j.1365-2915.1989.tb00229.x.

Farah, S., Tsach, T., Bentolila, A., Domb, A.J., 2014. Morphological, spectral and chromatography analysis and forensic comparison of PET fibers. Talanta 123, 54–62. https://doi.org/10.1016/j.talanta.2014.01.041.

Gagliano-Candela, R., Aventaggiato, L., 2001. The detection of toxic substances in entomological specimens. Int. J. Legal Med. 114, 197–203. https://doi.org/10.1007/s004140000181.

Gerace, E., Veronesi, A., Martra, G., Salomone, A., Vincenti, M., 2017. Study of cocaine incorporation in hair damaged by cosmetic treatments. Forensic Chem. 3, 69–73. https://doi.org/10.1016/j.forc.2017.02.004.

Goldstein, J.I., Newbury, D.E., Echlin, P., Joy, D.C., Lyman, C.E., Lifshin, E., Sawyer, L., Michael, J.R., 2003. The SEM and its modes of operation. In: Goldstein, J.I., Newbury, D.E., Echlin, P., Joy, D.C., Lyman, C.E., Lifshin, E., Sawyer, L.,

Michael, J.R. (Eds.), Scanning Electron Microscopy and X-ray Microanalysis, third ed. Springer US, Boston, MA, pp. 21–60, https://doi.org/10.1007/978-1-4615-0215-9_2.
Hagmann, D.F., Kruge, M.A., Goodey, N.M., Krumins, J.A., 2020. Characterization of coal particles in the soil of a former rail yard and urban brownfield: Liberty State Park, Jersey City (NJ), USA. Int. J. Coal Geol. 217, 103328. https://doi.org/10.1016/j.coal.2019.103328.
Hawkes, P.W., Spence, J.C.H., 2019. Springer Handbook of Microscopy. Springer International Publishing.
Hore, G., Maity, A., Naskar, A., Ansar, W., Ghosh, S., Saha, G.K., Banerjee, D., 2017. Scanning electron microscopic studies on antenna of *Hemipyrellia ligurriens* (Wiedemann, 1830) (Diptera: Calliphoridae)—a blow fly species of forensic importance. Acta Trop. 172, 20–28. https://doi.org/10.1016/j.actatropica.2017.04.005.
Hore, G., Saha, G.K., Banerjee, D., 2018. Sensory organs of forensically important fly *Ophyra capensis* (Wiedemann, 1818) (Diptera: Muscidae): a scanning electron microscopic study. Acta Trop. 185, 400–411. https://doi.org/10.1016/j.actatropica.2018.06.023.
Inkson, B.J., 2016. Scanning electron microscopy (SEM) and transmission electron microscopy (TEM) for materials characterization. In: Materials Characterization Using Nondestructive Evaluation (NDE) Methods. Elsevier, pp. 17–43, https://doi.org/10.1016/B978-0-08-100040-3.00002-X.
Joy, D.C., 2019. Scanning electron microscopy: theory, history and development of the field emission scanning electron microscope. In: Biological Field Emission Scanning Electron Microscopy. John Wiley & Sons, Ltd, Chichester, UK, pp. 1–6, https://doi.org/10.1002/9781118663233.ch1.
Kim, J., Kim, M., An, J., Kim, Y., 2016. Determination of the sequence of intersecting lines using focused ion beam/scanning electron microscope. J. Forensic Sci. 61, 803–808. https://doi.org/10.1111/1556-4029.13076.
Kuo, J., 2014. Electron microscopy: methods and protocols. In: Methods in Molecular Biology, third ed. Humana Press, New York.
Kwiecińska, B., Pusz, S., Valentine, B.J., 2019. Application of electron microscopy TEM and SEM for analysis of coals, organic-rich shales and carbonaceous matter. Int. J. Coal Geol. 211, 103203. https://doi.org/10.1016/j.coal.2019.05.010.
Melki, J.A.D., Martin, C.C.S., Zerbini, T., 2011. Scanning electron microscopy as an auxiliary method in the study of exhumed bones. Forensic Sci. Int. 206, e67–e70. https://doi.org/10.1016/j.forsciint.2010.09.004.
Mendonça, P.M., dos Santos-Mallet, J.R., de Mello, R.P., Gomes, L., Queiroz, M.M.D.C., 2008. Identification of fly eggs using scanning electron microscopy for forensic investigations. Micron 39, 802–807. https://doi.org/10.1016/j.micron.2008.01.014.
Montoriol, R., Guilbeau-Frugier, C., Payre, B., Telmon, N., Savall, F., 2019. Easy preservation protocol for SEM-EDX analysis of skin specimens in forensic context. La Rev. Méd. Lég. 10, 82–87. https://doi.org/10.1016/j.medleg.2019.02.003.
Palenik, C.S., Palenik, S., Groves, E., 2018. Microscopy applications: forensic microscopy. In: Reference Module in Chemistry, Molecular Sciences and Chemical Engineering. Elsevier, https://doi.org/10.1016/B978-0-12-409547-2.11426-X. B978012409547211426X.
Peter, N.D., 2019. Scanning transmission electron microscopy. In: Handbook of Microscopy. Springer Nature, Switzerland.
Pirrie, D., 2018. Testing the efficiency of soil recovery from clothing for analysis by SEM-EDS. Forensic Sci. Int. 289, 83–91. https://doi.org/10.1016/j.forsciint.2018.05.026.
Pirrie, D., Pidduck, A.J., Crean, D.E., Nicholls, T.M., Awbery, R.P., 2019. Identification and analysis of man-made geological product particles to aid forensic investigation of provenance in the built environment. Forensic Sci. Int. 305, 109974. https://doi.org/10.1016/j.forsciint.2019.109974.

Pye, K., Croft, D., 2007. Forensic analysis of soil and sediment traces by scanning electron microscopy and energy-dispersive X-ray analysis: an experimental investigation. Forensic Sci. Int. 165, 52–63. https://doi.org/10.1016/j.forsciint.2006.03.001.

Schwerdt, I.J., Olsen, A., Lusk, R., Heffernan, S., Klosterman, M., Collins, B., Martinson, S., Kirkham, T., McDonald, L.W., 2018. Nuclear forensics investigation of morphological signatures in the thermal decomposition of uranyl peroxide. Talanta 176, 284–292. https://doi.org/10.1016/j.talanta.2017.08.020.

Sehgal, V.N., Singh, S.R., Dey, A., Kumar, M.R., Jain, C.K., Grover, S.K., Dua, D.K., 1988. Tool marks comparison of a wire cut ends by scanning electron microscopy—a forensic study. Forensic Sci. Int. 36, 21–29. https://doi.org/10.1016/0379-0738(88)90211-3.

Sharma, S.K., 2018. Handbook of Materials Characterization. Springer Science+Business Media, New York, NY.

Spence, J.C.H., 2009. High-Resolution Electron Microscopy. Oxford University Press, New York.

Valle, F., Bianchi, M., Tortorella, S., Pierini, G., Biscarini, F., D'Elia, M., 2012. Nanotechnology for forensic sciences: analysis of PDMS replica of the case head of spent cartridges by optical microscopy, SEM and AFM for the ballistic identification of individual characteristic features of firearms. Forensic Sci. Int. 222, 288–297. https://doi.org/10.1016/j.forsciint.2012.07.005.

Visonà, S.D., Chen, Y., Bernardi, P., Andrello, L., Osculati, A., 2018. Diagnosis of electrocution: the application of scanning electron microscope and energy-dispersive X-ray spectroscopy in five cases. Forensic Sci. Int. 284, 107–116. https://doi.org/10.1016/j.forsciint.2018.01.001.

Williams, D.B., Carter, B.C., 1996. The Transmission Electron Microscope. Springer, Boston.

Woods, B., Paul Kirkbride, K., Lennard, C., Robertson, J., 2014. Soil examination for a forensic trace evidence laboratory. Part 2. Elemental analysis. Forensic Sci. Int. 245, 195–201. https://doi.org/10.1016/j.forsciint.2014.10.018.

Zhao, J., Wang, Y., Wang, G., Ma, Y., Shi, H., Wen, J., Li, X., Hu, S., Chen, F., Liu, C., 2015. Application of the microwave digestion-vacuum filtration-automated scanning electron microscopy method for diatom detection in the diagnosis of drowning. J. Forensic Legal Med. 33, 125–128. https://doi.org/10.1016/j.jflm.2015.03.009.

CHAPTER 14

Atomic force microscopy for forensic samples

1 Introduction

Optical and electron microscopes have been widely used for the analysis of different types of forensic evidence. Recently, further advancements have been made in developing scanning probe microscopy (SPM) techniques that, when combined with optical and electron microscopes, can provide image analysis of the sample much more accurately. SPM techniques use a specialized tip or probe that is capable of imaging the surface of the sample through proximal contour analysis. There are mainly three types of SPM: scanning tunneling microscopy (STM), near-field scanning optical microscopy (NSOM), and atomic force microscopy (AFM). STM employs a conducting metallic nanosized probe that is capable of tunneling currents between the tip and the surface. Based on the current changes, the gap distance between the sample and the tip can be determined, thereby allowing the three-dimensional (3D) imaging of a conducting sample. Meanwhile, the probe in NSOM is an optical fiber that is tapered at the end to a range of a few nanometers. This optical fiber then further allows the transmission of the light on the sample, thus allowing its visualization. AFM, just like all other SPM techniques, has two main components: a nanoscale probe or tip and a piezoelectric actuator. These two components are often assisted by several more components that precisely control the movements of the probe and help in image analysis. The difference between AFM and other SPM techniques is that it has a sharp conducting or nonconducting tip that moves along the surface of the sample. There are several modes involved during the operation of AFM that shall be discussed in detail in this chapter.

AFM has been widely used in the analysis of various physical or biological samples. During a crime scene investigation, numerous trace evidence such as hair, fiber, blood, soil, plant specimens, or biological fluids from humans or animals is collected (Pandey et al., 2017a). Several techniques have been developed for the analysis of such trace evidence. Combining AFM analysis with the existing techniques allows the addition of a new

Handbook of Analytical Techniques for Forensic Samples
https://doi.org/10.1016/B978-0-12-822300-0.00014-8

dimension for the characterization of the samples. AFM allows the high-resolution microscopic analysis of the sample and can analyze even the bonds involved at a molecular level. In this chapter, the principle of AFM, its instrumentation, working modes, and applications for the analysis of different types of forensic evidence are discussed in detail. Also, the advantages and drawbacks associated with the use of this technique for forensic evidence analysis are also discussed.

2 Principle, working, and instrumentation of atomic force microscopy

2.1 Principle of atomic force microscopy

Analysis through AFM is similar to a blind man walking on a path with a stick, wherein the blind man analyzes with his stick the path in front of him through the different bumps and holes. Due to this analogy, AFM has also been called many times a "blind" microscopy technique. Such imaging allows accurate 3D scanning of the surface of the samples. Unlike the different microscopy techniques that employ line-of-sight visualizations, or analyze the reflections or shadows on a sample, AFM analyzes the height of the irregularities present on the surface of the sample (Haugstad, 2012). These irregularities are very small and therefore can be easily analyzed with the nanoscale probe in AFM. The height calculated is further subjected to algorithmic and mathematical processing, due to which the final 3D image of the sample is presented by the software.

2.2 Instrumentation and modes of atomic force microscopy

As mentioned in the introduction section, AFM has two main components: the probe and the piezoelectric actuator or cantilever that carefully controls the movements of the needle with high accuracy (Fig. 1). The tip is attached to the microcantilever and when the tip comes in contact with the sample surface, the microcantilever bends a little. This bending or deflection is measured by the deflections in the laser that is pinpointed at the tip. This reflected laser then reaches the photodiode, which along with the entire feedback system carefully gauges the position of the tip and thereby controls the entire analysis. There are many modes of detection in AFM such as piezoelectric detection that is capable of directly detecting the deflections in the cantilever, optical interferometry, capacitive detection, and piezoresistive detection. The most widely used mode of detection is piezoelectric

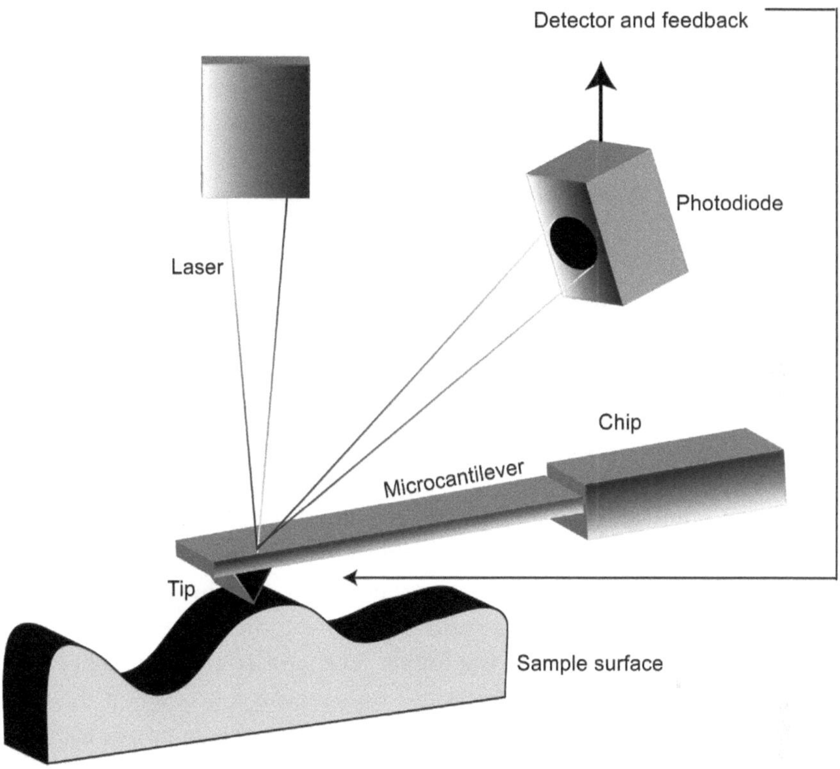

Fig. 1 Instrumentation of atomic force microscopy.

detection (Jumpertz et al., 1998; Kunicki et al., 2019; Wang and Chu, 2013; Xiong et al., 2014).

AFM works in three main modes: direct contact mode, noncontact mode, or tapping mode (Dombrowski, 2013). In contact mode, the probe is in constant physical contact with the sample and the sample topography is developed as a result of the vertical movements of the cantilever and the probe. The feedback loop system in the instruments allows the maintenance of a preset load force while the movement of the probe on the sample occurs. This mode is highly useful for analyzing soft samples such as biological samples, as they are less likely to provide counter force that may damage the probe during the analysis. Contact mode achieves the highest resolution in a liquid medium. In the presence of a liquid, the capillary forces between the probe and the sample surface are eliminated. However, if the preset force is not properly maintained, then it may cause damage to the soft or friable biological samples that are commonly encountered at a crime scene

(Binnig et al., 1986; Hölscher, 2012a; Junno et al., 1995; Le Grimellec et al., 1998; Manfredotti, 2013; Walter et al., 2017).

In noncontact mode, the probe is in very close proximity to the surface of the sample; however, it is not touching the sample. The main driving force between the interaction of the probe and the sample surface is the attractive van der Waals force. In this mode, the tip is kept in a constant oscillation mode in order to detect the extreme low magnitude of van der Waals and other attractive forces between the tip and the sample. Any interaction between the probe and the sample is measured by measuring the change in the amplitude of the oscillation or by measuring the resonant frequency via the feedback system. The noncontact mode of AFM achieves images with high resolution and can perform even submolecular imaging (Hölscher, 2012b; Smallman and Ngan, 2014).

The tapping mode is the most recent and the most advanced mode of AFM. In this technique, the probe is kept very close to the sample surface, and at alternating short intervals of time (approximately 50,000–500,000 cycles per second), it touches the sample surface. Just like the noncontact mode, the probe and the microcantilever are kept oscillating. This oscillation is kept at the lowest level and at a constant speed by the feedback system. When the probe comes across a groove on the surface, the oscillation of the probe increases while the opposite happens when the probe encounters any bump on the sample surface. The feedback loop system carefully monitors the interaction of the probe with the sample surface and automatically adjusts the tip-surface separation such that the magnitude of the preset force is maintained. Based on the changes in the amplitude of the oscillation, the surface topography can be easily analyzed. Due to the intermittent touching of the sample surface by the probe, the energy loss is significantly reduced. The tapping mode has several advantages over other modes of AFM, as it alternatively touches the surface. This technique allows high-resolution scanning on a myriad of samples and overcomes limitations such as friction, electrostatic forces, and adhesion. This mode of AFM allows sample analysis in air, vacuum, or fluid and is the most suitable technique for the analysis of very fragile and soft surfaces (Hansma et al., 1994; Ishida et al., 2000; Kueng et al., 2003; Magonov et al., 1997; Thomson, 2005). Fig. 2 depicts the changes in the forces with the change in distance between the tip and the in the different modes of AFM.

In brief, contact mode is the most damaging mode for analyzing samples in AFM, as it can not only damage or break the probe but can also distort the

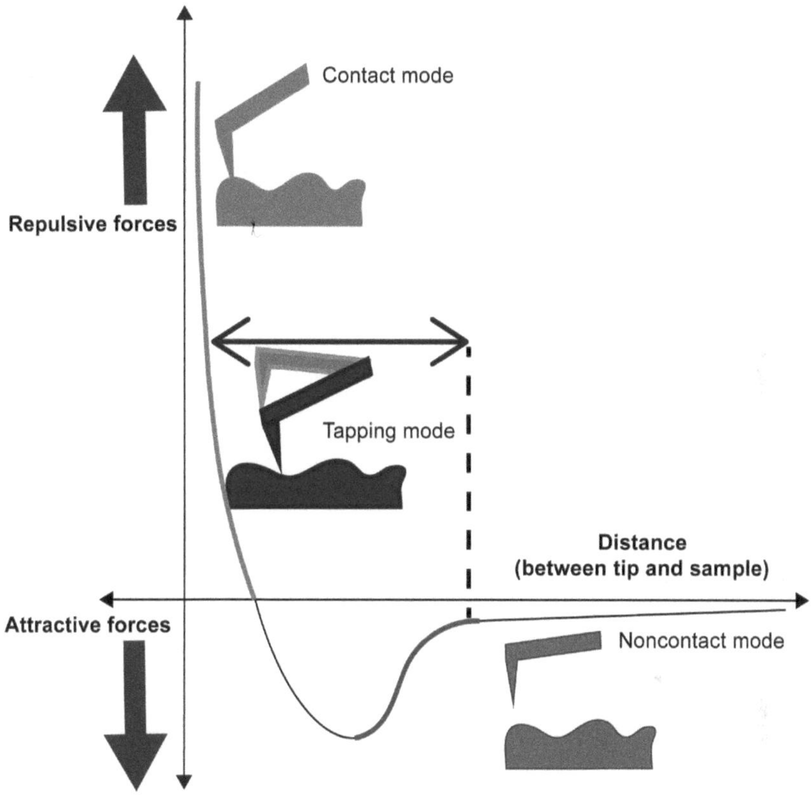

Fig. 2 Modes of atomic force microscopy.

sample surface. Meanwhile, the noncontact mode is significantly better than the contact mode, especially in terms of probe or sample damage. However, when compared to the tapping mode, the resolution provided by the noncontact mode is relatively less. A major drawback associated with the noncontact mode is that it is rendered useless in the presence of a contaminant layer that can hinder its oscillation. This drawback is overcome by the tapping mode, which is capable of achieving high resolution while causing minimal or negligible damage to the sample surface. Myriad samples can be analyzed through this mode in air, vacuum, or liquid.

3 Atomic force microscopy for forensic investigations

In past years, AFM has emerged as an efficient characterization tool for myriad samples of forensic importance. These samples usually have a physical, chemical, or biological origin. This section illustrates the various research works carried out by scientists using AFM for the analysis of forensic samples such as blood, hair, plastic wrapping material, questionable documents, fingerprints, fibers, SIM cards, explosives, and gunshot residue (GSR). The use of AFM for different types of forensic samples has been presented in Table 1.

3.1 Age determination of bloodstains

The examination of health and the diagnosis and treatment of disease usually include the analysis of blood. This analysis has become a very common practice for forensic scientists and investigators as well. The examination of bloodstain scattering patterns contribute to understanding the criminal dynamics and thus provide necessary crime-related elements during investigation (Gardner and Krouskup, 2018). It forms a major chunk of the crime investigation, and thus needs to be very accurate and precise. Innumerable and reliable tests are available to detect and identify bloodstains such as the benzidine test, the phenolphthalein test, the Takayama test, and the Teichmann test (Cox, 1991). Apart from these tests, the estimation of the age of bloodstains becomes crucial in order to determine the probable time the crime was committed. When there is an injury during a crime, and blood gets splattered outside and comes in contact with different substrates, various types of degradation occur on the exposed blood. The oxidation of Fe^{2+} (in hemoglobin) to Fe^{3+} (in methemoglobin) becomes responsible for the reddish-brown color of the exposed and dried bloodstains. The transformation of hemoglobin to methemoglobin changes the structure of red blood cells. These changes can be noticed with the aging of bloodstains, and therefore their analysis can be correlated to the probable time the crime was committed. However, the age determination of bloodstains has remained an issue for forensic investigators in the past. Worldwide, techniques such as electron paramagnetic resonance (Fujita et al., 2005), high-performance liquid chromatography (Andrasko, 1997), quantitative RNA degradation analysis (Anderson et al., 2005), and hyperspectral imaging (HSI) (Edelman et al., 2012) have been used for this purpose. However, apart from HIS, other techniques have proved to be invasive and are thus less attractive in the eyes of forensic experts. The drawback of HSI was the deviation in the bloodstain's age determination with increasing time (Smijs and Galli, 2019). With the advent of nanotechnology-based instruments such as AFM, the analysis of bloodstains for their age determination has become a comparatively easy task. AFM assists in providing the topographical data of the RBC's surface such as perimeter, area, height, and volume as well as in determining their viscoelasticity (Kuznetsova et al., 2007).

Table 1 AFM analysis of different forensic samples (Pandey et al., 2017b).

S. no.	Aim of study	Type of samples	Environmental conditions	Mode of cantilever	Frequency resonance	Applied force by probe (N/m)	Type of cantilever/ cantilever thickness	Spring constant (N/m)	References
1.	Analysis of PDMS replica	Firearms	Ambient	Semicontact	90–230 kHz		–	–	(Valle et al., 2012)
2.	Cause of fire	Copper molten mark	Typical	–	–	–	Si_3N_4 cantilever	–	(Ao et al., 2011)
3.	Age determination of bloodstains	Blood spots	Ambient	Contact	405 kHz	0.3	Si cantilever (4.8–5.5 mm)	74	(Strasser et al., 2007)
4.	Time-dependent surface adhesive force and morphology study	Red blood cells (RBCs)	Room temperature	Tapping	72–96 kHz	–	Si tips	3	(Wu et al., 2009)
5.	Time-dependent surface adhesive force and morphology study	Red blood cells (RBCs)	Room temperature	Contact	255–315 kHz	–	Si tips	0.9	(Wu et al., 2009)
6.	Time lapse investigation	Blood film	Room temperature	Tapping	0.5 Hz	–	Si cantilevers	2.8	(Chen and Cai, 2006)

Continued

Table 1 AFM analysis of different forensic samples (Pandey et al., 2017b)—cont'd

S. no.	Aim of study	Type of samples	Environmental conditions	Mode of cantilever	Frequency resonance	Applied force by probe (N/m)	Type of cantilever/ cantilever thickness	Spring constant (N/m)	References
7.	Quantitative analysis and classification of human hair	Human hair	Atmospheric	Contact	–	0.06	Si_3N_4 pyramidal (200 μm)	–	(Gurden et al., 2004)
8.	Examination of line crossing	Plain paper with dot matrix	Room temperature	Tapping	–	–	120 mm	–	(Friedbacher and Fuchs, 1999)
9.	Investigation of surface textures	Quartz grain	Ambient	Noncontact/ Tapping			Si cantilevers	–	(Konopinski et al., 2012)
10.	Investigation of commercial pressure sensitive adhesives	Packaging and adhesive tapes		Semicontact	330 kHz	–	Si cantilevers	–	(Canetta and Adya, 2011)
11.	Morphological, spectral, and chromatography analysis and forensic comparison	PET fibers	–	Tapping	300 kHz	–	Oscillating Si probe	–	(Farah et al., 2014)
12.	Morphological changes	–	Room temperature	Contact	0.5 Hz	–	Au-coated Si_3N_4	0.03	(Canetta et al., 2009a, 2009b)

One of the pioneer works related to the age determination of bloodstains was carried out by Strasser et al., and was published in 2007. For analysis, blood was collected from the fingertip of a volunteer and was dropped onto a glass slide, and was further dried at room temperature. The analysis was done in noncontact mode. The typical doughnut-like shape of the RBCs was observed under the microscope. The morphological investigation of the RBCs was carried out for 4 weeks, and there was no alteration observed in the structure of RBCs, even after 31 days, as shown in Fig. 3. However, the elasticity of the blood decreased over this period of time. This change was probably because of the morphological changes the RBCs underwent during the coagulation and drying of the bloodstain (Strasser et al., 2007).

Another study focused on analyzing the morphological changes in the RBCs at different environmental conditions as well as at high and low temperature (Wu et al., 2009). In a recent research work, the examination of time since deposition (TSD) of RBCs from bloodstains was carried out using AFM. The researchers investigated the changes in blood smears with time as well as variations in the analysis of TSD for smears deposited on three types of surfaces: glass, metal, and ceramic. For 28 days, there was no noticeable variation observed in the morphology of the RBCs. However, the quantitative analysis through AFM revealed an increasing drift in the perimeter, volume, height, and area of the RBCs during the aforementioned time duration. However, significant differences were observed in the morphology of the RBCs from the blood deposited on metal, ceramic, and glass. This study thus highlighted the importance of the type of surface on which the bloodstain was deposited as a crucial factor in the age determination of bloodstains (Cavalcanti and Silva, 2019). In spite of these kinds of encouraging results obtained, the application of AFM in routine forensic investigations still needs more development, expansion, and research.

3.2 Hair examination

Hair is one of the most common types of evidence found at a crime scene. Hair is a favorable site for the accumulation of chemicals, drugs, and biomolecules. The importance of hair analysis is very high in crime investigations because of the unique nature of hair in retaining its chemical profile and show only minute variations. The analysis can give an idea about drug intake and abuse, even if the hair sample has been collected after several days, months, or years (Palmeri et al., 2000). AFM has proved to be an excellent tool for the examination of hair because of its capability to function in liquid,

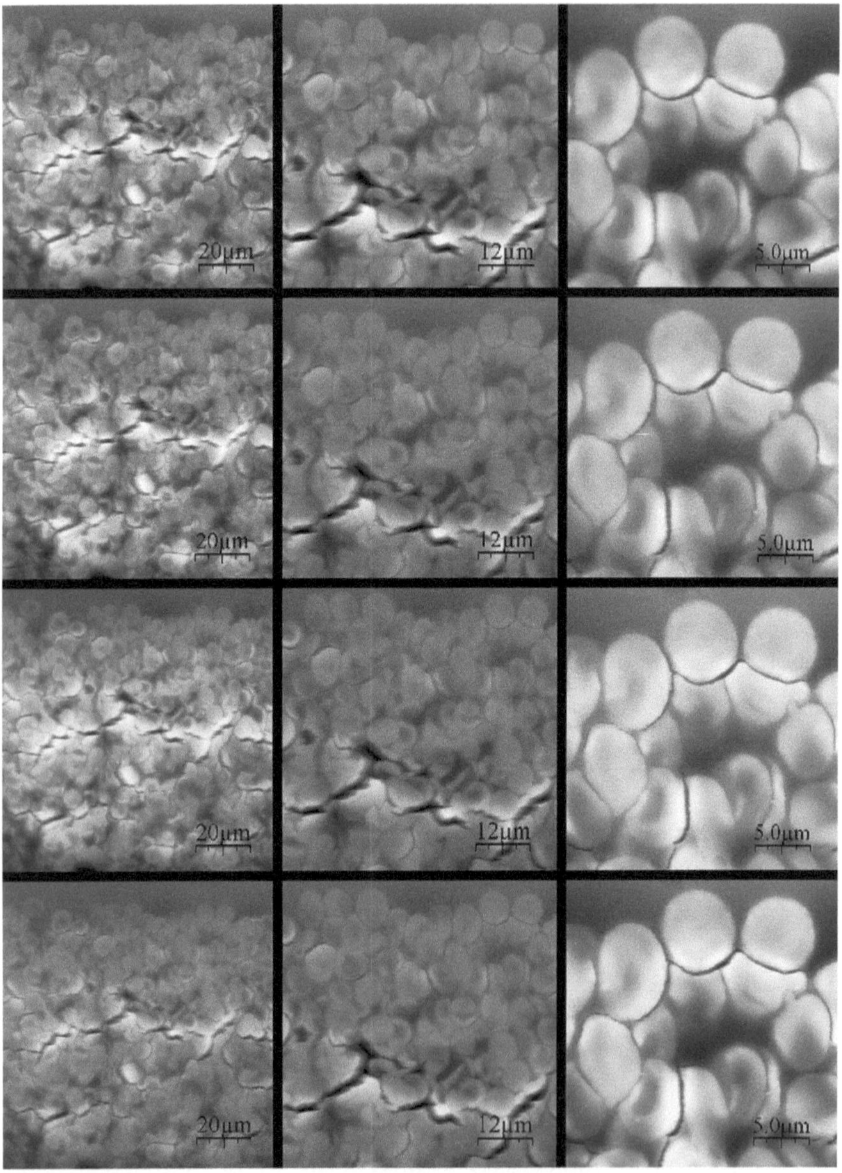

Fig. 3 Morphological analysis of bloodstains through AFM. Rows are related to the measuring days (individual) and columns are related to the scan area (identical) of the blood spot on different days. The time after which each row was scanned was 1.5 h (1st row), 1 week (2nd row), 2 weeks (3rd row), and 4 weeks (4th row). No morphological change was observed with time (Strasser et al., 2007).

air, and vacuum (Pandey et al., 2017b). The effect of environmental factors on the nanomechanical properties and structure of the surface of hair can be easily studied through AFM (Bhushan and Chen, 2006). However, in order to perform the toxicological hair investigations, AFM needs to be coupled with Raman spectroscopy to identify different drugs, chemicals, and toxins on the surface of the hair. AFM imaging of hair for forensic investigations often faces issues related to lack of reference samples of hair with similar physical assets (Smijs and Galli, 2019).

In the past, a research work employed AFM to study the influence of treatments and ethnicity on structural variations and tensile reaction (in situ) of human hair. Caucasian, African, and Asian volunteers gave parts of their hair for the AFM analysis. The tensile response was evaluated for changes due to soaking, ethnicity, and fatigue. It was observed that hair samples from different ethnicities had different mechanical properties (Seshadri and Bhushan, 2008). In a recent study, an advanced AFM was developed via robot technology with the capability of multiparametric imaging for 3D reconstruction and 360-degree mapping. Human hair was taken as the anisotropic material of biological origin and its adhesion, topography, energy dissipation, and Young's modulus were studied using the developed AFM system (Lu et al., 2019).

3.3 Plastic wrapping material analysis

Burial of evidence in soil related to a crime, such as cadavers, clothes, weapons, etc., has always remained a common practice among criminals. This evidence is usually found to be wrapped in materials made of plastics such as polyvinyl chloride (PVC) and polyethylene (PE) (Schotsmans et al., 2017). These plastic-based materials are usually thought to be nondegradable and inert to the environment; however, at the microscopic level, these materials get damaged. The decomposition of these materials depends upon the nature of the soil in which they are buried and their chemical structure (Singh and Sharma, 2008). Certain environmental conditions favor the growth of microbial flora and fauna, which ultimately become responsible for the biological degradation of plastics (Shah et al., 2008). The structure of PE gets damaged due to the mycelial growth on its surface while the leaching out or damage caused by fungi and bacteria to the platicizers added in PVC-based materials becomes responsible for their biodegradation (Pradeep and Benjamin, 2012). The forensic investigation of such wrapping materials is done to analyze the effect of environmental factors in soil that stimulate the degradation of these polymeric materials.

In a recent study, AFM was used to analyze the surface alterations in polymeric materials with forensic importance that were buried in different soil environments at the nanometer level. PE bags and PVC sheets were concealed in different model soil environments such as loam, sand, clay, and lime for a period of 24 months. Further, the pH, moisture content, and temperature of these soil types were varied. The analysis was done in tapping mode while maintaining the vibration amplitude of the cantilever to be constant. Over a period of 2 years, it was found that the high pH of soil and elevated moisture content favor the degradation of PE bags. A continuous leaching of plasticizer from the PVC sheeting was observed over the period of burial, which was aggravated in the moist environment. AFM also helped in measuring the nanoscale changes in the roughness of these materials, which shows the potential of this technique that can be utilized to study even the minute details of the degradation of wrapping materials used to bury evidence during forensic investigations (Sullivan et al., 2019).

3.4 Examination of textile fibers

Textile fibers are often encountered at the crime scene that could be directly correlated to the clothing belonging to the suspect or victim. The forensic examination of textile fibers usually involves chemical, microscopic, and mechanical investigations. The basic structure of fibers change upon exposure to different environmental conditions, which could be determined using microscopy (Brüschweiler and Grieve, 1997).

AFM was first used by Canetta et al. in 2009 to study the morphological parameters of textile fibers at the nanoscale level. Two natural fibers, cotton and wool, and one synthetic fiber, viscose, were exposed to various environmental conditions for variable intervals of time. The imaging was done in contact mode using a gold-coated Si_3N_4 cantilever. The nanoscale variations in the structure of all three fibers were vividly visible in the microscopic analysis (Canetta et al., 2009a, 2009b). In a recent study, AFM coupled with infrared spectroscopy (IR) was used to study the photodegradation of the polyester fiber. The microscopy was performed in contact mode using a gold-plated Si_3N_4 cantilever. AFM coupled with IR assisted in studying the structural and chemical changes in the polyester fiber due to changes in humidity, temperature, and solar radiation (Nguyen-Tri and Prud'homme, 2019). These analyses show the ability of AFM to be used for structural studies related to textile fibers collected from a crime scene.

3.5 SIM card analysis

SIM cards are often recovered in damaged condition from crime scenes, especially where any cybercrime has been committed. These cards are important evidence for cybercrime and mobile forensics. The data recovered from these cards often give insight about the identity of the victim, suspect, or both (Pandey et al., 2017b).

A group of scientists worked in the past to advance and characterize a technique that can be validated forensically for the recovery and processing of data from damaged SIM cards. De Nardi et al. combined the studies published in past years to develop a technique based on AFM to view underneath the data embedded in the EEPROM/flash memory arrays, which normally exist in the microcontrollers of the smart card (De Nardi et al., 2005; DeNardi et al., 2006).

3.6 Explosives and gunshot residue analysis

Terrorism has spread in today's world, and explosives are very commonly used for this purpose. The identification of the type of explosive used at a bombing site becomes important in forensic investigations. Terrorist activities, mainly bomb explosions, are dynamic and therefore need accurate and precise techniques to characterize and detect the explosives used during explosions (Zhang et al., 2018). The surface analysis of explosives such as ammonium perchlorate, PBX, and triamino-trinitro-benzene have been studied using AFM. The analysis of explosives through AFM provides fresher perceptions for understanding various properties of existing explosives (Kumari et al., 2013; Xu et al., 2018). The phenomenon of local heating leading to the induction of chemical detonation in explosives was also investigated in a study in which the hotspot formation occurred on the crystal of picric acid (2,4,6-trinitrophenol) when it came in contact with the tip of AFM (Kovalev and Sturm, 2011).

In addition to explosives, GSR analysis is also an important examination during crime investigations, especially those that involve the use of firearms. GSR analysis tells the shooting distance and angle from which bullet was shot and the type of firearm used. This information helps in reconstructing crimes related to firearms (López-López and García-Ruiz, 2014). GSR is usually collected from the clothing of the suspect and victim, the hands of the suspect, and the entry point of the bullet into the victim's body. In a study done by Mou et al., AFM was used to determine the shooting distance, the manufacturer of the cartridge, and the powder present in it. AFM

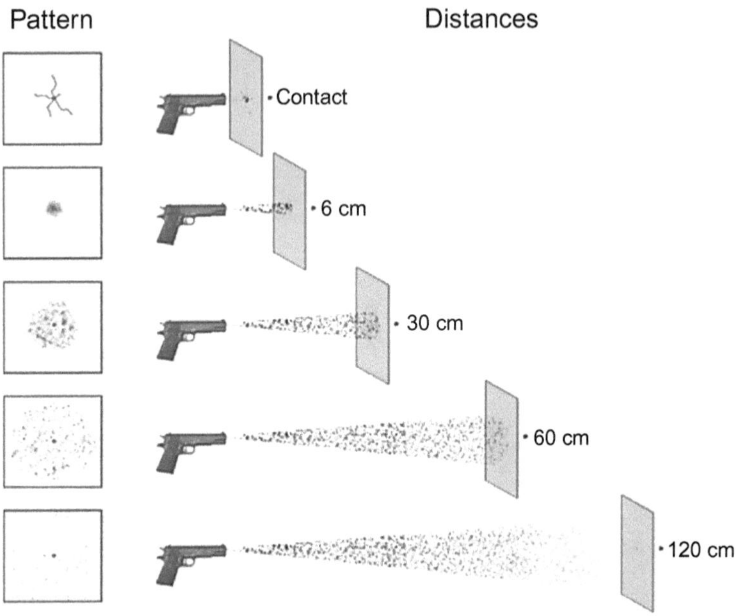

Fig. 4 Difference in GSR pattern obtained due to variable shooting distance by firing ammunition from same firearm (López-López and García-Ruiz, 2014).

imaging revealed various GSR shapes such as irregular, spherical, rectangular, heart rod, and boomerang. An inverse relation was concluded between the shooting distance and the size distribution of the GSR through this study (Mou et al., 2008). This study later received a commentary stating the drawback of AFM for analyzing fine GSR powder because these fine particles attach to the tip of the AFM's probe, resulting in drastic variations in the shape and size of GSR powders (Jones, 2009). Fig. 4 shows the variations in patterns formed by the GSR spread from the ammunition fired from the same gun.

3.7 Fingermark examination

Fingerprints are highly crucial evidence as they are responsible for incriminating any suspects who may have been present at the crime scene. The techniques involved for detecting latent fingerprints target the moisture content or grease of the ridges and the chemical composition of the fingerprints. Because latent fingerprints are barely visible and consist of minute details, AFM can be used to magnify the ridges to a higher scale and thus allow the visualization of the fingerprints in much detail (Frick et al., 2013; Rawtani et al., 2019; Thomas, 1978). Over the years, various studies, albeit limited, have employed AFM for the detection of fingerprints. This usage comes after certain challenges in detecting aged and contaminated fingerprints. In one study, AFM was used in tapping mode to analyze fingerprints on individual features on a polished brass surface. With AFM, it was possible to visualize ridges and other minutiae of the fingerprints with much more detail. This technique was also able to visualize even the microscopic sweat droplets present between the ridges (Goddard et al., 2010). In another study, a new mode of AFM was developed that was known as PeakForce Quantitative Nanomechanical Mapping (PF QNM). Through this mode, it was possible to develop images of high-resolution topography and other factors such as adhesion, stiffness, deformation, and elastic modulus. The researchers took into consideration the adhesion forces, electrostatic attractions, capillary forces, chemical interactions, and van der Waals forces between the probe and the fingerprint surface. Different fingers had different fluid (sweat and moisture) compositions that influenced these forces and thereby allowed the imaging of the latent fingerprint (Dorakumbura et al., 2016). AFM was also used to detect changes in the composition of the latent fingerprints with time with much accuracy. These changes were observed through measuring changes in the height of the fingerprint ridge and by observing the horizontal movement of the fingerprint components across the substrate (Popov et al., 2017).

3.8 Analysis of questionable documents

Questionable documents are those that contain information that may be considered potentially disputed in a court of law. They can include fraudulent, stolen, or counterfeit papers, boards, or even glass. Documents such as checks and other bank documents, currency, wills, land deeds and titles, and stamps are often widely encountered in questionable document cases. Conventionally, questionable documents are often analyzed for crossed out lines, overwriting, or forged signatures through microscopes, digital imaging instrumentation, electrostatic detection apparatus (ESDA), and techniques involving analytical chemistry. It is highly crucial to employ nondestructive techniques for the analysis of the questionable documents (Calcerrada and García-Ruiz, 2015; Dirwono et al., 2010; Hilton, 1992). The AFM analysis of questionable documents enhances the repertoire of the preexisting nondestructive techniques for the examination of questionable documents. It offers high resolution even with such low orders of force. In a study, AFM in tapping mode was employed to analyze the ink deposition, line crossing, and papers used in a questionable document. The nondestructive and soft nature of the tapping mode allowed the detection of the topography, amplitude, and the height profile of the samples and thereby analyzed the samples; it was able to differentiate between the different types of papers, line crossings, and inks (Chen et al., 2012). Similarly, AFM was also used to determine the sequence of crossed lines. AFM was set in the tapping mode and was used to image heterogenous lines that were randomly produced on plain paper with different writing media such as ballpoint pens, typewriter carbon film ribbon, fibertip pens, floating ball pens, toner, and dot-matrix fabric ribbons. It was concluded that it is possible to image ink deposits of different origins and the sequence of line crossing. Because AFM does not require any prior sample preparation as observed in the case of scanning electron microscopy (SEM) and transmission electron microscopy (TEM), it is highly useful for the preservation of questionable documents while analyzing them (Khanmy-Vital et al., 2001).

4 Conclusion

AFM is a type of SPM technique that uses a conducting or nonconducting probe to visualize the 3D form of the sample surface. This probe is attached to a microcantilever that deflects or bends a little whenever the probe comes across a bump or groove on the sample surface. This deflection is analyzed through the deviations in the transmitted laser and the data are processed through a feedback loop system. Through such processing, it is possible to obtain the height profile of the sample surface and thus procure a 3D image. AFM commonly works in three different modes: contact mode, tapping or intermittent mode, and noncontact mode. The modes are a representation of the probe to the sample separation. In contact mode, the probe is in constant contact with the sample surface while in noncontact mode, the opposite occurs. In the intermittent or tapping mode, the probe alternatively touches the samples at regular intervals of time. This mode offers the highest topographic resolution with the least risk of damaging the probe or the sample surface.

All three modes have been used for the analysis of different forensic samples such as aged bloodstains, hair, plastic wrapping materials, explosives, GSR, damaged SIM cards, textile fibers, fingerprints, and questionable documents. AFM is a versatile tool that can also be combined with other characterization techniques such as Raman spectroscopy and electron microscopes. Despite the high resolution and nondestructive analysis of the sample by this technique, there are certain drawbacks that limit its frequent use in analyzing different forensic evidence. These limitations include the requirement of expertise, time, and cost. However, recently several advancements and several modes of operation have been made in the modification of this technique. Therefore, in the future, AFM may become a common technique for the analysis of different forensic samples.

References

Anderson, S., Howard, B., Hobbs, G.R., Bishop, C.P., 2005. A method for determining the age of a bloodstain. Forensic Sci. Int. 148, 37–45. https://doi.org/10.1016/j.forsciint.2004.04.071.

Andrasko, J., 1997. The estimation of age of bloodstains by HPLC analysis. J. Forensic Sci. 42, 601–607.

Ao, G., Chang-zheng, Z., Man, D., Wei, G., Ming, Z., Da-Wei, X., 2011. Microscopic investigation of a copper molten mark by optical microscopy (OM) and atomic force microscopy (AFM). Procedia Engineering. In: The 5th Conference on Performance-Based Fire and Fire Protection Engineering, vol. 11, pp. 100–106, https://doi.org/10.1016/j.proeng.2011.04.633.

Bhushan, B., Chen, N., 2006. AFM studies of environmental effects on nanomechanical properties and cellular structure of human hair. In: Ultramicroscopy, Proceedings of the Seventh International Conference on Scanning Probe Microscopy, Sensors and Nanostructures. vol. 106, pp. 755–764, https://doi.org/10.1016/j.ultramic.2005.12.010.

Binnig, G., Quate, C.F., Gerber, C., 1986. Atomic force microscope. Phys. Rev. Lett. 56, 930–933. https://doi.org/10.1103/PhysRevLett.56.930.

Brüschweiler, W., Grieve, M.C., 1997. A study on the random distribution of a red acrylic target fibre. Sci. Justice 37, 85–89. https://doi.org/10.1016/S1355-0306(97)72152-X.

Calcerrada, M., García-Ruiz, C., 2015. Analysis of questioned documents: a review. Anal. Chim. Acta 853, 143–166. https://doi.org/10.1016/j.aca.2014.10.057.

Canetta, E., Adya, A.K., 2011. Atomic force microscopic investigation of commercial pressure sensitive adhesives for forensic analysis. Forensic Sci. Int. 210, 16–25. https://doi.org/10.1016/j.forsciint.2011.01.029.

Canetta, E., Montiel, K., Adya, A.K., 2009a. Morphological changes in textile fibres exposed to environmental stresses: atomic force microscopic examination. Forensic Sci. Int. 191, 6–14. https://doi.org/10.1016/j.forsciint.2009.05.022.

Canetta, E., Montiel, K., Adya, A.K., 2009b. Morphological changes in textile fibres exposed to environmental stresses: atomic force microscopic examination. Forensic Sci. Int. 191, 6–14. https://doi.org/10.1016/j.forsciint.2009.05.022.

Cavalcanti, D.R., Silva, L.P., 2019. Application of atomic force microscopy in the analysis of time since deposition (TSD) of red blood cells in bloodstains: a forensic analysis. Forensic Sci. Int. 301, 254–262. https://doi.org/10.1016/j.forsciint.2019.05.048.

Chen, S.-Z., Tsai, T.-L., Chen, Y.-F., 2012. Forensic application of atomic force microscopy—questioned document. J. Chin. Chem. Soc. 59, 283–288. https://doi.org/10.1002/jccs.201100739.

Chen, Y., Cai, J., 2006. Membrane deformation of unfixed erythrocytes in air with time lapse investigated by tapping mode atomic force microscopy. Micron 37, 339–346. https://doi.org/10.1016/j.micron.2005.11.011.

Cox, M., 1991. A study of the sensitivity and specificity of four presumptive tests for blood. J. Forensic Sci. 36, 1503–1511. https://doi.org/10.1520/JFS13170J.

De Nardi, C., Desplats, R., Perdu, P., Beaudoin, F., Gauffier, J.-L., 2005. Oxide charge measurements in EEPROM devices. In: Microelectronics Reliability, Proceedings of the 16th European Symposium on Reliability of Electron Devices, Failure Physics and Analysis, vol 45, pp. 1514–1519, https://doi.org/10.1016/j.microrel.2005.07.055.

DeNardi, C., Desplats, R., Perdu, P., Gauffier, J.-L., Guérin, C., 2006. Descrambling and data reading techniques for flash-EEPROM memories. Application to smart cards. In: Microelectronics Reliability, Proceedings of the 17th European Symposium on Reliability of Electron Devices, Failure Physics and Analysis. Wuppertal, Germany 3rd–6th October 2006 46, pp. 1569–1574, https://doi.org/10.1016/j.microrel.2006.07.022.

Dirwono, W., Park, J.S., Agustin-Camacho, M.R., Kim, J., Park, H.-M., Lee, Y., Lee, K.-B., 2010. Application of micro-attenuated total reflectance FTIR spectroscopy in the forensic study of questioned documents involving red seal inks. Forensic Sci. Int. 199, 6–8. https://doi.org/10.1016/j.forsciint.2010.02.009.

Dombrowski, R.T., 2013. Microscopy techniques for analyzing the phase nature and morphology of biomaterials. In: Jaffe, M., Hammond, W., Tolias, P., Arinzeh, T. (Eds.), Characterization of Biomaterials, Woodhead Publishing Series in Biomaterials. Woodhead Publishing, pp. 1–33, https://doi.org/10.1533/9780857093684.1 (Chapter 1).

Dorakumbura, B.N., Becker, T., Lewis, S.W., 2016. Nanomechanical mapping of latent fingermarks: a preliminary investigation into the changes in surface interactions and topography over time. Forensic Sci. Int. 267, 16–24. https://doi.org/10.1016/j.forsciint.2016.07.024.

Edelman, G., van Leeuwen, T.G., Aalders, M.C.G., 2012. Hyperspectral imaging for the age estimation of blood stains at the crime scene. Forensic Sci. Int. 223, 72–77. https://doi.org/10.1016/j.forsciint.2012.08.003.

Farah, S., Tsach, T., Bentolila, A., Domb, A.J., 2014. Morphological, spectral and chromatography analysis and forensic comparison of PET fibers. Talanta 123, 54–62. https://doi.org/10.1016/j.talanta.2014.01.041.

Frick, A.A., Fritz, P., Lewis, S.W., 2013. Chemistry of print residue. In: Siegel, J.A., Saukko, P.J., Houck, M.M. (Eds.), Encyclopedia of Forensic Sciences, second ed. Academic Press, Waltham, pp. 92–97, https://doi.org/10.1016/B978-0-12-382165-2.00262-2.

Friedbacher, G., Fuchs, H., 1999. Classification of scanning probe microscopies. Pure Appl. Chem. 71, 1337–1357.

Fujita, Y., Tsuchiya, K., Abe, S., Takiguchi, Y., Kubo, S., Sakurai, H., 2005. Estimation of the age of human bloodstains by electron paramagnetic resonance spectroscopy: long-term controlled experiment on the effects of environmental factors. Forensic Sci. Int. 152, 39–43. https://doi.org/10.1016/j.forsciint.2005.02.029.

Gardner, R., Krouskup, D., 2018. Practical crime scene processing and investigation. In: Practical Aspects of Criminal and Forensic Investigations, third ed. CRC Press, Boca Raton.

Goddard, A.J., Hillman, A.R., Bond, J.W., 2010. High resolution imaging of latent fingerprints by localized corrosion on brass surfaces. J. Forensic Sci. 55, 58–65. https://doi.org/10.1111/j.1556-4029.2009.01217.x.

Gurden, S.P., Monteiro, V.F., Longo, E., Ferreira, M.M.C., 2004. Quantitative analysis and classification of AFM images of human hair. J. Microsc. 215, 13–23. https://doi.org/10.1111/j.0022-2720.2004.01350.x.

Hansma, P.K., Cleveland, J.P., Radmacher, M., Walters, D.A., Hillner, P.E., Bezanilla, M., Fritz, M., Vie, D., Hansma, H.G., Prater, C.B., Massie, J., Fukunaga, L., Gurley, J., Elings, V., 1994. Tapping mode atomic force microscopy in liquids. Appl. Phys. Lett. 64, 1738–1740. https://doi.org/10.1063/1.111795.

Haugstad, G., 2012. Atomic Force Microscopy: Understanding Basic Modes and Advanced Applications. John Wiley & Sons, Hoboken, NJ.

Hilton, O., 1992. Scientific Examination of Questioned Documents, revised ed. CRC Press.

Hölscher, H., 2012a. AFM, tapping mode. In: Bhushan, B. (Ed.), Encyclopedia of Nanotechnology. Springer Netherlands, Dordrecht, p. 99, https://doi.org/10.1007/978-90-481-9751-4_33.

Hölscher, H., 2012b. AFM, non-contact mode. In: Bhushan, B. (Ed.), Encyclopedia of Nanotechnology. Springer Netherlands, Dordrecht, pp. 93–99, https://doi.org/10.1007/978-90-481-9751-4_32.

Ishida, N., Inoue, T., Miyahara, M., Higashitani, K., 2000. Nano bubbles on a hydrophobic surface in water observed by tapping-mode atomic force microscopy. Langmuir 16, 6377–6380. https://doi.org/10.1021/la000219r.

Jones, B.J., 2009. Commentary on: Mou Y, Lakadwar J, Rabalais JW. Evaluation of shooting distance by AFM and FTIR/ATR analysis of GSR. J Forensic Sci 2008;53:1381-6. J. Forensic Sci. 54, 502. author reply 503 https://doi.org/10.1111/j.1556-4029.2009.00989.x.

Jumpertz, R., Hart, A., Ohlsson, O., Saurenbach, F., Schelten, J., 1998. Piezoresistive sensors on AFM cantilevers with atomic resolution. Microelectron. Eng. 41–42, 441–444. https://doi.org/10.1016/S0167-9317(98)00102-6.

Junno, T., Anand, S., Deppert, K., Montelius, L., Samuelson, L., 1995. Contact mode atomic force microscopy imaging of nanometer-sized particles. Appl. Phys. Lett. 66, 3295–3297. https://doi.org/10.1063/1.113735.

Khanmy-Vital, A., Kasas, S., Dietler, G., 2001. The use of atomic force microscopy to determine the sequence of crossed lines. Probl. Forensic Sci. 46, 401–412.

Konopinski, D.I., Hudziak, S., Morgan, R.M., Bull, P.A., Kenyon, A.J., 2012. Investigation of quartz grain surface textures by atomic force microscopy for forensic analysis. Forensic Sci. Int. 223, 245–255. https://doi.org/10.1016/j.forsciint.2012.09.011.

Kovalev, A., Sturm, H., 2011. Observation of nanoscale hot-spot generation on a 2,4,6-trinitrophenol (TNP) single crystal. Surf. Sci. Graphene Surf. Interfaces 605, 1747–1753. https://doi.org/10.1016/j.susc.2011.06.012.

Kueng, A., Kranz, C., Lugstein, A., Bertagnolli, E., Mizaikoff, B., 2003. Integrated AFM–SECM in tapping mode: simultaneous topographical and electrochemical imaging of enzyme activity. Angew. Chem. Int. Ed. 42, 3238–3240. https://doi.org/10.1002/anie.200351111.

Kumari, A., Mehilal, Jain, S., Jain, M.K., Bhattacharya, B., 2013. Nano-ammonium perchlorate: preparation, characterization, and evaluation in composite propellant formulation. J. Energetic Mater. 31, 192–202. https://doi.org/10.1080/07370652.2012.694576.

Kunicki, P., Angelov, T., Ivanov, T., Gotszalk, T., Rangelow, I., 2019. Sensitivity improvement to active piezoresistive afm probes using focused ion beam processing. Sensors (Basel) 19. https://doi.org/10.3390/s19204429.

Kuznetsova, T.G., Starodubtseva, M.N., Yegorenkov, N.I., Chizhik, S.A., Zhdanov, R.I., 2007. Atomic force microscopy probing of cell elasticity. Micron Microsc. Nanobiotechnol. 38, 824–833. https://doi.org/10.1016/j.micron.2007.06.011.

Le Grimellec, C., Lesniewska, E., Giocondi, M.-C., Finot, E., Vié, V., Goudonnet, J.-P., 1998. Imaging of the surface of living cells by low-force contact-mode atomic force microscopy. Biophys. J. 75, 695–703. https://doi.org/10.1016/S0006-3495(98)77559-9.

López-López, M., García-Ruiz, C., 2014. Recent non-chemical approaches to estimate the shooting distance. Forensic Sci. Int. 239, 79–85. https://doi.org/10.1016/j.forsciint.2014.03.023.

Lu, H., Wen, Y., Zhang, H., Xie, H., Shen, Y., 2019. 360° multiparametric imaging atomic force microscopy: a method for three-dimensional nanomechanical mapping. Ultramicroscopy 196, 83–87. https://doi.org/10.1016/j.ultramic.2018.09.013.

Magonov, S.N., Elings, V., Whangbo, M.-H., 1997. Phase imaging and stiffness in tapping-mode atomic force microscopy. Surf. Sci. 375, L385–L391. https://doi.org/10.1016/S0039-6028(96)01591-9.

Manfredotti, C., 2013. Imaging at the nanoscale: scanning probe microscopies applied to semiconductors. In: Lamberti, C., Agostini, G. (Eds.), Characterization of Semiconductor Heterostructures and Nanostructures, second ed. Elsevier, Oxford, pp. 467–507 (Chapter 11) https://doi.org/10.1016/B978-0-444-59551-5.00011-X.

Mou, Y., Lakadwar, J., Rabalais, J.W., 2008. Evaluation of shooting distance by AFM and FTIR/ATR analysis of GSR. J. Forensic Sci. 53, 1381–1386. https://doi.org/10.1111/j.1556-4029.2008.00854.x.

Nguyen-Tri, P., Prud'homme, R.E., 2019. Nanoscale analysis of the photodegradation of polyester fibers by AFM-IR. J. Photochem. Photobiol. A Chem. 371, 196–204. https://doi.org/10.1016/j.jphotochem.2018.11.017.

Palmeri, A., Pichini, S., Pacifici, R., Zuccaro, P., Lopez, A., 2000. Drugs in nails: physiology, pharmacokinetics and forensic toxicology. Clin. Pharmacokinet. 38, 95–110. https://doi.org/10.2165/00003088-200038020-00001.

Pandey, G., Rawtani, D., Agrawal, Y.K., 2017a. Future aspects of Halloysite nanotubes in forensic investigations. J. Nanomed. Res. 6, 2.

Pandey, G., Tharmavaram, M., Rawtani, D., Kumar, S., Agrawal, Y., 2017b. Multifarious applications of atomic force microscopy in forensic science investigations. Forensic Sci. Int. 273, 53–63. https://doi.org/10.1016/j.forsciint.2017.01.030.

Popov, K.T., Sears, V.G., Jones, B.J., 2017. Migration of latent fingermarks on non-porous surfaces: observation technique and nanoscale variations. Forensic Sci. Int. 275, 44–56. https://doi.org/10.1016/j.forsciint.2017.02.015.

Pradeep, S., Benjamin, S., 2012. Mycelial fungi completely remediate di(2-ethylhexyl) phthalate, the hazardous plasticizer in PVC blood storage bag. J. Hazard. Mater. 235–236, 69–77. https://doi.org/10.1016/j.jhazmat.2012.06.064.

Rawtani, D., Tharmavaram, M., Pandey, G., Hussain, C.M., 2019. Functionalized nanomaterial for forensic sample analysis. TrAC Trends Anal. Chem. 120, 115661. https://doi.org/10.1016/j.trac.2019.115661.

Schotsmans, E., Márquez-Grant, N., Forbes, S., 2017. Taphonomy of Human Remains: Forensic Analyses of the Dead and the Depositional Environment. Wiley, Chichester.

Seshadri, I.P., Bhushan, B., 2008. Effect of ethnicity and treatments on in situ tensile response and morphological changes of human hair characterized by atomic force microscopy. Acta Mater. 56, 3585–3597. https://doi.org/10.1016/j.actamat.2008.03.039.

Shah, A.A., Hasan, F., Hameed, A., Ahmed, S., 2008. Biological degradation of plastics: a comprehensive review. Biotechnol. Adv. 26, 246–265. https://doi.org/10.1016/j.biotechadv.2007.12.005.

Singh, B., Sharma, N., 2008. Mechanistic implications of plastic degradation. Polym. Degrad. Stab. 93, 561–584. https://doi.org/10.1016/j.polymdegradstab.2007.11.008.

Smallman, R.E., Ngan, A.H.W., 2014. Characterization and analysis. In: Smallman, R.E., Ngan, A.H.W. (Eds.), Modern Physical Metallurgy, eighth ed. Butterworth-Heinemann, Oxford, pp. 159–250. Chapter 5 https://doi.org/10.1016/B978-0-08-098204-5.00005-5.

Smijs, T., Galli, F., 2019. Forensic potential of atomic force microscopy with special focus on age determination of bloodstains. In: Tański, T., Staszuk, M., Ziębowicz, B. (Eds.),

Atomic-Force Microscopy and its Applications. IntechOpen, https://doi.org/10.5772/intechopen.77204.
Strasser, S., Zink, A., Kada, G., Hinterdorfer, P., Peschel, O., Heckl, W.M., Nerlich, A.G., Thalhammer, S., 2007. Age determination of blood spots in forensic medicine by force spectroscopy. Forensic Sci. Int. 170, 8–14. https://doi.org/10.1016/j.forsciint.2006.08.023.
Sullivan, C., Thomas, P., Stuart, B., 2019. An atomic force microscopy investigation of plastic wrapping materials of forensic relevance buried in soil environments. Aust. J. Forensic Sci. 51, 596–605. https://doi.org/10.1080/00450618.2018.1450893.
Thomas, G.L., 1978. The physics of fingerprints and their detection. J. Phys. E: Sci. Instrum. 11, 722–731. https://doi.org/10.1088/0022-3735/11/8/002.
Thomson, N.H., 2005. Imaging the substructure of antibodies with tapping-mode AFM in air: the importance of a water layer on mica. J. Microsc. 217, 193–199. https://doi.org/10.1111/j.1365-2818.2005.01399.x.
Valle, F., Bianchi, M., Tortorella, S., Pierini, G., Biscarini, F., D'Elia, M., 2012. Nanotechnology for forensic sciences: analysis of PDMS replica of the case head of spent cartridges by optical microscopy, SEM and AFM for the ballistic identification of individual characteristic features of firearms. Forensic Sci. Int. 222, 288–297. https://doi.org/10.1016/j.forsciint.2012.07.005.
Walter, B., Mairiaux, E., Faucher, M., 2017. Atomic force microscope based on vertical silicon probes. Appl. Phys. Lett. 110, 243101. https://doi.org/10.1063/1.4985125.
Wang, H., Chu, P.K., 2013. Surface characterization of biomaterials. In: Bandyopadhyay, A., Bose, S. (Eds.), Characterization of Biomaterials. Academic Press, Oxford, pp. 105–174 (Chapter 4) https://doi.org/10.1016/B978-0-12-415800-9.00004-8.
Wu, Y., Hu, Y., Cai, J., Ma, S., Wang, X., Chen, Y., Pan, Y., 2009. Time-dependent surface adhesive force and morphology of RBC measured by AFM. Micron 40, 359–364. https://doi.org/10.1016/j.micron.2008.10.003.
Xiong, Z., Mairiaux, E., Walter, B., Faucher, M., Buchaillot, L., Legrand, B., 2014. A novel dog-bone oscillating AFM probe with thermal actuation and piezoresistive detection. Sensors (Basel) 14, 20667–20686. https://doi.org/10.3390/s141120667.
Xu, W., Pang, Z., Wang, J., Ping, C., Wang, J., Peng, J., 2018. Preparation and characterization of TATB/VitonA nanocomposites. J. Nanomater. 2018, 9721730. https://doi.org/10.1155/2018/9721730.
Zhang, W., Tang, Y., Shi, A., Bao, L., Shen, Y., Shen, R., Ye, Y., 2018. Recent developments in spectroscopic techniques for the detection of explosives. Materials (Basel) 11, 1364. https://doi.org/10.3390/ma11081364.

CHAPTER 15

Energy dispersive X-ray (EDX) coupled microscopy in forensic science

1 Introduction

Energy dispersive X-ray spectroscopy (EDX) is an X-ray based technique often used in conjunction with electron microscopy techniques such as scanning electron microscopy (SEM), and transmission electron microscopy (TEM) to obtain the elemental information of the sample. Over the years, EDX has been used to analyze elements in semiconductors, medicine, cement, pollution, and nanomaterials. Such myriad applications have been possible because of their rapid data collection, ease of use, and easy coupling with electron microscopy techniques. It is also a nondestructive technique, and because it can be coupled with electron microscopy techniques, it is possible to get elemental information along with the structure and morphology of the sample.

In forensic science, the elemental composition and X-ray mapping of samples such as paints, drugs, and heavy metals are highly necessary in order to perform their confirmatory analysis (Pandey et al., 2017; Rawtani et al., 2019). Due to such reasons, there is a widespread use of this technique in the analysis of various forensic evidence. In this chapter, a brief overview of the principle and theory of EDX, its instrumentation, and its coupling with SEM and TEM will be discussed. Also, the application of these techniques for the analysis of forensic evidence is also discussed.

2 Principle and theory of EDX

Because EDX is an X-ray technique, it is important to first know the fundamentals of X-ray generation. X-rays are generated when an inner-shell electron is excited to a higher level upon being hit by a high energy incident electron that creates a hole in the ground level. When an electron's energy

Handbook of Analytical Techniques for Forensic Samples
https://doi.org/10.1016/B978-0-12-822300-0.00015-X

level corresponds to the difference between the quantum number of the quantum angular momentum, it falls back to fill that hole and generates X-rays. Apart from the X-rays generated, another set of X-rays is also generated due to the interaction of the primary incident electrons with the electric field generated by the electrons around the nucleus of an element; these are known as Bremsstrahlung X-rays. Depending on the transition of the electron from the K, L, M, or N shells, the typical characteristic X-ray is released. For instance, when the electron transitions from the L to the K shell, it is known as Kα. If it transitions from the M shell to the K shell, then it is known as a Kβ X-ray. The generated X-rays are characteristic to every element because each element possesses a different electronic configuration and thus has different energy levels. Apart from this, a very important parameter known as the ionization energy plays a key role in the generation of X-rays. Ionization energy signifies the amount of energy required to remove the electron from its shell. Therefore, based on the ionization energy of the electron, the energy of the source can be manipulated as well. The ionization energy is measured in terms of the ionization cross-section whose unit is in Barn (1 Barn $= 10^{-28}$ m^2). This cross-section increases with an increase in the energy of the primary electron and after a certain increase in the level, it falls down. Therefore, it is important to choose an electron beam of optimal energy. The Bremsstrahlung is detected as a continuum of X-rays that possesses energies ranging from the lowest to the energy of the incident primary electron. Factors such as the accelerating of the primary electron beam, and the atomic number of the element play a crucial role in the generation of such radiation. This radiation is also different for conductive and nonconductive samples. In conductive samples, the continuum occurs in all frequencies up to the excitation energies of the primary electrons. Therefore, any nonconductive samples are typically coated with a conducting material prior to use in SEM (Hodoroaba, 2020). Also, sometimes auger electrons are also generated when an electron from the outer shell with a defined kinetic energy is released at the same time the excited electron hit by the primary electrons fills the hole at the lower level. This is more common in elements of lower atomic numbers, which makes this technique highly suitable for the analysis of heavy elements. Apart from auger electrons, fluorescence is also emitted during the relaxation process of the electron from the higher level to the ground level.

During the detection of the characteristic X-rays, several processes tend to happen that may interfere with the detection of the characteristic X-rays. Once such process is X-ray scattering and absorption, which happens during its travel to the detector. The X-ray absorption here follows the Beer Lambert law and is represented in terms of the mass adsorption coefficient. Techniques such as X-ray photoelectron spectroscopy take into account such processes for detection. The detector takes into account the energy loss due to such processes as well.

SEM-EDX is the most commonly used tool in the analysis of elements. SEM is typically used to identify the surface morphology by detecting the secondary X-rays and backscattered electrons. Secondary electrons are generated when the primary incident beam hits the sample and ejects electrons from their shells. Backscattered electrons are those in which the scattering angle is more than 90 degrees. The image formed depends on the type of electrons detected and secondary electrons are the most commonly detected electrons for SEM image formation. In the case of SEM-EDX, the EDX detector is placed along with the SEM detector so that the electron signals can be easily captured, and the elements can be mapped. SEM-EDX is a relatively simple technique, both in terms of the placement of the detector and the sample preparation. Because SEM only analyzes the surface morphology, a sample of any size can be used for analysis as long as it can be made conductive. However, in TEM, the samples prepared are ultrathin and need a complex sample preparation technique. However, what this technique loses in sample preparation, it makes up for in providing a highly spatially resolved analysis of the sample. Through this technique, it is possible to analyze variable areas of the sample.

In SEM, the quantitative analysis of the samples is done through *ZAF* correction in which the intensities of the characteristic peaks of the sample (*Is*) to that of the standard or pure sample (*Ir*) are compared while multiplying it with the mean atomic number (*Z*), absorption correction (*A*), and fluorescence correction (*F*). It is commonly denoted as:

$$Ci = [ZAF](Is/Ir)$$

where *Ci* is the concentration of the sample. Modifications in this method include the use of stored standard signals, thus eliminating the need for standards every time the analysis is done. Also, semiquantitative analysis can be done through a method called fingerprinting in which the analysis is not done by considering every single parameter, but rather by focusing on certain key features that are characteristic of an element.

Meanwhile, in TEM, the samples that are often treated as thin films are quantitatively analyzed by using the Cliff-Lorimer method, which basically takes into account the intensities of the characteristic X-ray peaks (I_A, I_B)

obtained from the sample and the concentrations of the elements (C_A, C_B). It is denoted as the following:

$$I_A/I_B = (C_A/C_B)K_{AB}$$

where K_{AB} is the Cliff-Lorimer k-factor. Apart from the Cliff-Lorimer method, the Hall method is also used that was developed to overcome the uncertainties that result from the lack of useful X-rays. In this method, the intensities of the characteristic peaks of the samples are compared with the Bremsstrahlung rays of each sample (Geiss, 1992; Hodoroaba, 2020; Mishra et al., 2017; Ngo, 1999; Russ, 1984; Shindo and Oikawa, 2002).

3 Instrumentation of EDX

3.1 Detectors in EDX

The primary electron beam required for the excitation of the electrons for the generation of X-rays from the sample is typically supplied by SEM and TEM. Therefore, in EDX, the detector plays the most crucial role in detecting the generated X-rays. The detectors are composed of a collimator that is responsible for the collection of X-rays in a single channel. From the collimator, it is sent to the heart of the EDX, the solid-state detector in which a silicon crystal doped with lithium is used. When the emitted X-rays from the sample hit the crystal, this generates a photoelectron of high energy. This electron later hits other atoms in the crystal until it loses energy. Due to this process, the photoelectron ends up releasing more electrons from the atoms, due to which an electric current is generated. This current or charge generated is directly proportional to the amount of X-rays received from the sample as long as factors such as lattice imperfections and carrier mobility do not hamper the movement. Once the electric charge is generated, it is converted to pulse voltage whose height is proportional to the charges or current produced. In order to ensure smooth functioning of the detector, it is usually covered with detector windows that are responsible for maintaining optimal conditions within the detector. The detector windows commonly used are the beryllium window type and the ultrathin window type. Beryllium-type windows are used for the transmission of the X-ray spectra in a vacuum tight environment. Beryllium has a low atomic mass and density, due to which it can remain transparent to the incoming X-rays and therefore serve as a suitable window. Meanwhile, ultrathin window types are typically used to analyze light elements up to atomic number 6 through an aluminum-doped organic film with a thickness less than 1 μm; it is more commonly used in TEM. Also, any elements that provide weak signals can also be detected using this type of detector window. A modification to this type of detector in which there is no window film is also used to analyze lighter elements in SEM.

Sample preparation is a highly crucial step before performing the SEM-EDX and TEM-EDX analysis. SEM and TEM both have different modes of sample preparation and supports. In SEM, the support must be able to provide stability and should secure it from the environment. Apart from this, it should also be capable of holding a sample of any size and shape. Typically, in both electron microscopes, primary and secondary supports are used. Primary supports are those that are capable of holding the sample, and its features are designed to enhance the microscope and EDX performance. In SEM, metallic stubs made of beryllium or pyrolytic carbon are used in case of analysis with EDX. They are highly conductive and can be used for any number of samples. Liquid samples are typically coated and dried on the stub while a small amount of the solid sample is loaded on adhesive tape, and the samples are sprinkled on it. Secondary supports in SEM are designed specifically according to the type of sample to be studied and are made of both conducting and nonconducting elements. Thin metal foil can be easily cut with scissors and fitted on the metallic stub. In TEM, thin metal disks of variable shapes with a diameter less than 4 mm are used. Here too, the grids can be conducting or nonconducting and are made of copper, stainless steel, beryllium, carbon, or any polymer. These disks are composed of finely divided mesh on which the particles can be easily supported. However, if EDX analysis is to be performed, beryllium is preferred (Abd Mutalib et al., 2017; Brodsky, 1988; Echlin, 2009; Frank, 1992; Goldstein, 2003). Fig. 1 shows the instrumental setup of EDX.

4 EDX-coupled microscopic analysis of forensic samples

Coupling of EDX along with a microscope helps in telling the elemental composition of the sample that is undergoing microscopic analysis. It is generally coupled with microscopes such SEM, TEM, and AFM. However, as far as forensic science is concerned, SEM coupled with EDX is the most commonly employed technique for the investigation of evidence such as biological samples, explosives, glass, gunshot residue (GSR), paint, soil, and toxins (Zadora and Brożek-Mucha, 2003). The research works related

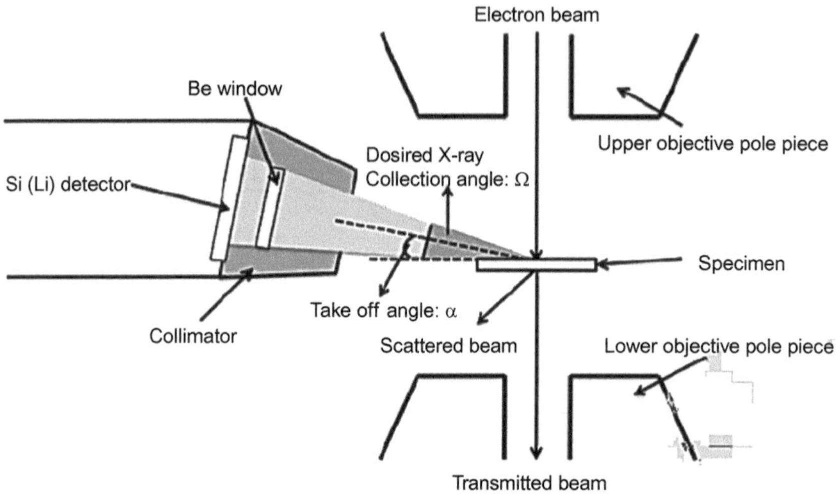

Fig. 1 Instrumental setup of energy dispersive X-ray spectroscopy (li, 2012).

to forensic sample analysis via EDX are elaborated in this section. The section has more emphasis on the EDX part than the microscopic analysis.

4.1 Biological samples

Biological samples are among the most common evidence present at a crime scene. These samples include body fluids, tissues, cremains, etc. The EDX-assisted microscopic analysis of such types of evidence is elaborated in this section.

In a research work from the late 1990s, EDX coupled with SEM was used for the analysis of elements present in semen stains. The cloth sample with semen stains was coated with a thin layer of carbon for the SEM EDX analysis. For EDX analysis, a Si-Li detector with a distance of 24 mm at an angle of 35 degrees from the sample was used. The X-rays were generated at 20 keV. The elemental analysis showed the presence of zinc, which is known to be a characteristic element in sperm (Lászik et al., 1999). Another study focused on the study of injured wounds that have received any kind of implement. Such injury implements are among the most crucial types of evidence that can be submitted in courts. Kitchen implements made of carbon steel and with a Zn-Cu or Ni-Cr coating were used to make wounds. For the EDX analysis, the angle between the detector (Si/Li) and the specimen was 25 degrees. Elements such as Cr, Ni, Zn, Cu, Si, Mn, Fe, Al, and Ca were detected in the wounds (Bai et al.,

2007). Brewer et al. used EDX attached to SEM as well as STEM (scanning transmission electron microscope) for the elemental analysis of bioagents. The model organism taken in the study was *Bacillus thuringiensis israelensis*, whose deactivated spores (via gamma irradiation) were used. The spectrum for SEM-EDX was recorded from an image with 192*256 pixels while for STEM-EDX, it was 128*128 pixels. Biologically relevant elements such as P, Ca, K, and Si were identified in the samples. The study can be helpful for other bioagents as well (Brewer et al., 2008). The analysis of cremains and burned bones was also done by Ellingham et al. using SEM-EDX. The samples were initially burned at a very high temperature, and then analyzed to check their elemental composition. Ca and P, the major components in bone, were identified. It was concluded that the Ca/P ratio in bones remains almost stable even after burning. The presence or absence of any tissue on the bone had no effect on its elemental composition (Ellingham et al., 2018).

4.2 Explosives

Explosives are commonly found in crime scenes related to bomb blasts or firearm incidents. The analysis of elements present in the explosive or its residue as well as the type of explosive used during the blast can be traced. Forensic scientists have used EDX analysis coupled with microscopes to trace the elemental composition of different explosives.

The team of Garner and Fultz published an article regarding the postblast investigation of explosives, in which EDX coupled with SEM was discussed regarding the instrumental analysis of the debris collected from the blast. The elements present in the debris as well as its microscopic appearance could be analyzed by this technique (Garner et al., 1986). In a study by Otieno-Alego, the examination of elements in explosives was carried out using EDX coupled with SEM. The microscopic analysis was done in low vacuum conditions, therefore it did not require any kind of coating on the sample. Different particles observed in the microscopic image were marked and their EDX data were collected. The EDX spectra of two of those particles are shown in Fig. 2. Particle 3 was identified as an explosive having potassium chlorate and sulfur because of their higher percentage while particle 3 was identified as an aluminum flake. Based on the EDX analysis along with Raman microscopy, it was concluded that the explosive mixture was based on chlorate and consisted of trinitrotoluene and tetryl (Otieno-Alego, 2009).

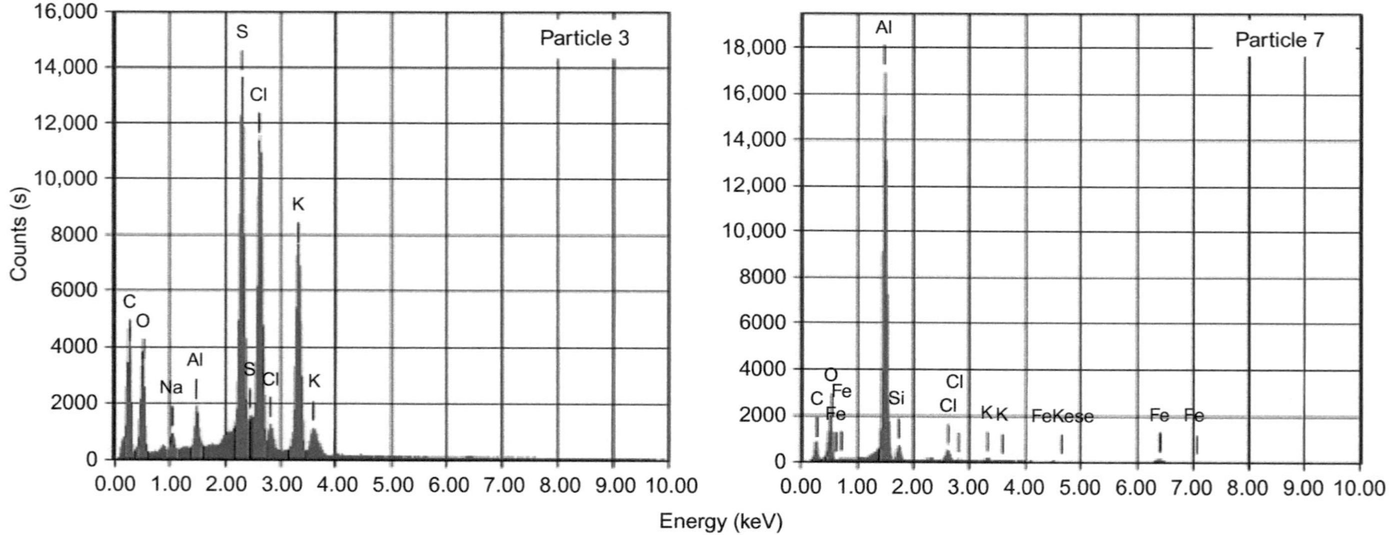

Fig. 2 EDX analysis of debris with explosive residues (Otieno-Alego, 2009).

4.3 Glass

Glass fragments are commonly encountered in cases of vehicular accidents, street fights, and robberies. The analysis of these fragments directly relates them to their original source, especially in the case of vehicles. The microscopic examination assisted by elemental analysis has contributed in such investigations over the years. This section discusses the research works regarding the forensic analysis of glass via EDX-assisted microscopy.

The article by Buscaglia in the early 1990s showed the advantages of EDX over other elemental analysis techniques such as atomic absorption spectroscopy (AAS) or atomic emission spectroscopy (AES) for the analysis of glass fragments. The merits include nondestructive sample analysis, no sample preparation, high analysis speed, and the requirement of very small amount of the sample (Buscaglia, 1994). In a research work by Falcone et al. that focused on the chemical analysis (quantitatively) of glass samples, EDX coupled with SEM was used. The accelerating voltage for the generation of electrons was 20 keV while the source of X-rays was the Kα shell. For the analysis of low energy line elements such as Na and Mg, the energy window was kept at 0.2 keV while for higher energy line elements such as Fe and Cr, it was kept at 0.34 keV. It was observed that for the detection of elements in minor quantities such as Fe and Cr, a higher acquisition time was required (up to 400 s) (Falcone et al., 2006). Zadora also performed the chemical profiling of glass fragments in order to categorize them according to their source such as the headlight of a car, a window, or any container. EDX coupled with SEM assisted in this purpose. The accelerating voltage for the generation of X-rays was 20 keV with a lifetime of 50 s. Elements such as O, Na, Mg, Al, Si, K, Ca, and Fe were identified in the glass fragments. The study proved to helpful in properly classifying these fragments based on their composition to different source groups (Zadora, 2007). A similar kind of work regarding the classification of traces of glass was done by Ramos and Zadora using SEM-EDX analysis. The elemental composition of glass detected via EDX along with the analysis of the refractive index using glass refractive index measurements helped in the classification of the collected glass traces (Ramos and Zadora, 2011). Another research work by El-Defter et al. performed the elemental analysis of Australian window glass collected from schools as well as residential and commercial buildings using EDX coupled with SEM. The accelerating voltage was 15 keV, and the EDX assembly in SEM had an Si-Li detector. Elements such as Na, Mg, Si, and Ca were detected in EDX analysis, and their ratios in six different combinations were used for discriminating the glass fragments collected from different locations.

However, the study proved that LIBS (laser-induced breakdown spectroscopy) has more advantages for elemental analysis over conventional techniques such as EDX (El-Deftar et al., 2014).

4.4 Gunshot residue

GSR is commonly encountered in crime scenes that involved firearm incidents. Spectroscopic and chromatographic techniques are used for its detection and quantification. However, for elemental analysis, EDX is commonly practiced. These are among the most common types of forensic samples analyzed via EDX-assisted microscopy. The research works related to such analyses are discussed in this section.

In a research work, EDX coupled with SEM was used to examine GSR in order to differentiate between the various kinds of ammunition. The residues of primers were collected from the hands of the shooters. Chemicals with element combinations such as PbSbBa, SbBa, SbSn, PbSbSn, Ba, Hg, etc., were observed through the elemental analysis. It was concluded that different kinds of ammunition had different compositions and ratios of elements in the primer residues collected (Brozek-Mucha and Jankowicz, 2001). Brozek-Mucha again performed SEM-EDX analysis of GSR in order to compare the cartridge cases. Similar kinds of elements were detected in these GSR as well. However, in order to compare their composition with the primer present in the inner cartridge, it was concluded that the composition varies a lot from the primer ignition to the last step of the projectile leaving the muzzle. It was recommended to carefully study these composition changes for proper comparison of cartridges (Brożek-Mucha, 2007). In a proficiency test conducted by ENFSI regarding the sensing and identification of GSR, EDX coupled with SEM was used. Laboratories from 28 nations participated in this test. The EDX analysis focused on the detection of Pb, Sb, and Ba in the GSR samples. The test results upon comparison with the previous tests revealed an improved capability of the new test for GSR detection (Niewoehner et al., 2008). In a case where a policeman fired at a person, the GSR from the discharged bullet as well as the intermediate target (concrete pole) were compared using EDX-coupled SEM. The analysis revealed that the elemental composition of GSR from the bullet and pole were almost similar, which helped in concluding that the death of the person was accidental. Fig. 3 shows the EDX spectra along with the SEM image of the particles collected from the concrete pole (Hu et al., 2009). EDX coupled with SEM has also been used by Amadasi et al. to prove

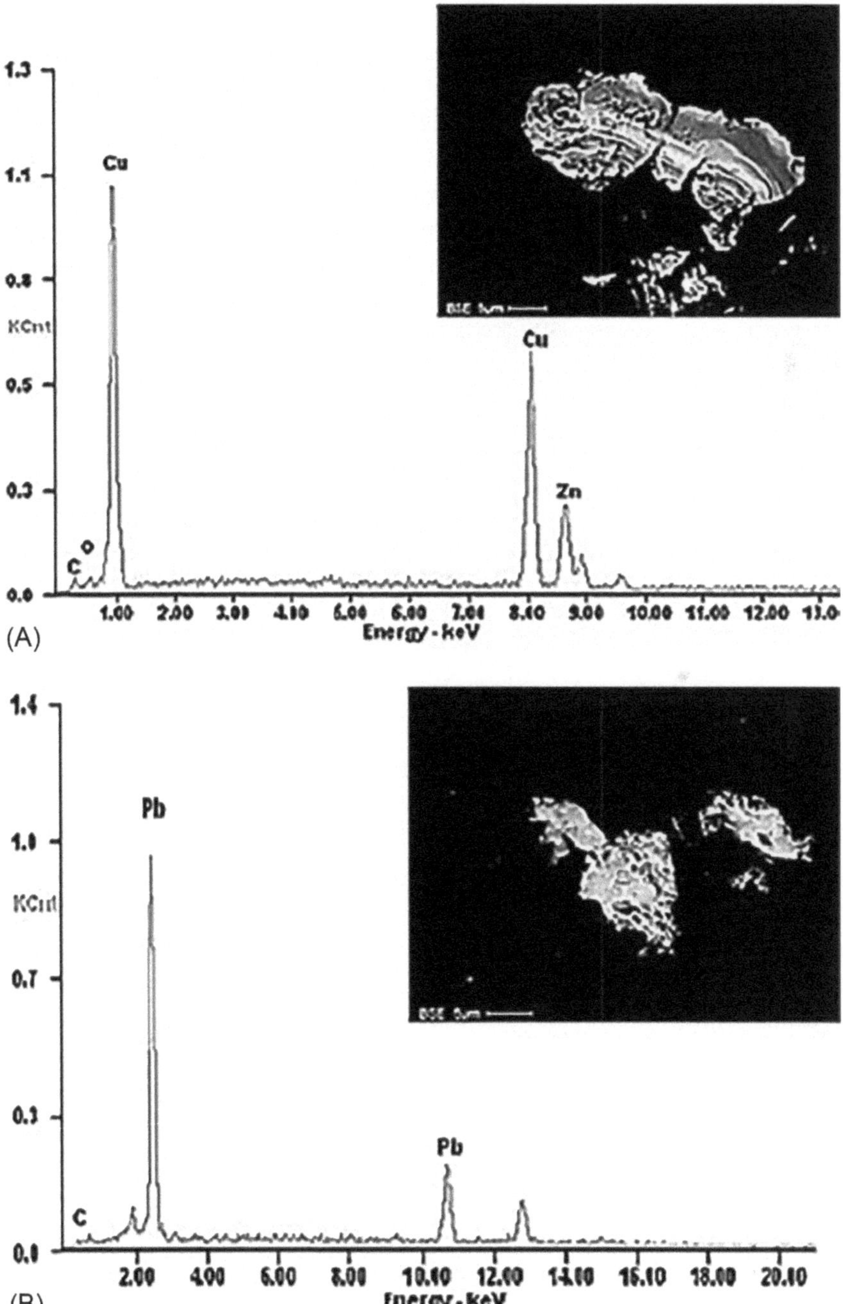

Fig. 3 EDX spectra and SEM images of GSR with Cu and Zn (A); and (B) Pb (Hu et al., 2009).

the existence of GSR present in gunshot wounds of even cremated bone specimens. The bone samples were shot and then charred in an oven at very high temperatures. The GSR was analyzed via SEM-EDX before and after charring. Elements such as Ba, Pb, and Sb were present in bones before and after charring (Amadasi et al., 2012). French et al. focused on the secondary transfer of GSR, which is the transfer of GSR from the shooter to any other person or object not involved in the shooting. Two scenarios were considered: a handshake after shooting and an exchange of a gun. Through the EDX and SEM analysis, GSRs of similar size and composition were observed in the person who did the handshake and exchanged the gun with the shooter. The study therefore recommended to carefully investigate such cases because the presence of GSR does not mean that the person has fired the bullet (French et al., 2014). French and Morgan also did a similar study for the tertiary transfer of GSR, in which samples were collected from people standing near the firearm discharge. GSR was found on them as well through the SEM-EDX analysis. The results again emphasized a proper investigation of firearm cases, considering in mind the transfer patterns for GSR.

4.5 Paints

Paint chips are often exchanged when two object collide with each other. This is crucial evidence in cases of hit and run as well as robberies. Forensic analysts examine the paint fragments using spectroscopic, chromatographic, and microscopic techniques. EDX coupled with microscopes has assisted in knowing the elemental composition of different paint samples.

Barium sulfate, a common extender used in paints, was examined by Haswell et al. using EDX coupled with SEM as well as TEM. The team examined three paintings in order to check whether the barium sulfate used was the same. EDX coupled with SEM assisted in characterizing the interparticle composition while EDX with TEM helped in characterizing the intraparticle composition. The particles were initially imaged via SEM-EDX, and the particles with average strontium content were further visualized via TEM-EDX. The presence of strontium indicated that the barium sulfate used was of natural origin. It was concluded on the basis of EDX analysis that all three paintings had different barium sulfates (Haswell et al., 2008). EDX coupled with SEM with a TES (transition edge sensor) microcalorimeter was used for the analysis of automotive paints. The use of TES helped in removing the problem of overlapping peaks observed in conventional SEM-EDX analysis with a solid-state detector (SSD). The analysis was done using TES as well as SSD at an accelerating voltage of 15 keV. The lifetime for TES was 300 s while for SSD, it was 100 s. The elements identified in the paint sample had peaks of Ba and S, indicating the presence of barium sulfate in paints (Fig. 4) (Nakai et al., 2010). Gum binders that are utilized in watercolor paints were also examined by Sano and Cumpson using SEM-EDX. An accelerating voltage of 15 keV was used during the analysis. The EDX analysis helped in knowing the surface composition of the samples at a sampling depth of around 1 μm. The analysis showed the presence of Cl and O elements in the binder samples. Fig. 5 shows the microscopic images along with EDX mapping for O and Cl of different samples (Sano and Cumpson, 2016). Malek et al. combined EDX and SEM along with other techniques such as Fourier transform infrared spectroscopy and Raman spectroscopy for the complete analysis of the

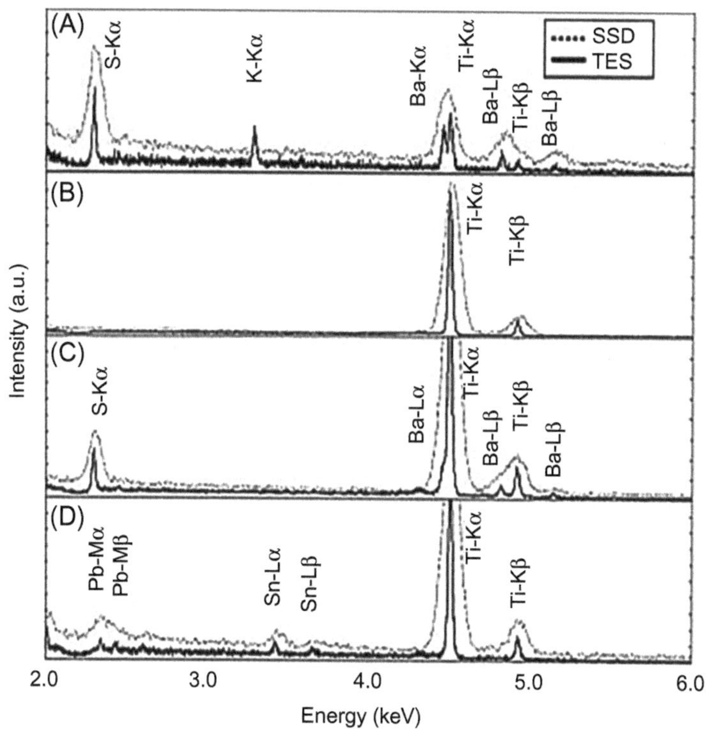

Fig. 4 EDX analysis of automotive paints using TES (transition edge sensor) and SSD (solid state detector) (Nakai et al., 2010).

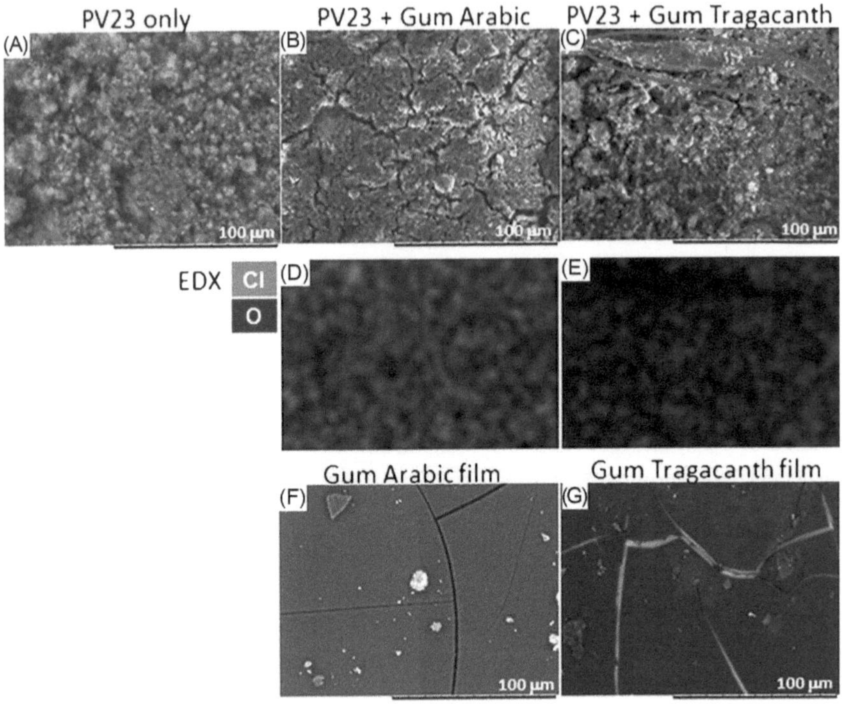

Fig. 5 SEM analysis of different gum binders (A, B, C, F, G) and EDX mapping for Cl and O (D and E) on binders shown in B and C (Sano and Cumpson, 2016).

composition of paint chips collected from automobiles. The accelerating voltage for the analysis was 15 keV while the resolution of the EDX detector was 133 eV for the X-rays from the Kα shell of Mn. It was concluded that while EDX analysis was not sufficient for the speciation of the molecules, it could still provide data that could be corroborated with the FTIR and Raman results (Malek et al., 2019).

4.6 Soil

Soil analysis becomes important in forensic science because it directly connects the evidence from which it has been collected to the location where the crime was possibly committed. It may also indicate a particular location or area where the victim or the suspect might have visited. The elemental analysis of soil gives an idea about the mineral composition of soil, and thus relates it to a particular location (Pirrie et al., 2009).

In a study by McVicar and Graves, EDX coupled with SEM was used for the forensic assessment of soil samples. The accelerating voltage was 20 keV, and a minimum count of 1500 was maintained for the X-rays. The elemental analysis helped in identifying and classifying the soil samples into different types based on their mineral composition (Mcvicar and Graves, 1997). Pye published an article showing the potential of EDX coupled with SEM for the forensic examination of different kinds of soils, rocks, dusts, and sediments. For the analysis, an EDX instrument with an Si-Li detector was used. Elements such as O, Al, Fe, Si, S, Zr, Ti, Ca, and Ni were detected in different samples with variable weight percentages. The study could be helpful for forensic experts that need to examine soils from different sources (Pye, 2004). In another research work, EDX coupled with SEM was used for the mineralogical analysis of soils, and further the speciation of arsenic in such soil samples. The accelerating voltage for the microscopic analysis was 20 keV, and the EDX analysis was performed from the area that had a clearer view of the particles. Many elements such as O, Al, Si, P, K, Ca, Ti, Mn, Fe, Cu, and As were present in most of the soil samples taken. However, elements such as Pb and Ba were present in only a few samples, and Mg, S, and Zn were present in only one soil sample (Gómez-Parrales et al., 2011). Woods et al. examined soil samples from Australia to study their elemental composition. Fig. 6 shows the EDX analysis of Australian soil from different locations. The study could be helpful in solving cases from Australia where soil is collected as evidence (Woods et al., 2014).

4.7 Toxins

The ingestion of toxins such as pesticides can be fatal at higher doses and have adverse health effects at milder doses. Many cases related to poisoning use the elemental analysis of stomach contents to understand the possible reason for poisoning. EDX analysis has played an important role in this area.

A research work by Kinoshita et al. used EDX analysis for the examination of the stomach contents in poisoning through the ingestion of methomyl, belonging to the carbamate class of pesticides. An Si/Li detector was used for the instrument that was generating X-rays via an Rh anode at a voltage of 50 keV. Through the analysis, high peaks of Si and S were identified, which confirmed the presence of methomyl in the stomach (Fig. 7) (Kinoshita et al., 2013). The research team also performed a similar kind of work by analyzing the stomach contents in fenitrothion poisoning. EDX analysis showed the presence of P and S in the stomach. This result

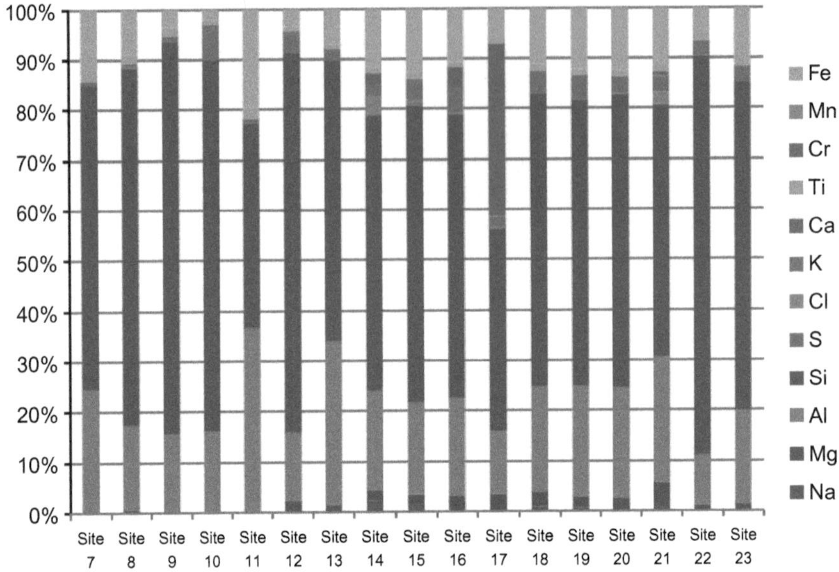

Fig. 6 EDX analysis data for Australian soil from different sites (Woods et al., 2014).

along with other analyses helped in concluding that the ingestion of fenitrithion was the reason behind the poisoning (Tanaka et al., 2015).

5 Conclusion

In this chapter, we presented a brief overview of the principle and theory behind EDX and how it can be coupled with electron microscopy techniques such as SEM and TEM, along with their applications in the detection of various forensic science evidence such as biological samples, explosives, glass, gunshot residue, paint, soil, and toxins. EDX is a technique commonly used for the elemental mapping of a substance by detecting the characteristic X-ray peaks and Bremsstrahlung radiation. It is also possible to perform a quantitative analysis of the elements present. When this technique is combined with electron microscopy techniques, it provides information about structure and morphology. SEM is commonly used to analyze the surface morphology of samples while TEM is used to provide an internal view of the sample. In forensic science, it is important to obtain structural as well as the compositional proof in order to establish the positive identity of the substance. The images obtained from these techniques can also be easily

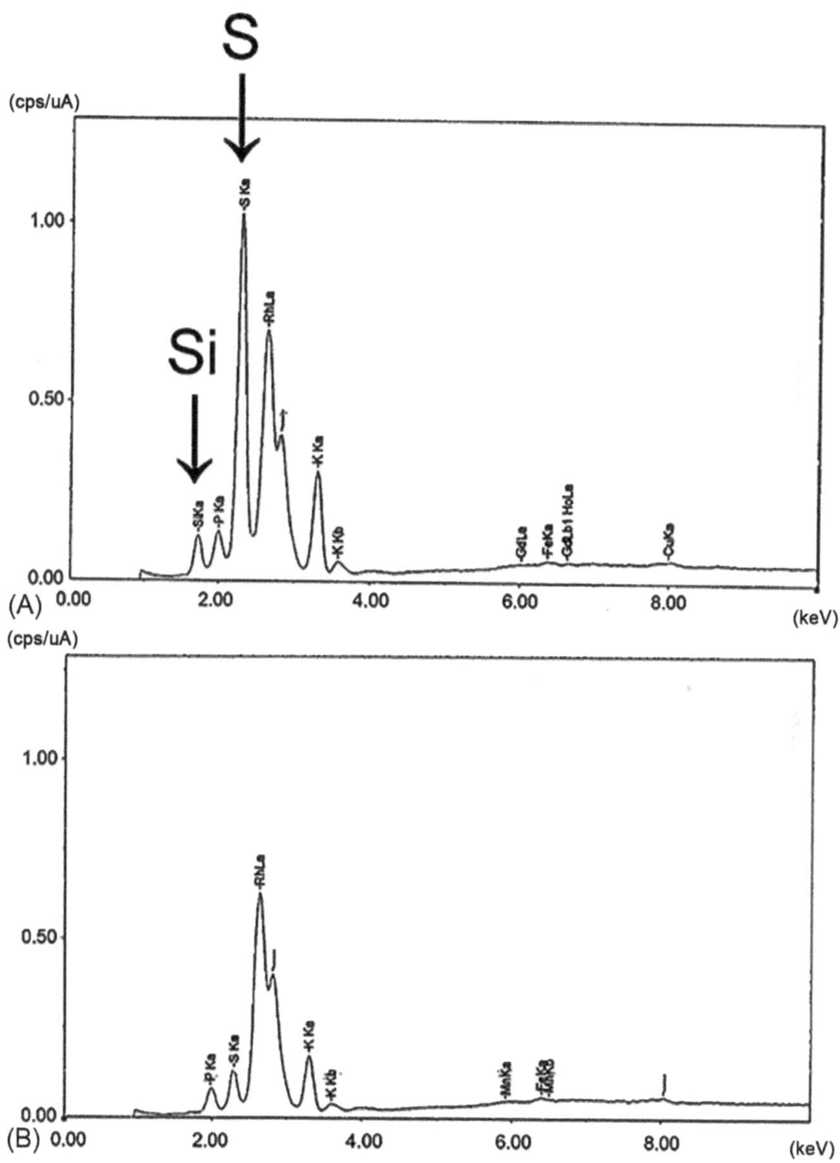

Fig. 7 EDX analysis of stomach contents from (A) victim; and (B) control (Kinoshita et al., 2013).

provided in court as testimony. There are, however, certain drawbacks to these techniques, especially in terms of the cost, maintenance, and sample preparation. However, despite the drawbacks, these techniques are an indispensable tool for the confirmatory analysis of evidence.

References

Abd Mutalib, M., Rahman, M.A., Othman, M.H.D., Ismail, A.F., Jaafar, J., 2017. Scanning electron microscopy (SEM) and energy-dispersive X-ray (EDX) spectroscopy. In: Membrane Characterization. Elsevier, pp. 161–179, https://doi.org/10.1016/B978-0-444-63776-5.00009-7.

Amadasi, A., Brandone, A., Rizzi, A., Mazzarelli, D., Cattaneo, C., 2012. The survival of metallic residues from gunshot wounds in cremated bone: a SEM–EDX study. Int. J. Legal Med. 126, 525–531. https://doi.org/10.1007/s00414-011-0661-7.

Bai, R., Wan, L., Li, H., Zhang, Z., Ma, Z., 2007. Identify the injury implements by SEM/EDX and ICP-AES. Forensic Sci. Int. 166, 8–13. https://doi.org/10.1016/j.forsciint.2006.03.008.

Brewer, L.N., Ohlhausen, J.A., Kotula, P.G., Michael, J.R., 2008. Forensic analysis of bioagents by X-ray and TOF-SIMS hyperspectral imaging. Forensic Sci. Int. 179, 98–106. https://doi.org/10.1016/j.forsciint.2008.04.020.

Brodsky, E.L., 1988. The mechanical design of thin beryllium windows for synchrotron radiation. Nucl. Inst. Methods Phys. Res. A 266, 358–361. https://doi.org/10.1016/0168-9002(88)90411-1.

Brożek-Mucha, Z., 2007. Comparison of cartridge case and airborne GSR—a study of the elemental composition and morphology by means of SEM-EDX. X-Ray Spectrom. 36, 398–407. https://doi.org/10.1002/xrs.990.

Brozek-Mucha, Z., Jankowicz, A., 2001. Evaluation of the possibility of differentiation between various types of ammunition by means of GSR examination with SEM–EDX method. Forensic Sci. Int. 123, 39–47. https://doi.org/10.1016/S0379-0738(01)00518-7.

Buscaglia, J.A., 1994. Elemental analysis of small glass fragments in forensic science. Anal. Chim. Acta 288, 17–24. https://doi.org/10.1016/0003-2670(94)85112-3.

Echlin, P., 2009. Handbook of Sample Preparation for Scanning Electron Microscopy and X-ray Microanalysis. Springer US, Boston, MA, https://doi.org/10.1007/978-0-387-85731-2.

El-Deftar, M.M., Speers, N., Eggins, S., Foster, S., Robertson, J., Lennard, C., 2014. Assessment and forensic application of laser-induced breakdown spectroscopy (LIBS) for the discrimination of Australian window glass. Forensic Sci. Int. 241, 46–54. https://doi.org/10.1016/j.forsciint.2014.04.040.

Ellingham, S.T.D., Thompson, T.J.U., Islam, M., 2018. Scanning electron microscopy-energy-dispersive X-ray (SEM/EDX): a rapid diagnostic tool to aid the identification of burnt bone and contested cremains. J. Forensic Sci. 63, 504–510. https://doi.org/10.1111/1556-4029.13541.

Falcone, R., Sommariva, G., Verità, M., 2006. WDXRF, EPMA and SEM/EDX quantitative chemical analyses of small glass samples. Microchim. Acta 155, 137–140. https://doi.org/10.1007/s00604-006-0531-z.

Frank, J. (Ed.), 1992. Electron Tomography. Springer US, Boston, MA, https://doi.org/10.1007/978-1-4757-2163-8.

French, J., Morgan, R., Davy, J., 2014. The secondary transfer of gunshot residue: an experimental investigation carried out with SEM-EDX analysis: the secondary transfer of GSR studied experimentally using SEM-EDX. X-Ray Spectrom. 43, 56–61. https://doi.org/10.1002/xrs.2498.

Garner, D.D., Fultz, M.L., Byall, E.B., 1986. The ATF approach to post-blast explosives detection and identification. J. Energ. Mater. 4, 133–148. https://doi.org/10.1080/07370658608011337.

Geiss, R.H., 1992. EDS. In: Encyclopedia of Materials Characterization. Elsevier, pp. 120–134, https://doi.org/10.1016/B978-0-08-052360-6.50015-1.

Goldstein, J. (Ed.), 2003. Scanning Electron Microscopy and X-ray Microanalysis, third ed. Kluwer Academic/Plenum Publishers, New York.

Gómez-Parrales, I., Bellinfante, N., Tejada, M., 2011. Study of mineralogical speciation of arsenic in soils using X ray microfluorescence and scanning electronic microscopy. Talanta 84, 853–858. https://doi.org/10.1016/j.talanta.2011.02.010.

Haswell, R., Zeile, U., Mensch, K., 2008. Van Gogh's painting grounds: an examination of barium sulphate extender using analytical electron microscopy – SEM/FIB/TEM/EDX. Microchim. Acta 161, 363–369. https://doi.org/10.1007/s00604-007-0910-0.

Hodoroaba, V.-D., 2020. Energy-dispersive X-ray spectroscopy (EDS). In: Characterization of Nanoparticles. Elsevier, pp. 397–417, https://doi.org/10.1016/B978-0-12-814182-3.00021-3.

Hu, S., Shen, H., Wang, S., Fang, C., 2009. Trajectory reconstruction through analysis of trace evidence in bullet-intermediate target interaction by SEM/EDX. J. Forensic Sci. 54, 1349–1352. https://doi.org/10.1111/j.1556-4029.2009.01158.x.

Ii, S., 2012. Nanoscale chemical analysis in various interfaces with energy dispersive X-ray spectroscopy and transmission electron microscopy. In: X-Ray Spectroscopy., https://doi.org/10.5772/31645.

Kinoshita, H., Tanaka, N., Jamal, M., Kumihashi, M., Okuzono, R., Tsutsui, K., Ameno, K., 2013. Application of energy dispersive X-ray fluorescence spectrometry (EDX) in a case of methomyl ingestion. Forensic Sci. Int. 227, 103–105. https://doi.org/10.1016/j.forsciint.2012.08.026.

Lászik, A., Weisser, H.-J., Keresztury, L., Pollak, S., Papp, G., Pozsgai, I., 1999. PCR typing of human semen stains after SEM-EDX examination. Int. J. Legal Med. 112, 376–379. https://doi.org/10.1007/s004140050017.

Malek, M., Nakazawa, T., Kang, H.-W., Tsuji, K., Ro, C.-U., 2019. Multi-modal compositional analysis of layered paint chips of automobiles by the combined application of ATR-FTIR imaging, Raman microspectrometry, and SEM/EDX. Molecules 24, 1381. https://doi.org/10.3390/molecules24071381.

Mcvicar, M.J., Graves, W.J., 1997. The forensic comparison of soils by automated scanning electron microscopy. J. Can. Soc. Forensic Sci. 30, 241–261. https://doi.org/10.1080/00085030.1997.10757104.

Mishra, R.K., Zachariah, A.K., Thomas, S., 2017. Energy-dispersive X-ray spectroscopy techniques for nanomaterial. In: Microscopy Methods in Nanomaterials Characterization. Elsevier, pp. 383–405, https://doi.org/10.1016/B978-0-323-46141-2.00012-2.

Nakai, I., Ono, Y., Li, Q., Nishiwaki, Y., Tanaka, K., Nakayama, S., 2010. Capability of aTES microcalorimeter SEM-EDS system for forensic analysis of automotive paint and gunshot residue. Surf. Interface Anal. 42, 402–410. https://doi.org/10.1002/sia.3113.

Ngo, P.D., 1999. Energy Dispersive Spectroscopy., p. 11.

Niewoehner, L., Andrasko, J., Biegstraaten, J., Gunaratnam, L., Steffen, S., Uhlig, S., Antoni, S., 2008. GSR2005—continuity of the ENFSI proficiency test on identification of GSR by SEM/EDX. J. Forensic Sci. 53, 162–167. https://doi.org/10.1111/j.1556-4029.2008.00594.x.

Otieno-Alego, V., 2009. Some forensic applications of a combined micro-Raman and scanning electron microscopy system. J. Raman Spectrosc. 40, 948–953. https://doi.org/10.1002/jrs.2206.

Pandey, G., Rawtani, D., Agrawal, Y.K., 2017. Future aspects of halloysite nanotubes in forensic investigations. J. Nanomed. Res. 6.

Pirrie, D., Power, M.R., Rollinson, G.K., Wiltshire, P.E.J., Newberry, J., Campbell, H.E., 2009. Automated SEM-EDS (QEMSCAN®) mineral analysis in forensic soil investigations: testing instrumental reproducibility. In: Ritz, K., Dawson, L., Miller, D. (Eds.),

Criminal and Environmental Soil Forensics. Springer Netherlands, Dordrecht, pp. 411–430, https://doi.org/10.1007/978-1-4020-9204-6_26.
Pye, K., 2004. Forensic examination of rocks, sediments, soils and dusts using scanning electron microscopy and X-ray chemical microanalysis. Geol. Soc. Lond., Spec. Publ. 232, 103–121. https://doi.org/10.1144/GSL.SP.2004.232.01.11.
Ramos, D., Zadora, G., 2011. Information-theoretical feature selection using data obtained by scanning electron microscopy coupled with and energy dispersive X-ray spectrometer for the classification of glass traces. Anal. Chim. Acta 705, 207–217. https://doi.org/10.1016/j.aca.2011.05.029.
Rawtani, D., Tharmavaram, M., Pandey, G., Hussain, C.M., 2019. Functionalized nanomaterial for forensic sample analysis. TrAC Trends Anal. Chem. 120, 115661. https://doi.org/10.1016/j.trac.2019.115661.
Russ, J.C., 1984. Fundamentals of Energy Dispersive X-ray Analysis, Butterworths Monographs in Materials. Butterworths, London; Boston.
Sano, N., Cumpson, P.J., 2016. Surface analysis characterisation of gum binders used in modern watercolour paints. Appl. Surf. Sci. 364, 870–877. https://doi.org/10.1016/j.apsusc.2015.12.162.
Shindo, D., Oikawa, T., 2002. Energy dispersive X-ray spectroscopy. In: Analytical Electron Microscopy for Materials Science. Springer, Tokyo.
Tanaka, N., Kinoshita, H., Takakura, A., Jamal, M., Kumihashi, M., Uchiyama, Y., Tsutsui, K., Ameno, K., 2015. Combination of energy-dispersive X-ray fluorescence spectrometry (EDX) and head-space gas chromatography mass spectrometry (HS-GC/MS) is a useful screening tool for stomach contents. Rom. J. Leg. Med. 23, 43–44. https://doi.org/10.4323/rjlm.2015.43.
Woods, B., Paul Kirkbride, K., Lennard, C., Robertson, J., 2014. Soil examination for a forensic trace evidence laboratory—part 2: elemental analysis. Forensic Sci. Int. 245, 195–201. https://doi.org/10.1016/j.forsciint.2014.10.018.
Zadora, G., 2007. Glass analysis for forensic purposes—a comparison of classification methods. J. Chemom. 21, 174–186. https://doi.org/10.1002/cem.1030.
Zadora, G., Brożek-Mucha, Z., 2003. SEM–EDX—a useful tool for forensic examinations. Mater. Chem. Phys. 81, 345–348. https://doi.org/10.1016/S0254-0584(03)00018-X.

CHAPTER 16

Mass spectrometry in forensic chemistry

1 Introduction

Forensic science, especially forensic toxicology, requires appropriate characterization techniques to properly detect compounds with high reliability, accuracy, and validity. In order to achieve this, chromatography techniques are widely used to separate the analytes; they are usually paired with a powerful technique known as mass spectrometry. This technique allows the highly sensitive and accurate determination of a myriad of compounds and provides rich elemental information, even in a mixture of compounds. It also has unparalleled detection limits and speed and thus is highly suitable for the detection of forensic samples such as pesticides, drugs, toxins, and poisons (Maurer, 1998; Rawtani et al., 2019).

Mass spectrometers target the mass-to-charge ratio of any molecule and detects it on this basis. The theory and principle of mass spectrometers will be discussed further in this chapter. It can not only detect the molecules, but can also perform its quantitative detection. This is done through a series of steps that mainly involves the ionization of the molecules, due to which fragmentation occurs that is further detected. In this chapter, the principle, theory, and instrumentation of mass spectrometry are discussed. Also, the applications of the technique in the analysis of various forensic evidence is discussed.

2 Principle and theory of mass spectrometry

As mentioned in the introduction, mass spectrometry is based on the generation of the positive ion of a molecule. During ionization, a molecule breaks into ions containing either an odd or even number of electrons. These ions have different chemical properties. The positively charged ion is known as the parent ion or the molecular ion (M+). These molecular ions, due to their high instability, further break into a single or multiple daughter ion. These ions are separated on the basis of their mass-to-charge ratio (m/z)

Handbook of Analytical Techniques for Forensic Samples
https://doi.org/10.1016/B978-0-12-822300-0.00016-1

and detected by a detector to form a spectrum that consists of varying intensities of the ions. The intensity of the ions in the spectra is directly proportional to their abundance in the molecule or the mixture. A typical mass spectrum consists of the m/z ratio in the *x*-axis and the intensity in the *y*-axis. The ions also provide information on the nature and structure of the molecule. If the M+ ion formed has a high molecular weight and is single, it means that the compound is pure. However, if its contaminated, then multiple M+ ions are found on the spectrum. The M+ ion peak with the highest intensity is known as the base peak and corresponds directly to the molecular weight of the compound.

Conventionally, mass spectrometers work under high vacuum in order to allow the facile movement of ions to the detector without undergoing any elastic or inelastic collision, as they not only influence the trajectory of the ions but also make them lose their energy. In order to produce this, various ion sources are used that will be discussed further in this chapter.

3 Overview of mass spectrometry

A typical mass spectrometer consists of an ionization source, a mass analyzer, and a detector (Fig. 1). There are numerous ionization sources that are responsible for the ionization of the samples and these sources are dependent

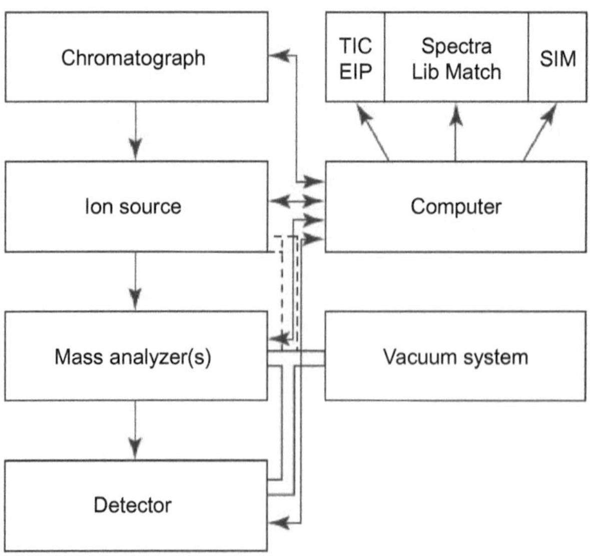

Fig. 1 Instrumentation of mass spectrometry (Rockwood et al., 2018).

on the type of sample used. In this section, a brief overview of the ionization sources, the mass analyzers, and the detectors used is discussed.

3.1 Sources of ionization

The ionization sources are highly crucial, and the sources developed take into consideration the transfer of internal energy during the process and the physicochemical properties of the sample. The ionization sources are mainly divided into two parts based on the nature of the sample, that is, whether they are solid or liquid samples. In the analysis of liquid samples, the solution is converted into droplets through nebulization and later ionized at atmospheric pressure. In the case of solid samples, they are usually subjected to a series of sample preparation techniques that allow them to be embedded in a matrix, followed by the bombardment of photons that ionize the sample particles.

Electron ionization is a source in which a heated electron source such as a filament radiates electrons and accelerates them toward the ions. Through their path, they collide with the gas-phase samples. When the electrons reach the sample molecules, a transfer of energy between the electron and the sample takes place when the energy of the electron matches the transitional energy of the electrons in the molecule. In order to ensure proper ionization of the samples, the entire setup is kept in a vacuum chamber under a high accelerating voltage. However, the high energy of the electron can also cause overfragmentation of the molecule, due to which there may be multiple peaks but with low intensities in the spectra.

Another technique that is also useful for the analysis of liquid samples is the chemical ionization technique, also known as the soft ionization technique. Here, a lower amount of energy as compared to the electron ionization is required and is highly dependent on the sample. A great advantage with this technique is that the fragmentation is less extensive, due to which a simpler and clearer spectrum is formed. Chemical ionization takes places via multiple processes including proton transfer, charged species transfer, or the formation of an adduct from the condensed phase to the gas phase. In adduct formation, due to the ionization, there is a formation of a reaction product due to the direct addition of some distinct molecules. In the charge transfer, gases that have high ionization potential react with the samples. Chemical desorption ionization is also a variation of the chemical ionization technique in which the temperature of the chamber is reduced to prevent the pyrolysis of the samples (Hoffmann and Stroobant, 2007). These samples are then placed on a heated filament and for a few seconds, they get desorbed and ionized. Field ionization is also another technique in which very strong electric fields produce ions. This is also a soft ionization technique. Under the presence of an intense electric field, the samples that are usually in a gas phase undergo ionization to produce positive ions through quantum tunneling. These ions are later passed onto the mass analyzer through a hole in the ionization chamber, which is similar to the setup used in electron ionization and chemical ionization.

An interesting technique that is preferably used for the analysis of solid samples is secondary ion mass spectrometry or SIMS. In this technique, secondary ions that are emitted when high-intensity electrons hit a surface are detected. These secondary ions are capable of providing a detailed chemical analysis of the sample. Static SIMS is a technique in which a very low dose of primary ions is impinged on the sample such that only one primary electron hits a particular sample surface at a time. Due to this, it is possible to analyze only one area of the sample while the remaining areas of the sample remain unaffected. Another technique is the use of field desorption, which is highly useful for thermally sensitive compounds and polymers. The mechanism of this technique is similar to that of chemical desorption in that the sample is deposited on a heated filament and is passed through an electric field. Plasma and laser desorption techniques are also used. However, these techniques require a high level of expertise and are currently used with modifications. Matrix-assisted laser desorption ionization (MALDI) is one such technique in which the sample is dissolved in a suitable solvent known as the matrix and the sample is dried while the solvent is removed. Due to this, the matrix molecules get embedded with the sample molecules. They are, however, still isolated and can be easily analyzed further through the ablation of large parts of the dried sample-analyte solution with the help of a laser. Due to the excessive heat of the laser, portions of the matrix get ablated while simultaneously ionizing the samples. Due to the isolation of the sample molecules by the matrix, there is a very minimal change of formation of sample clusters, due to which the sensitivity of the analysis is very high (Hoffmann and Stroobant, 2007; Rockwood et al., 2018).

3.2 Mass analyzers

Mass analyzers are used to separate the ions produced during the ionization stage via the m/z ratio. These work by using a combination of an electric field that is either kept constant or variable and a magnetic field. There are mainly two types of mass analyzers that are classified on the basis of their functioning. Mass analyzers that allow the transmission of ions of different masses along a time scale are known as scanning analyzers. There are also mass analyzers that allow the simultaneous transmission of the ions at a time such as quadrupole, ion trap, time-of-flight, Fourier transform ion cyclotron resonance (FTICR), and Fourier transform orbitrap.

Quadrupole analyzers separate ions based on their m/z ratios according to the stability of the trajectories of the ions. Typically, a quadrupole consists of four circular or hyperbolic rods that are parallel to each other. When a positive ion enters between the quadrupoles, it faces a surge of either positive or negative potential. Based on the manipulation of the potentials, the movement and the path of the ions can be controlled according to the m/z ratio. Ion trap analyzers are another set of mass analyzers in which a radio frequency quadrupole serves as a trap to capture ions in two or three dimensions. In three-dimensional (3D) ion traps, there is a circular electrode in which two electrodes, usually ellipsoidal, are fixed on the top and bottom of the ion. In two-dimensional (2D) ion traps, instead of ellipsoidal electrodes, rod-shaped quadrupoles are used that are responsible for the forward and backward reflection of the ions. Another type of mass analyzer is the FTICR, which traps the ions with the help of electric trapping plates and excitations in between a magnetic field, and separates the trapped ions as per their cyclotron frequency. The ions pass close to the electrodes and emit a signal known as the free induction decay. This signal is used to develop a mass spectrum through Fourier transformation. The electrostatic trap mass analyzer uses two sets of electrodes: one outer barrel-shaped electrode and an inner spindle-shaped electrode. The signals emitted from the trapped ions are converted into a mass spectrum through Fourier transformation (Gratz et al., 2006). Another commonly used mass analyzer is the time-of-flight analyzer, which is based on the principle that heavier ions take a longer time to travel from the ionization source to the detector. Therefore, when the ions are released from the ionization source, they enter a high-vacuum environment so that the velocities of the ions are solely determined by their masses (Lockyer and Vickerman, 2001).

3.3 Detectors

The detectors are responsible for the generation of an electric current obtained from the incident ions; this is further used to generate the mass spectra. The detector used for an instrument is dependent on the application and the type of sample. The detectors work either by counting the mass of individual ions at a time or by counting the masses of multiple ions.

Detectors commonly used in mass spectrometers are photographic plates, Faraday cups, electron multipliers, and electrooptical ion detectors. A photographic plate is responsible for detecting ions with the same m/z ratio at the same time. When the ions hit the photographic plate, they create a dark spot on the plate at different positions. The darker the spot, the greater the abundance of the ion in the compound. However, this technique has been replaced by other alternative detectors.

A Faraday cup is a detector in which a small cup with a minute hole is connected to a resistor. When the ions hit the inside of the cup, they undergo multiple reflections, due to which they lose their energy and become neutralized due to the gain or loss of electrons that further generates an electric current. The resistor serves to control the flow of this current, which is further amplified for detection. The greater the current, the greater the abundance of ions in the sample. However, this detector is not sensitive and has a slow response time. These detectors are used to identify the precise ratio of specific ions in a compound.

Electron multipliers work by subjecting the ions to a high accelerating voltage that strikes a dynode to set up a cascade of such strikes. Due to this, secondary ions are generated of multiple cages that are further converted to electrons and amplified to produce a current that is further detected. Conventionally, this detector usually comprises 10–20 dynodes in order to achieve a significant emission of the secondary particles. There have been several modifications to this detector in which the individual dynodes that are placed separately have been replaced by a single continuous dynode to achieve high efficiency. Lastly, the electrooptical ion detectors are responsible for converting the ions into photons for detection. The Daly detector is the most common detector and consists of dynodes, a scintillation screen, and a photomultiplier. These dynodes can be used either in the positive or negative potential. In positive potential, the ions are attracted toward the negative dynode and the vice versa happens in case of negative potential mode. The secondary electrons generated hit the photomultiplier screen, due to which photons are emitted that are further converted to an electric current and amplified. The detectors discussed are supported by computers that serve as an interface to control various factors that enable the user to achieve a high efficiency of separation (Baghel et al., 2017; Hoffmann and Stroobant, 2007; Kaklamanos et al., 2012).

4 Applications of mass spectrometry for forensic investigations

MS has been widely used in different industries such as pharmaceuticals, textiles, food, and many more. Forensic scientists have also been using this technology, either alone or in conjugation with other analytical techniques, for various types of evidence (Correa et al., 2016; Orellana et al., 2013). This section discusses the applications of MS for forensic investigations such as the analysis of biological samples, the examination of drugs and foods, explosives and gunshot residue analysis, fiber analysis, the detection of inks in questionable documents, and the investigation of oil spills and petroleum products.

4.1 Analysis of biological samples

Different types of biological samples of forensic importance such as hair, body fluids, and toxins of biological origin have been analyzed through MS either alone or in combination with other techniques that are mostly related to chromatography. This section discusses some research works that utilized MS for analyzing such biological samples.

In a study carried out by Kharbouche et al., MS was used in combination with gas chromatography (GC) for the analysis of hair with traces of ethyl glucuronide, which is a metabolite formed during ethanol metabolism. Negative chemical ionization was used in the tandem MS (NCI-MS) for the analysis. The hair samples were collected from corpses with a history of alcohol consumption; hair spiked with analytes was also analyzed. The SRM (selected reaction monitoring) mode was adopted for the analysis using m/z transitions for quantification (352-163), identification (347-119), and the internal standard (352-163). The NCI-MS spectra of ethyl glucuronide is shown in Fig. 2 (Kharbouche et al., 2009). MS has also been used for the detection of different compounds and drugs in body fluids. In a study, MS has been used in combination with liquid chromatography (LC) to determine illicit drugs and their metabolites such as opiates, cocaine, and lysergic acid in body fluids such as urine, blood, and serum. Spiked body fluid samples with drugs were used for the study. Atmospheric pressure chemical ionization MS was used in the positive ionization mode to analyze the drugs. The study concluded that this variant of MS can analyze illicit drugs in body fluid without any requirement of derivatization (Bogusz et al., 1998). In another study, time-of-flight MS was used for the identification of drugs in urine samples. The ionization source for MS was electrospray ionization while the mode of operation was positive ion. This variant of MS with LC efficiently analyzed drugs in urine with an automatic target database (Ojanperä et al., 2006). A similar kind of study was also carried out by Gergov et al. that

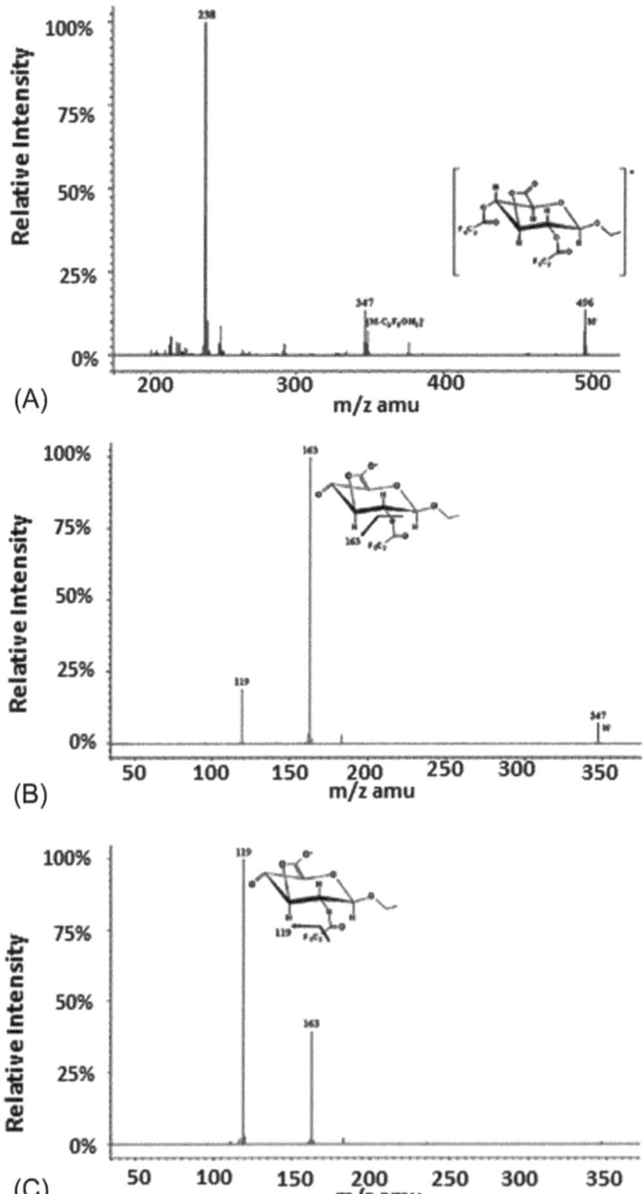

Fig. 2 NCI-MS spectra of ethyl glucuronide (Kharbouche et al., 2009). (A) Full scan mode mass spectra of negative chemical ion of EtG postderivatization with perfluoropentanoic anhydride. (B) PFP derivative of EtG formed after collision induced fragmentation (5 eV) (precursor ion: m/z 347). (C) PFP derivative of EtG formed after collision induced fragmentation (25 eV) (precursor ion: m/z 347).

employed an automatic target database search for the identification of drugs in urine (Gergov et al., 2001). In another study, ethyl glucuronide was analyzed in urine through MS with electrospray ionization (Weinmann et al., 2004). Drugs have also been analyzed in blood serum. In a study by Bogusz et al., amphetamines (underivatized) and associated phenethylamines were detected in serum using atmospheric pressure chemical ionization MS preceded by high-performance LC (HPLC). The molecular ions and remains (in the protonated state) of drugs were observed in the MS spectra (Bogusz et al., 2000). In another study by Dresnan et al., tandem MS was used for the examination of methanandamide and cannabinoids (synthetic) in serum. A total of 101 serum samples were analyzed during the study, and 56.4% were found to be positive for the aforementioned drugs (Dresen et al., 2011). Another study by Fredriksson et al. showed the potential of MS for the analysis of ricin (protein toxin) from the extracts of castor beans. The sample was prepared for MS analysis by enzymatic digestion using trypsin. Matrix-assisted laser desorption ionization MS with a UV ionization lazer was used for this purpose. The analysis was able to identify the intact ricin as well as its peptides (Fredriksson et al., 2005).

4.2 Examination of drugs and foods

Forensic chemistry mostly utilizes MS for the analysis of unlawful drugs, and this practice has been going on for many years. In a study, MS was used in combination with HPLC for the detection of scopolamine and atropine (chemicals responsible for poisoning by Datura). The electrospray ionization mode was used for MS analysis. The drugs were analyzed from the seed extracts and from the contents of the stomach. The chemical profile of the drugs as analyzed by MS was similar in the stomach contents and the seed extracts (Steenkamp et al., 2004). Ecstasy tablets have also been detected using desorption electrospray ionization MS. Linear ion trap triple quadrupole MS was used for the analysis. The spectra obtained revealed the protonated and deprotonated ions of ecstasy (Leuthold et al., 2006). The designer drug analogs (counterfeited drugs) of vardenafil, tadalafil, and sildenafil have also been analyzed through MS. The analysis was carried out using electrospray ionization. Fourier transform ion cyclotron resonance MS used in this study had high resolution and accuracy in predicting the structures of the analogs through the spectra obtained (Gratz et al., 2006). Another study has also utilized MS for the analysis of various kinds of drugs such as Vicodin, methamphetamine, prednisone, Xanax, and cannabis extract (Rodriguez-Cruz, 2006).

Food samples have also been investigated for the presence of any contaminants or foreign materials through MS. Nielen et al. showed the ability of desorption electrospray ionization MS for the examination of food contaminants of chemical origin such as pesticides, natural toxins, veterinary drugs, and food additives (Nielen et al., 2011). In a study, high-resolution MS was used for the analysis of pesticides that were collected from the surfaces of apples, grapes, and oranges using swabbing. The collected swabs were investigated for pesticide residues using MS coupled with direct analysis in real-time ionization. Almost 86% of the residual pesticides were detected, even at a very low concentration of 10 ng/g in the grapes or 2 ng/g in the apples and oranges (Edison et al., 2011). In another study, paper spray MS was used to forensically discriminate among beers. A total of 141 samples of 8 different brands of lager beers were taken for the analysis. The mass range for the analysis was 100–1800 m/z while the voltage for the paper spray was 40 kV. The peaks denote those maltooligosaccharide ions which had resulted from the hydrolytic action of the polysaccharides in barley. Fig. 3 shows the MS spectra of lager beer (Pereira et al., 2016).

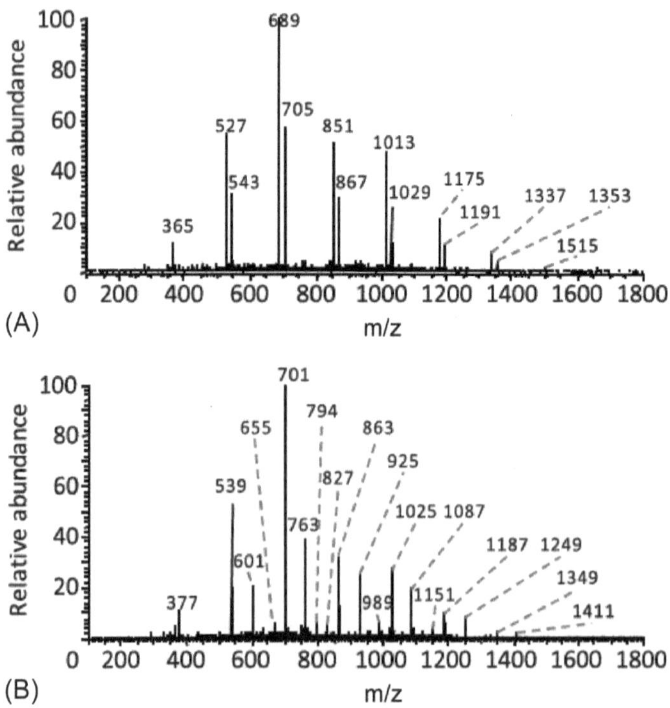

Fig. 3 (A) Paper spray (+) and (B) paper spray (−) of lager beer (Pereira et al., 2016).

4.3 Explosives and gunshot residue analysis

Explosives that are commonly found at terrorist attack scenes need to be precisely analyzed. MS has been widely used by researchers for the detection of explosives. In a study by Zhao and Yinon, inorganic oxidizers that are used in explosives were detected using electrospray ionization MS. The mass spectra were obtained in positive and negative ion modes. The voltage in the ion modes was 5 kV and 4 kV, respectively. A cluster ion series was observed in the MS spectra in both ion modes, which helped in the identification of oxidizers (Zhao and Yinon, 2002). In another study, adduct ions of trinitrotriazacyclohexane (RDX) were also detected using MS in combination with GC. Electrospray ionization and atmospheric pressure chemical ionization conditions were used for the analysis, in which characteristic ions were produced by the attachment of the negative ion to the molecule of the analyte. It was observed that in electrospray ionization, adduct ion formation was not governed by the RDX's self-decomposition while in atmospheric pressure chemical ionization, self-decomposing RDX was responsible for the adduct ion formation (Gapeev et al., 2003). Triacetone triperoxide (TATP) has also been detected using MS with different kinds of ionization such as methane negative ion chemical ionization, electron ionization, and ammonia positive chemical ionization (Sigman et al., 2006). TATP along with PETN (pentaerythritoltetranitrate) was also analyzed in a study carried out by Benson et al. using isotope ratio MS. The values of isotopes of carbon and hydrogen were measured for TATP while for PETN, isotopes of carbon and nitrogen were measured (Benson et al., 2009).

Gunshot residue (GSR) investigations have also been carried out using MS. In a study, GSR from rimfire ammunition of 0.22 caliber was analyzed by secondary ion MS. The elemental sensitivity of MS was higher as compared to the scanning electron microscope analysis coupled with an energy dispersive X-ray spectrometer. The study showed the applicability of this variant of MS for analyzing GSR collected in this micrometer size range (Coumbaros et al., 2001). GSR fired from a 0.4 caliber pistol was analyzed for the identification of metals such as barium, antimony, and lead using inductively coupled plasma MS. The analysis was able to detect metal concentrations even lower than 1 ppm. The study helped in discriminating between GSR collected from the hands of nonshooters and shooters (Sarkis et al., 2007). MS coupled with inductively coupled plasma has also been used to differentiate different types of bullets through the analysis of GSR (Udey et al., 2011). Morelato et al. also utilized desorption electrospray ionization MS for the screening of GSR. The quadrupole time-of-flight MS was used during the study to check for the presence of organic and inorganic GSR (Morelato et al., 2012).

4.4 Fiber analysis

The forensic analysis of fibers is an important step during crime investigations. Various types of physical and chemical characterization of fibers are carried out. The analysis of dyes in the fibers is one such analysis, and this examination extensively utilizes MS analysis. Dyed fibers have been extensively studied using different variants of MS such as LC-MS, capillary electrophoresis MS (CE-MS), and desorption techniques (Goodpaster and Liszewski, 2009). This section discusses the research works that utilized MS and its various variants for the analysis of dyes extracted from various kinds of fibers.

In a study by Petrick et al., electrospray ionization MS was used for the analysis of dyes in fibers made up of polyester and acrylic. The dyes were extracted from fibers which were 0.5 cm long. In addition to the extracted dyes, 13 dispense and 15 basic dye standards were also taken. Dye analysis using MS was able to differentiate fibers that had similar chromatograms and UV spectra (Petrick et al., 2006). In another study, basic dyes coming from acrylic fibers such as azo and methane were detected using electrospray ionization MS; the voltage for electrospray ionization was 3.72 kV. The length of the acrylic fiber used was 2 mm. Three basic dyes were extracted from this length of fiber. The study helped in quantifying the dyes present in fibers as well as characterized them at the molecular level (Stefan et al., 2009). In another study, real time-time-of-flight MS was used for the direct detection of dyes in textile fibers. The study focused on the detection of indigotin, quercetin, and alizarin in reference samples in solution form as well as in textile fibers that were dyed with them. Direct analysis in real-time ionization MS with high resolution was used for this purpose, in which no sample preparation was required. This variant of MS detected these organic dyes in animal fibers such as wool and silk as well as in plant fibers such as cotton and linen (Selvius DeRoo and Armitage, 2011). In another study related to the direct detection of dyes in textile fibers, matrix-assisted laser desorption electrospray ionization MS was used. The direct analysis of dye removed the step of dye extraction from the fabric material, thereby omitting the tedious sample preparation steps. In this variant of MS, dye was desorbed from the fabric using the IR laser with a wavelength of

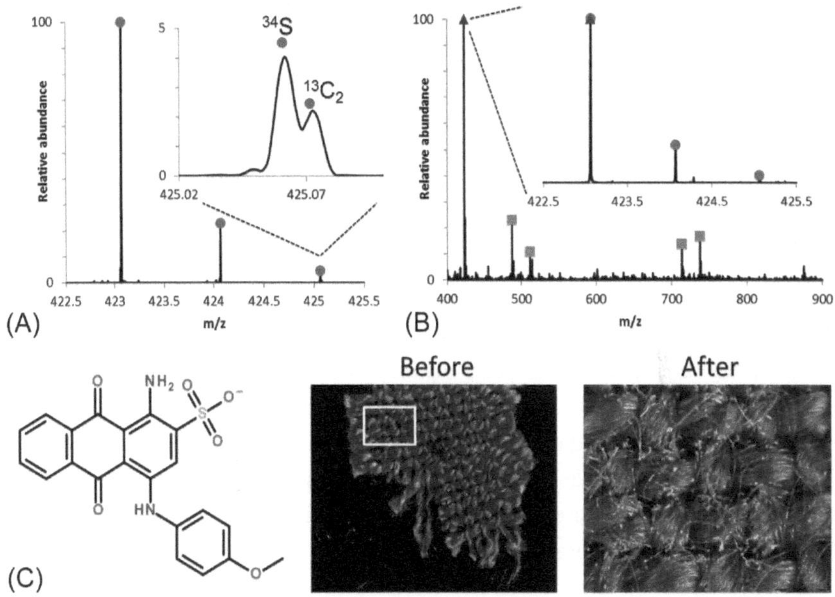

Fig. 4 MS spectra of nylon fiber dyed with acid blue 40 using (A) electrospray ionization, and (B) matrix-assisted laser desorption electrospray ionization; (C) structure of acid blue with nylon fiber area that was analyzed (Cochran et al., 2013).

2.94 μm. The MS measurements were taken in both positive and negative ion modes. In addition to the merit of no sample preparation, another advantage of direct analysis was the minimal damage caused to the fiber during analysis, proving this technique to be almost nondestructive. Fig. 4 depicts the MS spectra and structure of the dye analyzed in fiber (Cochran et al., 2013).

4.5 Detection of inks in questionable documents

The analysis of questionable documents includes the examination of inks and their constituents present in that paper. This examination is necessary to estimate the correct age and source of written documents as well as their authentication. MS has been widely utilized by forensic analysts for this purpose. Here are some of the research works that have employed MS and its combinations for the analysis of inks present in questionable documents.

Jones et al. used real-time MS for direct discrimination among writing inks. As discussed in the previous section, there was no requirement of sample preparation in this case as well because of direct detection. For the analysis, black and blue inks from ballpoint, gel, and ink pens were collected. Atmospheric pressure ionization was used in time-of-flight MS, which was operated in the positive ion mode; the spectra were obtained in the m/z range of 60–1000. The protonated molecules obtained in the MS spectra were because of the thermally unstable constituents of inks (Jones et al., 2006). Another research work used matrix-assisted laser desorption ionization MS for the analysis of dyes in pen inks straight from the paper. The laser used was a nitrogen pulsed laser with a wavelength of 337 nm. The study was helpful in detecting dye mixtures with multiple charges, which are commonly used in pen inks (Dunn and Allison, 2007). Desorption MS with electrospray ionization has also been utilized for the examination of inks on questionable documents. Writing samples were prepared using blue ballpoint pens that possessed dyes such as basic violet 3, basic blue 7, solvent blue 2, and solvent blue 26, as identified by the MS spectra (Ifa et al., 2007). In a similar kind of study, laser desorption ionization MS in combination with HPLC was used to differentiate blue inks from ballpoint pens. The detected inks were classified into 26 classes because the MS spectra provided data about the molecular weight and quantity of dyes in inks (Weyermann et al., 2007). Electrospray ionization MS has been used by Williams et al. for the analysis of documents written with black ink. The sources of black inks were gel, ballpoint, and rollerball pens. The ions that were observed in the MS spectra were responsible for the differentiation of inks from different pens (Williams et al., 2009). Matthews et al. presented a case report of dye detection in inks from ballpoint pens by utilizing time-of-flight MS with laser desorption ionization. The study focused on causing minimal damage to the document, and collected single fiber bearing ink from its surface. The MS spectra were obtained in reflection mode. Fig. 5 shows the MS spectra of inks collected from blue, black, and red pens (Matthews et al., 2011). Paper spray MS has also been used for ink analysis from ballpoint pens (da Silva Ferreira et al., 2015).

4.6 Investigation of oil spills and petroleum products

Oil spills and cases related to arson usually contain evidence of crude oil or petroleum products. MS in combination mostly with chromatographic techniques has been utilized by forensic analysts for the investigation of oil spills and petroleum products. A compilation by Wang et al. focused on the characterization and identification of biomarkers from oil spills through GC-MS. Biomarkers in petroleum do not degrade easily in the

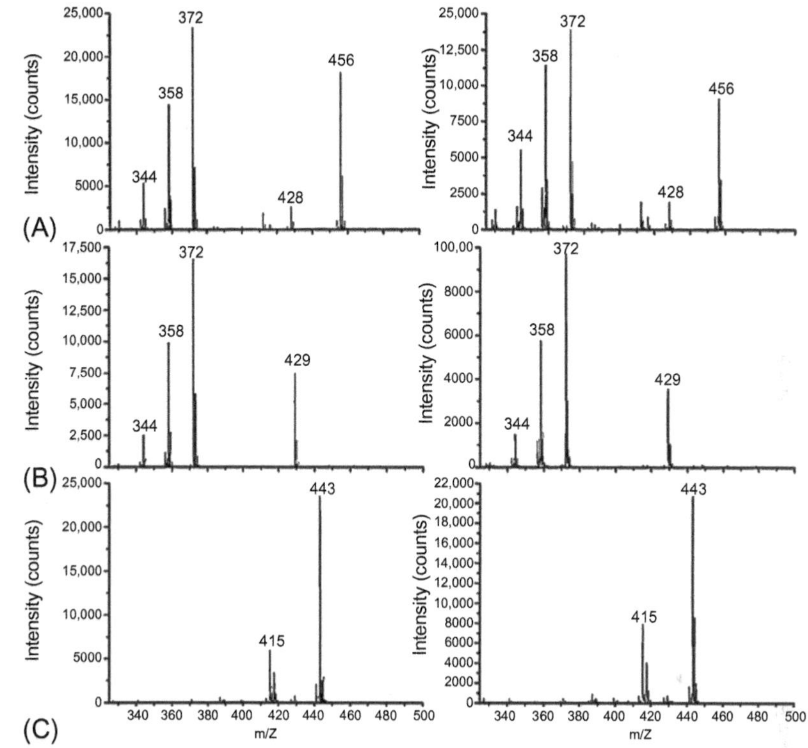

Fig. 5 MS spectra of (A) blue; (B) black; (C) red pen inks collected from single fiber from paper *(right)* and solid line *(left)* (Matthews et al., 2011).

environment and exhibit unique properties under different geological conditions (Wang et al., 2006a).

In a study by Keto and Wineman, accelerants based on petroleum collected from very contaminated extracts of fire debris were analyzed by MS in combination with GC. The MS spectra was obtained in the range of 50–200 a.m.u. using electron impact ionization, with an electron energy of 70 eV. The analysis was helpful in getting signals of gasoline as well as medium and heavy petroleum distillates in highly contaminated samples of fire debris (Keto and Wineman, 1991). In another study, an oil spill was identified using MS combined with GC. The spiked samples consisted of petroleum biomarkers and PAH (polycyclic aromatic hydrocarbon). The real samples for the oil spill were comprised of crude oil, spill samples, refined products, and oil mixtures that were collected from the forensic oil spill laboratory. The sources of the oil spills were identified using this analysis (Christensen et al., 2004). Oil spills and petroleum products have also been detected by the fingerprinting of five different diamondoids (sesquiterpane, steranes, terpane, diamantane, adamantane) using GC-MS (Wang et al., 2006b). Alberta oil sands, which are known to contain huge bitumen reserves, were studied along with petroleum products using GC-MS. Compounds such as PAHs and diamondoids (terpane, sesquiterpane, sterane) were identified in the samples on the MS spectra that were taken in the selected ion monitoring mode (Yang et al., 2011). Triple quadrupole MS has also been used for the analysis of petroleum products in samples collected from oil spills at Deepwater Horizon (Adhikari et al., 2017).

5 Conclusion

Forensic science requires the rapid and accurate analysis of samples. This is easily possible due to mass spectrometry in which compounds can be analyzed through this technique by the determination of their molecular weight and elemental composition. This technique mainly consists of four crucial parts: the ionization source, the mass analyzer, the detector, and the computer. The ionization source is responsible for the ionization of the molecules and is different for solid and liquid samples. The mass analyzers are responsible for separating the ions according to their mass-by-charge ratio. Once separated, these ions are sent further to the detector where every signal received from the ions is converted to an electric signal and then amplified. The amplified electric signals are later displayed on the computer as the mass spectra plotted against the intensity and the mass-to-charge ratio. In forensic science, this technique has been used for the analysis of biological samples as well as the examination of drugs and foods, explosives, gunshot residue, fibers, the detection of inks in questionable documents, and oil spills and petroleum products. Such applications were discussed in this chapter.

References

Adhikari, P.L., Wong, R.L., Overton, E.B., 2017. Application of enhanced gas chromatography/triple quadrupole mass spectrometry for monitoring petroleum weathering and forensic source fingerprinting in samples impacted by the deepwater horizon oil spill. Chemosphere 184, 939–950. https://doi.org/10.1016/j.chemosphere.2017.06.077.

Baghel, U.S., Singh, A., Singh, D., Sinha, M., 2017. Application of mass spectroscopy in pharmaceutical and biomedical analysis. In: Sharmin, E., Zafar, F. (Eds.), Spectroscopic Analyses—Developments and Applications. InTech, https://doi.org/10.5772/intechopen.70655.

Benson, S.J., Lennard, C.J., Maynard, P., Hill, D.M., Andrew, A.S., Roux, C., 2009. Forensic analysis of explosives using isotope ratio mass spectrometry (IRMS)—preliminary

study on TATP and PETN. Sci. Justice 49, 81–86. https://doi.org/10.1016/j.scijus.2009.03.004.
Bogusz, M.J., Maier, R.-D., Krüger, K.-D., Kohls, U., 1998. Determination of common drugs of abuse in body fluids using one isolation procedure and liquid chromatography-atmospheric-pressure chemical-ionization mass spectrometry*. J. Anal. Toxicol. 22, 549–558. https://doi.org/10.1093/jat/22.7.549.
Bogusz, M.J., Kruger, K.-D., Maier, R.-D., 2000. Analysis of underivatized amphetamines and related phenethylamines with high-performance liquid chromatography-atmospheric pressure chemical ionization mass spectrometry. J. Anal. Toxicol. 24, 77–84. https://doi.org/10.1093/jat/24.2.77.
Christensen, J.H., Hansen, A.B., Tomasi, G., Mortensen, J., Andersen, O., 2004. Integrated methodology for forensic oil spill identification. Environ. Sci. Technol. 38, 2912–2918. https://doi.org/10.1021/es035261y.
Cochran, K.H., Barry, J.A., Muddiman, D.C., Hinks, D., 2013. Direct analysis of textile fabrics and dyes using infrared matrix-assisted laser desorption electrospray ionization mass spectrometry. Anal. Chem. 85, 831–836. https://doi.org/10.1021/ac302519n.
Correa, D.N., Santos, J.M., Eberlin, L.S., Eberlin, M.N., Teunissen, S.F., 2016. Forensic chemistry and ambient mass spectrometry: a perfect couple destined for a happy marriage? Anal. Chem. 88, 2515–2526. https://doi.org/10.1021/acs.analchem.5b02397.
Coumbaros, J., Kirkbride, K.P., Klass, G., Skinner, W., 2001. Characterisation of 0.22 caliber rimfire gunshot residues by time-of-fight secondary ion mass spectrometry (TOF-SIMS): a preliminary study. Forensic Sci. Int. 119 (1), 72–81.
da Silva Ferreira, P., Silva, D.F.D.A., Augusti, R., Piccin, E., 2015. Forensic analysis of ballpoint pen inks using paper spray mass spectrometry. Analyst 140, 811–819. https://doi.org/10.1039/C4AN01617C.
Dresen, S., Kneisel, S., Weinmann, W., Zimmermann, R., Auwärter, V., 2011. Development and validation of a liquid chromatography-tandem mass spectrometry method for the quantitation of synthetic cannabinoids of the aminoalkylindole type and methanandamide in serum and its application to forensic samples. J. Mass Spectrom. 46, 163–171. https://doi.org/10.1002/jms.1877.
Dunn, J.D., Allison, J., 2007. The detection of multiply charged dyes using matrix-assisted laser desorption/ionization mass spectrometry for the forensic examination of pen ink dyes directly from paper. J. Forensic Sci. 52, 1205–1211. https://doi.org/10.1111/j.1556-4029.2007.00535.x.
Edison, S.E., Lin, L.A., Gamble, B.M., Wong, J., Zhang, K., 2011. Surface swabbing technique for the rapid screening for pesticides using ambient pressure desorption ionization with high-resolution mass spectrometry. Rapid Commun. Mass Spectrom. 25, 127–139. https://doi.org/10.1002/rcm.4831.
Fredriksson, S.-Å., Hulst, A.G., Artursson, E., de Jong, A.L., Nilsson, C., van Baar, B.L.M., 2005. Forensic identification of neat ricin and of ricin from crude castor bean extracts by mass spectrometry. Anal. Chem. 77, 1545–1555. https://doi.org/10.1021/ac048756u.
Gapeev, A., Sigman, M., Yinon, J., 2003. Liquid chromatography/mass spectrometric analysis of explosives: RDX adduct ions. Rapid Commun. Mass Spectrom. 17, 943–948. https://doi.org/10.1002/rcm.1006.
Gergov, M., Boucher, B., Ojanperä, I., Vuori, E., 2001. Toxicological screening of urine for drugs by liquid chromatography/time-of-flight mass spectrometry with automated target library search based on elemental formulas: LC/TOFMS screening for drugs in urine. Rapid Commun. Mass Spectrom. 15, 521–526. https://doi.org/10.1002/rcm.260.
Goodpaster, J.V., Liszewski, E.A., 2009. Forensic analysis of dyed textile fibers. Anal. Bioanal. Chem. 394, 2009–2018. https://doi.org/10.1007/s00216-009-2885-7.
Gratz, S.R., Gamble, B.M., Flurer, R.A., 2006. Accurate mass measurement using Fourier transform ion cyclotron resonance mass spectrometry for structure elucidation of

designer drug analogs of tadalafil, vardenafil and sildenafil in herbal and pharmaceutical matrices. Rapid Commun. Mass Spectrom. 20, 2317–2327. https://doi.org/10.1002/rcm.2594.

Hoffmann, E.D., Stroobant, V., 2007. Mass Spectrometry: Principles and Applications, third ed. John Wiley, Chichester, West Sussex, England; Hoboken, NJ.

Ifa, D.R., Gumaelius, L.M., Eberlin, L.S., Manicke, N.E., Cooks, R.G., 2007. Forensic analysis of inks by imaging desorption electrospray ionization (DESI) mass spectrometry. Analyst 132, 461. https://doi.org/10.1039/b700236j.

Jones, R.W., Cody, R.B., McClelland, J.F., 2006. Differentiating writing inks using direct analysis in real time mass spectrometry. J. Forensic Sci. 51, 915–918. https://doi.org/10.1111/j.1556-4029.2006.00162.x.

Kaklamanos, G., Aprea, E., Theodoridis, G., 2012. Mass spectrometry. In: Chemical Analysis of Food: Techniques and Applications. Elsevier, pp. 249–283, https://doi.org/10.1016/B978-0-12-384862-8.00009-1.

Keto, R.O., Wineman, P.L., 1991. Detection of petroleum-based accelerants in fire debris by target compound gas chromatography/mass spectrometry. Anal. Chem. 63, 1964–1971.

Kharbouche, H., Sporkert, F., Troxler, S., Augsburger, M., Mangin, P., Staub, C., 2009. Development and validation of a gas chromatography–negative chemical ionization tandem mass spectrometry method for the determination of ethyl glucuronide in hair and its application to forensic toxicology. J. Chromatogr. B 877, 2337–2343. https://doi.org/10.1016/j.jchromb.2008.11.046.

Leuthold, L.A., Mandscheff, J.-F., Fathi, M., Giroud, C., Augsburger, M., Varesio, E., Hopfgartner, G., 2006. Desorption electrospray ionization mass spectrometry: direct toxicological screening and analysis of illicit Ecstasy tablets. Rapid Commun. Mass Spectrom. 20, 103–110. https://doi.org/10.1002/rcm.2280.

Lockyer, N.P., Vickerman, J.C., 2001. SIMS, time-of-flight, and surface analysis. In: Reference Module in Materials Science and Materials Engineering, pp. 8624–8628, https://doi.org/10.1016/B978-0-12-803581-8.03310-5.

Matthews, B., Walker, G.S., Kobus, H., Pigou, P., Bird, C., Smith, G., 2011. The analysis of dyes in ball point pen inks on single paper fibres using laser desorption ionisation time of flight mass spectrometry (LDI-TOFMS). Forensic Sci. Int. 209, e26–e30. https://doi.org/10.1016/j.forsciint.2011.03.033.

Maurer, H.H., 1998. Liquid chromatography–mass spectrometry in forensic and clinical toxicology. J. Chromatogr. B Biomed. Sci. Appl. 713, 3–25. https://doi.org/10.1016/S0378-4347(97)00514-8.

Morelato, M., Beavis, A., Ogle, A., Doble, P., Kirkbride, P., Roux, C., 2012. Screening of gunshot residues using desorption electrospray ionisation–mass spectrometry (DESI–MS). Forensic Sci. Int. 217, 101–106. https://doi.org/10.1016/j.forsciint.2011.10.030.

Nielen, M.W.F., Hooijerink, H., Zomer, P., Mol, J.G.J., 2011. Desorption electrospray ionization mass spectrometry in the analysis of chemical food contaminants in food. TrAC Trends Anal. Chem. 30, 165–180. https://doi.org/10.1016/j.trac.2010.11.006.

Ojanperä, S., Pelander, A., Pelzing, M., Krebs, I., Vuori, E., Ojanperä, I., 2006. Isotopic pattern and accurate mass determination in urine drug screening by liquid chromatography/time-of-flight mass spectrometry. Rapid Commun. Mass Spectrom. 20, 1161–1167. https://doi.org/10.1002/rcm.2429.

Orellana, F.A., Gálvez, C.G., Orellana, F.A., Gálvez, C.G., Roldán, M.T., García-Ruiz, C., Roldán, M.T., García-Ruiz, C., 2013. Applications of laser-ablation-inductively-coupled plasma-mass spectrometry in chemical analysis of forensic evidence. TrAC Trends Anal. Chem. 42, 1–34. https://doi.org/10.1016/j.trac.2012.09.015.

Pereira, H.V., Amador, V.S., Sena, M.M., Augusti, R., Piccin, E., 2016. Paper spray mass spectrometry and PLS-DA improved by variable selection for the forensic discrimination of beers. Anal. Chim. Acta 940, 104–112. https://doi.org/10.1016/j.aca.2016.08.002.

Petrick, L.M., Wilson, T.A., Ronald Fawcett, W., 2006. High-performance liquid chromatography-ultraviolet-visible spectroscopy-electrospray ionization mass spectrometry method for acrylic and polyester forensic fiber dye analysis*. J. Forensic Sci. 51, 771–779. https://doi.org/10.1111/j.1556-4029.2006.00155.x.

Rawtani, D., Tharmavaram, M., Pandey, G., Hussain, C.M., 2019. Functionalized nanomaterial for forensic sample analysis. TrAC Trends Anal. Chem. 120, 115661. https://doi.org/10.1016/j.trac.2019.115661.

Rockwood, A.L., Kushnir, M.M., Clarke, N.J., 2018. Mass spectrometry. In: Principles and Applications of Clinical Mass Spectrometry. Elsevier, pp. 33–65, https://doi.org/10.1016/B978-0-12-816063-3.00002-5.

Rodriguez-Cruz, S.E., 2006. Rapid analysis of controlled substances using desorption electrospray ionization mass spectrometry. Rapid Commun. Mass Spectrom. 20, 53–60. https://doi.org/10.1002/rcm.2267.

Sarkis, J.E.S., Neto, O.N., Viebig, S., Durrant, S.F., 2007. Measurements of gunshot residues by sector field inductively coupled plasma mass spectrometry—further studies with pistols. Forensic Sci. Int. 172, 63–66. https://doi.org/10.1016/j.forsciint.2006.12.007.

Selvius DeRoo, C., Armitage, R.A., 2011. Direct identification of dyes in textiles by direct analysis in real time-time of flight mass spectrometry. Anal. Chem. 83, 6924–6928. https://doi.org/10.1021/ac201747s.

Sigman, M.E., Clark, C.D., Fidler, R., Geiger, C.L., Clausen, C.A., 2006. Analysis of triacetone triperoxide by gas chromatography/mass spectrometry and gas chromatography/tandem mass spectrometry by electron and chemical ionization. Rapid Commun. Mass Spectrom. 20, 2851–2857. https://doi.org/10.1002/rcm.2678.

Steenkamp, P.A., Harding, N.M., van Heerden, F.R., van Wyk, B.-E., 2004. Fatal datura poisoning: identification of atropine and scopolamine by high performance liquid chromatography/photodiode array/mass spectrometry. Forensic Sci. Int. 145, 31–39. https://doi.org/10.1016/j.forsciint.2004.03.011.

Stefan, A.R., Dockery, C.R., Baguley, B.M., Vann, B.C., Nieuwland, A.A., Hendrix, J.E., Morgan, S.L., 2009. Microextraction, capillary electrophoresis, and mass spectrometry for forensic analysis of azo and methine basic dyes from acrylic fibers. Anal. Bioanal. Chem. 394, 2087–2094. https://doi.org/10.1007/s00216-009-2897-3.

Udey, R.N., Hunter, B.C., Smith, R.W., 2011. Differentiation of bullet type based on the analysis of gunshot residue using inductively coupled plasma mass spectrometry*: differentiation of bullet type using ICP-MS. J. Forensic Sci. 56, 1268–1276. https://doi.org/10.1111/j.1556-4029.2011.01836.x.

Wang, Z., Stout, S.A., Fingas, M., 2006a. Forensic fingerprinting of biomarkers for oil spill characterization and source identification. Environ. Forensic 7, 105–146. https://doi.org/10.1080/15275920600667104.

Wang, Z., Yang, C., Hollebone, B., Fingas, M., 2006b. Forensic fingerprinting of diamondoids for correlation and differentiation of spilled oil and petroleum products. Environ. Sci. Technol. 40, 5636–5646. https://doi.org/10.1021/es060675n.

Weinmann, W., Schaefer, P., Thierauf, A., Schreiber, A., Wurst, F.M., 2004. Confirmatory analysis of ethylglucuronide in urine by liquid-chromatography/electrospray ionization/tandem mass spectrometry according to forensic guidelines. J. Am. Soc. Mass Spectrom. 15, 188–193. https://doi.org/10.1016/j.jasms.2003.10.010.

Weyermann, C., Marquis, R., Mazzella, W., Spengler, B., 2007. Differentiation of blue ballpoint pen inks by laser desorption ionization mass spectrometry and high-performance thin-layer chromatography. J. Forensic Sci. 52, 216–220. https://doi.org/10.1111/j.1556-4029.2006.00303.x.

Williams, M.R., Moody, C., Arceneaux, L.-A., Rinke, C., White, K., Sigman, M.E., 2009. Analysis of black writing ink by electrospray ionization mass spectrometry. Forensic Sci. Int. 191, 97–103. https://doi.org/10.1016/j.forsciint.2009.07.003.

Yang, C., Wang, Z., Yang, Z., Hollebone, B., Brown, C.E., Landriault, M., Fieldhouse, B., 2011. Chemical fingerprints of Alberta oil sands and related petroleum products. Environ. Forensic 12, 173–188. https://doi.org/10.1080/15275922.2011.574312.

Zhao, X., Yinon, J., 2002. Forensic identification of explosive oxidizers by electrospray ionization mass spectrometry. Rapid Commun. Mass Spectrom. 16, 1137–1146. https://doi.org/10.1002/rcm.692.

CHAPTER 17

X-ray diffraction for forensic samples

1 Introduction

X-rays for a long time have served as a suitable tool to analyze the chemical and structural morphology of any element because of their high energies, which enable them to penetrate any material. Over the years, this high energy part of the electromagnetic spectrum has been employed in the analysis of crystal structure, particle size, chemical composition, nanoscale imaging, and even in the visualization of internal structures through techniques such as X-ray diffraction (XRD), X-ray microscopy, scanning X-ray fluorescence (XRF), and X-ray computed tomography (XCT).

X-ray beams, upon interacting with a sample, undergo a series of scattering and absorption processes that shall be discussed in this chapter. Based on these phenomena, different characterization tools have been developed. X-ray diffraction is a technique that is useful for the analysis of solid crystalline substances. This technique produces a distinct pattern based on the diffraction of the X-rays on the sample. The pattern is characteristic to every sample and indicates the lattice parameters, space group, crystallite size, etc. This technique is useful for analyzing amorphous substances such as DNA, vitamins, and medicines. This technique is also nondestructive in nature and easy to use (Chatterjee, 2001; Waseda et al., 2011).

In forensic science, XRD does not find applications in the analysis of common evidence such as fingerprints and body fluids (Pandey et al., 2017; Rawtani et al., 2019). However, it is highly useful in the analysis of evidence such as fibers, fabrics, explosives, and archaeological evidence such as cremains. Because it is nondestructive in nature, it is highly useful in the analysis of archaeological evidence. In this chapter, X-ray diffraction, its principle and theory, instrumentation, and applications in the examination of numerous forensic evidence are elaborated.

Handbook of Analytical Techniques for Forensic Samples
https://doi.org/10.1016/B978-0-12-822300-0.00017-3

2 Principle and theory of X-ray diffraction

2.1 Theory of X-ray

X-rays are a part of the electromagnetic spectrum with a wavelength of 0.1–100 Å; they were first discovered by William Roentgen. They can be classified into two types: hard X-rays and soft X-rays. Hard X-rays have significantly higher energies than the soft X-rays. X-rays are usually generated through X-ray tubes or through a synchrotron. In an X-ray tube, a cathodic element such as tungsten is heated, which causes it to emit electrons that are accelerated with a high voltage. When these high-energy electrons hit the sample, the electrons in different shells in the atom get excited and are emitted as X-rays. X-rays are high-energy electrons that can penetrate any sample. Therefore, it is necessary that the energy of the electrons sent to the target is very high. Also, factors such as the material used and the initial and final shell positions of the electron also influence the energy of the X-rays (Kasai et al., 2005; Misture and Snyder, 2001).

In the case of a synchrotron, the electrons are made to move in a circular orbit under a constant voltage. Due to the circular motion of the electrons under the influence of accelerating voltage, they acquire extremely high momentum, and start emitting X-rays. The synchrotrons are a great alternative over the traditionally used X-ray tubes as they have minimum thermal loss. Also, this technique is time saving as there is no need for the cathodic filament to cool before the next usage.

X-rays are generally emitted in the form of white radiation or Bremsstrahlung, which is basically a continuous spectrum of X-rays. This continuous spectrum consists of different wavelengths of X-rays that are emitted when the photon-electron collision takes place. The white radiation is dependent on the kinetic energy of the incident electron. It is continuous until all the kinetic energy of the electrons is used in the generation of short and long wavelengths.

Typically, in an atom's emission spectra, there are four main characteristic lines—K, L, M, N—that correspond to $n = 1,2,3,4$ quantum levels of the electron's energy state, respectively. When an incident electron hits the target metal with an energy higher than the excitation potential of the electron, it makes a transition from a higher level to lower level. Depending on the shell from which the electron was emitted, the characteristic lines appear on the spectra that are stronger than the white radiation. The characteristic lines that always appear are Kα and Kβ because the K line is the lowest quantum energy level, and therefore it is the only state that all electrons go to once they lose energy during the transition. Based on this, the X-ray sources in XRD are chosen, which will be discussed in the next session.

2.2 Theory of X-ray diffraction

When the X-ray beam hits a sample, it undergoes a series of scattering and absorption events. X-ray absorption takes place because of their ability to easily penetrate solids. The amount of X-rays absorbed can be easily estimated by using the Beer-Lambert law, in which the intensity of the transmitted X-ray photons is compared with the initial intensity. Clearly, the denser the sample, the lesser the intensity of the transmitted photons and thus there is more absorption. In X-ray scattering, the presence of the atoms in an order in the sample causes the X-rays to be scattered. Here, there are two types of scattering events: elastic and inelastic scattering. In elastic scattering, the energy of the X-ray photons is conserved when they hit the sample. However, in inelastic scattering, the energy is lost and converted into some other form. Elastic scattering also preserves the information of the source from which the X-rays were emitted while inelastic scattering doesn't. The elastic scattering that X-rays commonly undergo is known as Thomson scattering. In inelastic scattering, the X-rays undergo Compton scattering in which there is a decrease in the energy of the X-ray upon collision with electrons.

Commonly, apart from scattering in which there is a transfer of energy from the X-ray photon to the electron or vice versa, X-rays also undergo diffraction in which their direction changes with respect to the positioning of the atom in crystalline structures. XRD preferably uses Thomson scattering to characterize the crystallinity of any material. Also, the diffracted X-rays interact with each to form either constructive or destructive interference, and the pattern generated is characteristic of the material's structure. This is known as Bragg's diffraction and this phenomenon is described in Bragg's law. The equation of Bragg's Law is as follows (Kasai et al., 2005):

$$n\lambda = 2d\sin\theta$$

Here, n can be any integer, λ is the incident X-rays' wavelength, d corresponds to the interplanar space responsible for generating the diffraction, and θ is the diffraction angle (Fig. 1) (Bunaciu et al., 2015). In XRD, the sample is scanned through two methods: the Laue method and the powder-diffraction method. These methods will be discussed further in this chapter.

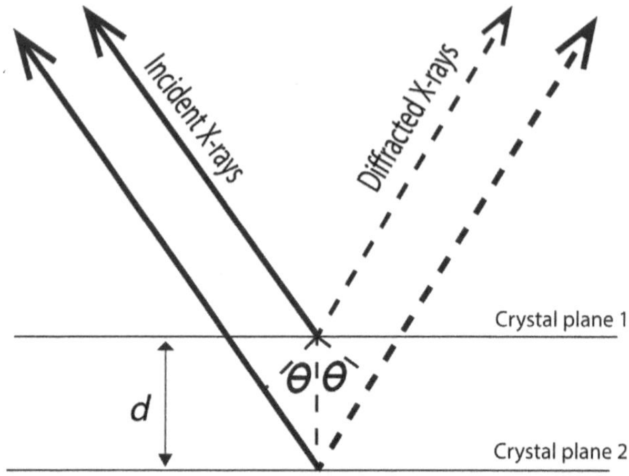

Fig. 1 X-ray diffraction.

3 Instrumentation of X-ray diffraction

XRD consists of three main components: the X-ray source, the sample holder, and the X-ray detector. The Bragg angle in the instrumentation is between the plane of the sample holder and the source. Apart from this, the angle between the projection source and the detector is 2θ. As mentioned above, a sample in XRD can be analyzed either through the Laue method or the powder diffraction method. In the Laue method, the diffraction angle is kept constant while the diffraction is varied. In the powder method, the diffraction angle is varied while the wavelength is kept constant.

The instrumentation of XRD varies according to the type of sample analysis. For instance, in the Laue method, the X-ray source, the sample holder, and the detector such as photographic film are kept in a straight line. Prior to the sample holder, there are slits that help in switching to different wavelengths of the X-ray. When the X-rays hit the sample, diffraction spots appear on the photographic film. These spots are characteristic of the crystal orientation of the sample. The placement of the crystal in this technique must be perfect, as any distortion may lead to the formation of incorrect diffraction spots, thereby contributing to false information. There are several variations of the Laue method such as the back-reflection method and the transmission method. In the back-reflection method, the photographic film is positioned between the source and the sample and the diffraction occurs in a backward direction. In the transmission method, the film is positioned behind the sample and detects diffraction beams transmitted through the sample. This method is typically used for orienting single crystals (Bishnoi et al., 2017).

In the powder diffraction method, the instrumentation is set up in a geometric arrangement known as a Bragg-Brentano parafocusing system. This is a circular arrangement where the sample is kept between the source and the detector. Between the source and the detector, there is a divergence slit that controls the intensity of the X-ray beams and the angle at which it falls on the sample. Therefore, smaller diffraction angles can be easily achieved using this technique, although at the cost of intensity. The sample is also movable and based on various tilt angles, the diffraction angle is decided. The various crystal planes in the sample show up as peaks with variable intensities in the XRD spectra. A typical XRD spectra consists of a list of angles (2θ) on the x-axis with the intensities on the y-axis. This data can also be combined with SEM-EDX to obtain confirmation on the elemental composition (Bunaciu et al., 2015) Fig. 2 shows the instrumentation for the Laue and powder diffraction methods.

The detectors in XRD are classified as three types: photon counting detectors, integrating detectors, and spatial sensitive detectors. The integrating detectors, also known as analog detector, work by integrating the charge produced by the conversion of the photons to electrons. Photon counting detectors, also known as digital detectors, are sensitive to the energy of each individual converted photon. Spatial sensitive detectors are those that detect photons on the basis of the area covered by them. They can be one-dimensional (1D) or line detectors, two-dimensional (2D) or area detectors, and point detectors. These detectors require a goniometer that is capable of measuring the angular position of the sample. Modern instruments use a solid-state detector such as germanium or Si (Li) detectors. Factors such

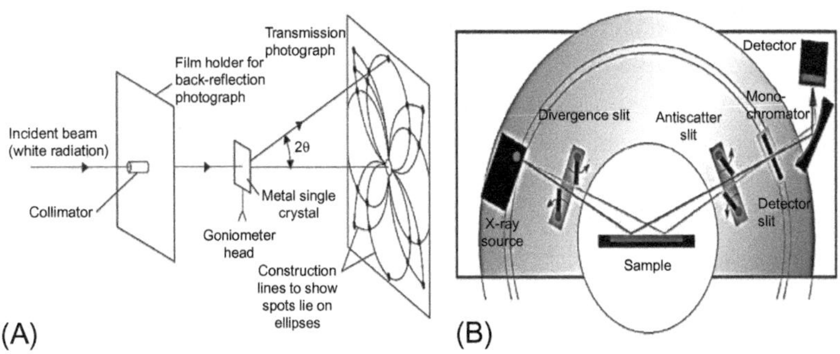

Fig. 2 Instrumentation of XRD (A) Laue method (Smallman and Ngan, 2014); (B) powder diffraction method (de Villiers and Lu, 2015).

as analysis time and better image quality are considered while deciding the detector.

A very crucial aspect of XRD analysis, just like any other analytical technique, is sample preparation. Solids as well as finely divided colloidal samples can be analyzed using XRD as long as they are properly prepared. Factors such as sample treatment to remove the impurities or unwanted substances, particle size, thickness, and orientation play crucial roles in the analysis of the sample. In the powder diffraction method, finely ground powders are required to achieve a high signal-to-noise ratio and a clean spectrum. Random and oriented mounts for samples are employed for the analysis (de Villiers and Lu, 2015; Smallman and Ngan, 2014).

3.1 Types of X-ray diffraction

There are many types of XRD that are classified on the basis of the diffraction angle, resolution, and type of samples. In this chapter, however, only small angle X-ray scattering (SAX) and wide-angle X-ray diffraction (WXRD) will be discussed, as these two techniques are the most widely used for the analysis of forensic evidence.

SAX is a technique that is used to analyze objects less than 100 nm. In this case, the samples are made ultrathin so that the X-rays can transmit through them. A typical SAX instrumentation consists of an X-ray source, a sample holder, a beam stopper, and an area detector. Sometimes, the dimensions of the objects themselves are more than the wavelength of the X-ray; therefore, the diffraction angles are kept extremely small. Also, because in this case the samples are made ultrathin, a majority of the X-rays with primary photons get transmitted through the sample onto the detector, due to which the detector may get damaged or saturated. To prevent this, a beam stopper is used that blocks such primarily transmitted X-ray beams. The X-rays are collected multiple times in order to overcome any loss due to weak scattering of the X-rays. The crystallinity of samples such as biological compounds, nanomaterials, and polymers is often characterized using this technique (Brügemann and Gerndt, 2004).

WXRD is another technique that is similar to SAX; however, in this the diffraction angles are kept wide. This is achieved by shortening the distance between the detector and the sample. This technique is widely used to analyze crystallinity and the chemical or phase composition of the sample. This technique, however, cannot be used for analyzing amorphous samples (Bishnoi et al., 2017).

4 X-ray diffraction for forensic evidence analysis

This technique has assisted in knowing the amorphous and crystalline behavior of any material, which can be a unique signature for that material (Bunaciu et al., 2015). Forensic scientists have employed XRD for decades to study these unique features of forensic evidence. The different kinds of forensic samples studied through XRD can be classified into anthropogenic, geological, and biological evidence, as discussed in this section.

4.1 Anthropogenic evidence

Anthropogenic evidence that can be encountered at a crime scene includes drugs, explosives, paints, chemicals, papers, fibers, counterfeit products, gunshot residue (GSRs), etc. In this section, research works that employed XRD for the analysis of this anthropogenic evidence are discussed.

Drugs, mainly illicit and recreational drugs (cocaine, heroin, nicotine, morphine, marijuana), are widely used at rave parties. They are transported in different types of storage boxes, bags, containers, etc. In this regard, Causin et al. carried out a study on plastic bags for investigating the traces of unlawful drugs using wide angle XRD (WAXRD). A total of 33 shopping bags that were collected from Venice, Italy, that had been used as packaging material for drugs was examined for the XRD analysis. The study differentiated among these bags without any requirement for sample preparation. The 2θ range for the WAXRD was from 15 to 50 degrees. The source of X-ray radiation was $CuK\alpha_1$. The analysis was helpful in discriminating among visually indistinguishable bags from different sources on the basis of their degree of crystallinity, signals from additive used, and the intensity ratios of peaks of polyethylene (Causin et al., 2007). In another study, single crystal XRD was used to investigate hydrochlorides of pentedrone and metaphedrone. These are "legal high" designer drugs, that is, drugs whose structures are modified to bypass the provisions of laws. The study was performed through collaboration between the forensic laboratory of police and an academic institution from Poland. The radiation source for the XRD analysis was $MoK\alpha$. The dimensions and volume of unit cells were used to differentiate between the crystals of both drugs. The study was the first of its kind that employed single crystal XRD for the identification of derivatives of cathinone (Trzybiński et al., 2013). Psychoactive drugs have also been examined through XRD, which served as a nondestructive analytical technique. Powder XRD was used for the analysis of eight synthetic cathinones that had a similar molecular assembly. Four drugs were also identified in the street samples. The 2θ range for the analysis was 5–60 degrees, and the radiation source was CuKα. The 2θ range between 10 and 30 degrees, which is the region for organic compounds, became the fingerprint region for the cathinones taken for the study because each drug had its own unique peak in this range. Fig. 3 shows the XRD pattern of drugs analyzed in this study (Jurásek et al., 2019). GSR and explosives can also be detected using XRD. In a recent study by Miranda et al., XRD along with multivariate analysis was used to estimate the distance of shooting through GSR. Cardboard boxes covered with cotton fabric were the target from which GSR was collected. The voltage for XRD analysis was 40 kV, with a current of 40 mA and a final angle of 90 degrees. The study helped in creating a calibration curve by examining GSR from targets hit at different shooting distances. Fig. 4 shows the combined XRD pattern of GSR on fabric due to variable shooting distance (Miranda et al., 2019). In another recent study that focused toward developing a forensic database for explosives, XRD along with other characterization techniques was used for detecting and discriminating between military and commercial explosives. Powder XRD with $CuK\alpha_1$ as the radiation source was used in the 2θ range of 7–80 degrees. SO_2 reflexes were detected for hexahydrotrinitrotriazine (RDX), trinitrotoluene (TNT), and pentaerythritol tetranitrate (PETN) in the XRD analysis. The analysis provided information about the molecular constituents in the explosive samples (Schachel et al., 2020).

XRD analysis has also been employed by forensic experts for the analysis of papers. In a study, the technique helped in determining the mineral composition and crystallinity of cellulose present in the paper. CuKα was used as the radiation source, and the analysis was carried out at a voltage of 40 kV with 20 mA current. The scanning 2θ range was 17–25 degrees. Minerals such as kaolinite, talc, anatase, montmorillonite, and illite were found in the papers through XRD analysis (Foner and Adan, 1983). In another study, the effects of ninhydrin and printing on paper were studied using XRD analysis in combination with other characterization techniques. The aim of the study was to link the paper evidence from its source, which is a hectic process when comparing printed or treated paper with blank paper. CuKα was used as radiation source, with a voltage of 45 kV and current of 200 mA. The scanning range was 5–40 degrees. The fillers used in the paper, such as chlorite, talc, kaolin and calcium carbonate, were detected through the XRD pattern of the papers. The pattern also remained the same after ninhydrin treatment and the printing of papers, as shown in Fig. 5 (Itamiya and Sugita, 2015). Causin et al. also carried out the distinction of paper of forensic

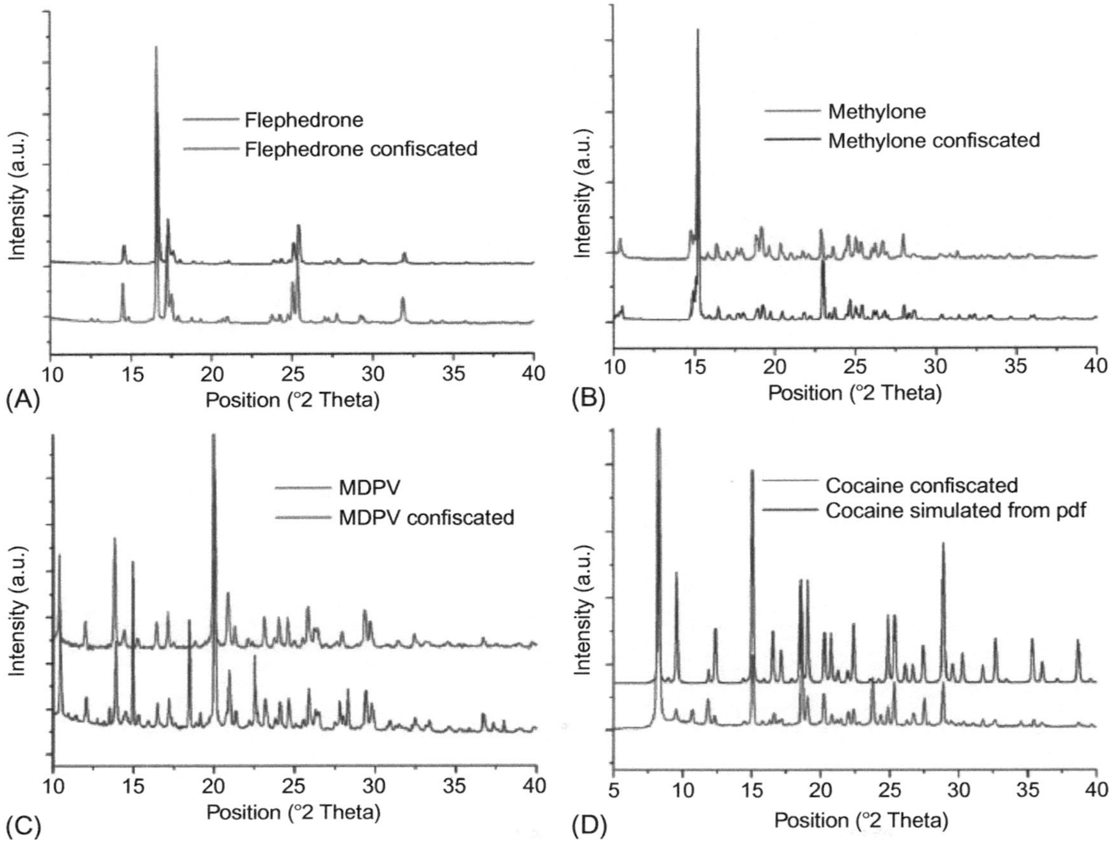

Fig. 3 XRPD pattern of pure and seized cathinone drugs (Jurásek et al., 2019). (A) XRPD pattern of pure and confiscated flephedrone. (B) XRPD pattern of pure and confiscated methylone. (C) XRPD pattern of pure and confiscated MPDV. (D) XRPD pattern of pure and confiscated cocaine.

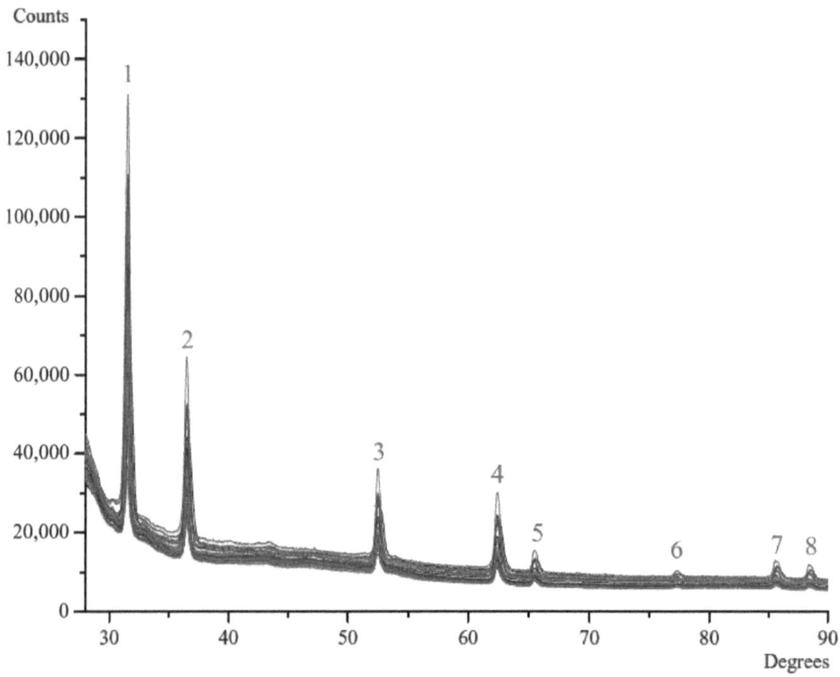

Fig. 4 XRD pattern of GSR with variable shooting distance (Miranda et al., 2019).

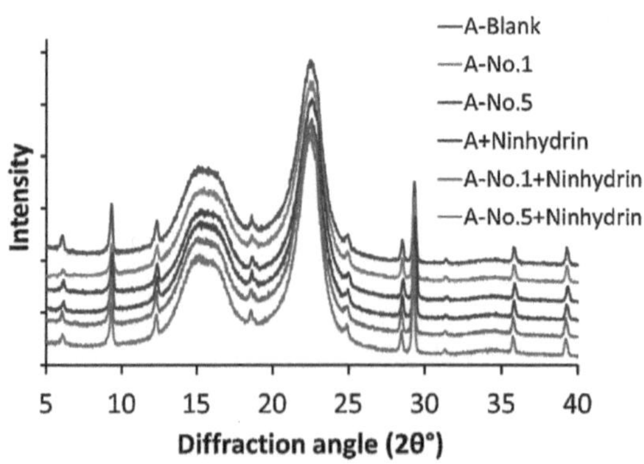

Fig. 5 XRD pattern of blank, printed, and ninhydrin treated paper (Itamiya and Sugita, 2015).

importance using XRD in combination with infrared spectroscopy. Analogous types of 19 office papers were taken for the WAXRD analysis done using $CuK\alpha_1$ as the source of radiation. The focus of the XRD analysis was on the inorganic and cellulose content of papers. The diffraction pattern was very broad as well as diffused, and the main peaks were located at 22.7 and 22 degrees with secondary peaks at 20.3, 16.3, and 14.8 degrees (Causin et al., 2010).

Wooden safety matches have also been analyzed through XRD. The study was related to a murder case in which evidence was attempted to be burned, but due to damp conditions, some unburned materials and matches were collected that were later also recovered from the suspect's house. Both matches were analyzed through XRD for comparison. Through the analysis, it was found that the matches recovered from the site and the suspect's house were similar, thus showing the importance of XRD in the investigation of such cases as well (Farmer et al., 2007). Fibers have also been analyzed through XRD. This potential of XRD was proved in the early 1980s by Lynch and Kerrigan. The radiation source used for the analysis was CoKα. Fibers of jute, nylon 6, wool, acrylic, linen, viscose rayon, polypropylene, and polyester were analyzed. It was observed that the photographs taken through XRD could assist in distinguishing between these fibers (Lynch and Kerrigan, 1981). Counterfeiting in coins was also analyzed through XRD by Hida et al. For the analysis, the coins were used without any pretreatment and the analysis was done using Cu filament with a voltage of 40 kV and a current of 100 mA. The scanning range for the study was from 20 to 160 degrees. Principal component analysis and component analysis were used for the XRD patterns, which helped in the classification of counterfeit coins into two groups: one made by nickel plating and the other made by a pressed method. Eight peaks were observed for real and pressed method coins while two additional peaks for nickel-plated coins were observed in the diffraction pattern (Hida et al., 2001). In another study from the early 1980s, a forensic examination of paints for the analysis of pigments was carried out by Curry et al. using powder XRD. A total of 71 pigments was collected for the analysis. A CoKα radiation source was used for the analysis with a voltage of 35 kV and a current of 34 mA. The study suggested that the XRD pattern of pigments in paint samples could be compared with that of pure pigments for their identification. However, the limitation of the analysis was the requirement of a high concentration of pigment in the paint sample (Curry et al., 1982). Chemicals such as perchlorate have also been detected using XRD in combination with other analytical techniques. The study was relevant to the field of environmental forensics for finding the sources of perchlorate. The results from XRD analysis revealed the presence of smectite, calcite, orthoclase feldspar, quartz, plagioclase, and muscovite in the analyzed samples (sediment containing perchlorate collected from the road cut) (Duncan et al., 2005).

4.2 Geological evidence

Geological evidence is often collected from the crime scene because it gives an idea about the location relevant to the crime as well as the status of the building in the case of concrete-related materials. This section discusses research works that employed XRD for the analysis of geological evidence such as soil, concrete, and minerals.

The qualitative and quantitative forensic analysis of soil and rock samples was carried out by Ruffell and Wiltshire in 2004 using XRD. The mineral abundance in the samples was determined by quantitative XRD (QXRD). The samples were taken from the tire tracks of cars present at the crime scene. The radiation source for the QXRD analysis was CuKα. The analysis revealed that a mineral proportion of soil from the tire matched with the location. However, the soil samples from two different locations were discriminated only by conventional XRD, and not by QXRD. The study therefore recommended the combined use of both techniques for effective quantitative and qualitative analysis (Ruffell and Wiltshire, 2004). In another study that focused on the discrimination of different soil samples as well as the same soil from different locations, XRD was used for the mineralogical analysis of clay. Samples were extracted with CBD (citrate/bicarbonate/dithionite) and NaOH. CuKα was used as the radiation source with 30 mA/40 kV/10–70 degrees 2θ as instrument operating parameters. CBD-extracted samples were analyzed for gibbsite and kaolinite while NaOH-extracted samples were analyzed for hematite and goethite. Among all the techniques used in the study, XRD emerged as the only technique that accurately discriminated between soil samples that were collected even within a kilometer range (Corrêa et al., 2018). In another research focusing on the next-generation sequencing of soils for a forensic comparison, XRD was used for their geological analysis. The analysis was carried out using CuKα as the radiation source, which generated X-rays at a voltage of 40 kV and a current of 40 mA. The scanning range for the samples was 5–80 degrees. The XRD analysis showed that the crystallinity of the unfertilized and naturally fertilized soil was similar while for soil with natural fertilizer, the crystallinity was different. Soil from a country road and that close to houses were also not distinguishable by XRD (Giampaoli et al., 2014). A compilation related to the application of XRD in mineralogy and pedology for solving a double murder was also published a few years back (Fitzpatrick and Raven, 2012). In a technical note presented by Prandel et al., XRD was used for the forensic analysis of clays. The Rietveld method was used with XRD (RM-XRD) for this purpose. RM refines the data generated by XRD such as the proportion of minerals and their crystal structure. CuKα as the radiation source at 40 kV and 30 mA was used for the analysis in a step scan manner for the range from 5 to 20 to 70 degrees. The analysis revealed the dominance of kaolinite in most of the clay samples collected while quartz dominated in the sample from the first site (Prandel et al., 2018). Rawlins et al. also employed XRD for the analysis of samples related to the earth. The XRD instrument had an X-ray tube equipped with cobalt as the target that generated X-rays at 45 kV and 40 mA while the scanning range was fixed at 3–65 degrees. The analysis revealed the presence of ferruginous, feldspar, and quartz in the collected samples (Rawlins et al., 2006).

Apart from soil and clay, XRD has also been used for the analysis of concrete. In a study by Alqassim et al., powder XRD was used for the forensic analysis of concrete that was affected by fire in the United Arab Emirates. The scanning range was 5–50 degrees while the radiation source was CuKα. A change in the intensity of peaks for the materials used in concrete such as calcite, gypsum, portlandite, and quartz was observed in the diffraction pattern upon changing the temperature (ambient to 900°C) of the concrete as per the fire's temperature (Alqassim et al., 2016). Uranium dioxide (UO_2), naturally found in uraninite mineral, has also been studied for its surface oxidation using grazing incidence XRD (GIXRD). The study was relevant to nuclear forensics, which investigates the production and trafficking of such materials. The oxidation of the surface of UO_2 is a unique indication toward its exposure to atmosphere during transport and handling, which was investigated through GIXRD that is sensitive toward the surface. The analysis revealed a contraction of the unit cells of UO_2 near the surface (Tracy et al., 2018).

4.3 Biological evidence

Biological evidence is one of the most common types found at a crime scene. However, not all types of biological evidence need XRD analysis. This section discusses the applications of XRD for the analysis of biological evidence such as burnt remains and cremains of humans as well as for the analysis of the postmortem interval through bone analysis (PMI).

Piga and coresearchers have shown the potential of XRD for the analysis of burnt remains with context to forensic science in many of their publications. In one of their works, human teeth and bone were analyzed by powder XRD to study the behavior of their microstructure in the temperature range of 200–1000°C. The analysis was done using CuKα as the radiation source while 40 kV and 30 mA were the operational parameters. The angular range for the study was 15–120 degrees. The diffraction pattern indicated toward escalation in the hydroxylapatite crystal content with increasing temperature (Piga et al., 2009). A similar kind of analysis was published before by Piga et al. that focused on calibrating the changes in bones as a function of time and temperature. Fig. 6 shows the XRD pattern of bone at different temperatures. In the figure, the increasing content of calcium oxide is evident with increasing temperature (Piga et al., 2008). However, another study published by the same research group showed the inability of XRD

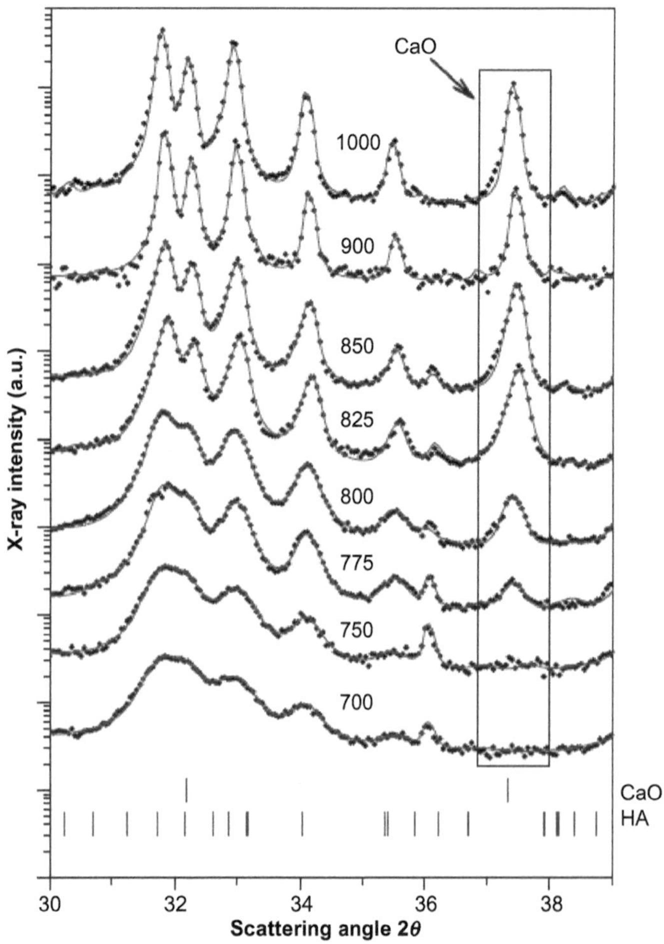

Fig. 6 XRD pattern of bone at variable temperature (Piga et al., 2008).

to distinguish between animal and human bones. The operation parameters were 40 kV, 40 mA, and 9–140 degrees of angular range. The XRD analysis of animal and human bones showed that as far as lattice parameters are concerned, there was complete overlapping of the results for both types of bones (Piga et al., 2013). This research group also carried out a study to understand the pattern of the crystallinity index of burnt bones and teeth by mixing bioapatite with carbonates and phosphates using XRD. The analysis displayed the formation of whitlockite (β-tricalcium phosphate) in teeth, thereby showing the higher frequency of the crystallinity index (Piga et al., 2016).

Bergslien et al. used XRD to differentiate between the filler materials and human cremains. The analysis was done for an angular range of 5–120 degrees using Cu as the target, generating X-rays at 45 kV and 40 mA. The apatite present in the biological sample (bone/teeth) had a dissimilar XRD pattern with that of the apatite present in the geological filler materials (Bergslien et al., 2008). In a study, PMI was estimated from the analysis of the remains of bones using XRD. The analysis was done to assess the crystallinity of the minerals present in the cortical and medullar region of the bone. The 2θ range for the analysis was 10–65 degrees with CuKα as the source of radiation. It was concluded that with increasing PMI, the crystallinity in the bone increases due to the growth of hydroxyapatite, which was evident because of the sharp peaks obtained in the diffraction pattern (Prieto-Castelló et al., 2007).

5 Conclusion

XRD is a nondestructive technique that is used to analyze crystallinity, mineral and textural identification, chemical and phase composition, lattice mismatch, and dislocation density. This technique is widely used as it is easy to use; requires minimal sample preparation, especially in the case of powder-diffraction method; and the spectra obtained are easy to read and interpret. The data obtained from XRD can also be compared with SEM-EDX to gain information on the elemental analysis. This technique works on the basis of Bragg's law in which the crystallinity of any material is observed through constructive and destructive interference of the X-rays diffracting upon hitting the sample. Samples that are in the nanometer range and samples that are larger such as a single crystal can be easily analyzed with this technique. This is possible due to the two modes of analysis, the Laue method and the powder diffraction method. In forensic science, the powder diffraction method has been widely used in which the diffraction angle is varied while the wavelength of the X-ray beam is kept constant. The use of SAX and WXRD modes allows a comprehensive analysis of the material.

While there are several advantages to this technique, there are certain limitations as well. For instance, the sample that is to be analyzed should preferably be in a single phase. While XRD can analyze multiphasic compounds, the chances of misinterpreting the data are high. Sometimes, there may be multiple peak overlays due to the presence of similar compounds in the sample.

In forensic science, XRD has been used in the analysis of biological, geological, and anthropogenic samples. Different modes of XRD are employed to analyze these substances. However, XRD is still lacking in the analysis of trace evidence that is found in extremely small quantities. With proper sample preparation techniques, modifications to the sample holder that will be capable of holding microvolumes of the sample, and a detector that is highly sensitive to even weak X-ray scattering, it is possible to overcome even this hindrance. Nonetheless, XRD is still a highly valued technique for the analysis of various forensic evidence.

References

Alqassim, M.A., Jones, M.R., Berlouis, L.E.A., Nic Daeid, N., 2016. A thermoanalytical, X-ray diffraction and petrographic approach to the forensic assessment of fire affected concrete in the United Arab Emirates. Forensic Sci. Int. 264, 82–88. https://doi.org/10.1016/j.forsciint.2016.03.015.

Bergslien, E.T., Bush, M., Bush, P.J., 2008. Identification of cremains using X-ray diffraction spectroscopy and a comparison to trace element analysis. Forensic Sci. Int. 175, 218–226. https://doi.org/10.1016/j.forsciint.2007.07.004.

Bishnoi, A., Kumar, S., Joshi, N., 2017. Wide-angle X-ray diffraction (WXRD). In: Microscopy Methods in Nanomaterials Characterization. Elsevier, pp. 313–337, https://doi.org/10.1016/B978-0-323-46141-2.00009-2.

Brügemann, L., Gerndt, E.K.E., 2004. Detectors for X-ray diffraction and scattering: a user's overview. Nucl. Inst. Methods Phys. Res. A 531, 292–301. https://doi.org/10.1016/j.nima.2004.06.019.

Bunaciu, A.A., Udriştioiu, E.G., Aboul-Enein, H.Y., 2015. X-ray diffraction: instrumentation and applications. Crit. Rev. Anal. Chem. 45, 289–299. https://doi.org/10.1080/10408347.2014.949616.

Causin, V., Marega, C., Carresi, P., Schiavone, S., Marigo, A., 2007. A quantitative differentiation method for plastic bags by wide angle X-ray diffraction for tracing the source of illegal drugs. Forensic Sci. Int. 168, 37–41. https://doi.org/10.1016/j.forsciint.2006.06.029.

Causin, V., Marega, C., Marigo, A., Casamassima, R., Peluso, G., Ripani, L., 2010. Forensic differentiation of paper by X-ray diffraction and infrared spectroscopy. Forensic Sci. Int. 197, 70–74. https://doi.org/10.1016/j.forsciint.2009.12.056.

Chatterjee, A.K., 2001. X-ray diffraction. In: Ramachandran, V.S., Beaudoin, J.J. (Eds.), Handbook of Analytical Techniques in Concrete Science and Technology. William Andrew Publishing, Norwich, NY, pp. 275–332 (Chapter 8) https://doi.org/10.1016/B978-081551437-4.50011-4.

Corrêa, R.S., Melo, V.F., Abreu, G.G.F., Sousa, M.H., Chaker, J.A., Gomes, J.A., 2018. Soil forensics: how far can soil clay analysis distinguish between soil vestiges? Sci. Justice 58, 138–144. https://doi.org/10.1016/j.scijus.2017.09.003.

Curry, C.J., Rendle, D.F., Rogers, A., 1982. Pigment analysis in the forensic examination of paints. I. Pigment analysis by X-ray powder diffraction. J. Forensic Sci. Soc. 22, 173–177. https://doi.org/10.1016/S0015-7368(82)71467-7.

de Villiers, J.P.R., Lu, L., 2015. XRD analysis and evaluation of iron ores and sinters. In: Iron Ore. Elsevier, pp. 85–100, https://doi.org/10.1016/B978-1-78242-156-6.00003-4.

Duncan, P.B., Morrison, R.D., Vavricka, E., 2005. Forensic identification of anthropogenic and naturally occurring sources of perchlorate. Environ. Forensic 6, 205–215. https://doi.org/10.1080/15275920590952883.

Farmer, N.L., Ruffell, A., Meier-Augenstein, W., Meneely, J., Kalin, R.M., 2007. Forensic analysis of wooden safety matches—a case study. Sci. Justice 47, 88–98. https://doi.org/10.1016/j.scijus.2007.04.001.

Fitzpatrick, R.W., Raven, M.D., 2012. How pedology and mineralogy helped solve a double murder case: using forensics to inspire future generations of soil scientists. Soil Horiz. 53, 14. https://doi.org/10.2136/sh12-05-0016.

Foner, H.A., Adan, N., 1983. The characterization of papers by X-ray diffraction (XRD): measurement of cellulose crystallinity and determination of mineral composition. J. Forensic Sci. Soc. 23, 313–321. https://doi.org/10.1016/S0015-7368(83)72269-3.

Giampaoli, S., Berti, A., Di Maggio, R.M., Pilli, E., Valentini, A., Valeriani, F., Gianfranceschi, G., Barni, F., Ripani, L., Romano Spica, V., 2014. The environmental biological signature: NGS profiling for forensic comparison of soils. Forensic Sci. Int. 240, 41–47. https://doi.org/10.1016/j.forsciint.2014.02.028.

Hida, M., Sato, H., Sugawara, H., Mitsui, T., 2001. Classifcation of counterfeit coins using multivariate analysis with X-ray diffraction and X-ray fluorescence methods. Forensic Sci. Int. 115, 129–134.

Itamiya, H., Sugita, R., 2015. Effects of printing and ninhydrin treatment on forensic analysis of paper. Forensic Sci. Int. 255, 38–42. https://doi.org/10.1016/j.forsciint.2015.05.026.

Jurásek, B., Bartůněk, V., Huber, Š., Kuchař, M., 2019. X-ray powder diffraction—a non-destructive and versatile approach for the identification of new psychoactive substances. Talanta 195, 414–418. https://doi.org/10.1016/j.talanta.2018.11.063.

Kasai, N., Kakudo, M., Kasai, N., 2005. X-ray diffraction by macromolecules. In: Springer Series in Chemical Physics. Kodansha/Springer, Tokyo/Berlin, New York.

Lynch, B.J., Kerrigan, G.C., 1981. Identification of single fibres by X-ray diffraction in forensic analysis. X-Ray Spectrom. 10, 196–197. https://doi.org/10.1002/xrs.1300100411.

Miranda, K.L., Ortega-Ojeda, F.E., García-Ruíz, C., Martínez, P.S., 2019. Shooting distance estimation based on gunshot residues analyzed by XRD and multivariate analysis. Chemom. Intell. Lab. Syst. 193, 103831. https://doi.org/10.1016/j.chemolab.2019.103831.

Misture, S.T., Snyder, R.L., 2001. X-ray diffraction. In: Buschow, K.H.J., Cahn, R.W., Flemings, M.C., Ilschner, B., Kramer, E.J., Mahajan, S., Veyssière, P. (Eds.), Encyclopedia of Materials: Science and Technology. Elsevier, Oxford, pp. 9799–9808, https://doi.org/10.1016/B0-08-043152-6/01778-2.

Pandey, G., Rawtani, D., Agrawal, Y.K., 2017. Future aspects of halloysite nanotubes in forensic investigations. J. Nanomed. Res. 6. https://doi.org/10.15406/jnmr.2017.06.00153.

Piga, G., Malgosa, A., Thompson, T.J.U., Enzo, S., 2008. A new calibration of the XRD technique for the study of archaeological burned human remains. J. Archaeol. Sci. 35, 2171–2178. https://doi.org/10.1016/j.jas.2008.02.003.
Piga, G., Thompson, T.J.U., Malgosa, A., Enzo, S., 2009. The potential of X-ray diffraction in the analysis of burned remains from forensic contexts. J. Forensic Sci. 54, 534–539. https://doi.org/10.1111/j.1556-4029.2009.01037.x.
Piga, G., Solinas, G., Thompson, T.J.U., Brunetti, A., Malgosa, A., Enzo, S., 2013. Is X-ray diffraction able to distinguish between animal and human bones? J. Archaeol. Sci. 40, 778–785. https://doi.org/10.1016/j.jas.2012.07.004.
Piga, G., Gonçalves, D., Thompson, T.J.U., Brunetti, A., Malgosa, A., Enzo, S., 2016. Understanding the crystallinity indices behavior of burned bones and teeth by ATR-IR and XRD in the presence of bioapatite mixed with other phosphate and carbonate phases. Int. J. Spectrosc. 2016, 1–9. https://doi.org/10.1155/2016/4810149.
Prandel, L.V., Melo, V.D.F., Brinatti, A.M., Saab, S.D.C., Salvador, F.A.S., 2018. X-ray diffraction and Rietveld refinement in deferrified clays for forensic science. J. Forensic Sci. 63, 251–257. https://doi.org/10.1111/1556-4029.13476.
Prieto-Castelló, M.J., Hernández del Rincón, J.P., Pérez-Sirvent, C., Álvarez-Jiménez, P., Pérez-Cárceles, M.D., Osuna, E., Luna, A., 2007. Application of biochemical and X-ray diffraction analyses to establish the postmortem interval. Forensic Sci. Int. 172, 112–118. https://doi.org/10.1016/j.forsciint.2006.12.014.
Rawlins, B.G., Kemp, S.J., Hodgkinson, E.H., Riding, J.B., Vane, C.H., Poulton, C., Freeborough, K., 2006. Potential and pitfalls in establishing the provenance of earth-related samples in forensic investigations. J. Forensic Sci. 51, 832–845. https://doi.org/10.1111/j.1556-4029.2006.00152.x.
Rawtani, D., Tharmavaram, M., Pandey, G., Hussain, C.M., 2019. Functionalized nanomaterial for forensic sample analysis. TrAC Trends Anal. Chem. 120, 115661. https://doi.org/10.1016/j.trac.2019.115661.
Ruffell, A., Wiltshire, P., 2004. Conjunctive use of quantitative and qualitative X-ray diffraction analysis of soils and rocks for forensic analysis. Forensic Sci. Int. 145, 13–23. https://doi.org/10.1016/j.forsciint.2004.03.017.
Schachel, T.D., Stork, A., Schulte-Ladbeck, R., Vielhaber, T., Karst, U., 2020. Identification and differentiation of commercial and military explosives via high performance liquid chromatography—high resolution mass spectrometry (HPLC-HRMS), X-ray diffractometry (XRD) and X-ray fluorescence spectroscopy (XRF): towards a forensic substance database on explosives. Forensic Sci. Int. 308, 110180. https://doi.org/10.1016/j.forsciint.2020.110180.
Smallman, R.E., Ngan, A.H.W., 2014. Characterization and analysis. In: Modern Physical Metallurgy. Elsevier, pp. 159–250, https://doi.org/10.1016/B978-0-08-098204-5.00005-5.
Tracy, C.L., Chen, C.-H., Park, S., Davisson, M.L., Ewing, R.C., 2018. Measurement of UO2 surface oxidation using grazing-incidence x-ray diffraction: implications for nuclear forensics. J. Nucl. Mater. 502, 68–75. https://doi.org/10.1016/j.jnucmat.2018.01.052.
Trzybiński, D., Niedziałkowski, P., Ossowski, T., Trynda, A., Sikorski, A., 2013. Single-crystal X-ray diffraction analysis of designer drugs: hydrochlorides of metaphedrone and pentedrone. Forensic Sci. Int. 232, e28–e32. https://doi.org/10.1016/j.forsciint.2013.07.012.
Waseda, Y., Matsubara, E., Shinoda, K., 2011. X-ray Diffraction Crystallography: Introduction, Examples and Solved Problems. Springer, Heidelberg [Germany]/New York.

CHAPTER 18

Lab-on-chip devices

1 Introduction

Lab-on-chip devices are miniaturized analytical devices that are independently capable of carrying out the synthesis or analysis of various components and analytes. They consist of integrated electromechanical systems that are capable of carrying out repetitive and complex laboratory tasks. They can also be portable and handheld and therefore can be used for onspot detection with increased speed, efficiency, accuracy, and sensitivity (Giannitsis, 2011). These devices also require much less of a sample while also increasing the overall throughput and enhancing analysis automation. Due to such features, they are also known as microfluidic devices. The integrated circuit of the lab-on-chip device is developed as a result of merging electronics, optics, biosensors, and fluidics. The analysis in a lab-on-chip device is done through a series of processes that is assisted by various application-oriented components in the integrated circuit. The process includes the extraction of the sample followed by the conversion of the analyte signal to a readable form with the help of a transducer. This is followed by amplification of the signal and the detection and processing of the data received. While there are several advantages of lab-on-chip technology, there are certain challenges to this technology, including designing constraints and the requirement of ultrapristine conditions for its fabrication. However, with constant improvements over the years, the time and cost for the fabrication of such devices have been drastically reduced. Over the years, lab-on-chip devices have been used in molecular biology, environmental, and diagnostic applications (Didar and Tabrizian, 2010; Li and Lin, 2009; Srinivasan et al., 2004).

In forensic science, the most crucial aspect of carrying out a crime scene investigation is the analysis of evidence present at the crime scene. Quite often, forensic science analysis is also carried out outside a crime scene such as in airports where incriminating substances such as narcotics and counterfeit substances need to be detected. The application of lab-on-chip devices in such cases is therefore necessary to perform the onspot, efficient, and facile analysis of the analyte in order to obtain prompt results. Such immediate

Handbook of Analytical Techniques for Forensic Samples
https://doi.org/10.1016/B978-0-12-822300-0.00018-5

analysis drastically reduces the risk of contamination of the sample (Rawtani et al., 2019). The requirement of micro- or nanoamounts of a sample for analysis also allows them to be used even in tampered crime scenes where the evidence is either removed or contaminated significantly. The use of lab-on-chip devices in forensic science is still in its infancy. However, this technology holds immense potential for applications in this field. In this chapter, the overall architecture of the lab-on-chip devices and their fabrication processes will be discussed, followed by their applications in forensic science.

2 Architecture of a lab-on-chip device

2.1 Components of a lab-on-chip device

The components in a lab-on-chip device are often application-oriented and are solely based on the process of analysis and the analyte present. For instance, the development of nucleic acid-based lab-on-chip devices for diagnostic applications requires components that are capable of performing cell lysis for the extraction of DNA and DNA purification. In biomedical lab-on-chip devices, emphasis is given on cell and peptide analysis and on enzymatic assays (Lee et al., 2005; Lim et al., 2010).

Most of the lab-on-chip devices consist of an injector or a sample chamber that is responsible for the collection of micro- or nanoamounts of the sample fluid. They can be either amalgamated with the internal circuitry or can be present as an interface or a bridge between the device and the sample environment. Commonly used injectors are syringe pumps, robotic pumps, or electronic pipettes (Haeberle and Zengerle, 2007). Modifications to these injectors are often made to ensure the disbursement of minute amounts of the sample to the device. Modifications are often made on the shape and dimension of the injector (Foote et al., 2005; Haeberle and Zengerle, 2007). Once the sample has entered the circuit, it is crucial that its flow is carefully controlled so that it enters in minute quantities to the remaining parts of the circuit. The duration, sequence, and amount of the flow are carefully controlled through pumps that operate on different principles. Some of the transport controllers are pneumatic, thermopneumatic, piezoelectric, hydrodynamic, and hydrostatic pumps. These pumps can be either active, in which an external force is required to operate, or passive, in which the fluid automatically flows through it. Once the sample is properly transported, it has to be prepared before any further processing. For example, as mentioned before, cell lysis and purification are some of the steps that are responsible for the preparation of DNA extraction from the cell. Conventionally, ultrafiltration, preconcentration, and derivatization are some of the methods used to make them ready for further analysis (Foote et al., 2005; Underberg and Waterval, 2002). This step is highly crucial as improper sample preparation can increase the risk of contamination or decrease the efficiency of sample analysis.

The next process involves mixing the reagents with the samples. The reagents and the samples are transported through multiple microfluidic channels and are later mixed at one spot through magnetic, ultrasonic, pneumatic, electrical, or chaotic mixing. The mixing depends on the type of reagents and samples used and can be either passive or active (Lee et al., 2009; Lim et al., 2010).

Once the mixing is done, the mixture is sent to a reactor chamber where the final reaction assay takes place in a controlled environment. This chamber is supplemented by sensors and actuators that constantly monitor the changes in the reaction. During fabrication, it is important that this chamber be made with utmost care, as any failure in this chamber may result in the contamination of the sample or leaking of the mix. These chambers are designed for the gas or liquid phase, but they may also have a packed bed reactor. In the gas phase reactor chambers, the reaction is supplemented by inert gases such as argon or with water vapor. In liquid phase reactors, the reagents and the samples are liquids and react together in this chamber. In packed-bed reactors, the reaction takes place on the surface of the catalyst instead of the empty space in the reactor (Lim et al., 2010).

The reaction products formed are later collected in a separate chamber. This process is done with the help of separators that are capable of separating the samples and reagents once the reaction has occurred. The separator used varies for each sample. For instance, magnetic microbeads can be magnetically separated while protein cells can be separated though dielectrophoresis, isotachophoresis, or a simple laminar flow. Homogenous mixtures or colloidal solutions can be separated with the help of optical tweezers in which minute amounts of force are exerted on the molecules with the help of a laser.

After the separation of the reaction products, the formed products are detected using a series of transducers that converts the physical signals from the product to readable electrical signals: Over the years, several detection methods have been developed. They include optical, electrochemical, capacitive, electrochemiluminescence, piezoresistive, and mass spectrometric techniques. Apart from the separator, a controller and data processor

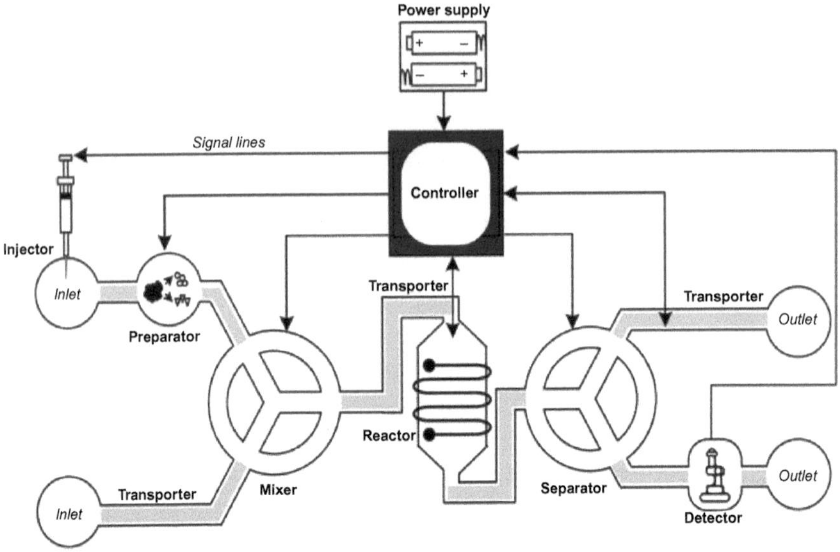

Fig. 1 Components in a lab-on-chip device (Lim et al., 2010).

receive the signal and display the data to the user for further analysis. Fig. 1 depicts the different components present in a lab-on-chip device (Auroux et al., 2002; Lee et al., 2005; Lee and Lee, 2004; Lim et al., 2010; Pamme and Wilhelm, 2006; Popovtzer et al., 2006; Renzi et al., 2005; Sabounchi et al., 2008; Schulte et al., 2000; Underberg and Waterval, 2002).

2.2 Fabrication techniques of lab-on-chip devices

The components discussed in the previous section are brought together in a small integrated chip comprising thousands of microfluidic channels capable of conducting a miniaturized laboratory-scale experiment with high throughput and efficiency. The fabrication of all these components on a substrate usually takes place through two approaches: front end and back end. In front-end methods, a single substrate, usually silica or silicon, is taken and the circuitry of different components is fabricated on it. Techniques such as photolithography come under front-end technology. In back-end technology, individual circuits of the components are fabricated separately and then packed together to be put in a more complex system. The fabrication of such microelectronic lab-on-chip devices takes place in specialized rooms called clean rooms. The clean rooms are designed and maintained such that there are very low levels of particulates, organisms, or cells and the cleanliness is usually determined on the basis of the number of particles per cubic meter (Iniewski, 2015).

In front-end methods, planar technology is widely used for the fabrication of the chips. This process is called planar because of the requirement of planar masks or sheets that are two-dimensional and repeatedly deposited on the substrate to fabricate the circuit (Iniewski, 2015). Photolithography is a commonly used technique that involves creating a desired shape on the substrate, followed by deposition of the material film on the developed shape. After this, a photoresistor is applied on the film and the substrate is exposed to ultraviolet (UV) light. Once UV light is exposed on the film, the parts that are not covered by the photoresistor undergo chemical changes and are removed, leaving behind only the desirable shape with the film present (Bharadwaj, 2015). After this, the materials are etched. If required, the steps are repeated until the circuitry is fabricated. Another lithography technique is electron beam lithography in which electron beams instead of light are used to fabricate desirable small shapes by covering the desirable parts with an electrosensitive film. Apart from lithography techniques, etching is also used for the fabrication of these devices. There are mainly two types of etching: dry and wet etching. In wet etching, through a chemical reaction, a part of the film or substrate is removed while in dry etching, the parts are removed through mechanical processes. In wet etching, the substrate is kept in a chemical bath and the reaction is supplemented by plasma. Meanwhile, dry etching has three types of processes: sputtering, reactive ion etching, and chemical dry etching (Nguyen, 2012). Apart from etching, various deposition techniques such as chemical vapor deposition and physical vapor deposition are also used. In these techniques, the precursors are themselves volatilized or are mixed with volatile substances, after which their vapors are carefully deposited on the surface of the substrate. Molecular beam epitaxy is another process in which the ionic form of the precursors is deposited on the substrate in an ultrahigh vacuum (Emrich et al., 2002; Giannitsis, 2011).

The fabrication processes as discussed above are suitable for small-scale fabrication. However, at a large scale, these techniques become very cost-intensive. Also, it is very difficult to develop the complex infrastructure required for the large-scale fabrication of the chips using these techniques. Over the years, polymers have been used as base materials for fabrication as they are cost-effective and easy to handle. Using polymers for the fabrication process allows the easy detection of biological macromolecules such as proteins, DNA, and RNA that require complex architecture. Fabrication using polymers includes techniques such as soft lithography in which the polymers serve as the base structural material and are micropatterned or molded using specific dies or molds. Unlike photolithography, this requires significantly less energy and can be easily molded. Spin coating, Langmuir Blodgett deposition, Lithographie, Galvanomung, Abformung or in short LIGA are also widely used in fabrication using polymers. Apart from these techniques, electroplating and laser ablation are also used for fabrication.

Once the components are fabricated through the front-end process, the circuits are broken into individual components in a process called dicing. They are then are attached on a die that will serve as the base of the final product through a process called die attach. However, before the components are diced, they are subjected to preliminary testing known as on-wafer testing to ensure that no major or minor defects are present that may hamper the overall integration of the components in a circuit. The attachment of the die also enhances the robustness of the component and protects it from any mishandling or chipping. Once the die is attached, the wires are bonded and joined. Visual, optical, and electrical tests are usually done during on-wafer testing. After these processes, the components are sent to back-end technologies that deal with the integration of all the individual components and set them up in an enclosed package. The final testing of the chips is held once the back-end process is completed (Iniewski, 2015). Fig. 2 highlights the fabrication process of a lab-on-chip device.

3 Forensic applications of lab-on-chip devices

Lab-on-chip devices, otherwise known as microfluidic devices, have proved their utility in different fields of science. They are preferred by scientists due to the requirement of trace quantities of samples and reagents for analysis. Forensic samples are usually recovered in minute quantities from the crime scene, and from that, a number of analyses need to be carried out for effective investigation (Pandey et al., 2017a,b; Rawtani et al., 2019). Therefore, the importance of microfluidic devices in forensic investigations increases several fold. A number of forensic applications have utilized lab-on-chip devices for the analysis of different forensic samples such as DNA, explosives, biological fluids, drugs, etc. This section focuses on the use of lab-on-chip devices by forensic scientists and experts for investigation and analyses purposes.

3.1 Analysis of DNA

The process of DNA analysis usually involves five major steps: sampling, cell lysis, extraction and purification, amplification, and detection and storage. These steps require a lot of time to complete when performed at forensic

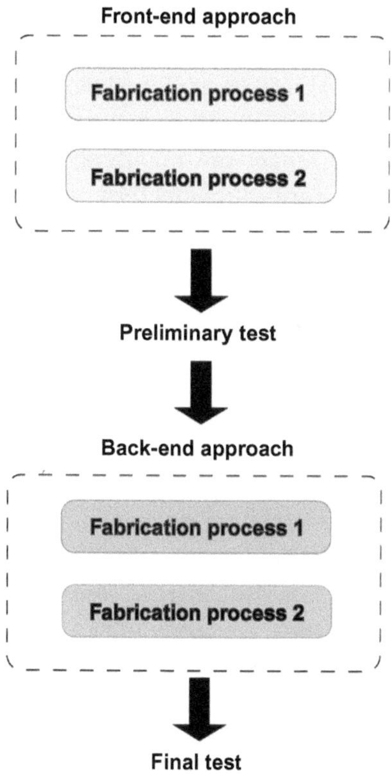

Fig. 2 Fabrication process of a lab-on-chip device.

laboratories. The analysis may take days to produce results, due to which its relevancy and contribution may get diminished with time in a crime investigation (Mapes et al., 2015). An effective crime scene scenario can be developed through the use of techniques that are fast, accurate, and specific such as lab-on-chip devices. Regarding DNA examination, a number of developments have been made in past years to integrate almost all the aforementioned steps of DNA analysis on a single lab-on-chip device (Bruijns et al., 2016).

As far as sampling of DNA evidence is concerned, lab-on-chip devices usually prefer samples as either pure DNA in a buffer or a cell lysate as the input material. DNA samples collected through swabbing should not be used in these microfluidic devices because there are very narrow flow channels in these devices that may get clogged because of the fibers coming from the swab. The loading of DNA samples from a swab to the device has remained a matter of concern for experts working in this field (Bruijns et al., 2016).

The next step involves, DNA extraction, and its purification. On the chip, the latter two steps are combined. Cell lysis can be carried out on a chip in three ways: thermal, mechanical, and electrochemical (Kim et al., 2009). After cell lysis, the DNA needs to be separated from the remaining cell debris because it may hinder the amplification process (Reedy et al., 2011). The extraction and purification of DNA are carried out using micro-solid phase extraction, differential extraction, and magnetic beads (Bruijns et al., 2016).

A further step is the multiplication or amplification of the extracted DNA sample from the evidence. The amplification is carried out through PCR chips, which are of two types: well-based and continuous flow (Zhang and Ozdemir, 2009). In a well-based PCR chip, the sample is heated and cooled to PCR temperatures while in a continuous-flow PCR chip, the sample is allowed to flow from different zones with a fixed temperature. Continuous-flow chips can be further divided into fixed-loop, closed-loop, and oscillatory chips (Zhang and Ozdemir, 2009). Apart from conventional PCR, digital droplet PCR has also emerged as an alternative for amplification purposes. Here, the sample is initially mixed with a PCR mixture and divided into various reaction compartments, in which position-based statistics are applied (Basova and Foret, 2014).

In the later stage, DNA detection is carried out. In a laboratory experiment, DNA can be detected usually through its absorbance value. However, as far as on-chip analysis is concerned, fluorescence-based detection, occasionally in combination with capillary electrophoresis (CE), is used to detect DNA. Fluorescence detection can be done via the use of fluorescent intercalating dyes, fluorescently labeled deoxynucleotides, or fluorescently labeled primers. CE is combined with fluorescent detection when an STR (short tandem repeat) profile is needed. However, the incorporation of the detection step on a chip is not an easy task. So, many devices usually go for off-chip DNA detection (Bruijns et al., 2016). Regarding the storage of the DNA sample, on-chip storage conditions are yet to be integrated. The sample remains on the chip in a closed environment. However, the cross-contamination of the sample needs to be prevented during transportation and handling.

In a study, polymethyl methacrylate (PMMA) and polydimethylsiloxane (PDMS) in combination with silicon and glass were used as chip material to develop a microfluidic device for DNA analysis. The lysis was carried out in a guanidine lysis buffer, and the extraction was solid-phase extraction. SYBR green dye was used in the detection of DNA (Chen et al., 2007).

In another study, a microfluidic device was used to analyze ancient DNA samples for the determination of the sex of the individual (Fig. 3). The device was made up of glass, and was fabricated using techniques such as wet etching and photolithography. The selective extraction of damaged DNA was carried out by labeling certain sites in DNA with biotin, and extracting the labeled DNA through superparamagnetic particles coated with streptavidin. The sex determination was done by the analysis of the Y chromosome and amelogenin (Parton et al., 2013). The technique can

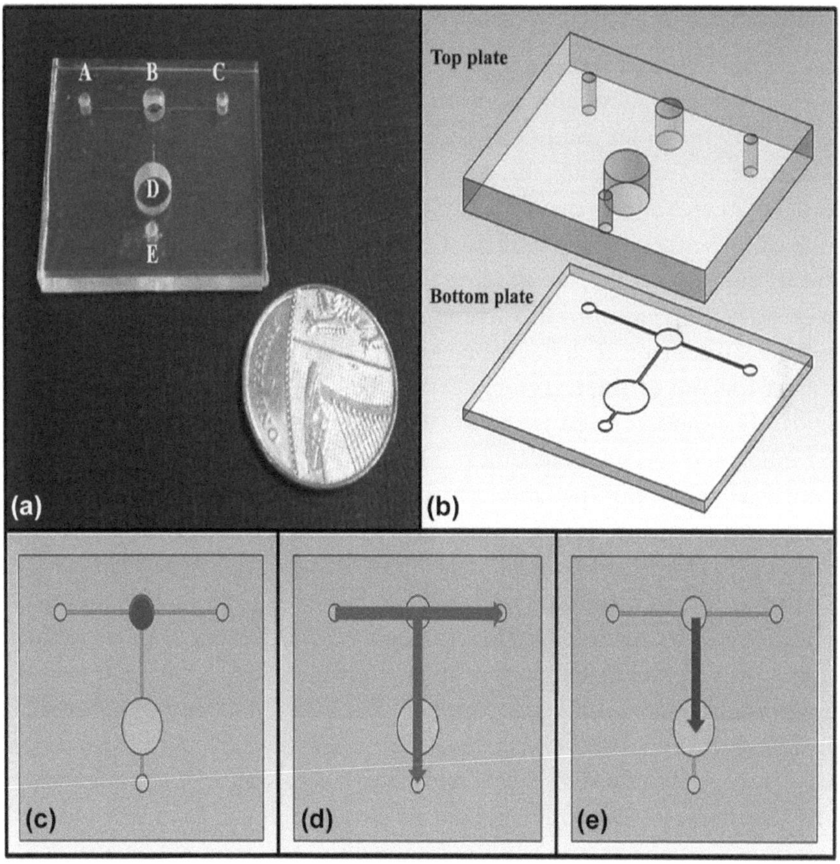

Fig. 3 (a) Microfluidic device for DNA extraction and amplification. Here, A, C, and E are the in/out holes for electrodes and reagents, and B and D are the extraction chamber and amplification chamber for DNA, respectively; (b) structure of a two-layered microfluidic device; (c) site for introduction of particles/beads and binding of sample DNA; (d) wash solution flow path; (e) drive of eluted DNA or particles toward the amplification chamber (Parton et al., 2013).

be helpful in cases where a cadaver has been recovered after many years, and gender identification is difficult through conventional techniques.

3.2 Identification of biological fluids

Forensic serology deals with the analysis of biological fluids such as blood, semen, saliva, and urine. The analysis of these fluids in forensic laboratories is carried out in two steps. The first step involves a preliminary test to confirm the presence of biological fluids, and the second step involves a confirmatory test to specifically identify the type of biological fluid recovered as evidence from the crime scene (Mozayani and Noziglia, 2011). However, the conventional strategies and methods employed at forensic laboratories for the identification and confirmation of biological fluids are time-consuming, complex, and destructive in nature, which means the sample cannot be recovered after the completion of the analysis. There is a need to develop techniques that reduce the amount of sample requirement and time while being simple to operate. In recent years, spectroscopic techniques and immunoassays that are commonly used in forensic serological analysis have emerged (Harbison and Fleming, 2016). However, the drawbacks such as the requirement of skilled labor and the immobility of the instruments restrict the use of these techniques for the onsite detection of biological fluids. The current need is to develop a portable lab-on-chip device for the quick and specific detection of biological fluids at a crime scene using a minimal amount of the sample without employing any expensive instrumentation or chemicals. A number of scientists across the globe have been working in this regard, which will help in bringing justice at a faster pace.

In a recent study, a group of researchers developed a microfluidic analytical device based on paper for the detection of biological fluids such as blood, saliva, semen, and urine. The study provides a very convenient and presumptive way of detecting, identifying, and differentiating between the aforementioned biological fluids at the crime scene. A chromatographic paper was used to print a pattern with four lanes using a wax printer. Each lane was designated for only one type of biological fluid. The mechanism of detection was colorimetric sensing. The traditional colorimetric methods for these fluids, such as an iodine test for saliva, a Nessler reagent test for urine, a Kastle-Mayer test for blood, and an acid phosphatase enzyme test for detecting semen, were modified for compatibility with the chromatographic paper and used as the detection method in the developed microfluidic device. The device had a shelf life of 2 weeks when stored in dark and dry conditions

(Cromartie et al., 2019). In another recent study, a portable device was developed for the biochemical examination of blood (finger pricked). The device has been developed to detect glucose, cholesterol, and triglycerides present in the blood. For the analysis, the blood can be directly transferred onto the chip, where all the processes are done. The results were produced within 15 min of sample injection. The study provides a user friendly, self-service device to monitor the biochemical aspects of blood, especially during chronic diseases (Zhu et al., 2020).

3.3 Detection of explosives

The detection of explosives is of utmost importance in today's world, which is facing a huge number of terrorist activities that usually involve the application of different explosives. Innumerable techniques have been developed for the detection of explosives and their residues, such as high-performance liquid chromatography, ion mobility spectrometry, thin-layer chromatography, etc. (Douse, 1982; Ewing et al., 2001; Rawtani et al., 2019). However, the long time for analysis, high cost, requirement of expert manpower, and bulky instrumentation have limited the use of these techniques only to laboratories. In cases where a large number of people or surfaces need to be scanned for explosives, a portable lab-on-chip device is needed that can specifically detect even trace quantities of explosives and their residues (Erçağ et al., 2011).

In a study, an immunosensing-based microfluidic device was developed for the detection of explosives under water. An immunosensor had antibodies for the nitro-aromatic compounds that were responsible for detecting the explosives. In addition to immune-sensing, fluorescence detection was also employed to sense the explosives. The lab-on-chip microfluidic device was fabricated using micromilling and hot embossing. The sample processing was 60 times more than the conventional immunoassay, and was able to detect up to 0.01 ng/mL of trinitrotoluene (TNT). The device also decreased the total analysis time for such compounds by 60-fold (Adams et al., 2011). The same concept of high-throughput microfluidic immunosensors (HTMI) was used in another study to integrate this device onto a REMUS100 autonomous underwater vehicle (AUV). The study was done with an aim to provide onsite monitoring of explosives in marine water and to protect waterways, people, and the military from any terrorist activities. Concentrations of TNT as low as 20 ppb were detected through this system (Adams et al., 2013). In another study, the explosive dinitrotoluene (DNT) was monitored in real time using surface-enhanced Raman spectroscopy (SERS) and free-surface microfluidics. The concept of the device took inspiration from canine olfaction. The system was able to detect trace quantities of DNT up to 1 ppb in the vapor phase. The SERS integrated in the device helped in enhancing the signal, thereby becoming responsible for the high sensitivity of the device toward trace quantities of the explosive. The minimal sample requirement, continuous monitoring, and reproducibility were the highlights of the lab-on-chip device developed in this study (Piorek et al., 2012). A paper-based microfluidic analytical device has also been developed for the detection of improvised explosives using colorimetric principles. For this, hydrophobic channels were created on the chromatography paper using wax ink that had five lanes. Each lane was comprised of a colorimetric reagent that, upon reaction with a specific explosive, produced a specific color. The study developed two such devices. The first one was to detect inorganic explosives such as ammonium nitrate, flash powder, and black powder. The second was used for sensing high and military explosives such as TNT and RDX (research department explosive), urea nitrate, and TATP (triacetone triperoxide). The detection time was approximately 5 min with the limit of detection up to 0.39 μg of the explosive (Peters et al., 2015). TNT, TNB (trinitrobenzene), and tetryl have also been detected using a paper-based microfluidic device. The paper had a deposition of potassium hydroxide that changed color upon reacting with different nitro-aromatic explosives. The detection limit was 12.5 ng for TNT, 15 ng for tetryl, and 7.5 ng for TNB. The device was used with a bioanalyzer lab-on-chip device that was able to detect 7 ng of TNT, 3.8 ng of TNB, and 4.7 ng of tetryl. The coupling of a paper-based microfluidic device with a lab-on-chip device could be able to detect even 1 μg of explosive that is distributed over an area of about 100 cm^2 (Pesenti et al., 2014).

3.4 Anticounterfeiting

Counterfeiting has spread its tentacles in almost all areas. Counterfeit currency notes directly impact the economy of any country while counterfeit edible products and drugs adversely affect the public health. A number of techniques and strategies have been developed in the past to prevent the counterfeiting of currency notes and drugs. In a study, anticounterfeit labels were developed using photonic structures. These structures were printed through the tuning of orientation. The labels had many axially symmetrical orientations of photonic structures that switched color upon tilting the sample or changing the incident angle for light. The technology could be used to fabricate invisible or colorimetric labels that can be decoded either spectrophotometrically or visually (Xuan and Ge, 2011).

In another study, a microtaggant was developed by lithographically encoding a polymer using an error-correctable QR code with high capacity. The technology was utilized to prevent the counterfeiting of drugs. The microtaggant was fabricated with PDMS by using a soft lithographic technique. The QR code pattern was generated using software called OR code generator. These QR-coded microtaggants were detected using a charge-coupled camera attached to an optical microscope. The image captured can be decoded using any smartphone application to scan the QR codes. This enabled the drugs to be uniquely tagged, thereby avoiding their counterfeiting (Han et al., 2012). The authentication procedure for the anticounterfeiting of drugs is shown in Fig. 4.

Barcode particles can also be used as microtaggants for anticounterfeiting. These particles contain the data related to the composition and identification of that material (Zhao et al., 2015). In another study, nanocrystals of rare earth metals that are distinct in their spectra were used to generate specific particle barcodes through the use of micropatterning. A defined population of these particles or a barcode with these particles was attached on the surface of different materials. The particles remained invisible in normal light. But with near-infrared light, the microtaggant was illuminated. Such technology can be used to tag almost all the manufactured products of this world (Lee et al., 2014).

3.5 Detection of illicit drugs

Narcotics and other drugs of abuse are a growing menace due to their ability to weaken an individual's mental and physical health as well as the overall functioning of society. In order to curb this menace, it is important to develop techniques that their allow rapid and onspot detection. Through this, it is possible to constantly monitor criminal activities and ensure quick deliverance of justice. Over the years, drugs have been usually detected through chromatographic, spectroscopic, and spectrometric techniques after extraction from biological fluids such as blood, saliva, or urine. Apart from this, immunoassay techniques have also been developed that are capable of rapidly eliminating negative samples. In recent years, lab-on-chip devices have been developed for the detection of drugs. These devices are capable

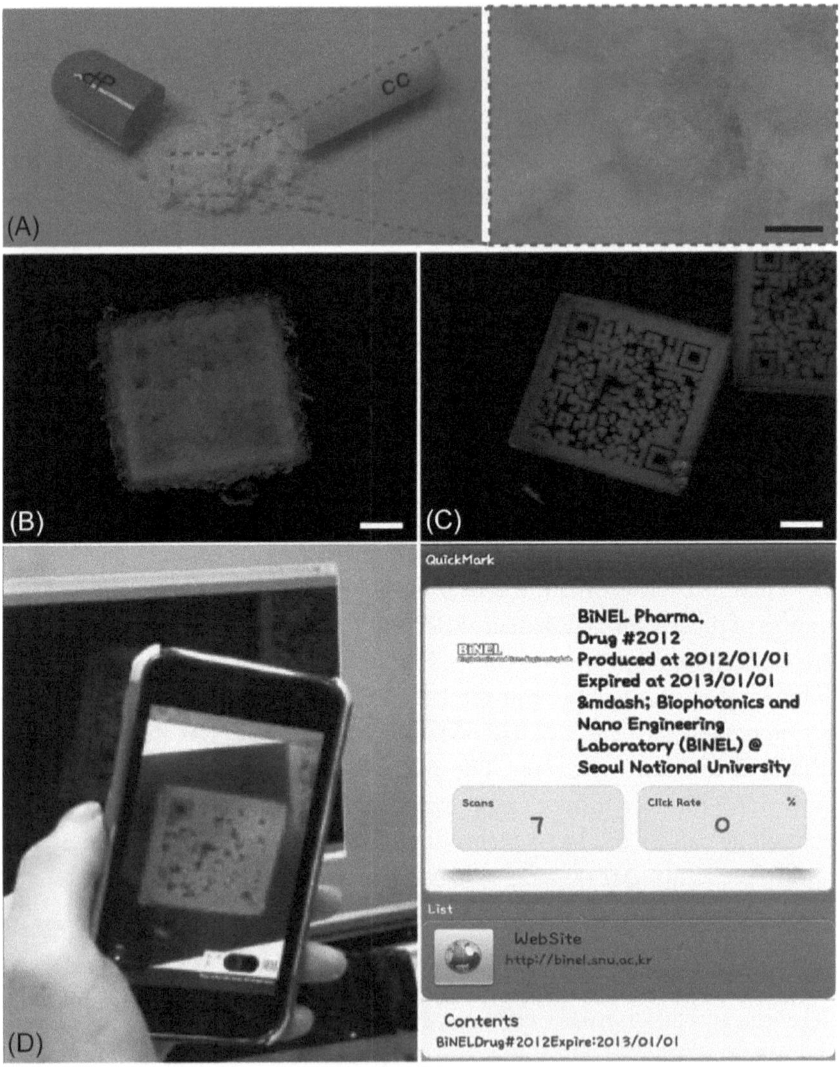

Fig. 4 (A) Capsule with the microtaggant; (B) microtaggant under fluorescence microscope, image was blurred because of the drug powder; (C) drug dissolved in water and microtaggant under fluorescence microscope; (D) reading of QR code through a smartphone application (Han et al., 2012).

of screening, extraction, separation, and detection. These devices are highly suitable for such applications because of their ability to accurately perform quantitative as well as qualitative analysis with increased accuracy, selectivity, and sensitivity. They also require micro- or nanoamounts of the sample and can be easily carried or even disposed. Screening of the drugs has been done through microfluidic devices that employ an enzyme linked immunosorbent assay (ELISA). Extraction has been done through filtration and various solid-phase or liquid-phase extraction techniques. Separation was usually conducted using capillary electrophoresis or high-pressure liquid chromatography while detection was done using electrochemical, optical, colorimetric, or FRET methods. In a study, a microfluidic device was developed in which methamphetamine was detected from saliva through surface enhanced Raman spectroscopy. The device had microfluidic channels that connected different spots where the saliva, silver ions, and salts were stored. A channel led the drug-containing saliva to the silver ions where it was adsorbed. When the salt was added, the silver ions were aggregated, resulting in the development of surface enhanced Raman spectroscopy. This signal was detected to sense the presence of the drug (Andreou et al., 2013). In another study, a microfluidic electronic nose was developed for the detection of the opiates heroin and cocaine. This device employed quartz crystal microbalance (QCM) on which antibodies specific to the opiates were immobilized. The antibodies interacted with the drugs, which caused an increase in the weight of the quartz microbalance. Through this change, the presence of the drug was detected. Apart from this, the resonance frequency of the sensor was also observed. In the presence of the drugs, this frequency showed changes (Frisk et al., 2005). Another detection method was through using micellar electrokinetic capillary chromatography (MECC) in which the narcotics are detected through the presence of surfactants in an electrophoresis buffer. This technique is cost-effective, consumes fewer reagents, and is less time-consuming. The device consisted of a circuit board with PMMA as the substrate and silica capillaries as the microchannel. Once high voltage was applied, the separation of the drugs started to occur. Through this, they were able to detect from 1 to 1500 μg/mL.

4 Conclusion

Lab-on-chip devices are small integrated microfluidic devices that are capable of carrying out a miniaturized version of large-scale laboratory analysis. These devices are capable of independently carrying out the quantitative and qualitative analysis of micro- or nanovolumes of analytes through different detection methods. A lab-on-chip device comprises different components that are responsible for extracting the sample, transporting them in the device through microfluidic channels, and detection. Certain lab-on-chip devices also consist of sample preparation and purification components and a reaction chamber as well. These devices are fabricated through front-end and back-end processes. In the front end, various fabrication processes such as lithography, etching, molecular beam epitaxy, chemical and physical vapor deposition, laser ablation, and LIGA are used. However, these techniques can become quite cumbersome and cost-intensive if these components need to be fabricated in a large scale. Therefore, polymer-based technologies are widely used to overcome this drawback. Polymers such as polydimethylsiloxane are widely used as a base material on which the microfluidic channels are microprinted. In back-end processes, the components are diced into individual parts through a process known as dicing, then attached to dies. These are later subjected to different integration processes for the whole-scale fabrication of the chip.

As far as the applications of lab-on-chip devices in forensic science is concerned, this chapter showed their potential to be used for the analysis of DNA, biological fluids, explosives, counterfeiting, and drugs. The five major steps of DNA analysis have been researched and most of the steps have been integrated on the device for the onsite analysis of DNA. Devices that can detect, identify, and differentiate between the commonly found biological fluids at a crime scene such as blood, urine, semen, and saliva have also been developed. Lab-on-chip devices that can detect glucose, cholesterol, and triglycerides in blood have also been researched. Explosives such as TNT, TNB, RDX, tetryl, etc., have been specifically detected using such devices. Barcode particles have also been incorporated to develop microtaggants that can be used as anticounterfeiting labels. These taggants have a specific QR code that can be decoded using any smartphone application for scanning the QR code. Further, the detection of illicit drugs such as heroin, cocaine, methamphetamine, etc., has also been done using such lab-on-chip devices.

Currently, the fabrication of lab-on-chip devices for forensic applications is still in the research stages. In the future, it will be necessary to improve the existing semiconductor and polymer technologies so that the chips can be fabricated in a large scale in a facile manner. Through such innovations, it is possible to promote the ubiquitous use of lab-on-chip devices in various forensic science laboratories.

References

Adams, A.A., Charles, P.T., Deschamps, J.R., Kusterbeck, A.W., 2011. Demonstration of submersible high-throughput microfluidic immunosensors for underwater explosives detection. Anal. Chem. 83, 8411–8419. https://doi.org/10.1021/ac2009788.

Adams, A.A., Charles, P.T., Veitch, S.P., Hanson, A., Deschamps, J.R., Kusterbeck, A.W., 2013. REMUS100 AUV with an integrated microfluidic system for explosives detection. Anal. Bioanal. Chem. 405, 5171–5178. https://doi.org/10.1007/s00216-013-6853-x.

Andreou, C., Hoonejani, M.R., Barmi, M.R., Moskovits, M., Meinhart, C.D., 2013. Rapid detection of drugs of abuse in saliva using surface enhanced Raman spectroscopy and microfluidics. ACS Nano 7, 7157–7164. https://doi.org/10.1021/nn402563f.

Auroux, P.-A., Iossifidis, D., Reyes, D.R., Manz, A., 2002. Micro total analysis systems. 2. Analytical standard operations and applications. Anal. Chem. 74, 2637–2652. https://doi.org/10.1021/ac020239t.

Basova, E.Y., Foret, F., 2014. Droplet microfluidics in (bio)chemical analysis. Analyst 140, 22–38. https://doi.org/10.1039/C4AN01209G.

Bharadwaj, P., 2015. Microfabrication via Photolithography. https://doi.org/10.13140/RG.2.2.33218.58565.

Bruijns, B., van Asten, A., Tiggelaar, R., Gardeniers, H., 2016. Microfluidic devices for forensic DNA analysis: a review. Biosensors 6, 41. https://doi.org/10.3390/bios6030041.

Chen, X., Cui, D., Liu, C., Li, H., Chen, J., 2007. Continuous flow microfluidic device for cell separation, cell lysis and DNA purification. Anal. Chim. Acta 584, 237–243. https://doi.org/10.1016/j.aca.2006.11.057.

Cromartie, R.L., Wardlow, A., Duncan, G., McCord, B.R., 2019. Development of a microfluidic device (μPADs) for forensic serological analysis. Anal. Methods 11, 587–595. https://doi.org/10.1039/C8AY02080A.

Didar, T.F., Tabrizian, M., 2010. Adhesion based detection, sorting and enrichment of cells in microfluidic lab-on-chip devices. Lab Chip 10, 3043–3053. https://doi.org/10.1039/C0LC00130A.

Douse, J.M.F., 1982. Trace analysis of explosives in handswab extracts using Amberlite XAD-7 porous polymer beads, silica capillary column gas chromatography with electron-capture detection and thin-layer chromatography. J. Chromatogr. A 234, 415–425. https://doi.org/10.1016/S0021-9673(00)81879-8.

Emrich, C.A., Tian, H., Medintz, I.L., Mathies, R.A., 2002. Microfabricated 384-lane capillary array electrophoresis bioanalyzer for ultrahigh-throughput genetic analysis. Anal. Chem. 74, 5076–5083. https://doi.org/10.1021/ac020236g.

Erçağ, E., Uzer, A., Eren, S., Sağlam, S., Filik, H., Apak, R., 2011. Rapid detection of nitroaromatic and nitramine explosives on chromatographic paper and their reflectometric sensing on PVC tablets. Talanta 85, 2226–2232. https://doi.org/10.1016/j.talanta.2011.07.080.

Ewing, R.G., Atkinson, D.A., Eiceman, G.A., Ewing, G.J., 2001. A critical review of ion mobility spectrometry for the detection of explosives and explosive related compounds. Talanta 54, 515–529. https://doi.org/10.1016/s0039-9140(00)00565-8.

Foote, R.S., Khandurina, J., Jacobson, S.C., Ramsey, J.M., 2005. Preconcentration of proteins on microfluidic devices using porous silica membranes. Anal. Chem. 77, 57–63. https://doi.org/10.1021/ac049136w.

Frisk, T., Ronnholm, D., van der Wijngaart, W., Stemme, G., 2005. Fast narcotics and explosives detection using a microfluidic sample interface. In: The 13th International Conference on Solid-State Sensors, Actuators and Microsystems, 2005. Digest of Technical Papers. TRANSDUCERS'05. IEEE, Seoul, Korea, pp. 2151–2154. https://doi.org/10.1109/SENSOR.2005.1497530.

Giannitsis, A.T., 2011. Microfabrication of biomedical lab-on-chip devices. A review. Estonian J. Eng. 17, 109. https://doi.org/10.3176/eng.2011.2.03.

Haeberle, S., Zengerle, R., 2007. Microfluidic platforms for lab-on-a-chip applications. Lab Chip 7, 1094–1110. https://doi.org/10.1039/B706364B.

Han, S., Bae, H.J., Kim, J., Shin, S., Choi, S.-E., Lee, S.H., Kwon, S., Park, W., 2012. Lithographically encoded polymer microtaggant using high-capacity and error-correctable QR code for anti-counterfeiting of drugs. Adv. Mater. 24, 5924–5929. https://doi.org/10.1002/adma.201201486.

Harbison, S., Fleming, R., 2016. Forensic body fluid identification: state of the art. Res. Rep. Forensic Med. Sci. 11. https://doi.org/10.2147/RRFMS.S57994.

Iniewski, K., 2015. Devices, Circuits, and Systems. p. 1166.

Kim, J., Johnson, M., Hill, P., Gale, B.K., 2009. Microfluidic sample preparation: cell lysis and nucleic acid purification. Integr. Biol. 1, 574–586. https://doi.org/10.1039/b905844c.

Lee, S.J., Lee, S.Y., 2004. Micro total analysis system (μ-TAS) in biotechnology. Appl. Microbiol. Biotechnol. 64, 289–299. https://doi.org/10.1007/s00253-003-1515-0.

Lee, S.-H., Cho, S.I., Lee, C.-S., Kim, B.-G., Kim, Y.-K., 2005. Microfluidic chip for biochemical reaction and electrophoretic separation by quantitative volume control. Sensors Actuators B Chem. 110, 164–173. https://doi.org/10.1016/j.snb.2005.01.030.

Lee, Y.-F., Lien, K.-Y., Lei, H.-Y., Lee, G.-B., 2009. An integrated microfluidic system for rapid diagnosis of dengue virus infection. Biosens. Bioelectron. 25, 745–752. https://doi.org/10.1016/j.bios.2009.08.020.

Lee, J., Bisso, P.W., Srinivas, R.L., Kim, J.J., Swiston, A.J., Doyle, P.S., 2014. Universal process-inert encoding architecture for polymer microparticles. Nat. Mater. 13, 524–529. https://doi.org/10.1038/nmat3938.

Li, H.-F., Lin, J.-M., 2009. Applications of microfluidic systems in environmental analysis. Anal. Bioanal. Chem. 393, 555–567. https://doi.org/10.1007/s00216-008-2439-4.

Lim, Y.C., Kouzani, A.Z., Duan, W., 2010. Lab-on-a-chip: a component view. Microsyst. Technol. 16, 1995–2015. https://doi.org/10.1007/s00542-010-1141-6.

Mapes, A.A., Kloosterman, A.D., de Poot, C.J., 2015. DNA in the criminal justice system: the DNA success story in perspective. J. Forensic Sci. 60, 851–856. https://doi.org/10.1111/1556-4029.12779.

Mozayani, A., Noziglia, C., 2011. The Forensic Laboratory Handbook Procedures and Practice. Humana Press, New York.

Nguyen, N.-T., 2012. Fabrication technologies. In: Nguyen, N.-T. (Ed.), Micromixers, Micro and Nano Technologies. second ed. William Andrew Publishing, Oxford, pp. 113–161. (Chapter 4) https://doi.org/10.1016/B978-1-4377-3520-8.00004-8.

Pamme, N., Wilhelm, C., 2006. Continuous sorting of magnetic cells via on-chip free-flow magnetophoresis. Lab Chip 6, 974–980. https://doi.org/10.1039/b604542a.

Pandey, G., Rawtani, D., Agrawal, Y.K., 2017a. Future aspects of halloysite nanotubes in forensic investigations. J. Nanomed. Res. 6https://doi.org/10.15406/jnmr.2017.06.00153.

Pandey, G., Tharmavaram, M., Rawtani, D., Kumar, S., Agrawal, Y., 2017b. Multifarious applications of atomic force microscopy in forensic science investigations. Forensic Sci. Int. 273, 53–63. https://doi.org/10.1016/j.forsciint.2017.01.030.

Parton, J., Abu-Mandil Hassan, N., Brown, T.A., Haswell, S.J., Brown, K.A., Shaw, K.J., 2013. Sex identification of ancient DNA samples using a microfluidic device. J. Archaeol. Sci. 40, 705–711. https://doi.org/10.1016/j.jas.2012.07.012.

Pesenti, A., Taudte, R.V., McCord, B., Doble, P., Roux, C., Blanes, L., 2014. Coupling paper-based microfluidics and lab on a chip technologies for confirmatory analysis of trinitro aromatic explosives. Anal. Chem. 86, 4707–4714. https://doi.org/10.1021/ac403062y.

Peters, K.L., Corbin, I., Kaufman, L.M., Zreibe, K., Blanes, L., McCord, B.R., 2015. Simultaneous colorimetric detection of improvised explosive compounds using microfluidic paper-based analytical devices (μPADs). Anal. Methods 7, 63–70. https://doi.org/10.1039/C4AY01677G.

Piorek, B.D., Lee, S.J., Moskovits, M., Meinhart, C.D., 2012. Free-surface microfluidics/surface-enhanced Raman spectroscopy for real-time trace vapor detection of explosives. Anal. Chem. 84, 9700–9705. https://doi.org/10.1021/ac302497y.

Popovtzer, R., Neufeld, T., Ron, E.Z., Rishpon, J., Shacham-Diamand, Y., 2006. Electrochemical detection of biological reactions using a novel nano-bio-chip array. Sensors Actuators B Chem. 119, 664–672. https://doi.org/10.1016/j.snb.2006.01.037.

Rawtani, D., Tharmavaram, M., Pandey, G., Hussain, C.M., 2019. Functionalized nanomaterial for forensic sample analysis. TrAC Trends Anal. Chem. 120, 115661. https://doi.org/10.1016/j.trac.2019.115661.

Reedy, C.R., Price, C.W., Sniegowski, J., Ferrance, J.P., Begley, M., Landers, J.P., 2011. Solid phase extraction of DNA from biological samples in a post-based, high surface area poly(methyl methacrylate) (PMMA) microdevice. Lab Chip 11, 1603–1611. https://doi.org/10.1039/c0lc00597e.

Renzi, R.F., Stamps, J., Horn, B.A., Ferko, S., VanderNoot, V.A., West, J.A.A., Crocker, R., Wiedenman, B., Yee, D., Fruetel, J.A., 2005. Hand-held microanalytical instrument for chip-based electrophoretic separations of proteins. Anal. Chem. 77, 435–441. https://doi.org/10.1021/ac049214f.

Sabounchi, P., Morales, A.M., Ponce, P., Lee, L.P., Simmons, B.A., Davalos, R.V., 2008. Sample concentration and impedance detection on a microfluidic polymer chip. Biomed. Microdevices 10, 661–670. https://doi.org/10.1007/s10544-008-9177-4.

Schulte, T., Bardell, R., Weigl, B.H., 2000. Sample acquisition and control on-chip microfluidic sample preparation. J. Assoc. Lab. Autom. 5, 83–86. https://doi.org/10.1016/S1535-5535-04-00089-9.

Srinivasan, V., Pamula, V.K., Fair, R.B., 2004. An integrated digital microfluidic lab-on-a-chip for clinical diagnostics on human physiological fluids. Lab Chip 4, 310–315. https://doi.org/10.1039/B403341H.

Underberg, W.J.M., Waterval, J.C.M., 2002. Derivatization trends in capillary electrophoresis: an update. Electrophoresis 23, 3922–3933. https://doi.org/10.1002/elps.200290010.

Xuan, R., Ge, J., 2011. Photonic printing through the orientational tuning of photonic structures and its application to anticounterfeiting labels. Langmuir 27, 5694–5699. https://doi.org/10.1021/la200571y.

Zhang, Y., Ozdemir, P., 2009. Microfluidic DNA amplification—a review. Anal. Chim. Acta 638, 115–125. https://doi.org/10.1016/j.aca.2009.02.038.

Zhao, Y., Cheng, Y., Shang, L., Wang, J., Xie, Z., Gu, Z., 2015. Microfluidic synthesis of barcode particles for multiplex assays. Small 11, 151–174. https://doi.org/10.1002/smll.201401600.

Zhu, Y., Meng, X., Chen, Y., Li, J., Shao, H., Lu, Y., Pan, L., Xu, Y., Cheng, J., 2020. Self-served and fully automated biochemical detection of finger-prick blood at home using a portable microfluidic analyzer. Sensors Actuators B Chem. 303, 127235. https://doi.org/10.1016/j.snb.2019.127235.

CHAPTER 19

Nanotechnology in forensic science

1 Introduction

Dr. Richard Feynman's talk in 1959 gave birth to the revolutionary field of nanotechnology, and his quote, "There is plenty of room at the bottom," quite aptly summarized the essence and potential of nanotechnology. This field deals with the design, synthesis, manipulation, and applications of nanoscale (10^{-9} m) materials in different fields. Prior to 1959, Romans had used nanoscale materials for the decoration of different items. After that, nanomaterials found applications in photography as well. Over the years, nanotechnology has seen unsurmountable leaps in the development of nanomaterials, their characterization techniques, and their applications. The small size of the nanomaterials imparts a high surface area to the materials, due to which they have high reactivity and a tunable surface chemistry. The same nanoscale dimension allows the nanomaterials to be different from their bulk counterparts. Due to such properties, they have found applications in nanomedicine, nanoelectronics, agriculture, pollutant remediation, clean energy synthesis, energy storage, and even in defense and cosmetics. There are numerous nanomaterials that are either directly synthesized or are naturally procured from the nature. Every nanomaterial has its own property, toxicity, and fate in the environment.

In forensic science, evidence such as fingerprints, DNA, hair, semen, drugs, and even explosives are often found in trace amounts at a crime scene. Nanomaterials, due to their small size and high surface area, are capable of attaching to such trace evidence with higher selectivity (Rawtani et al., 2019). This ensures that such compounds can be detected with high sensitivity and contrast. In this chapter, a brief overview of the different synthetic approaches, the types of nanomaterials, and their characterization techniques will be discussed. Also, the different functionalization strategies for nanomaterials will be discussed in brief. Further, light will be shed on the applications of nanomaterials in forensic science.

Handbook of Analytical Techniques for Forensic Samples
https://doi.org/10.1016/B978-0-12-822300-0.00019-7

2 An insight into the world of nanotechnology

2.1 Types of nanomaterials and their characterization techniques

Nanomaterials are any material whose size ranges from 1 to 100 nm. Conventionally, nanomaterials are divided on the basis of their dimensions, that is, on the basis of their x, y, and z axis. Based on these parameters, nanomaterials are classified into zero, one, two, and three dimensions. Nanomaterials in the zero dimension have all their axes in the nanoscale range of 1–100 nm. Nanomaterials such as nanoparticles, quantum dots, fullerenes, and hollow spheres come under this class (Tiwari et al., 2012; Yu et al., 2015). One-dimensional (1D) nanomaterials are those whose any two axes occur in the nanoscale range while the third axis is larger than the nanoscale. This class includes nanomaterials such as nanowires, nanorods, and nanotubes. Meanwhile, two-dimensional (2D) nanomaterials comprise nanomaterials with only one axis is in the nanoscale dimension. The remaining axes are larger than the nanoscale. Examples of two-dimensional nanomaterials include nanosheets, nanofilms, or thin films. The last class, three-dimensional (3D) nanomaterials, consists of materials in which all the axes are beyond the nanoscale dimensions. However, their structure is similar to a nanocrystalline structure, that is, they consist of individual nanosized crystals that are arranged in an orderly manner, albeit in different orientations. Examples of three-dimensional nanomaterials include fullerites, membranes with nanoparticles embedded in them, or powder fibers. Apart from the dimensional classification of nanomaterials, there are several more classes that divide nanomaterials based on their origin, chemical composition, shape, size, and properties (Saleh and Gupta, 2016).

Nanomaterials can be either synthesized through different approaches or can be naturally procured from nature (Pandey et al., 2017a,b; Rawtani et al., 2017; Rawtani and Agrawal, 2012a; Tharmavaram et al., 2018); they will be discussed in brief in the next section. Once synthesized, the nanomaterials are characterized using various microscopy and spectroscopy techniques. Microscopy approaches such as electron microscopy and scanning probe microscopy (SPM) are used. Electron microscopy includes techniques such as transmission electron microscopy (TEM) and scanning electron microscopy (SEM) while SPM consists of atomic force microscopy (AFM), near-field scanning microscopy (NSOM), and scanning tunnelling microscopy (STM). TEM and SEM use an electron beam to visualize the interior morphology and the outer area of the nanomaterials, respectively.

AFM offers 3D imaging of the nanomaterial surface (Pandey et al., 2017c) while STM is widely used for characterizing the surface of the conducting materials. NSOM allows visualization through light. A commonality between the three techniques is that they use a probe or tip that can be conducting, nonconducting, or optically transparent for the visualization of the surface of the nanomaterials. Apart from the microscopy and spectroscopy techniques, X-ray diffraction (XRD) is also used to characterize the crystalline or amorphous structure of a nanomaterial by measuring the d-space between the atoms. Spectroscopy techniques such as UV-visible, infrared, Raman, and fluorescence spectroscopy are widely used. Mass spectrometry is also used to identify the molecular weight of the nanomaterial.

The characterization techniques are selected based on the purpose. For instance, in order to perform structural characterization, SEM, TEM, SPM, and the spectroscopy techniques are used. However, if the composition of the nanomaterial must be found out, then mass spectroscopy, infrared spectroscopy, and even nuclear magnetic resonance (NMR), X-ray photoelectron spectroscopy (XPS) and X-ray fluorescence are used. Electron microscopy and SPM techniques are often combined with dynamic light scattering (DLS) to determine the size of the nanomaterial and its dispersion. DLS measures the hydrodynamic radius of the nanomaterial while electron microscopy and SPM offer the exact size of the nanomaterial. Every nanomaterial is always subjected to a series of characterization studies to get an overall idea of its structure and composition (Salame et al., 2018).

2.2 Types of synthetic approaches of nanomaterials

Nanomaterials can be synthesized through solid, liquid, or even gaseous state precursors. Conventionally, the approaches can be summarized as two basic approaches: top down and bottom up. In the top-down approach, the nanomaterial processing starts with a bulk material that is broken down into smaller nanoscale pieces through mechanical or chemical methods. The second approach is the bottom-up approach in which synthesis occurs through the self-assembly of the atomic or molecular form of the precursor. In theory, these approaches may seem very simple. Practically, though, significant investment in maintaining the environmental conditions such as pressure, temperature, and cleanliness is required. Proper maintenance of such conditions ensures that nanomaterials of optimal shape, size, composition, and degree of agglomeration are fabricated. These approaches are often modified through the use of green synthetic routes, self-assembly, and biomimetic approaches (Sasidharan et al., 2019).

Top-down approaches involves processes such as ball milling, etching, lithography, and laser ablation. Ball milling is a process in which the precursor powder is subjected to high-energy compressive forces. Grinding of the precursor under such high forces causes it to break down into smaller pieces until it reaches a constant size that is usually in the nanometer range. This minimum range of size differs for different precursors. A major advantage with this technique is that conditions such as temperature are kept at a constant minimum. Due to this, the process of nanomaterial formation is slow and uniform. Metal alloys and composites can also be synthesized through ball-milling approaches (Baláž, 2018; Casati and Vedani, 2014; Koch, 1998). Etching is another top-down technique in which a certain portion of the precursor is removed in a controlled manner to synthesize nanomaterials. It can be performed through chemicals, electric arc discharge, or plasma. Etching, when done on uniform substrates, produces nanomaterials of optimal size and shape (Chen et al., 2015; Puliyalil and Cvelbar, 2016). Another technique for nanomaterial synthesis is laser-ablation, which is similar to etching because it involves the removal of a certain part of a material. However, this is done through a laser. The extremely high energy of the laser heats the surface of the precursor, due to which it ablates and sublimates. The ablation can be made uniform by controlling the laser parameters such as its energy and pulse (Phipps, 2007). Apart from the mentioned techniques, lithography is a technique that is widely used in the semiconductor industry to fabricate IC circuits. This technique involves a series of processes that removes materials from a substrate in a selective manner. Photolithography, scanning lithography, and soft lithography are some of lithography types (Kretz, 2004; Tran and Nguyen, 2017).

Bottom-up approaches involve vapor-phase deposition, molecular beam epitaxy (MBE), chemical reduction and precipitation reactions, sol-gel, sonochemical synthesis, and self-assembly techniques. Physical vapor deposition (PVD) (Ali et al., 2005) and chemical vapor deposition (CVD) (Carlsson and Martin, 2010) are two types of vapor phase deposition techniques in which a uniform coating of films is performed. These techniques are highly used in the semiconductor industry to produce thin films. PVD involves the gasification of the precursor material, after which the material's vapor goes through a reacting chamber and gets deposited on the surface of the substrate to be coated. CVD is similar to PVD in the way it coats the substrate. However, in this technique, the precursor is usually mixed with volatile compounds such as halides or hydrides so that they can easily transform into the vapor phase and get deposited on the surface of the substrate present in the reaction chamber. Meanwhile, MBE is a complex evaporation technique that is carried out in ultrahigh vacuum to produce uniform single-crystal thin films (Alavi, 2001). One of the most common synthetic routes of nanoparticles is through chemical reduction and precipitation techniques. Precipitation is a type of exchange reaction method in which the solids are precipitated out from a solution of the precursor. Careful control of the experimental parameters and the environmental conditions allows the formation of nanoscale particles. In the chemical reduction technique, powerful reducing agents such as sodium borohydride and ascorbic acid are added to a solution of the precursor in order to reduce the precursor to its ionic state and allow the formation of nanoparticles. Very often, the precursor solution also consists of capping agents that are responsible for increasing the stability of the nanoparticles while maintaining uniform size distribution. The sonochemical technique is a method in which the high-intensity ultrasound waves are passed through a precursor solution to form nanoparticles. Such high-intensity soundwaves are known to create cavitation in the solvent, which causes bubble formation. Each bubble then serves as a minireactor with extreme high pressure and temperature, due to which the formation of the nanoparticles is promoted. Lastly, nanomaterials can be functionalized in such a way that they induce self-assembly in the presence of other nanomaterials. This technique is known as the self-assembly method of forming nanomaterials. However, this synthetic route is not devoid of major challenges. Finding the precise location for a particular moiety and interconnecting the nanomaterials and their applications in different fields are some of the commonly encountered challenges (Sasidharan et al., 2019). Fig. 1 depicts the overall representation of the different synthetic approaches.

2.3 Functionalization strategies of nanomaterials

The functionalization of nanomaterials involves the incorporation of functional groups on the surface of a nanomaterial. Proper functionalization not only imparts desired properties such as enhanced conductivity as well as optical, magnetic, antimicrobial, or antioxidant properties to the nanomaterial, but it also helps in controlling their growth in synthesis (Tharmavaram et al., 2018). Appropriate functionalization techniques allow a nanomaterial of optimal shape, size dispersion, and stability to be developed. Over the years,

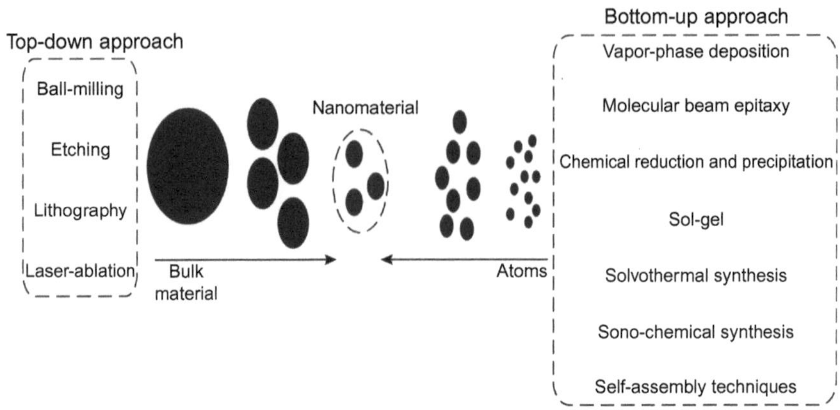

Fig. 1 Synthetic approaches of nanomaterials.

several functionalization strategies for nanomaterials have been developed. Some functionalization strategies focus on modifying the characteristic of the surface of the nanomaterial while some strategies directly attach a particular molecule on it (Sperling and Parak, 2010). For example, strategies such as ligand exchange and modification, polymeric coatings, and silanization involve phase transfer reactions in which hydrophobic nanomaterials are converted to hydrophilic molecules in order to promote their applications in water-based solvent systems. In ligand exchange reactions, a particular moiety on the surface of the nanomaterial is exchanged by a ligand with the required properties. For example, a negatively charged nanomaterial may be subjected to ligand exchange wherein a positively charged nanomaterial gets attached on it, thereby maintaining the nanomaterial stability and size (Caragheorgheopol and Chechik, 2008; Twigg, 1994). Polymers that are amphiphilic in nature are highly useful in the functionalization of a nanomaterial as they are capable of surviving in two phases at once. Therefore, coating with polymers allows the applicability of the nanomaterial in two different phases and there is no dependence on the actual nature of the nanoparticle (Kirschling et al., 2011; Petruska et al., 2004). Silanization is another technique in which after a series of hydrolysis, condensation, hydrogen, and other bond formations, a functional group is imparted on the surface of the nanomaterial (Brzoska et al., 1994; Krasnoslobodtsev and Smirnov, 2002). Strategies where direct functional groups are attached on the surface include attaching chemical functional groups such polyethylene glycol and mercaptocarboxylic acids and biomolecules such as peptides and nucleic acids (Mo et al., 2017; Zhu et al., 2018).

3 Nanotechnology-mediated forensic investigations

Recent years have seen tremendous growth in the field of nanotechnology, which has now started spreading into forensic science as well. Different types of nanomaterials, either in their pristine state or postsurface functionalization, as well as nanotechnology-based devices and sensors have found applications in the investigation of crimes. This section discusses the utilization of nanomaterials and nanotechnology-based devices in the world of forensics. Some of the research works involving various kinds of nanomaterials for different forensic applications are discussed in Table 1.

3.1 Detection of explosives and gunshot residues

Terrorist activities often involve the use of explosives such as improvised explosive devices (IEDs), bombs, grenades, etc., that cause destruction on a very large scale. The detection of even trace quantities of these explosives becomes very crucial as far as forensic investigations are concerned. The conventional techniques used for sensing explosives are based on spectrometry or chromatography (Rawtani et al., 2019). However, concerns such as high cost, the requirement of skilled manpower, and high time consumption have forced forensic scientists to look for alternatives to these conventional strategies. Nanomaterials have started contributing to the detection of explosive residues. Various types of nanomaterial-based sensors have also been developed for this purpose (Liu et al., 2019). Nanomaterials such as AuNP (Riskin et al., 2008), SiO_2NP (Fu et al., 2010), carbon NP hybrids (Guo et al., 2010), curcumin NPs (Pandya and Shukla, 2018), and carbon nanotubes (CNT) (Lu et al., 2011) have been used by various research groups for the detection of different explosives. The mechanism of detection is usually based on the redox properties of the explosives or the luminescent taggants to provide luminescence to even traces of explosives (Rawtani et al., 2019). Explosives such as dinitrobenzene (DNB) and trinitrotoluene (TNT) are common explosives detected using nanomaterials. In a research study, TNT was detected using carbon dots that were capped with molecularly imprinted polymers. Mesoporous silicon dioxide served as the imprinting matrix. The lowest concentration of TNT detected was 17 nM, and the mechanism of detection was based on fluorescence (Xu and Lu, 2016). The schematic representation of TNT sensing through the aforementioned system is shown in Fig. 2. Devices with these types of detection technologies can help security personnel to stop many terrorist activities.

Table 1 Nanomaterials employed in different applications of forensic science (Rawtani et al., 2019).

S. no.	Nanomaterial	Functionalization agent	Functionalization method	Application	Target/product	Detection method	Reference
1	YVO_4:Eu nanocrystals	PEI	Hydrothermal method	Fingerprint identification	Latent fingerprint	Fluorescence	Wang et al. (2015b)
2	$LaPO_4$:Ce, Tb nanobelts	Ethylene glycol	Hydrothermal method	Fingerprint identification	Latent fingerprint	Fluorescence	Wang et al. (2015b)
3	CNT	MHTPP THTPP	Electrolytic method	Explosives detection	DNB	DPS voltammetry	Lu et al. (2011)
4	AuNPs	2-Mercaptoethane sulfonate *p*-Aminothiophenol	Wet synthesis method	Explosives detection	TNT	Voltammetry	Riskin et al. (2008)
5	Si nanoshells	3-APTS	Reflux hydrolysis	Explosives detection	TNT, TNB, DNT	Voltammetry	Fu et al. (2010)
6	PNEGHNs	PMMA	Microwave-assisted heating method	Explosives detection	TNT	Electrochemical	Guo et al. (2010)
7	HNT	Chloroauric acid	Onspot reduction method	DNA analysis	DNA	Fluorescence	Rawtani and Agrawal, 2012b
8	HNT	Silver nitrate	Onspot reduction method	DNA analysis	DNA	Fluorescence	Rawtani et al. (2013)
9	Graphene nanosheets	PSS	Hydrothermal reaction	Illicit drug sensing	Heroine, morphine, noscapine	Voltammetric	Navaee et al. (2012)

10	Ag, AuNP nano inks	MUA, DDAB	Wet synthesis method	Counterfeiting	Erasable inks	Colorimetric	Klajn et al. (2009)
11	Fe_3O_4 CNC	Silica layer	Wet synthesis method	Counterfeiting	–	Optical detection	Ge and Yin (2008) and Klajn et al. (2009)
12	AuNP	Ascorbic acid	Wet synthesis method	Homeland security	Dichlorvos	Colorimetric	D'souza et al. (2014)

PEI, polyethylene amine; *CNT*, carbon nanotubes; *HNT*, halloysite nanotubes; *MHTPP*, 5-(4-hydroxyphenyl)-10,15,20-triphenylporphyrin; *THTPP*, (meso-tetrakis (4-hydroxyphenyl)porphyrin); *DNB*, dinitrobenzene; *TNB*, trinitrobenzene; *TNT*, trinitrotoluene; *DNT*, dinitrotoluene; *3-APTS*, (3-aminopropyl)triethoxysilane; *MUA*, (11-mercaptoundcanoxy)azobenzene; *DDAB*, didodecyldimethylammonium bromide; *CNC*, colloidal nanocrystal clusters.

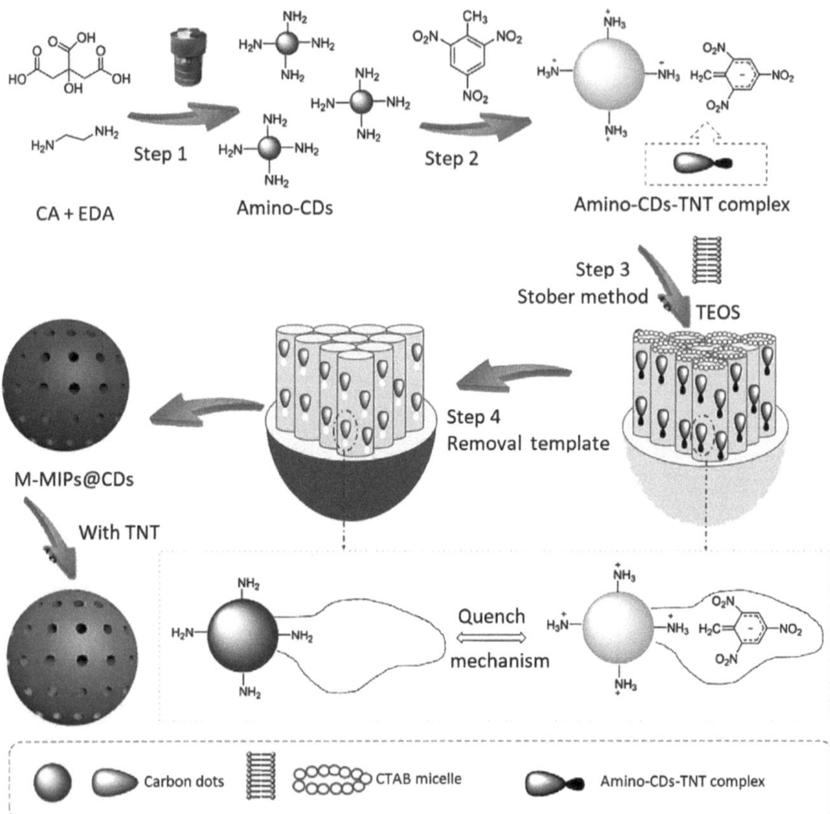

Fig. 2 Schematic representation of TNT sensing via molecularly imprinted carbon dots (Xu and Lu, 2016).

Apart from explosives, gunshot residue (GSR) has also been identified using nanotechnology. Techniques such as SEM and TEM have been utilized by a group of researchers from Brazil to characterize GSR from ammunition of various calibers (Melo et al., 2014). Another study though electron microscopy revealed the presence of metallic NPs in GSR (Yang et al., 2006). These nanotechnology-based advancements have eased the investigation process in crimes involving firearms and ammunition to a great extent.

3.2 Biosensors

Biosensors are among the recent and excellent innovations of nanotechnology that have started gaining the interest of forensic experts and investigators, mainly because of their ability for the onsite detection of forensic analytes with great accuracy and specificity. The utilization of biosensors can significantly increase the selectivity, specificity, and sensitivity of sensing even trace levels of evidence such as blood, sweat, poison, illicit drugs, alcohol, warfare agents, etc. Various kinds of nanomaterials, preferably those with good conductivity and optical properties, are used in the construction of biosensors. The nanomaterials lacking optical properties but having good conductivity, such as carbon nanotubes, are conjugated with different biomolecules such as nucleic acids or enzymes. The nanomaterial itself or the enzyme linked to the nanomaterial that is present in the biosensor are majorly responsible for the sensing of various analytes. Regarding the use of biosensors in forensic science, they have found applications in the detection of marine toxins, microbes, alcohol, forensic markers, and warfare agents. The different biosensors used for these analytes are discussed below:

- **Biosensors for marine toxins, microbes, and alcohol**: Marine toxins are used to kill or damage the health of people, especially in the coastal areas that have widespread consumption of seafood. Consumers of seafood usually eat shellfish that sometimes are contaminated with shellfish poisons. These toxins are known to cause paralysis in the muscles of the respiratory system. Shellfish poisons have been detected using surface plasmon resonance (SPR) biosensors employing polyclonal antibodies for saxitoxin (Haughey et al., 2011). Gold nanoparticles conjugated with antibodies for tetrodotoxin were used to develop immunosensors to sense tetrodotoxin that comes from the puffer fish (Zhou et al., 2010). Another immunosensor to detect okadeic acid has also been developed using the antibody against the toxin (Stewart et al., 2009). Microbes such as *Chlamydia trachomatis* (Parab et al., 2010) and *Listeria monocytogenes* (Banerjee and Bhunia, 2010) have also been detected using gene- and cell-based biosensors. These microbes are responsible for the spread of pandemics. Nanomaterials such as carbon nanotubes and gold nanoparticles have been used to immobilize alcohol oxidase (Das and Goswami, 2013) and alcohol dehydrogenase (Zhen et al., 2011), respectively, for the detection of alcohol, which is majorly abused and the cause of many criminal activities.
- **Biosensors for forensic markers**: Studies have been performed in the past regarding the detection of sweat and semen, which can be considered forensic markers. For the detection of sweat, a cascade system involving enzymes such as horseradish peroxidase (HRP) and lactate

dehydrogenase (LDH) was used to develop the biosensor. The study helped in differentiating between sweat and saliva by sensing the lactate present in sweat (Huynh et al., 2017). The identification of spermatozoa in a semen sample can give false results in case of sexual assaults if done by any aspermic or oligospermic individual. Therefore, a biosensor was developed for the prostate specific antigen (PSA) that can serve as a forensic marker in cases of sexual assaults (Koukouvinos et al., 2017).

- **Biosensors for warfare agents**: Warfare agents can be broadly categorized as chemical and biological. Chemical warfare agents majorly include organophosphates (OPs) and explosives while biological warfare agents have toxins originating from biological sources such as botulin, ricin, etc. The inhibitory action of OPs against acetylcholinesterase (AChE) has been utilized to develop AchE-based biosensors to detect such agrochemicals (Viveros et al., 2006). TNT (Singh, 2007) and TATP (triacetone triperoxide) (Girotti et al., 2011) have also been detected by using different biosensors. SPR-based biosensors have been developed to sense ricin, which comes from *Ricinus communis* (Uzawa et al., 2008). Other bioterrorism agents such as cholera toxin, staphylococcal enterotoxin B, and ricin have also been sensed through a multianalyte biosensing device (Lian et al., 2010).

3.3 Visualization of latent fingermarks

Fingerprints are among the most common evidence that an investigator can collect from a crime scene. Broadly, there exist three types of fingermarks: latent (invisible), patent (visible), and plastic (indentations). Among the three, latent fingerprints are difficult to find and are usually developed by experts using fingerprinting kits. A physical developer, ninhydrin, diazofluoren-9-one (DFO), iodine fuming, and superglue fuming are some of the widely used techniques for fingerprint development (Sodhi and Kaur, 2016). However, concerns related to these techniques such as toxicity, low preservation time, and less selectivity and sensitivity have led to the development of other novel techniques for the development of fingerprints (Wang et al., 2017).

Research related to nanomaterial-assisted fingerprint development has seen tremendous growth in past years. The major reason behind this is the nanoscale size of these materials that get accumulated between the ridges of the fingerprint and make visible even the fine details of the fingermarks (Rawtani et al., 2019). In order to enhance the contrast, these nanomaterials are tagged with colored and fluorescent dyes before developing the fingerprints. The availability of different colored dyes provides the benefit of developing different dye-nanomaterial complexes that have contrasting colors with the background. Fluorescent nanomaterials such as quantum dots, $LaPO_4$:Ce, and YVO_4:Eu have gained the interest of forensic researchers due to their optical properties (Wang et al., 2015b). The potential of clay-based nanomaterials such as bentonite (Fig. 3) (Chen et al., 2009) and halloysite (Pandey et al., 2017b) has also been discussed by researchers through the incorporation of fluorescent and colored dyes.

3.4 Detection of DNA and illicit drugs

The importance of DNA sensing and analysis is very crucial in forensic investigations because it directly relates the evidence with the victim and the suspect. Traditional techniques of DNA detection involve sensing systems labeled with fluorophores and real-time polymerase chain reaction (RT-PCR) assays. However, the drawbacks associated with these techniques such as high cost, the requirement of expertise, and time have started shifting the interest of researchers from these techniques toward more

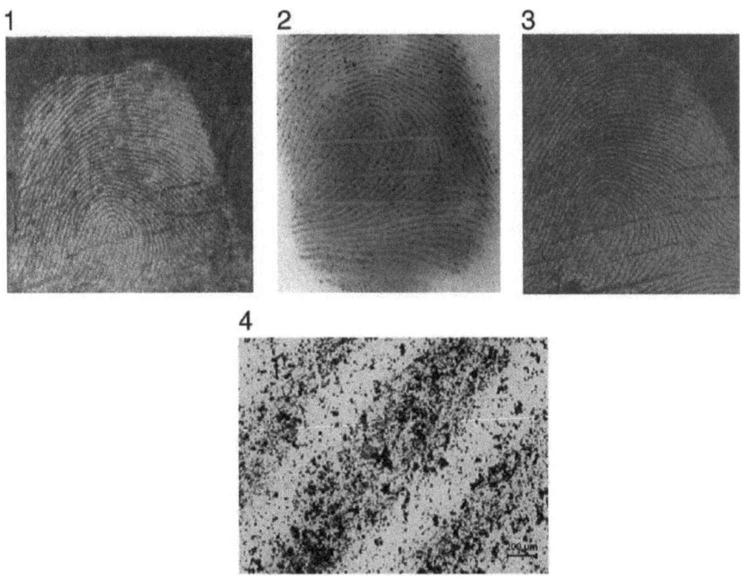

Fig. 3 Development of fingermarks on a glass slide using bentonite loaded with rhodamine 6G (1, 3) and methylene blue (2). (4) Shows optical microscopic image of the deposition of bentonite-rhodamine complex between the fingerprint ridges (Chen et al., 2009).

advanced nanotechnology-assisted detection methods. Nanomaterials such as halloysite nanotubes have been used to detect DNA via the colorimetric approach after functionalization with silver and gold nanoparticles (Rawtani et al., 2013; Rawtani and Agrawal, 2012b). Carbon-based materials such as CNT and graphene have also been utilized for this purpose (Li et al., 2013).

Illicit and illegal drugs such as heroin, cocaine, morphine, codeine, etc., are widely used, especially by youngsters, and these drugs become a major cause of criminal activities. The onsite detection of these drugs is needed for forensic experts and investigators. In this regard, various nanomaterial-based systems have been researched by scientists. The surface functionalization of nanomaterials with various agents in order to provide any reactive functional group on the nanomaterial's surface becomes important for sensing drugs. The interaction between the functional groups on the surface of nanomaterials with those present on the drug is the key behind sensing these drugs. In a study, CdSe/ZnS quantum dots were conjugated with antimorphine to sense morphine (Zhang et al., 2017). The high sensitivity and conductivity of carbon-based nanomaterials such as graphene oxide help them in differentiating between different unlawful drugs through voltammetry (Navaee et al., 2012). Other studies in this field include the use of nanocrystals for the detection of morphine (Arabali and Sadeghi, 2013) and nanodiamonds for sensing codeine (Simioni et al., 2017).

3.5 Counterfeiting prevention and homeland security

Copyright encroachment and damage to society are the major concerns related to the act of counterfeiting. Currency notes and drugs produced by counterfeiting become responsible for decreasing the country's economy and increasing the risks to public health, respectively (Rawtani et al., 2019). Water marks, holograms, radio frequency identification (RFID), metal threads, etc., are commonly used security techniques to identify counterfeiting in currency notes. However, the security levels can be increased severalfold by using fluorescent nanomaterials, quantum dots, metastable nanoinks, and colloidal nanocrystals. The safety of materials increases to a great extent upon using these nanostructures (Ge and Yin, 2008; Klajn et al., 2009).

Homeland security is majorly concerned with the detection of weapons of mass destruction (WMDs) because of their ability to cause heavy destruction to life and property (Rawtani et al., 2019). Nanomaterials have been used to develop nanomotors that help in the identification of warfare agents. These nanomotors are propelled due to the photoactivation property of nanoparticles and through bubble formation (Reynolds and Hart, 2004; Singh and Wang, 2015). Dichlorvas, a potent insecticide and warfare agent, has been detected using the colorimetric method with the help of gold nanoparticles (D'souza et al., 2014). This nanomaterial-based identification of WMDs truly has a bright future in forensic investigations due to the specificity of the method.

3.6 Nanotechnology-based tracker

Nanotechnology-based tracking devices, commonly known as nanotrackers, are a nanoscale dimensioned system that can be inserted or attached on any human or product in order to track it during movement. These trackers along with barcodes can be secretively riveted onto the products in a unique pattern, thereby preventing the shoplifting of the product (Paunescu et al., 2016). These nanodevices can also be inserted into the bodies of prisoners or criminals in order to keep a check on their movement if they are trying to escape the prison (Wang et al., 2015a). This technology is already gaining attention among common people, who are using this as a small GPS device for themselves, their belongings such as keyrings, or on the collars of their pets. These tracking nanodevices need more research in order to enhance their utilization in routine forensic work. With improvised technological aid, these devices can efficiently track the movement of criminals and products throughout the world. The devices can also assist in differentiating between counterfeit and real products.

4 Conclusion

Nanomaterials are nanosized materials that have a high surface area and increased reactivity. The small size of the nanomaterials causes them to have enhanced optical, electrical, and magnetic properties. Due to such properties, they find applications in many fields such as nanomedicine, pollutant remediation, nanoelectronics, and in the production and storage of clean energy. There are two basic modes of synthesizing nanomaterials: the top-down approach and the bottom-up approach. In the top-down approach, a bulk material is broken down to smaller parts through mechanical or chemical methods while in the bottom-up approach, the atomic form of the precursors is taken and then combined to form the nanomaterials. It is also important that nanomaterials be functionalized before being used in any application. Functionalization has several roles and some of them include imparting additional properties such as enhanced conductivity, optical transparency, and antimicrobial or antioxidant properties to the nanomaterial as well as controlling its growth and stability and reducing its overall toxicity.

In the field of forensic science, many types of nanomaterials have been utilized for a wide variety of applications. Explosives and gunshot residue that are collected by investigators from a crime scene have been detected using nanomaterials. The redox properties of explosives along with the luminescent taggants have been utilized for the detection of explosive residues. Nanotechnology-based electron microscopes have helped in studying GSR patterns to find the shooting distance and angle at which the bullet was fired. Nanomaterials with fluorescent properties or those conjugated with enzymes have been used to develop biosensors for the detection of marine toxins, microbes, alcohol, forensic markers such as sweat and semen, and warfare agents of chemical or biological origin. The nanoscale size and conjugation with colored and fluorescent dyes has also enabled nanomaterials to be used for visualizing latent fingerprints. These nanomaterials accumulate between the fingerprint ridges and display very minute details of fingermarks. The detection of DNA and unlawful drugs has also been carried out using different nanomaterials. These nanomaterials can also assist in preventing the counterfeiting of currency notes as well as the forgery of different products. The detection of WMDs has also been researched using these nanomaterials. Nanodevices known as nanotrackers have also shown potential to be used for tracking the movements of goods and criminals. The potential of nanomaterials can be explored to a much greater extent upon carrying out more research in the field. Collaborations between scientists and experts from the worlds of forensics and nanotechnology can produce excellent results for the investigation team, thereby assisting in bringing proper justice to the victims.

References

Alavi, K., 2001. Molecular beam epitaxy. In: Buschow, K.H.J., Cahn, R.W., Flemings, M.C., Ilschner, B., Kramer, E.J., Mahajan, S., Veyssière, P. (Eds.), Encyclopedia of Materials: Science and Technology. Elsevier, Oxford, pp. 5765–5780. https://doi.org/10.1016/B0-08-043152-6/01002-0.

Ali, M., Hamzah, E., Toff, M., 2005. Review of physical vapour deposition (PVD) techniques hard coating. J. Mekanikal 20.

Arabali, V., Sadeghi, R., 2013. Surface properties of nano-Al_2O_3 film and its application in the preparation of morphine electrochemical sensor. Ionics 19, 1775–1782. https://doi.org/10.1007/s11581-013-0929-7.

Baláž, M., 2018. Ball milling of eggshell waste as a green and sustainable approach: a review. Adv. Colloid Interface Sci. 256, 256–275. https://doi.org/10.1016/j.cis.2018.04.001.

Banerjee, P., Bhunia, A.K., 2010. Cell-based biosensor for rapid screening of pathogens and toxins. Biosens. Bioelectron. 26, 99–106. https://doi.org/10.1016/j.bios.2010.05.020.

Brzoska, J.B., Azouz, I.B., Rondelez, F., 1994. Silanization of solid substrates: a step toward reproducibility. Langmuir 10, 4367–4373. https://doi.org/10.1021/la00023a072.

Caragheorgheopol, A., Chechik, V., 2008. Mechanistic aspects of ligand exchange in Au nanoparticles. Phys. Chem. Chem. Phys. 10, 5029. https://doi.org/10.1039/b805551c.

Carlsson, J.-O., Martin, P.M., 2010. Chemical vapor deposition. In: Martin, P.M. (Ed.), Handbook of Deposition Technologies for Films and Coatings. third ed. William Andrew Publishing, Boston, pp. 314–363. https://doi.org/10.1016/B978-0-8155-2031-3.00007-7 (Chapter 7).

Casati, R., Vedani, M., 2014. Metal matrix composites reinforced by nano-particles—a review. Metals 4, 65–83. https://doi.org/10.3390/met4010065.

Chen, Q., Kerk, W.T., Soutar, A.M., Zeng, X.T., 2009. Application of dye intercalated bentonite for developing latent fingerprints. Appl. Clay Sci. 44, 156–160. https://doi.org/10.1016/j.clay.2009.01.011.

Chen, T.-H., Nieh, C.-C., Shih, Y.-C., Ke, C.-Y., Tseng, W.-L., 2015. Hydroxyl radical-induced etching of glutathione-capped gold nanoparticles to oligomeric AuI–thiolate complexes. RSC Adv. 5, 45158–45164. https://doi.org/10.1039/C5RA05159B.

D'souza, S.L., Pati, R.K., Kailasa, S.K., 2014. Ascorbic acid functionalized gold nanoparticles as a probe for colorimetric and visual read-out determination of dichlorvos in environmental samples. Anal. Methods 6, 9007–9014. https://doi.org/10.1039/C4AY01004C.

Das, M., Goswami, P., 2013. Direct electrochemistry of alcohol oxidase using multiwalled carbon nanotube as electroactive matrix for biosensor application. Bioelectrochemistry 89, 19–25. https://doi.org/10.1016/j.bioelechem.2012.08.007.

Fu, X.-C., Chen, X., Wang, J., Liu, J.-H., Huang, X.-J., 2010. Amino functionalized mesoporous silica microspheres with perpendicularly aligned mesopore channels for electrochemical detection of trace 2,4,6-trinitrotoluene. Electrochim. Acta 56, 102–107. https://doi.org/10.1016/j.electacta.2010.09.045.

Ge, J., Yin, Y., 2008. Magnetically tunable colloidal photonic structures in alkanol solutions. Adv. Mater. 20, 3485–3491. https://doi.org/10.1002/adma.200800657.

Girotti, S., Ferri, E., Maiolini, E., Bolelli, L., D'Elia, M., Coppe, D., Romolo, F.S., 2011. A quantitative chemiluminescent assay for analysis of peroxide-based explosives. Anal. Bioanal. Chem. 400, 313–320. https://doi.org/10.1007/s00216-010-4626-3.

Guo, S., Wen, D., Zhai, Y., Dong, S., Wang, E., 2010. Platinum nanoparticle ensemble-on-graphene hybrid nanosheet: one-pot, rapid synthesis, and used as new electrode material for electrochemical sensing. ACS Nano 4, 3959–3968. https://doi.org/10.1021/nn100852h.

Haughey, S.A., Campbell, K., Yakes, B.J., Prezioso, S.M., DeGrasse, S.L., Kawatsu, K., Elliott, C.T., 2011. Comparison of biosensor platforms for surface plasmon resonance based detection of paralytic shellfish toxins. Talanta 85, 519–526. https://doi.org/10.1016/j.talanta.2011.04.033.

Huynh, C., Brunelle, E., Agudelo, J., Halámek, J., 2017. Bioaffinity-based assay for the sensitive detection and discrimination of sweat aimed at forensic applications. Talanta 170, 210–214. https://doi.org/10.1016/j.talanta.2017.04.016.

Kirschling, T.L., Golas, P.L., Unrine, J.M., Matyjaszewski, K., Gregory, K.B., Lowry, G.V., Tilton, R.D., 2011. Microbial bioavailability of covalently bound polymer coatings on model engineered nanomaterials. Environ. Sci. Technol. 45, 5253–5259. https://doi.org/10.1021/es200770z.

Klajn, R., Wesson, P.J., Bishop, K.J.M., Grzybowski, B.A., 2009. Writing self-erasing images using metastable nanoparticle "inks" Angew. Chem. Int. Ed. 48, 7035–7039. https://doi.org/10.1002/anie.200901119.

Koch, C.C., 1998. Intermetallic matrix composites prepared by mechanical alloying—a review. Mater. Sci. Eng. A 244, 39–48. https://doi.org/10.1016/S0921-5093(97)00824-1.

Koukouvinos, G., Metheniti, A., Karachaliou, C.-E., Goustouridis, D., Livaniou, E., Misiakos, K., Raptis, I., Kondili, A., Miniati, P., Petrou, P., Kakabakos, S., 2017. White light reflectance spectroscopy biosensing system for fast quantitative prostate specific antigen determination in forensic samples. Talanta 175, 443–450. https://doi.org/10.1016/j.talanta.2017.07.074.

Krasnoslobodtsev, A.V., Smirnov, S.N., 2002. Effect of water on silanization of silica by trimethoxysilanes. Langmuir 18, 3181–3184. https://doi.org/10.1021/la015628h.

Kretz, J., 2004. Lithography for silicon nanotechnology. In: Siffert, P., Krimmel, E.F. (Eds.), Silicon: Evolution and Future of a Technology. Springer, Berlin, Heidelberg, pp. 399–414. https://doi.org/10.1007/978-3-662-09897-4_19.

Li, F., Pei, H., Wang, L., Lu, J., Gao, J., Jiang, B., Zhao, X., Fan, C., 2013. Nanomaterial-based fluorescent DNA analysis: a comparative study of the quenching effects of graphene oxide, carbon nanotubes, and gold nanoparticles. Adv. Funct. Mater. 23, 4140–4148. https://doi.org/10.1002/adfm.201203816.

Lian, W., Wu, D., Lim, D.V., Jin, S., 2010. Sensitive detection of multiplex toxins using antibody microarray. Anal. Biochem. 401, 271–279. https://doi.org/10.1016/j.ab.2010.02.040.

Liu, R., Li, Z., Huang, Z., Li, K., Lv, Y., 2019. Biosensors for explosives: state of art and future trends. TrAC Trends Anal. Chem. 118, 123–137. https://doi.org/10.1016/j.trac.2019.05.034.

Lu, X., Quan, Y., Xue, Z., Wu, B., Qi, H., Liu, D., 2011. Determination of explosives based on novel type of sensor using porphyrin functionalized carbon nanotubes. Colloids Surf. B: Biointerfaces 88, 396–401. https://doi.org/10.1016/j.colsurfb.2011.07.020.

Melo, L.G.A., Martiny, A., Pinto, A.L., 2014. Nano characterization of gunshot residues from Brazilian ammunition. Forensic Sci. Int. 240, 69–79. https://doi.org/10.1016/j.forsciint.2014.04.010.

Mo, L., Li, J., Liu, Q., Qiu, L., Tan, W., 2017. Nucleic acid-functionalized transition metal nanosheets for biosensing applications. Biosens. Bioelectron. 89, 201–211. https://doi.org/10.1016/j.bios.2016.03.044.

Navaee, A., Salimi, A., Teymourian, H., 2012. Graphene nanosheets modified glassy carbon electrode for simultaneous detection of heroine, morphine and noscapine. Biosens. Bioelectron. 31, 205–211. https://doi.org/10.1016/j.bios.2011.10.018.

Pandey, G., Munguambe, D.M., Tharmavaram, M., Rawtani, D., Agrawal, Y.K., 2017a. Halloysite nanotubes—an efficient 'nano-support' for the immobilization of α-amylase. Appl. Clay Sci. 136, 184–191. https://doi.org/10.1016/j.clay.2016.11.034.

Pandey, G., Rawtani, D., Agrawal, Y.K., 2017b. Future aspects of halloysite nanotubes in forensic investigations. J. Nanomed. Res. 6, 2.

Pandey, G., Tharmavaram, M., Rawtani, D., Kumar, S., Agrawal, Y., 2017c. Multifarious applications of atomic force microscopy in forensic science investigations. Forensic Sci. Int. 273, 53–63. https://doi.org/10.1016/j.forsciint.2017.01.030.

Pandya, A., Shukla, R.K., 2018. New perspective of nanotechnology: role in preventive forensic. Egypt. J. Forensic Sci. 8, 57. https://doi.org/10.1186/s41935-018-0088-0.

Parab, H.J., Jung, C., Lee, J.-H., Park, H.G., 2010. A gold nanorod-based optical DNA biosensor for the diagnosis of pathogens. Biosens. Bioelectron. 26, 667–673. https://doi.org/10.1016/j.bios.2010.06.067.

Paunescu, D., Stark, W.J., Grass, R.N., 2016. Particles with an identity: tracking and tracing in commodity products. Powder Technol. 291, 344–350. https://doi.org/10.1016/j.powtec.2015.12.035.

Petruska, M.A., Bartko, A.P., Klimov, V.I., 2004. An amphiphilic approach to nanocrystal quantum dot–titania nanocomposites. J. Am. Chem. Soc. 126, 714–715. https://doi.org/10.1021/ja037539s.

Phipps, C., 2007. Laser Ablation and Its Applications. Springer.

Puliyalil, H., Cvelbar, U., 2016. Selective plasma etching of polymeric substrates for advanced applications. Nanomaterials 6, 108. https://doi.org/10.3390/nano6060108.

Rawtani, D., Agrawal, Y.K., 2012a. Multifarious applications of halloysite nanotubes: a review. Rev. Adv. Mater. Sci. 30, 282–295.

Rawtani, D., Agrawal, Y.K., 2012b. Study the interaction of DNA with halloysite nanotube-gold nanoparticle based composite. J. Bionanosci. 6, 95–98. https://doi.org/10.1166/jbns.2012.1080.

Rawtani, D., Agrawal, Y.K., Prajapati, P., 2013. Interaction behavior of DNA with halloysite nanotube–silver nanoparticle-based composite. BioNanoScience 3, 73–78. https://doi.org/10.1007/s12668-012-0071-4.

Rawtani, D., Pandey, G., Tharmavaram, M., Pathak, P., Akkireddy, S., Agrawal, Y.K., 2017. Development of a novel 'nanocarrier' system based on Halloysite Nanotubes to overcome the complexation of ciprofloxacin with iron: an in vitro approach. Appl. Clay Sci. 150, 293–302. https://doi.org/10.1016/j.clay.2017.10.002.

Rawtani, D., Tharmavaram, M., Pandey, G., Hussain, C.M., 2019. Functionalized nanomaterial for forensic sample analysis. TrAC Trends Anal. Chem. 120, 115661. https://doi.org/10.1016/j.trac.2019.115661. https://doi.org/10.1016/j.trac.2019.115661.

Reynolds, J.G., Hart, B.R., 2004. Nanomaterials and their application to defense and homeland security. JOM 56, 36–39. https://doi.org/10.1007/s11837-004-0270-8.

Riskin, M., Tel-Vered, R., Bourenko, T., Granot, E., Willner, I., 2008. Imprinting of molecular recognition sites through electropolymerization of functionalized au nanoparticles: development of an electrochemical TNT sensor based on π-donor–acceptor interactions. J. Am. Chem. Soc. 130, 9726–9733. https://doi.org/10.1021/ja711278c.

Salame, P.H., Pawade, V.B., Bhanvase, B.A., 2018. Characterization tools and techniques for nanomaterials. In: Bhanvase, B.A., Pawade, V.B., Dhoble, S.J., Sonawane, S.H., Ashokkumar, M. (Eds.), Nanomaterials for Green Energy, Micro and Nano Technologies. Elsevier, pp. 83–111. https://doi.org/10.1016/B978-0-12-813731-4.00003-5 (Chapter 3).

Saleh, T.A., Gupta, V.K., 2016. Synthesis, classification, and properties of nanomaterials. In: Saleh, T.A., Gupta, V.K. (Eds.), Nanomaterial and Polymer Membranes. Elsevier, pp. 83–133. https://doi.org/10.1016/B978-0-12-804703-3.00004-8 (Chapter 4).

Sasidharan, S., Raj, S., Sonawane, S., Sonawane, S., Pinjari, D., Pandit, A.B., Saudagar, P., 2019. Nanomaterial synthesis: chemical and biological route and applications. In: Beeran Pottathara, Y., Thomas, S., Kalarikkal, N., Grohens, Y., Kokol, V. (Eds.), Nanomaterials Synthesis, Micro and Nano Technologies. Elsevier, pp. 27–51. https://doi.org/10.1016/B978-0-12-815751-0.00002-X (Chapter 2).

Simioni, N.B., Oliveira, G.G., Vicentini, F.C., Lanza, M.R.V., Janegitz, B.C., Fatibello-Filho, O., 2017. Nanodiamonds stabilized in dihexadecyl phosphate film for electrochemical study and quantification of codeine in biological and pharmaceutical samples. Diamond Relat. Mater. 74, 191–196. https://doi.org/10.1016/j.diamond.2017.03.007.

Singh, S., 2007. Sensors—an effective approach for the detection of explosives. J. Hazard. Mater. 144, 15–28. https://doi.org/10.1016/j.jhazmat.2007.02.018.

Singh, V.V., Wang, J., 2015. Nano/micromotors for security/defense applications. A review. Nanoscale 7, 19377–19389. https://doi.org/10.1039/c5nr06254c.

Sodhi, G.S., Kaur, J., 2016. Physical developer method for detection of latent fingerprints: a review. Egypt. J. Forensic Sci. 6, 44–47. https://doi.org/10.1016/j.ejfs.2015.05.001.

Sperling, R.A., Parak, W.J., 2010. Surface modification, functionalization and bioconjugation of colloidal inorganic nanoparticles. Proc. R. Soc. A 368, 1333–1383. https://doi.org/10.1098/rsta.2009.0273.

Stewart, L.D., Hess, P., Connolly, L., Elliott, C.T., 2009. Development and single-laboratory validation of a pseudofunctional biosensor immunoassay for the detection of the okadaic acid group of toxins. Anal. Chem. 81, 10208–10214. https://doi.org/10.1021/ac902084a.

Tharmavaram, M., Pandey, G., Rawtani, D., 2018. Surface modified halloysite nanotubes: a flexible interface for biological, environmental and catalytic applications. Adv. Colloid Interface Sci. 261, 82–101. https://doi.org/10.1016/j.cis.2018.09.001.

Tiwari, J.N., Tiwari, R.N., Kim, K.S., 2012. Zero-dimensional, one-dimensional, two-dimensional and three-dimensional nanostructured materials for advanced electrochemical energy devices. Prog. Mater. Sci. 57, 724–803. https://doi.org/10.1016/j.pmatsci.2011.08.003.

Tran, K.T.M., Nguyen, T.D., 2017. Lithography-based methods to manufacture biomaterials at small scales. J. Sci.: Adv. Mater. Devices 2, 1–14. https://doi.org/10.1016/j.jsamd.2016.12.001.

Twigg, M.V., 1994. Ligand exchange reactions of inert-metal complexes—coordination numbers 4 and 5. In: Twigg, M.V. (Ed.), Mechanisms of Inorganic and Organometallic Reactions. Springer US, Boston, MA, pp. 79–95. https://doi.org/10.1007/978-1-4899-1624-2_5.

Uzawa, H., Ohga, K., Shinozaki, Y., Ohsawa, I., Nagatsuka, T., Seto, Y., Nishida, Y., 2008. A novel sugar-probe biosensor for the deadly plant proteinous toxin, ricin. Biosens. Bioelectron. 24, 929–933. https://doi.org/10.1016/j.bios.2008.07.049.

Viveros, L., Paliwal, S., McCrae, D., Wild, J., Simonian, A., 2006. A fluorescence-based biosensor for the detection of organophosphate pesticides and chemical warfare agents. Sens. Actuators B 115, 150–157. https://doi.org/10.1016/j.snb.2005.08.032.

Wang, M., Duong, B., Fenniri, H., Su, M., 2015a. Nanomaterial-based barcodes. Nanoscale 7, 11240–11247. https://doi.org/10.1039/C5NR01948F.

Wang, M., Li, M., Yu, A., Wu, J., Mao, C., 2015b. Rare earth fluorescent nanomaterials for enhanced development of latent fingerprints. ACS Appl. Mater. Interfaces 7, 28110–28115. https://doi.org/10.1021/acsami.5b09320.

Wang, M., Li, M., Yu, A., Zhu, Y., Yang, M., Mao, C., 2017. Fluorescent nanomaterials for the development of latent fingerprints in forensic sciences. Adv. Funct. Mater. 27, 1606243. https://doi.org/10.1002/adfm.201606243.

Xu, S., Lu, H., 2016. Mesoporous structured MIPs@CDs fluorescence sensor for highly sensitive detection of TNT. Biosens. Bioelectron. 85, 950–956. https://doi.org/10.1016/j.bios.2016.06.020.

Yang, J., Gunn, A., Palmbach, T., Wei, D., Sinha, S., 2006. Nano-forensics—nanoparticles in gun-shot-residue. In: 2006 IEEE Conference on Emerging Technologies—Nanoelectronics. Presented at the 2006 IEEE Conference on Emerging Technologies—Nanoelectronics, pp. 269–272. https://doi.org/10.1109/NANOEL.2006.1609727.

Yu, Z., Tetard, L., Zhai, L., Thomas, J., 2015. Supercapacitor electrode materials: nanostructures from 0 to 3 dimensions. Energy Environ. Sci. 8, 702–730. https://doi.org/10.1039/C4EE03229B.

Zhang, C., Han, Y., Lin, L., Deng, N., Chen, B., Liu, Y., 2017. Development of quantum dots-labeled antibody fluorescence immunoassays for the detection of morphine. J. Agric. Food Chem. 65, 1290–1295. https://doi.org/10.1021/acs.jafc.6b05305.

Zhen, S., Wang, Y., Liu, C., Xie, G., Zou, C., Zheng, J., Zhu, Y., 2011. A novel microassay for measuring blood alcohol concentration using a disposable biosensor strip. Forensic Sci. Int. 207, 177–182. https://doi.org/10.1016/j.forsciint.2010.10.002.

Zhou, Y., Li, Y., Lu, S., Ren, H., Li, Z., Zhang, Y., Pan, F., Liu, W., Zhang, J., Liu, Z., 2010. Gold nanoparticle probe-based immunoassay as a new tool for tetrodotoxin detection in puffer fish tissues. Sens. Actuators B 146, 368–372. https://doi.org/10.1016/j.snb.2010.02.049.

Zhu, L., Zhao, H., Zhou, Z., Xia, Y., Wang, Z., Ran, H., Li, P., Ren, J., 2018. Peptide-functionalized phase-transformation nanoparticles for low intensity focused ultrasound-assisted tumor imaging and therapy. Nano Lett. 18, 1831–1841. https://doi.org/10.1021/acs.nanolett.7b05087.

CHAPTER 20

Ethics and legal issues of forensic analysis techniques

1 Introduction

One of the key distinguishing features of humanity is its morality and its ability to judge the good, bad, and in-between. Due to the existence of this consciousness and morality, rules, laws, and regulations have been developed across the world that keep humans from crossing their morality boundaries. These laws are derived from a branch of philosophy known as ethics that is basically designed to oversee the morality of a human being. Ethical laws take into consideration the human rights of a person and their goal is to protect the rights and requirements of a person when they are surrounded by different shades of morality. Ethics are responsible for dictating the terms and conditions for good behavior but are not in actual control of the behavior of a person. There is also a fine line between morality and ethical issues. A person who is honest and uncorrupted is said to be morally inclined; however, that does not prevent them from encountering ethical issues in their life.

Because human nature is highly versatile with everyone having different moral conduct, it is important that appropriate universal ethical standards be developed that ensure that the overall ethical behavior of humans is the same irrespective of each person's ethical standards. Apart from ethical standards governing human society, there are legal standards as well that every citizen in a country must follow. While not following ethical standards does not necessarily permit punishment, not following a legal standard does warrant punishment.

In forensic science too, there are ethical and legal issues surrounding several areas of this field, mostly because it largely deals with criminals whose moral ambiguity is often questioned and brought to trial. While this field is majorly scientific, it is important that every step in a criminal investigation right from visiting the crime scene to processing the case in courts is governed by legal standards. Every country has its own set of legal standards that must be abided by stringently. Apart from this, ethical issues such as making

Handbook of Analytical Techniques for Forensic Samples
https://doi.org/10.1016/B978-0-12-822300-0.00020-3

decisions between competing obligations or maintaining discretion also demand a set of ethical standards. In this chapter, a brief overview of different ethical and legal issues in forensic science will be discussed.

2 Legal issues in forensic science

Forensic investigations need to follow a well-defined set of rules during the investigation, collection, and submission of evidence in a court. These rules can act as standard procedures while investigating a crime (Pandey et al., 2017; Rawtani et al., 2019). This section discusses the different legal steps in forensics from crime to court as well as elaborates the legal issues related to computer and digital forensics as well as DNA forensics.

2.1 Legal aspects: From crime scene to court

There are a number of legal steps involved from a crime scene to court. This section elaborates the legal steps involved in crime investigation and the submission of evidence in court. The rules related to evidence such as its authenticity and admissibility are also discussed. Further, the legal aspects of reports coming from forensic laboratories as well as the testimony of experts are elaborated as well.

2.1.1 Crime investigation procedure: Legal aspects

In a crime investigation process, forensic scientists are involved in different steps such as responding to a crime scene as well as analyzing evidence, depositions, and court testimony.

As far as the response to a is concerned, legally one or more forensic scientists should attend the crime scene, especially in high-profile cases as well as homicides. The reason behind this is the expertise possessed by such scientists in recognizing, collecting, and preserving the different evidence present at the scene of a crime. In the analysis of evidence, forensic scientists should be employed to perform this task, keeping in mind the expertise they have. Legally, the analysis reports issued by forensic scientists should go through the evaluation process by the crime investigators and prosecutors (Houck and Siegel, 2015).

A deposition, which is testimony given by taking an oath but not in front of a judge or in court, should also be taken through legal procedures. It should be taken by an authorized reporter or recorder only. The questions asked to the witness during a deposition should be the same as those asked during a trial. On a legal basis for court testimony, forensic scientists should testify to important and scientific findings to the courts and judges. These testimonies can be given in preliminary court hearings as well. The scientists giving testimonies in their area of expertise should be regarded as expert witnesses (Houck and Siegel, 2015).

As far as discovery is concerned, that is a process in which both sides seek to collect data about the evidence the opposite side has. An earlier rule was that the defendant was not entitled to inspect different evidence belonging to the prosecution. However, now the rule has changed and in today's time, both parties can legally inspect each other's evidence. This is called reciprocal discovery. In searches and seizures, a legal search warrant should be obtained from the authorizing body before searching and seizing evidence form any person's property. All these searches should preferably be done in the presence of forensic scientists because their expertise can effectively help in searching the physical evidence present. Self-incrimination, which is a confession made by the suspect committing the crime, is legally not prohibited. However, compelled self-incrimination is not allowed in the legal system, as per the Fifth Amendment in the US Constitution (Edmond and Cole, 2013; Houck and Siegel, 2015).

2.1.2 Subpoena: Legal aspects of submission of evidence

Whenever any evidence has to be submitted by any witness or the witness has to be called to court to give the testimony, a legal notice has to be issued by the court to the witness stating the location and time for giving the testimony or providing the evidence. The subpoena is given to the witness in person. If the witness is found to ignore the court summons, then he or she may face charges of contempt of court (Saks, 2013).

A subpoena for forensic scientists is a bit different, and is termed a "subpoena duces tecum." In this, the witness as well as any kind of evidence or document that is relevant to the case and is possessed by the witness (forensic scientist) should also be submitted to the court. This kind of subpoena can be sent by mail or delivered to the scientist's laboratory as well as being delivered in person. The penalties for disobeying this subpoena are similar to those faced by any normal witness (Houck and Siegel, 2015)

2.1.3 Evidence: Rules and legal aspects

Evidence can be defined as something that can assist the legal system in proving or disproving any kind of material fact. Evidence can be real (recovered from the crime scene), demonstrative (assisting the real evidence but not recovered from the crime scene, such as sketches and crime scene photos), and testimonial (given by a lay or expert witness under oath). Legally, the lay or nonexpert witness is not allowed to give any kind of opinion. However, expert witnesses can give opinions and conclusions on the basis of their expertise (Franjić, 2018).

The evidence submitted in court should be authentic, and its authenticity can be maintained using a proper and legal channel of labels and sponsors from the time of seizing until submission in court. This channel of labels and sponsors is called the chain of custody, which has all the information of people and places where evidence went before reaching court. Even if the evidence is opened by forensic scientists before submitting it to court, it should be kept in mind that previous seals remain intact and new seals are given to the evidence regarding the evidence handler (Houck and Siegel, 2015).

Regarding the admissibility of evidence, it should be relevant as well as competent. If the evidence to be submitted fulfills only one of these conditions, then it is not admissible in court. Any evidence can be relevant if it is a combination of materiality and probativeness. The competence of evidence in the legal system is governed by certain rules and restrictions such as the evidence should not be prejudiced, time wasting, unreliable, or presented through improper procedure such as by a surprise witness. The admissibility of evidence examined by novel scientific techniques or scientific evidence has changed after the Frye case and the Daurbert V. Merrell Dow case. Such evidence is admissible in court if the scientific technique used has general acceptance for usage in the scientific fraternity. If any novel technique is presented, then the underlying principles behind that scientific technique must be demonstrated in the court (Edmond and Cole, 2013; Houck and Siegel, 2015; Saks, 2013).

2.1.4 Legal aspects of reports issued by forensic laboratories

The report produced by a public forensic laboratory goes directly to the investigating officer related to that case in a legal system. Legally, an additional copy of the generated report should be sent to the prosecution and defense as well. The laboratory report can be used as evidence if the case goes to trial. These reports can be submitted in the court without the testimony of the forensic scientist.

Most of the reports are just the certification of the analysis. This is, therefore, a major concern regarding what data and how much data should be in the report. Legally, these scientific reports should have complete data, results, interpretations, and conclusions that can assist the jury in the courtroom trial. Another issue relevant to the reports of forensic laboratories is the recording of the analysis carried out by the forensic scientist. If the trial is conducted shortly after the analysis, then the scientist can testify to everything by memory. However, if the trial is done after months or years, the report of the analysis becomes crucial evidence because it helps the scientist to recollect if the analysis was carried out by him/her or testify if the procedures done during the analysis were correct or not (Houck and Siegel, 2010, 2015).

2.1.5 Testimony by expert witnesses: Legal aspects

Expert witnesses are generally the forensic scientists who performed the analysis of evidence collected from the crime scene. Not every case requires a forensic scientist to be summoned by the court. However, whenever it does, a legal procedure has to be followed, as discussed earlier, by sending a subpoena duces tecum, which is a court summons to produce the evidence and give testimony in court. This subpoena has to be given to the scientist from the prosecutor's office. In the court, unlike ordinary witnesses, expert witnesses are allowed to give their opinions and conclusions based on their expertise and findings during the evidence analysis, which can assist the jury to better understand, especially for scientific evidence. However, in doing so, the scientist also has to show his/her credentials, degrees, experience, and publications in a legal manner in order to prove his/her credentials expertise in that area because his/her opinions can directly influence the verdict in the case (Edmond and Cole, 2013; Houck and Siegel, 2015).

2.2 Legal aspects in computer and digital forensics

The main purpose of computer forensics is to give digital or electronic evidence that can be accepted by a jury in a courtroom. The legal system greatly influences the methods that can be used in computer forensics. As discussed earlier, any evidence to be submitted in court should be relevant and competent. Digital evidence is relevant because it is material and probative. However because it can be easily manipulated, modified, and duplicated, the competency of such evidence becomes a concern. No test is available that can determine if the presented digital evidence is scientifically valid. Therefore, any digital evidence can be submitted in court if it fulfills two situations: (a) it is derived and validated using a scientific method; and (b) it must be relevant (Houck and Siegel, 2015).

2.2.1 Digital evidence: Admissibility in court

Digital evidence generally needs a good base of introduction upon being admitted in court. However, courts do not discriminate between digital and other evidence as far as the introduction of evidence is concerned. Whenever any digital evidence is submitted, then the opposite party is give the opportunity to question the accuracy of the computer. However, arguments that such evidence cannot be trusted because it can be easily modified are not generally acknowledged in the court (Stahl et al., 2010). In order to preserve the admissibility of digital evidence, the guidelines issued at the International Conference on High-Tech Crime in 1999 are as follows (Strydom, 2001):

- Any actions taken while seizing digital evidence should not alter or change the evidence.
- Any person wanting to access the original copy of the digital evidence needs to be forensically competent.
- There should be proper documentation and preservation of all the activities carried out related to seizing, accessing, storing, and transferring the digital evidence, and all the information should be present for review.
- During the time that the digital evidence is under possession, a person should be responsible for any kind of actions taken related to the digital evidence.
- Any public or private agency that is doing the investigation and is responsible for the confiscation, access, packing, and transfer of the digital evidence has to carry out all the activities in compliance with the aforementioned principles.

2.2.2 Computer forensics and legal issues

There are some important legal issues that restrict the development of the computer forensics field, and they are as follows (Brungs and Jamieson, 2005):

- The differences between state and the federal legislative bodies on jurisdiction related to computer forensics and laws related to cybercrimes are a matter of concern.
- The differences in the jurisdiction laws related to the admissibility of digital evidence in the court are also a concern because one form of digital evidence is acceptable in one state and not in another.
- The absence of a proper guide related to the methodology for the collection, preservation, examination, and interpretation of digital evidence in computer forensics.
- A lack of knowledge and understanding about the significance of digital and electronic evidence in the legal system, especially with lawyers and judges.

- Evidence collection from digital records sometimes infringes the confidential records of an individual or company, for which law enforcement should be there.
- Uncertainty over the interception of telecommunication data while in transit, which is an offense and considered privacy infringement in some countries.
- No guidelines, especially for private forensic companies, as to what data can be legally collected and what breaches an individual's privacy related to digital evidence collection.
- Issues related to communications, such as emails, if they are read or unread, the beginning and completion of the communication, etc.
- Uncertainty with telecommunication companies as to how much data should be given to police so that the privacy of the client is also safe while the police get enough information.
- Requirement of cooperation between legal systems at the international level to increase the impact of computer forensics.
- Lack of access to digital evidence in real time for investigators, mainly due to the absence of mutual assistance among countries.
- Communications and broadcasts are digital evidence. However, they are listed in different acts in countries such as Australia, which creates a contrast.
- Offenses related to cybercrime are limited, and these need to be updated with the increasing cases of cybercrime in today's world.
- It is impossible to launch an action against an unknown person by private companies and nongovernment organizations, which is a major constraint for them in obtaining digital evidence.
- The reliability of tools and techniques that are used to analyze digital evidence is always a question, and should be therefore validated by a third party for its acceptance in court.
- The requirement of guidelines for the qualifications of expert witnesses related to digital evidence should be there so that their testimony can be relied upon.

2.3 Legal aspects in DNA forensics

DNA evidence came into picture for investigation purposes in 1985 after Alec Jeffreys confirmed the polymorphic nature of DNA due to the presence of variable number tandem repeats (VNTRs). However, there have always been concerns and questions about the techniques used for matching two DNA samples. DNA databanking laws have also been enacted in countries such as Australia, the United Kingdom, the United States, Austria, China, Canada, Germany, and France. However, there are issues related to DNA forensics that involve the DNA typing procedures for matching two DNA samples (Reilly, 2001; Walsh, 2005).

2.3.1 DNA forensics: Legal issues

The issues related to DNA forensics are (Reilly, 2001):

- The first issue relates to the kind of DNA sample that should be typed. In earlier days, DNA typing was a mandate only in some cases where the possibility of DNA evidence was high, such as sexual assault cases. However, nowadays, trends have changed and DNA analysis has become an important part of the investigation in all types of felonies.
- There is an issue of social inequality in the collection of DNA samples for DNA databanking in that samples are not taken from the privileged part of society. There is an issue of social inequality in DNA databanking of those DNA samples that are not collected from the society's privileged. Therefore, the DNA collection and databanking should be made equal for all the citizens of any country as long as the genetic privacy is promised.
- Whether the DNA sample taken from criminals should be stored or destroyed after DNA typing is performed has also remained a matter of concern for legal systems.
- With the advent of DNA analysis in investigations, the issue that the criminal justice system is facing is the increasing demand by convicts (especially rape convicts) to reopen their cases and perform DNA typing to prove their innocence. The legal system is in a dilemma as to whether postconviction testing should be allowed.
- DNA evidence can identify any individual, even after several years have passed since the crime was committed and the statue of limitations has expired; an example of this is rape. So, the legal system must increase the validity of such complaints in which DNA evidence is involved.
- The capacity of crime labs is low to analyze all DNA samples recovered from the crime scene. This should be increased and the people collecting evidence from the crime scene should be given proper training to collect all the DNA evidence available during the crime scene investigation.

3 Ethics and theories of ethics

Ethics, as mentioned before, are responsible for the analysis of right, wrong, and in-between. There are several ethical theories responsible for the guidance of people during any moral crisis by providing a check sheet of guidelines responsible for the appropriate moral conduct of a person. There are three basic theories of ethics: normative ethics, metaethics, and applied ethics. Normative ethics are responsible for providing guidelines for normal human behavior wherein concepts of principles and moral standards are defined. Metaethics is responsible for analyzing the meaning behind morality and its concepts while applied ethics revolves around the dilemmas faced during conflicting moral situations and the application of ethical concepts in professional situations. In forensic science, theories of applied ethics are applied.

While studying ethics, there are certain steps to be taken to ensure their proper implementation in forensic science. Typically, the first step is to create awareness regarding the different moral and ethical issues. Quite often, criminalists fail to understand that there are certain questions and situations for which there is no correct answer. In such cases, it is important that the ethical standards are properly defined. The majority of forensic science experts come from a varied background and from different universities. Therefore, it is also important that appropriate standards for professional ethics be used. Once the awareness of ethical situations is created, the different morality concepts in this field are defined. For instance, certain situations can be created with good intentions; however, the way the situation was handled could be unethical. Once the concepts are properly defined, the next step is to ensure that the ethical responsibility is maintained, even when the professionals are personally attached to the situation. In the final step, an overall observance of the professional culture and how ethical standards can be implemented in it is done.

In forensic science, ethics are especially important, as this field is a mix of science and law. Lawyers, scientists, police, the media, and the public are integrally involved with this field. A criminalist typically faces two major surroundings, namely, the court and the environment outside the court. In the court, the suspect or witness is usually made to make a statement that ensures that they speak the truth lest they perjure themselves. However, once outside the court, the criminalist has to face a number of ethical situations such as in the laboratory or in front of the media. Therefore, in order to simplify all these situations, a code of ethics is followed that is similar to the legal standards but is based on principles and morality and allows the criminalist to differentiate between the right and wrong (Barnett, 2001; Bowen, 2010).

3.1 Models of codes of ethics

In forensic science, there are generally two main types of ethical models: the general and the specific type. The Code of Ethics and Conduct from the

American Academy of Forensic Science (AAFS), and the Code of Ethics of the California Association of Criminalists (CAC) are the two models that respectively denote the general and the specific type. The AAFS models were created to ensure that the ethics of different forensic professions were unified in a single group. While CAC was involved in bridging the distance between the code of ethics of different scientific professions with the law, AAFS took into account only those ethics that are suitable to all professions. However, CAC develops ethics that can connect forensic science with legal standards.

The general or the broad model, as the name suggests, covers a variety of situations. However, this very feature serves as a limitation to this model. Due to the lack of details covered by this model, proper guidance that allows a member of an organization to abide by the code is not properly developed. This model by AAFS provides very little guidance to the professional on how to deal with a situation where their fellow member has violated an ethical code. However, this limitation too has been overcome with the development of the Good Forensic Science Practices Committee. Through this committee, it is possible to develop a proper guideline that includes the professionals in forensic science. It is now possible to ensure that the action or judgment taken by a member is the most appropriate ethical response. The situation is carefully reviewed, and a consensus is reached according to the guidelines prescribed by the AAFS. Through the Good Forensic Science Practices Committee, it is now truly possible to develop a common ground for the implementation, review, and analysis of appropriate ethical protocols in different forensic situations.

The detailed or specific model developed by CAC covers the roles and ethical conduct of every professional involved in an investigation. The code of ethics developed here also prescribes appropriate behavioral conduct as well. In this organization, the code of ethics was developed according to the trained criminalists and their collection experience. The guidelines mentioned here are very intricate and are connected to several sections at the same time. However, this comprehensive development of guidelines prevents their revision because a single guideline is often linked with several sections. While the revision is not itself an impossible task, it is definitely a very daunting task. Also, another drawback with such comprehensive guidelines is that they cannot be applied to unusual circumstances. In such cases, the general model is more suitable. However, with the use of more general language during the development of guidelines, it is possible to overcome this limitation.

These models also ensure that the ethical requirements meet the job activity of the professional. Factors such as the competence, thoroughness, relevance, and reviewability of a professional are considered important criteria for meeting the ethical requirements. Competence ensures that the professional is capable enough to fulfill a particular responsibility while thoroughness ensures that the task is completed with special attention to detail. Relevance is a factor responsible for ensuring that only the tests that are bsolutely required for the investigation are performed and reviewability also serves as a legal requirement that allows the professional's work to be examined and reviewed (Barnett, 2001; Bowen, 2010).

4 Code of ethics in different fields of forensic science

In this section, the codes of ethics applied in different fields of forensic science, especially, in forensic examination, DNA analysis, and psychiatry, will be discussed.

4.1 Ethics in forensic examination

Forensic examiners are any professionals who are involved in the examination and interpretation of physical evidence in order to produce testimony in court. They include all professionals who work in a forensic laboratory as well as pathologists, toxicologists, criminalists, and mental health practitioners. A code of ethics is highly required for forensic examiners as they directly deal with the intricate details of any investigation and therefore may become emotionally invested in them. This may cause the development of biased or erroneous results. Forensic examiners are supposed to present the results of evidence as simple objects without the attachment of any personal belief. The use of an appropriate code of ethics serves as a filter in this case. An ethical forensic examiner thus has scientific integrity, impartiality, the ability to acquire new knowledge, and trustworthiness. These traits are highly required when presenting expert testimony in court.

Forensic science is a field that is highly integral to the law. However, the very nature of forensic science and the law is highly varied. Examiners are supposed to handle evidence with utmost impartiality and should not waver the result to one direction. However, the law requires that two sides of a case be presented, out of which only one is selected. Due to such conflicting natures, it is quite easy for the forensic examiner to sway toward one side. Also, most of the time, the evidence given to the examiner for analysis has already been picked up by the person who visited the crime scene. The examiner also takes this factor into account while preparing for testimony in court.

The analysis of evidence largely results in formulating the results in a report. The examiner is authorized to update the side that employs them about any possible good or bad outcomes. This is a very difficult situation and thus needs to be bound tightly by appropriate ethical codes (Chisum and Turvey, 2013).

4.2 Ethics in DNA analysis

In the previous section, the overall ethical responsibility of the forensic examiner was discussed. However, the ethical conduct becomes very precise when dealing with particular evidence. For instance, DNA is highly crucial evidence capable of incriminating or proving a person innocent. It may benefit various individuals because it foolproof evidence for establishing a person's identity. However, DNA analysis for forensic applications is often subjected to two highly opposing views. One such opinion is that because DNA is capable of providing confirmation of a person's identity, the information should be kept away from the public as it is highly private. The opposing opinion states that such information should be kept public in order to facilitate the investigation. The first opinion rises when a proactive approach to prevent a crime is employed. In the proactive approach, with the help of DNA information, possible offenders can be controlled. However, this is a very idealistic approach in which there is no corruption within the justice system. The second opinion rises when it is assumed that a person is innocent until proven guilty.

DNA analysis and its information come under human rights. Therefore, ethical conduct revolving around this also comes under human rights. Apart from this, forensic DNA databases also come under human rights. For this, the Forensic Genetics Policy Initiatives were developed that are responsible for setting human rights standards for DNA databases (Beyleveld, 1997; Prainsack and Aronson, 2015; Wallace et al., 2014).

4.3 Ethics in forensic psychiatry

Forensic psychiatry is a bidisciplinary field that involves a psychiatrist possessing both a thorough knowledge of psychiatry and the law. Forensic psychiatry as such operates within a very specific area of legal and social context. The psychiatrists also serve as expert witnesses and provide their testimony on issues that require their expertise. However, just like any other forensic examiner, here too the psychiatrist is responsible for giving an opinion without any bias and should leave it to the judge to decide. Normally, when asked their opinion on the perpetrator, the psychiatrist must first get the consent of the perpetrator and then ensure that he or she has the ability or capacity to understand why he or she is being subjected to analysis by a psychiatrist. The psychiatrist is also responsible for performing risk assessment and predicting a possible future regarding the criminal activity of the offender. However, this is not always accurate and there are strict ethical guidelines regarding this (Konrad, 2016; Shapiro, 2016).

5 Conclusion

In this chapter, various legal and ethical issues in forensic science have been discussed. The legal aspects related to the different steps involved from crime scenes to the court were discussed. Further, the legal aspects and issues related to computer and digital forensics were elaborated. It can be said that with increased mutual assistance between countries, a wide network can be formed to decrease the increasing rate of cybercrime. Digital and DNA forensics can play an important role to curb such crimes. Also, the issues related to DNA typing in DNA forensics need to be resolved, keeping in mind the importance of DNA evidence in almost all kinds of felonies. It is also highly necessary to develop ethical standards that consider the human rights aspects of every forensic responsibility. Forensic science mainly caters to the law and therefore the evidence presented by experts must be free of partiality and should be delivered with absolute unattachment. Also, forensic science is largely a scientific field, due to which there are professional ethics to be followed by the different professionals in a forensic laboratory.

References

Barnett, P.D., 2001. Ethics in forensic science: professional standards for the practice of criminalistics. In: Protocols in Forensic Science. CRC Press, Boca Raton.

Beyleveld, D., 1997. Ethical issues in the forensic applications of DNA analysis. Forensic Sci. Int. 88, 3–15. https://doi.org/10.1016/S0379-0738(97)00078-0.

Bowen, R.T., 2010. Ethics and the Practice of Forensic Science. CRC Press, Taylor & Francis Group, United States.

Brungs, A., Jamieson, R., 2005. Identification of legal issues for computer forensics. Inf. Syst. Manage. 22, 57–66. https://doi.org/10.1201/1078/45099.22.2.20050301/87278.7.

Chisum, W.J., Turvey, B.E., 2013. Ethical issues in forensic examination. In: Ethical Justice. Elsevier, pp. 219–255. https://doi.org/10.1016/B978-0-12-404597-2.00008-5.

Edmond, G., Cole, S.A., 2013. Legal aspects of forensic science. In: Encyclopedia of Forensic Sciences. Elsevier, pp. 466–470. https://doi.org/10.1016/B978-0-12-382165-2.00226-9.

Franjić, S., 2018. Legal aspects of forensics. Forensic Sci. Today 009–017. https://doi.org/10.17352/pjfst.000011.

Houck, M.M., Siegel, J.A., 2010. Fundamentals of Forensic Science, second ed. Academic Press, Burlington, MA.
Houck, M.M., Siegel, J.A., 2015. Legal aspects of forensic science. In: Fundamentals of Forensic Science. Elsevier, pp. 621–650. https://doi.org/10.1016/B978-0-12-800037-3.00024-8.
Konrad, N., 2016. Ethical issues on forensic-psychiatric examinations. Ethics Med. Public Health 2, 119–126. https://doi.org/10.1016/j.jemep.2016.01.016.
Pandey, G., Tharmavaram, M., Rawtani, D., Kumar, S., Agrawal, Y., 2017. Multifarious applications of atomic force microscopy in forensic science investigations. Forensic Sci. Int. 273, 53–63. https://doi.org/10.1016/j.forsciint.2017.01.030.
Prainsack, B., Aronson, J.D., 2015. Forensic genetic databases: ethical and social dimensions. In: International Encyclopedia of the Social & Behavioral Sciences. Elsevier, pp. 339–345. https://doi.org/10.1016/B978-0-08-097086-8.82062-0.
Rawtani, D., Tharmavaram, M., Pandey, G., Hussain, C.M., 2019. Functionalized nanomaterial for forensic sample analysis. TrAC Trends Anal. Chem. 120, 115661. https://doi.org/10.1016/j.trac.2019.115661.
Reilly, P., 2001. Legal and public policy issues in DNA forensics. Nat. Rev. Genet. 2, 313–317.
Saks, M.J., 2013. History of the law's reception of forensic science. In: Encyclopedia of Forensic Sciences. Elsevier, pp. 481–487. https://doi.org/10.1016/B978-0-12-382165-2.00229-4.
Shapiro, D.L., 2016. Ethical issues in forensic psychology and psychiatry. Ethics Med. Public Health 2, 45–58. https://doi.org/10.1016/j.jemep.2016.01.015.
Stahl, B., Elizondo, D., Carroll-Mayer, M., Zheng, Y., Wakunuma, K., 2010. Ethical and legal issues of the use of computational intelligence techniques in computer security and computer forensics. In: The 2010 International Joint Conference on Neural Networks (IJCNN). Presented at the 2010 International Joint Conference on Neural Networks (IJCNN). IEEE, Barcelona, Spain, pp. 1–8. https://doi.org/10.1109/IJCNN.2010.5596546.
Wallace, H.M., Jackson, A.R., Gruber, J., Thibedeau, A.D., 2014. Forensic DNA databases–ethical and legal standards: a global review. Egypt. J. Forensic Sci. 4, 57–63. https://doi.org/10.1016/j.ejfs.2014.04.002.
Walsh, S.J., 2005. Legal perceptions of forensic DNA profiling. Forensic Sci. Int. 155, 51–60. https://doi.org/10.1016/j.forsciint.2004.11.001.

CHAPTER 21

Accreditations for forensic science laboratories

1 Introduction

Forensic science is a highly crucial field in law enforcement and in the deliverance of proper justice. It is an amalgamation of science and law and is therefore an intricate mixture of police, courts, and the scientific community. Such an intricate organization requires proper coordination and communications in order to ensure smooth processes from the crime scene investigation to the deliverance of testimony and proofs at court trials. Also, because it is largely a science, every protocol and procedure must be standardized so that it does not vary from every individual or organization. This smooth functioning is ensured through a series of accreditation organizations that lay out defined rules and certifications that ensure that there is uniformity, validity, reliability, and transparency in the entire process. In the accreditation process, a third party is involved that lays out rules for the various management and technical systems in forensic science. Any accreditation program is a holistic risk management strategy and therefore requires cooperation from all levels. Appropriate accreditation of the organization allows an optimal quality management system and therefore ensures that the results are valid.

While accreditation in forensic science varies from country to country, most of the countries base their accreditation laws on three main ISO standards; we discuss that further in this chapter. In this chapter, a brief overview of the history of accreditation and the different organizations responsible for accreditation in forensic science is discussed. Also, we give special emphasis on accreditation in different fields of forensic science, especially in digital, DNA, and toxicological forensics.

2 History and need of accreditation in forensic science

Accreditation is basically a formal process responsible for the accreditation of organizations in order to ensure that the procedures, policies, certifications,

Handbook of Analytical Techniques for Forensic Samples
https://doi.org/10.1016/B978-0-12-822300-0.00021-5

and programs are in accordance with general standards. Because in forensic science, the majority of the analysis takes place in the laboratory, the standards are usually defined according to the accreditation of the laboratory. The accreditation of a forensic laboratory is responsible for maintaining appropriate documentation, policies, and protocols for records. It also ensures that the organization is accountable to a third-party accreditation party in order to ensure that the above conditions are being satisfied while providing optimal customer satisfaction.

The accreditation of forensic laboratories was first started by the American Society of Crime Laboratory Directors/Laboratory Accreditation Board (ASCLD/LAB) in 1973 in order to discuss and open various channels of communication between the directors and managers of various crime laboratories. In 1988, the ASCLD/LAB was designated as a nonprofit independent organization, and served as a premier accreditation body for a long time. However, the integrity of these organizations was also questioned, and several articles were published that questioned the course of accreditation by these organizations. Therefore, in order to overcome this, the International Organization for Standardization (ISO), which is also an independent nongovernmental organization, was developed to offer and approve strict international standards for local, national, and international commerce. The ISO is responsible for ensuring the most basic and advanced requirements of accreditation so that the results produced at forensic laboratories are completely valid and reliable while the techniques used to produce the results are also standardized. It also develops standards with the International Electrotechnical Commission (IEC). The ISO itself has a technical committee that is responsible for the development of standards and a project committee that, unlike the technical committee, is temporary and is removed as soon as the standard is created. The standards are developed such that they are a suitable response to the needs in the market. They are developed according to experts across the globe and are consensual. The ISO standards are developed through a series of steps involving proposals, preparation, committees, inquiries, approvals, and publications. The different standards in ISO relevant to accreditation in forensic science will be discussed further in the next section. The ISO also clearly defines the accreditation process as a "procedure by which an authoritative body gives formal recognition that a body or person is component to carry out specific tests." It also takes into account the competence of any organization as well as that of every individual within the organization because the individuals themselves directly handle the development of results. Also, it is important, that the processes used by individuals are also competent so that the competence of the employees doesn't get shadowed out. In short, there is a strict need for the accreditation of forensic organizations by third-party organizations in order to develop trustworthy results that can be used in a court of justice respective to their countries (Collins, 2018; Malkoc and Neuteboom, 2007; Palmbach, 2013; Rankin and Welsh, 2013).

3 Accreditation in forensic science

Accreditation in forensic science is provided by an independent accreditation body (AB) that is further accredited by the ISO/IEC 17011 international standard. Every country has its own accreditation body according to their laws. However, the overall authority governing these processes that take place during testing and analysis is itself defined by the ISO. In forensic science, the following standards are commonly used:

- ISO 17025, which defines the general responsibilities and requirements to check the competence of the testing and calibration laboratories (ISO/IEC 17025:2017, 2017).
- ISO 17020 states the different requirements and responsibilities of the organization that undertakes the inspection of the laboratories (ISO/IEC 17020:2012, 2012).
- ISO 15189 is responsible for the quality assurance and management of medical laboratories (ISO/IEC 15189:2012, 2012).

Also, yet another organization exists that is responsible for assessing the competency of any organization; it is known as the ISO Committee on Conformity Assessment (CASCO) CASCO assesses an organization and declares whether it is suitable or competent enough to carry out its responsibilities. These standards have been developed to work in fulfilling conditions such as ensuring appropriate management systems and technical activities and also their documentation. They can also be applied through self-regulation, certification, or accreditation. In forensic science, evidence such as DNA and body fluids as well as trace evidence such as hair, toxins, and other inorganic components are commonly encountered, and their testing primarily takes place in a laboratory (Pandey et al., 2017; Rawtani et al., 2019). A critical organization in the accreditation and certification process is the International Laboratory Accreditation Cooperation (ILAC) that runs according to the ISO/IEC 17011 guidelines. Apart from this, ISO/IEC 9001:2015 and ISO/IEC 14001 are also implemented in laboratories to certify the quality and environmental management systems, respectively. However, these standards do not certify the organization but rather a small part of it. In terms of forensic science, ILAC G19:08/2014 is responsible for providing guidance for crime investigation laboratories and other counterparts that assist in evidence analysis.

Another major criterion for any accreditation body in forensic science is that the conditions it specifies must be suitable to the requirements of the justice system. In short, the "fitness of purpose" must be considered. Therefore, the requirements specified by the guidelines must be such that they allow a delivery of results that is suitable for the crime investigation. All the protocols used by the designated scientists are also therefore decided and certified accordingly. Factors such as documentation and its control are highly crucial and therefore stringent rules are specified for the control and handling of documents. The documents produced right from the crime scene to the testimony to be produced in court are carefully regulated to ensure their integrity. Apart from documents, the samples or evidence collected at the crime scene are also carefully regulated. Evidence is placed in a chain of custody that highlights the path of the evidence right from the collection to its transportation and analysis; it also includes details about the personnel handling them. The chain of custody itself is carefully scrutinized and the evidence is kept in a high-security building to prevent any loss or contamination. The personnel collecting, handling, and analyzing the samples are also required to be certified; those rules are also mentioned in the ISO guidelines. Not only forensic scientists, but medical personnel who are responsible for the autopsy and forensic medicine also need to be certified; those guidelines are mentioned in the ISO. The protocols for the analysis of evidence that occurs in laboratories are also carefully controlled and the protocols are also validated. Here again, careful documentation is maintained in order to record everything that occurs during the analysis. Apart from this, strict regulations are also mentioned that prevent the occurrence of any errors. If any errors do occur, corrective actions are also clearly mentioned and appropriate measures to prevent the reoccurrence of the errors are also in place (Ross and Davey, 2016).

In the United States, the ANSI-ASQ National Accreditation Board or ANAB is currently responsible for forensic laboratories. It merged the ASCLD/LAB into its program. It also oversees accreditation in countries such as Mexico, Bangladesh, the Cayman Islands, Canada, and Brazil. In India, the National Accreditation Board for Testing and Calibration Laboratories (NABL, 2008) is the main governing board for the accreditation and certification of laboratory personnel. For forensic science, a specific document known as the "Specific Criteria for Accreditation of Forensic Science

Laboratories" has been developed by the technical committee. It works in conjunction with the ISO/IEC 17025:2005 and is in compliance with the ISO/IEC 17025:2005. In Europe, the 2009 European Union Act is responsible for accrediting forensic personnel while the United Kingdom Accreditation Service holds this responsibility in the United Kingdom (Collins, 2018; Palmbach, 2013; Rankin and Welsh, 2013; Watson and Jones, 2013).

Biological samples such as DNA, blood, hair, semen, and saliva are often encountered. ISO has a special section to deal with the handling of such biological evidence in forensic science. The ISO Committee TC272 was started as a project committee in 2013. It is responsible for minimizing the risk contamination of biological evidence with human DNA. Later, this was generalized to include all the techniques pertaining to physical evidence and has five working groups responsible for defining evidence, its collection, products, analysis, interpretation, and reporting. Apart from these, there are several subcommittees pertaining to digital forensics, biometrics, and forensic dentistry (Wilson-Wilde, 2018).

3.1 Process of accreditation

The accreditation process occurs through a series of processes that will be discussed in detail in this section. The steps mentioned here are according to ISO/IEC 17020 and ISO/IEC 17025 (ILAC G19). ISO 17025 has two primary requirements: management and technical requirements. In management requirements, tasks such as proper organizational structuring, effective communication, policies, and protocols are included. Also, the authority chain in the organization is clearly defined. The purchase of equipment and other laboratory supplies is also clearly defined. Customer satisfaction and other areas of documentation control are also given utmost importance. With technical requirements, there are 10 subsections that control the qualifications of personnel, the selection and validation of methods, equipment selection and calibration, sampling, the measurement of uncertainty and traceability, and the reporting and documentation of results. The technical requirement is, however, not unique to forensic laboratories but is applicable to every testing laboratory.

The accreditation process takes place through a series of steps:

- Preliminary application or proof of conformance by the forensic laboratory.
- Application of the laboratory to the accreditation body for accreditation.
- Review of the laboratory documents and records.

- Onsite assessment of the forensic laboratory by the accreditation team.
- Report and review delivery of the assessment and remediation of findings.
- Awarding the forensic laboratory with accreditation.

The first step occurs when the laboratory wants to get accredited for the first time or to renew its existing accreditation. In both cases, the accreditation body must be informed prior to the applications. In case the organization is renewing its accreditation, the laboratory must apply for the renewal 6 months or 1 year before its accreditation expires. The information of the requirement of accreditation by the laboratory is required in order to ensure a smooth process and prevent any hassles during the entire accreditation process. Factors such as the policies and protocols as well as the management and technical requirements will be looked over during this step. Once this preliminary step is finished, and all the requirements by the laboratory seem to be appropriate, it can give a formal application to the accreditation body. Once the application has been made, the accreditation body will ensure that the application is complete and will later work in conjunction with the laboratory personnel. The application itself is a form of assessment because once it is submitted, a team of assessors from the accreditation body will start its assessment. This assessment is taken into further detail when the team of assessors starts to assess the documents and records of the laboratory. This assessment is usually done offsite as it saves time and energy. The records and documents are uploaded electronically to the website of the accreditation body and are examined in detail by the team of assessors. In order to facilitate this, the laboratories themselves maintain the electronic or paperless records. Once the offsite assessment is done, a team of assessors makes a laboratory visit and reviews the organizational structure and whether the laboratory is in compliance with the standard guidelines. The employees and other laboratory personnel will also be interviewed. This typically lasts for approximately 5 days. The laboratory administration is also responsible for clearing any queries raised by the team of assessors. Once the assessment of the organization is done, the report is made by the assessors in which any errors or nonconformances from the side of the laboratory personnel are mentioned. Also, a definite deadline is also mentioned before which the nonconformances must be fulfilled. These are mentioned in order to optimize the functioning of the laboratory and make sure that the protocols and policies are properly accredited. The report is usually submitted to the head of the laboratory, who ensures that any discrepancies are immediately removed. Once the nonconformances are removed and if every management and technical requirement is in accordance with the standards prescribed by the accreditation body, the laboratory is awarded accreditation. The certificate provided contains information regarding the scope of accreditation and the expiration date. This marks a momentous occasion for the laboratory because it proves that the results obtained from that particular laboratory are reliable and thus can be easily produced in a court of justice. However, in order to maintain this reliability, it is important that the laboratory undergoes renewal of the accreditation process at regular intervals of time.

The entire accreditation process is a large process that requires extremely efficient communication, first between the laboratory personnel followed by communication between the laboratory personnel and the accreditation body. Once this task has been accomplished; it is imperative that the same standards are maintained, even after accreditation, to ensure that the standards are maintained and reliable results are produced (Collins, 2018).

3.2 Digital forensics division

Digital forensics can be regarded as the route for the extraction of data in a usable form from data storage devices as well as systems. The extracted data can be utilized to gain knowledge during crime investigations or as evidence during courtroom proceedings (Tully et al., 2020). Currently, the keystone in legal systems is the fact that evidence admissibility in court is determined by *trier of fact* or finder of fact. The trier of fact is any person or group of people who are responsible for determining facts during any court proceedings. Digital forensics assists the *trier of fact* by providing undeviating evidence of any event that might have or has not occurred (Beckett and Slay, 2011). A researcher named Casey warned about the increasing number of cybercrime cases in today's world, and the decreasing quality of results produced by digital forensic experts (Casey, 2019). This calls for the accreditation of laboratories and experts working in digital forensics. ISO therefore provides necessary accreditations and guidelines for all laboratories and personnel, including those in digital forensics, to improve the quality of the results produced because they are very crucial in crime investigations and courtroom proceedings. As per ISO/IEC 17025, there are a number of necessities that must be incorporated while giving accreditation. These necessities include specification of technical proficiency, implementation of a well-authenticated and documented methodology, and involvement in an analysis of proficiency (UNIDO, 2009). A comparatively recent standard, ISO/IEC 27037, was introduced especially for digital forensics. This international standard focuses on identifying, collecting, and preserving digital evidence collected from a crime scene.

This standard could help private companies effectively preserve digital evidence, which could help criminal investigators in solving cases related to cybercrime in that company (Beckett and Slay, 2011).

In order to get accreditation as per ISO 17025, the following things need to be done by any digital forensic laboratory (Beckett and Slay, 2011):

- Methods and processes for evidence analysis should be validated.
- Equipment used should be properly and regularly calibrated and maintained.
- Procedures should be in written format.
- Traceability.
- Irrelevant testing should be controlled.
- Well-qualified staff should be present.

However, the laboratories that want to apply for this accreditation often face several challenges. Among them, the level of education of personnel working in the laboratories as well as the authentication of tools and procedures are the major challenges encountered by laboratories. Taking account of the challenge of education, a bachelors degree or higher is considered good. However, some certificate training courses are also available, but these can't be considered efficient qualification. Philips and Bowen very effectively distinguish this issue between a technically accomplished policeman and a forensic scientist (Philips and Bowen, 1985).

As far as authentication of tools and procedures is concerned, it is said that complete understanding of the methodology should be there, and well-known samples that can be closely related to the evidence should be included while verifying any tool or procedure (Barbara, 2005). The generation of false data by the equipment used is also a challenge, for example, a higher alcohol level than that in the sample due to an error in gas chromatography. However, as per the Scientific Working Group on Digital Evidence, false positives are not present in the field of digital forensics (SWDGE, 2008). This argument overlooks the challenge of generating false results in cases of analysis of digital evidence.

3.3 Crime scene investigators

The quality of crime scene investigations is solely dependent on the investigators involved in the process. In cases of lack of knowledge or experience with the investigator, any innocent person can be punished, or the suspect cannot be identified. In Australia, Justice Trevor Morling conducted an inquiry in which the entire group involved in the legal system and the forensic practitioners were charged for not following proper procedures and making changes in the function, structure, and practice of forensic methods during investigations (Morling, 1987). Instances like these increased the need for properly accredited individuals who can work as crime scene investigators. In this regard, SMANZFL (Senior Managers of Australian and New Zealand Forensic Laboratories) formed a committee named F&ISAG (Field and Identification Specialist Advisory Group) comprising specialists from ballistics and fingerprints as well as senior crime scene investigators for the accreditation of crime scene investigators. F&ISAG later formed a committee called CSAC (Crime Scene Accreditation Committee), whose job was to develop concepts and issues of investigators in order to enhance their proficiency. This committee plays an important role in the accreditation of investigators from Australia (Horsewell and Edwards, 1997).

Internationally, ISO 17025 gives accreditation to crime scene investigators on the basis of the person's competency in applying crime scene principles to photography, scene investigation, and in other forensic disciplines such as evidence handling. For any organization seeking accreditation only in crime scene investigation through ISO 17025, it is necessary for that organization to show the competency of their calibration and testing laboratories (NATA, 2000).

In order to develop competency and proficiency in crime scene investigators, duties in seven following criteria were decided, which were extracted from the CSAC committee (Horsewell, 2004):

- **Preliminary assessment of the crime scene**: It includes evaluation of health and safety risks, determining the incident circumstances, and deciding the boundary of the crime scene to preserve the evidence.
- **Supervision of the scene**: It involves the maintenance of an entry/exit log of people for the scene, preserving the scene during investigation, and advising people to follow an entry and exit path to minimize evidence contamination.
- **Scene examination**: It includes following a suitable search pattern for evidence, recording every detail of the scene, locating and properly collecting evidence from the suspect and victim, and using guidance from other experts wherever required.
- **Evidence interpretation**: This involves the development of an appropriate sequence of events that took place and the significance of the evidence collected, then communicating those interpretations with the officer in charge for that crime.

- **Scene recording**: It includes maintenance of the location, date, and time of the crime scene.
- **Collection of exhibits**: In this, exhibits need to be carefully collected in order to avoid any kind of contamination, then properly labeling them for identification. Also a record of the collected exhibits has to be maintained.
- **Management of case**: It involves the security as well as continuity of collected exhibits and evidence, maintaining a relationship with the officer in charge and other experts, and preparing reports and statements for the purpose of documentation.

3.4 DNA division

DNA division is one of the most important divisions in forensic science because DNA analysis gives confirmatory results about the identity of an individual involved in the crime, be it the suspect or the victim. Cases related to wrongful DNA analysis and reporting are present in the literature (Murphy, 2015; Vincent, 2010). The section of results evaluation and their reporting in forensic science does not have any standards to follow; it has only recommendations and guidelines (Hicks et al., 2019). It is therefore required for the laboratories involved in forensic DNA analysis and its reporting to get proper accreditation and follow international standards. For laboratories seeking accreditation in DNA analysis reporting, it has been added in their scope to state how they will evaluate the unmixed as well as mixed DNA profiles with the help of dedicated software. ISO 17025:2017 gives the standards that need to be followed by laboratories during DNA reporting for getting accreditation.

The certification of individuals involved in DNA analysis and interpretation is also required for an efficient investigation. In some countries such as the Netherlands, individual certification is mandatory while in some countries such as the United States, the certification is voluntary (Decorte, 2013). As per ISO//IEC 17025, proper documentation of all the events starting from receiving the sample to report preparation should be done. As per DAB (DNA Advisory Board) and ISFG (International Society of Forensic Genetics), the personnel involved in signing the DNA reports should have at least a masters' degree in biology or human genetics, and the analysts or examiners for DNA testing should at least possess a bachelors' degree and a minimum of 6 months experience in a forensic DNA laboratory. For technicians, a training program and qualification test (passed) are required to get accreditation. DAB and ISPG also recommend that laboratories and personnel involved in DNA analysis and reporting participate in proficiency testing at least twice a year (Decorte, 2013). In order to efficiently evaluate any laboratory's performance, blind proficiency testing is always recommended (Rand et al., 2004). These actions help in getting accreditation in forensic DNA analysis for any laboratory. In Italy, ACCREDIA has been appointed as the official accreditation authority for the forensic genetics laboratories since 2009. These laboratories need to follow standards as per ISO 17025/2005 in order to get the accreditation (Ricci et al., 2013). However, in case of DNA polymorphisms, there is no rule related to following an ISO protocol in Italy. Therefore, all laboratories have to make their own protocols using internal validation, keeping in mind that the method is reliable as well as reproducible.

3.5 Forensic toxicology division

Improper evidence analysis and report preparation can lead to deviation from proper investigation and thereby in apprehending the suspect. Toxicological analysis in forensics is a crucial step during investigations. However, undependable testing in forensic toxicological laboratories in well-known cases such as the drug chemist A. Dookhan from the United States (Cunha, 2014) and the Guy Paul Morin case from Canada (Kaufman, 1998) show that even legal systems and forensic experts can use science in a bad way. Therefore, such laboratories working in forensic toxicology need to be get proper accreditation.

There are a number of subdivisions in forensic toxicology such as poisoning cases, sports doping cases, postmortem cases, etc. These subdivisions have their own specific accreditation programs as well. The designing and scope of the accreditation program for a forensic toxicology laboratory therefore is dependent on the subdivision, and the standards may vary for different subdivisions of forensic toxicology (Jones, 2016). Generally, accreditation programs access laboratory administration, the qualification and training of personnel, and the competence and existence of procedures for standard operations as well as analysis, reporting, and safety. Quality assurance and competence of the chain of custody are also analyzed. However, the standards regarding these assessments vary for toxicology laboratories in different parts of the world such as Europe, Canada, the United States, and Australia.

Usually, maximum accreditation programs in Australia, Canada, and Europe for forensic toxicology laboratories are based on the standards prescribed by the ISO, mainly those in ISO/IEC 17025. These standards are sometimes supplemented by additional guidelines as well as peer-written standards. In the United Kingdom, the National Accreditation of Measurement and Sampling (NAMAS) gives accreditation as per ISO/IEC 17025. Similarly, other parts of the world have different accreditation bodies such as the National Association of Testing Laboratories (NATA) from Australia; the American National Standards Institute (ANSI), the American Society of Crime Laboratory Directors Laboratory Accreditation Board (ASCLD/LAB), and the American Association for Laboratory Accreditation (A2LA) in the United States; and the Forensic Science Accreditation program of the Standards Council of Canada (Jones, 2016).

In the United States, regarding toxicological analysis, few accreditation programs are specific for FUDT (forensic urine drug testing). The College of American Pathologists (CAP) also gives accreditations to forensic toxicology laboratories. However, the majority of forensic science laboratories in the United States are accredited by ASCLD/LAB. In this program, some standards are very specific for forensic toxicology laboratories, which mostly depend on the experience and judging capability of onsite inspection leaders. Another specific accreditation program for forensic toxicology, the American Board of Forensic Toxicology (ABFT), is also present. It is aimed at private and government laboratories, coroners, medical examiners, and police involved in the postmortem and human-related forensic toxicological analysis. Accreditation by ABFT can be maintained by laboratories by providing all the results of proficiency tests on an annual basis, and submitting an onsite inspection report every 2 years. After 2013, it has become mandatory for forensic toxicology laboratories in the United States to meet all standards before applying for accreditation or reaccreditation. However, getting these accreditations in not mandatory for laboratories in the United States, rather it is on a voluntary basis (Jones, 2016). But, if made mandatory, the quality of results produced by such laboratories would be enhanced several fold because it will become necessary for laboratories to upgrade the level of their methods and the procedures for analysis and reporting to get the accreditation.

4 Conclusion

Accreditation is a highly crucial process for forensic laboratories across the world in order to get valid and reliable analytical results of various evidence. It provides recognition and takes into account the transparency of the technical and management systems. In forensic science, evidence such as DNA, hair, semen, saliva, explosives, and gunshot residue is often encountered and analyzed. Apart from this, evidence such as computer hardware or software and flash drives are also encountered. The analysis of evidence in laboratories is therefore highly crucial and that analysis is also needed as testimony in a court of law. The accreditation of forensic laboratories allows that the analysis is done according to international standards and the personnel and employees involved in making this happen are also trained properly and certified. In forensic science, ISO 17025, ISO 17020, and ISO 15189 are the standards used to define the different technical and management requirements. The process of accreditation as such is a tedious one and the laboratory has to undergo several processes that allow them to be accredited. In this chapter, a brief overview of accreditation, its need, and history have been discussed. Apart from this, the standards and process of accreditation of a forensic laboratory have also been discussed.

References

Barbara, J., 2005. Digital evidence accreditation in the corporate and business environment. Digit. Investig. 2, 137–146.

Beckett, J., Slay, J., 2011. Scientific underpinnings and background to standards and accreditation in digital forensics. Digit. Investig. 8, 114–121. https://doi.org/10.1016/j.diin.2011.08.001.

Casey, E., 2019. The checkered past and risky future of digital forensics. Aust. J. Forensic Sci. 51, 649–664.

Collins, J., 2018. Accreditation. In: HR Management in the Forensic Science Laboratory. Elsevier, pp. 457–475. https://doi.org/10.1016/B978-0-12-801237-6.00026-9.

Cunha, G.A., 2014. Massachusetts drug chemist. Report of the Inspector General, United States.

Decorte, R., 2013. Accreditation in forensic DNA analysis. In: Encyclopedia of Forensic Sciences. Elsevier, pp. 227–232. https://doi.org/10.1016/B978-0-12-382165-2.00041-6.

Hicks, T., Biedermann, A., Taroni, F., Champod, C., 2019. Problematic reporting in DNA cases: the need for accredited formats and certified reporting competence. Forensic Sci. Int. Genet. Suppl. Ser. 7, 205–207. https://doi.org/10.1016/j.fsigss.2019.09.079.

Horsewell, J. (Ed.), 2004. The Practice of Crime Scene Investigation. CRC Press, Boca Raton.

Horsewell, J., Edwards, M., 1997. Development of quality systems accreditation for crime scene investigators in Australia. Sci. Justice 37, 3–8.

ISO/IEC 15189:2012, 2012. Medical Laboratories—Requirements for Quality and Competence. Available from: https://www.iso.org/standard/56115.html. (Accessed 19 September 2017).

ISO/IEC 17020:2012, 2012. Conformity Assessment—Requirements for the Operation of Various Types of Bodies Performing Inspection. Available from: https://www.iso.org/standard/52994.html. (Accessed 19 September 2017).

ISO/IEC 17025:2017, 2017. General Requirements for the Competence of Testing and Calibration Laboratories. Available from: https://www.iso.org/stan-dard/66912.html. (Accessed 2 February 2018).

Jones, G.R., 2016. Accreditation: toxicology laboratories. In: Encyclopedia of Forensic and Legal Medicine. Elsevier, pp. 30–33. https://doi.org/10.1016/B978-0-12-800034-2.00004-5.

Kaufman, F., 1998. Commission on Proceedings Involving Guy Paul Morin.

Malkoc, E., Neuteboom, W., 2007. The current status of forensic science laboratory accreditation in Europe. Forensic Sci. Int. 167, 121–126. https://doi.org/10.1016/j.forsciint.2006.06.064.

Morling, T.R., 1987. Royal Commission of Inquiry into the Chamberlain Convictions, Report of the Commissioner. Northern Territory Government, Darwin, NT.

Murphy, E., 2015. Inside the Cell: The Dark Side of Forensic DNA. Nation Books, New York.

NABL, 2008. Specific Guidelines for Accreditation of Forensic Science Laboratories and Checklist for Assessors. National Accreditation Board for Testing and Calibration Laboratories.

NATA, 2000. National Association of Testing Authorities, ISO/IEC17025 Application Document—Supplementary Requirements for Accreditation in the Field of Forensic Science. National Association of Testing Authorities, Sydney, NSW.

Palmbach, T.M., 2013. Education and accreditation in forensic science. In: Encyclopedia of Forensic Sciences. Elsevier, pp. 171–174. https://doi.org/10.1016/B978-0-12-382165-2.00290-7.

Pandey, G., Rawtani, D., Agrawal, Y.K., 2017. Future aspects of halloysite nanotubes in forensic investigations. J. Nanomed. Res. 6, 00153.

Philips, J., Bowen, J., 1985. Forensic Science and the Expert Witness. The Law Book Company Limited.

Rand, S., Schurenkamp, M., Hohoff, C., Brinkmann, B., 2004. The GEDNAP blind trial concept part II. Trends and developments. Int. J. Legal Med. 116, 199–206.

Rankin, B.W.J., Welsh, C., 2013. Accreditation. In: Encyclopedia of Forensic Sciences. Elsevier, pp. 515–518. https://doi.org/10.1016/B978-0-12-382165-2.00235-X.

Rawtani, D., Tharmavaram, M., Pandey, G., Hussain, C.M., 2019. Functionalized nanomaterial for forensic sample analysis. TrAC Trends Anal. Chem. 120, 115661. https://doi.org/10.1016/j.trac.2019.115661.

Ricci, U., De Sanzo, C., Carboni, I., Iozzi, S., Nutini, A.L., Torricelli, F., 2013. Accreditation of a forensic genetics laboratory in Italy. Forensic Sci. Int. Genet. Suppl. Ser. 4, e294–e295. https://doi.org/10.1016/j.fsigss.2013.10.150.

Ross, A., Davey, A., 2016. Accreditation: forensic specialties. In: Encyclopedia of Forensic and Legal Medicine. Elsevier, pp. 17–22. https://doi.org/10.1016/B978-0-12-800034-2.00002-1.

SWDGE, 2008. Standards and Controls. Scientific Working Group on Digital Evidence.

Tully, G., Cohen, N., Compton, D., Davies, G., Isbell, R., Watson, T., 2020. Quality standards for digital forensics: learning from experience in England & Wales. Forensic Sci. Int. Dig. Invest. 32. https://doi.org/10.1016/j.fsidi.2020.200905.

UNIDO, 2009. Complying With ISO 17025: A Practical Guidebook for Meeting the Requirements of Laboratory Accreditation Schemes Based on ISO 17025:2005 or Equivalent National Standards. United Nations Industrial Development Organization, Vienna.

Vincent, F., 2010. Inquiry Into the Circumstances that Led to the Conviction of Mr Farah Abdulkadir Jama. Victorian Government Printer, Australia.

Watson, D., Jones, A., 2013. Accreditation and certification for a forensic laboratory. In: Digital Forensics Processing and Procedures. Elsevier, pp. 795–824. https://doi.org/10.1016/B978-1-59749-742-8.00019-4.

Wilson-Wilde, L., 2018. The international development of forensic science standards—a review. Forensic Sci. Int. 288, 1–9. https://doi.org/10.1016/j.forsciint.2018.04.009.

CHAPTER 22

Quality control and quality assurance in forensic science laboratories

1 Introduction

Forensic science is basically the application of different scientific techniques for the analysis of various evidence to aid an investigation. Just like any other scientific field, here too the process performed by trained personnel needs to carefully be monitored and optimized to ensure that only trustworthy and reliable results are produced. Previously, the accreditation process in forensic science was discussed. However, along with accreditation comes other important measures such as quality control and assurance. These two topics are brought up whenever good laboratory practices are discussed.

Quality assurance is basically a set of planned or systematic protocols developed to ensure that a product or service is of utmost quality while quality control refers to the day-to-day operational protocols and activities to ensure the fulfillment of the quality requirements of a product or service. First, the protocols for the maintenance of the product or service quality are developed, followed by their implementation. Quality assurance ensures that the existence and effectiveness of a procedure, product, or service is of the highest quality and quality control carries out those activities.

In forensic science, the results developed directly cater to the law where the victim receives justice and the offenders are punished (Pandey et al., 2017a,b; Rawtani et al., 2019). In this chapter, the various aspects of quality assurance and control in forensic science will be discussed. Also, the use of these measures in certain specific areas of forensic science will also be discussed.

2 Quality assurance in forensic science

Forensic science laboratories are mainly accredited by three main standards: ISO 17025, ISO 17020, and ISO 15189. ISO 17025 is responsible for

Handbook of Analytical Techniques for Forensic Samples
https://doi.org/10.1016/B978-0-12-822300-0.00022-7

defining the general responsibilities and requirements to check the competence of testing and calibration laboratories. The second standard is responsible for stating the different requirements of the organizations that undertake the inspection of forensic laboratories. ISO 15189 is responsible for analyzing the quality assurance and management of medical laboratories.

Typically, each forensic laboratory consists of an established QA program. These programs ensure that the results developed are valid and have been developed through credible and valid means. A well-established QA program consists of certain requirements that are mandatory for all staff and professionals in a laboratory. Requirements include an appropriate organizational structure, a basic educational level, an appropriate laboratory, and other technical, administrative, and security facilities that ensure proper quality management of the services provided and the results produced. Proper documentation procedures, regular audits, quality control, and proficiency tests are also included (Lennard, 2013; Levy et al., 1999).

In forensic science, there are two types of tests—analytical and functional—that are routinely performed in a laboratory for the analysis of evidence. These tests can be either qualitative, quantitative, or comparative in nature and can be used for the identification, quantitation, or classification of the evidence. Quality assurance ensures that the protocols governing the three aspects of forensic science, which can be basically summarized as identification, individuality, and comparison of the evidence, are appropriate and fulfill the fitness of purpose (Doyle, 2019a). There are several organizations that are responsible for the quality management of an organization.

2.1 Organizations for quality management in forensic science

A group of bodies that is responsible for quality management in various laboratories and organizations is known as standard development organizations (SDOs). These are usually categorized into three levels based on their objectives. Level 1 organizations are usually comprised of organizations that apply generic standards before performing any activities. Level 1 is also the highest level and a common example is the ISO/IEC 17025:2017. ISO was established in 1947 and is basically an independent nongovernmental organization. The ISO standards are applied worldwide and serve as the primary standard setting organization. Another crucial standard setting body is the technical committee or TC that carries out the entire work of setting the standards by ISO; it is governed by the technical management board. Level 2 organizations are more specific than level 1 as they are responsible for selecting, adapting, and interpreting the standards according to the requirements of the laboratory. In forensic science, the Internal Laboratory Accreditation Guide (ILAC) serves as the level 2 organization. ILAC operates according to the ISO/IEC 17011 standard. ILAC ensures the quality management of forensic science providers by incorporating the standards ISO/IEC 17025 and ISO/IEC 17020. Finally, level 3 organizations are responsible for ensuring the specifics of a forensic laboratory and are the lowest in the hierarchy level. These standards are more specific and application-oriented. However, this itself serves as a limitation in making standards for every specific role. Despite these flaws, the standards in forensic science are developed according to the international, national, regional, and field-specific requirements by developing a systematic approach. The approach consists of the processes, methods, fields, and professional bodies involved in forensic science (Doyle, 2019b).

2.2 Aspects of quality assurance in forensic science

In this section, the various aspects of a forensic science organization that are subjected to quality assurance are discussed. A forensic science facility is similar to any other scientific facility in that it also involves personnel who carry out a set of technical, administrative, and security tasks. Carrying out such tasks, especially the technical aspects, requires following a set of standard operating procedures (SOP) that specifies the validated protocol for analyzing evidence through different analytical techniques. Throughout the analysis, documentation is recorded at every stage to ensure reviewability. Quality assurance therefore looks through every single aspect of this entire process to ensure that it is producing results of utmost quality.

The most crucial aspect of any facility is the personnel. The staff hired must be educated, competent, and trained before they are allowed to proceed with their designated activities. Improperly training staff not only provide invalid results but also poses a risk to the other members of an organization, especially when the activities are linked to each other. Through quality assurance, guidelines for the regular monitoring of their proficiency and performance are also developed.

The proficiency and performance of the staff are those factors that are highly dependent on the mental and physical ability of the staff. Improper workspaces, poor accommodation, bad lighting, and the lack of fresh air or cleanliness severely hamper staff productivity. Therefore, quality assurance in such cases is necessary to ensure that such environmental factors do not affect the productivity of the staff and that they are able to deliver quality results. Apart from this, clear divisions should also be present for each field in order to separate any incompatible activities. Apart from this, the equipment used to carry out the activities also needs to be regularly calibrated and serviced. The maintenance, safety, and calibration checks are performed by the in-house staff. Everything during this process is carefully documented and checked by senior officials who are in charge of this task. Improperly calibrated equipment will result in erroneous results, thus hampering quality. Therefore, quality assurance also ensures that the measurements or the standards used for calibrating purposes are taken from certified bodies. Just like any other task, every step in the selection of standards for calibration and validation must be documented carefully.

Once the standards are selected and the equipment is calibrated, the SOPs are made. The SOPs are extremely detailed protocols for carrying out policies or protocols and even for operating equipment. Because the sole purpose of SOPs is to produce detailed steps for carrying out an activity that produces optimal results, it is important that the staff adhere to it stringently. Any reason for deviation from the SOP must also be documented.

A crime investigation is largely based on the analysis of evidence found at a crime scene. A myriad of evidence such as physical, chemical, or biological evidence is found, often in varying quantities. It is highly imperative that every sample obtained from a crime scene must be documented and kept in a chain of custody. The amount of evidence collected as well as its type, date, and origin of collection must also be collected and processed. Here too, stringent protocols for the collection of evidence from a crime scene must be followed by officials who visit the crime scene (Lennard, 2013).

3 Quality assurance in different fields of forensic science

By carefully following the guidelines of quality assurance standards, it is possible to produce high-quality results and services. In this section, emphasis is given on quality assurance in different fields of forensic science such as digital forensics, forensic biology, forensic toxicology, questionable documents, and even forensic pathology and the postmortem of disaster victims.

3.1 Digital forensics

Digital forensics deals with the investigation and recovery of evidence in digital devices. Lately, digital, cyber, and network crimes have seen a surge due to technological developments. This has now been established as a vital tool for criminal investigation and justice systems. However, there is still a lack of standardization, operating protocols, and quality management systems. There are also several areas that require stringent quality management systems. However, the most demanding area is the processes a forensic scientist chooses to reliably analyze digital evidence to produce results of utmost quality. Newly encountered, undocumented digital evidence or data is often the most difficult to identify and classify or compare. Therefore, in order to overcome this, a verification of digital evidence (VODE) framework was developed that was capable of analyzing such evidence. The VODE framework allows a digital forensics scientist to interpret the meaning and impact of any evidence found at a crime scene. Through this framework, it is possible for the scientist to provide basic facts of the case, any possible offense type, and a set of actions that may have led to the investigation while also obtaining information for the same (Graeme, 2020; Horsman, 2020). Apart from digital forensics, network forensics that involves the identification, collection, analysis, and examination of digital evidence from a network or security log is another crucial part of digital forensics. Here too, proper evidence collection methods and solutions need to be developed to ensure that there is minimal data loss during collection. Non-QA and QA models have been developed to make sure that maximum data are collected (Cheng and Chen, 2007).

3.2 Forensic biology and toxicology

Quality assurance in forensic biology and toxicology follows the same guidelines of quality assurance as mentioned before. Forensic biology mainly involves the analysis of biological samples such as blood, saliva, semen, hair, vomit, urine, feces, and DNA. Techniques such as rt-PCR and sequencing are commonly used for DNA analysis while in vitro tests such as the Takayama and Teichmann tests are used for blood analysis. The quality assurance of forensic biology has been well developed as the samples analyzed in this field are ubiquitous, due to which the need for quality assurance in these fields was developed long before. These fields also involve appropriate protocols for documentation, chain of custody to handle evidence, equipment calibration, maintenance, and regular monitoring. Also, validated methods are used for the analysis of various evidence. The staff is also carefully monitored for proficiency (Quarino et al., 1994). For a forensic toxicologist, a quality product refers to the appropriate chemical analytical results that have been developed by trained and certified forensic scientists. Both fields use good laboratory practices to ensure they produce results of the highest quality (Aderjan, 2008).

3.3 Questionable document analysis

Quality assurance in forensic document examination, just like other forensic science fields, involves the production of highly consistent and informative results that are based on validated and up-to-date methods. Typically, questionable-document examiners are highly qualified and therefore the analytical results are usually reliable. However, it is important that a proper control mechanism be developed that allows the control and monitoring of an examiner's performance. The guidelines of quality assurance are also applied to this field. Here too, it is imperative that the examiners are trained and have a minimal threshold of expertise with a sound reasoning mind. Once competent and trained examiners are selected, it is important that the organization employing them adopts policies that are relevant to the questionable-document examination. Here too, a common protocol is required in order to ensure that the examiners do not adopt a protocol of their own. In quality assurance for questionable documents, there are four main areas where the guidelines can be applied. These involve the handling of exhibits, the examination processes used for their examination, report writing and documentation, and preparing for court testimony. By ensuring that these processes are controlled by appropriate quality assurance guidelines, quality results can be produced (Purdy, 1985).

3.4 Clinical and forensic pathology

Clinical and forensic pathologists are responsible for determining the cause of death through the examination of the corpse. The pathologists are medical doctors trained in forensic pathology and are responsible for analyzing the injury or any underlying disease, the manner of death, and tissue specimens. In the practical areas of medicine, standard statements and audits started to become routine in 1990. It was first established that the quality in this field will be determined through terms such as efficacy, equity, acceptability, accessibility, and appropriateness. Just like the guidelines for quality assurance in the different laboratory set ups, similar guidelines are applied to forensic pathology. The processing and control of the specimens, the careful interpretation of their histological analysis, accurate postmortem diagnosis with high sensitivity and specificity are some of the activities which require appropriate guidelines for quality assurance (Ferlan-Marolt and Balažic, 1999).

4 Quality control for the analysis of forensic samples

QC in forensic science is very essential in order to make sure that the quality of the investigation and evidence analysis remains the same and up to the mark for all cases. Many of the research works have focused on the analysis of standard reference materials (SRMs) of different forensic samples for QC purposes. This section discusses the research works related to QC for forensic samples such as alcohol, cartridge cases, DNA, drugs, entomological evidence, hair, and inks.

4.1 Alcohol

Alcohol is one of the most common types of evidence found in the blood and breath of people involved in drinking and driving as well as people at rave parties. In order to maintain the quality of analysis, standards need to be followed during sample collection and the analysis of samples should be done with respect to the reference material.

In the analysis of blood alcohol, two methods need to be followed: (a) the enzymatic method using the alcohol dehydrogenase enzyme (ADH); and (b) the instrumental method using gas chromatography. Each sample needs to be analyzed twice with both techniques. To maintain internal QC, a calibration curve with correlation coefficients of more than 0.999 must be used. All sample measurement plots need to be present in the confidence interval, with a probability of not less than 95%. The reproducibility as well as correctness of the results obtained for the test samples need to be checked with the reference blood as well as serum samples in order to maintain the internal QC. For external QC, interlaboratory tests need to be performed by all the accredited test laboratories for blood alcohol analysis on an annual basis. External experts should also regularly inspect such laboratories to make certain the correct regulations are applied during the analysis (Sutter, 2002). For QC, measurement uncertainty and document traceability are essential and are to for get laboratory accreditation in the future. For breath alcohol measurement, when all QC parts are applied, then a combined uncertainty of 73% in sampling, 10% in analysis, 4% in water and the air partition coefficient, and 13% in traceability are the accepted estimates in forensic science, and show the standards for measuring breath alcohol (Gullberg, 2006).

4.2 Cartridge case

Cartridge cases are readily recovered from crime scenes that had firearm incidents. These cases are sent for laboratory analysis in order to check what type of firearm fired them. Different kinds of analytical techniques, especially microscopic techniques, help in comparing the marking present on the cartridge cases due to the events that take place during firing.

When a firearm is recovered from the crime scene, then comparing the impression marks of the firing pin, ejector, and breech face on the cartridge cases helps in identifying the firearm and the cartridge. However, when a firearm is not present, then the images of fired cartridge cases present in the database of the Integrated Ballistics Identification System (IBIS) can be used for comparison. In the United States, in order to maintain QC for the IBIS, the National Institute of Standards and Technology (NIST) started a project related to the production of SRM 2460/2461 cartridge cases and bullets (standards). These SRMs help in maintaining the QC during bullet and cartridge case analysis using a comparison with IBIS images in the absence of a firearm (Song et al., 2006). Morris et al. performed a comparison study of cartridge cases using the IBIS system with the help of five different SRMs for cartridge cases with serial numbers 2P6325, 2P4316, 2P2415, 2P2335, and 2P2333. It was observed that the use of SRMs for comparison between cartridge cases from the same and different guns helped in differentiating the fired cartridges on the basis of breech face while there were some overlaps related to the impressions due to the firing pin (Morris et al., 2017).

4.3 DNA

In the initial days of forensic investigations, DNA analysis was carried out in cases where abundant DNA evidence could be found, such as cases of sexual assault. However, in today's time, DNA analysis has been made almost mandatory in almost all kinds of cases because it directly relates to the person who was involved in the crime, either as a victim, suspect, or witness (Gill et al., 1991).

QC in DNA profiling deals with the verification of fit for purpose as well as quality levels on a day-to-day basis. In DNA profiling, QC not only deals with the use of standard controls and reagents, but also is required for LIMS (laboratory information management system) as well as the digital expert systems present in a DNA laboratory. Elimination DNA databases as well as DNA environmental monitoring are also integral parts of QC for DNA profiling, along with the handling and report preparation of the analysis. In Europe, the European DNA Profiling Group (EDNAP) and the European Network of Forensic Science Institutes (ENFSI) work on the preparation of guidelines for DNA profiling for QC purposes. Quality assurance programs by ENFSI DNA focus on the use of PCR blanks and extraction blanks to maintain the quality and reduce the variability of interlaboratory results (Ansell, 2013).

DNA is also present in the mitochondria apart from the nucleus, where the analysis becomes important in cases of maternity disputes. In order to maintain the QC in mitochondrial DNA (mtDNA) analysis, SRMs have been developed. In a study by Levin et al., SRM 2392 was used as the reference material for QC during the analysis of mtDNA. This SRM has more than 50 sets of primers for a polymerase chain reaction (PCR), due to which almost any region of the mtDNA can be amplified followed by sequencing. The reproducible results of amplification from different laboratories ensure the use of this SRM for QC purposes (Levin et al., 1999). Due to the importance of mtDNA in forensic analysis, several studies have been published in past years related to QC in mtDNA profiling (Irwin et al., 2007; Parson et al., 2014; Yao et al., 2004).

During DNA analysis, short tandem repeats (STRs) have assisted in the identification of people due to their variability in different DNA. Butler et al. published a research paper related to QC for PCR primers that are used during multiple amplification reactions for STRs. The study focused on the analysis of primer sets provided in different commercially available kits for STR typing using high-performance liquid chromatography (HPLC) and mass spectroscopy (MS). It was concluded that the use of these techniques by forensic laboratories can help in ensuring that the quality of kits used for STR typing remains consistent as well as reliable to produce reproducible results (Butler et al., 2001). The evaluation of the genotype of these STRs in a statistical manner is dependent on the allele frequencies, which are compiled in the form of STR databases. However, before the study by Bodner et al., there was no QC available for the STR databases, and therefore the data produced by individual research groups were not reliable. Therefore, Bodner et al. presented STRidER, which is a novel database system related to autosomal STRs and provides software tools related to QC in the case of STR typing (Bodner et al., 2016). Szibor et al. also contributed in creating the pool of QC reference DNA (both autosomal and mitochondrial) from human lymphoid cell lines. The study could be helpful in forensic genetics, which relies on a comparison of samples with reliable standards to reduce the variability in results (Szibor et al., 2003).

4.4 Drugs

Drugs are very commonly abused by people, especially youngsters at parties. The collected drug samples can be in the form of powder or tablets, or can be present in the blood of the consumer. Different spectroscopic and chromatographic techniques are employed for such analysis in laboratories.

In order to prevent interlaboratory variations, several sets of rules and guidelines need to be followed by laboratories involved in forensic drug analysis, mainly the forensic toxicology laboratories. In order to maintain the QC in such laboratories, the Division of Narcotic Drugs from the United Nations proposed a set of requirements for forensic drug analysis. These requirements were related to the educational qualifications of the laboratory staff as well as the basic instruments and reagents required for drug analysis of for forensic purposes. TIAFT (the International Association of Forensic Toxicologists) has given such guidelines for drug analysis, especially in developing countries (Penders and Verstraete, 2006). The QC programs control the variations in the results by controlling the procedures adopted, and the instruments used using external and internal control drug samples. However, for efficient QC, the best control samples are the blind drug samples, which are unknown to the person doing the analysis. It is also recommended that the final results of the analysis should be in the control of another technician or scientist (Christophersen and Mørland, 1994). Malcolm et al. performed a study for the analysis of drugs through gas chromatography (GC). The study focused on the development of internal controls for QC purposes in forensic toxicology laboratories. During the analysis, control drug samples were prepared in ethanol and then spiked into the blood at variable concentrations, followed by analysis in GC. However, the study did not succeed in its aim to analyze the mixture of drugs in a blood sample, suggesting more work is needed in developing the QC control (Malcolm et al., 1995). As far as drug analysis in hair samples is concerned, guidelines have been issued by the Society of Hair Testing (SOHT). However, QC in drug analysis from hair remains an issue due to the lack of SRMs that could be used as external controls by the laboratories (Cooper et al., 2012).

4.5 Entomological evidence

Forensic entomology applies the science of insects to solve legal cases. Entomological evidence such as insects and their different developmental stages are generally recovered from the dead and decaying cadavers of humans as well as animals. These forensic samples are generally analyzed via various microscopic techniques.

However, in order to maintain the quality of results produced by such analyses, certain standards and guidelines need to be followed. In this regard, in Australia and New Zealand, a standard from ANZFSS (Australian and

New Zealand Forensic Science Society) called the Code for Professional Practice for Members of ANFSS has been established. This standard deals with the basic expectations related to the use of the scientific technique, report preparation, and conduct during forensic investigations with entomological evidence. A recent development in forensic science related to forensic entomology is the addition of the following Australian forensic science standards: (a) identification, recording, retrieval, and storage of materials; (b) investigation and analysis of materials; (c) result interpretation; and (d) report preparation (Archer and Wallman, 2017).

In order to provide QC for entomological evidence analysis, the Medical Sciences Specialist Advisory Group (MS SAG) Forensic Entomology Scientific Working Group (SWG) took the initiative to form a set of standard questions related to testimony for entomological evidence, which can be asked to the expert witness while giving the testimony. The accreditation of laboratories with standard NF ISO/CEI 17025 can assist in ensuring the QC in the entomological evidence analysis (Gaudry and Dourel, 2013). However, a relatively low number of cases with entomological evidence is a major concern and is hampering the growth of this discipline with standards for QC and QA. With increased international cooperation, the guidelines and standards can be developed to maintain QC in this field (Archer and Wallman, 2017).

4.6 Hair

Hair samples are one of the most common kinds of evidence encountered in almost any kind of crime scene. These samples are analyzed for structural details using microscopic techniques and for the presence of drugs using chromatographic and spectroscopic techniques. This section sheds light on drug analysis in hair samples and the QC-related studies done in this field. Because the amount of hair sample recovered from a crime scene isn't very large, the detection limit for drugs is also low. Therefore, it has increased relative deviation.

The results of hair analysis have a lot of variations among different laboratories. The main reason behind this is the use of spiked hair samples as controls for drug analysis. These hair samples spiked with different drugs cannot be used for quality control purposes because in the case of real hair samples, drugs are attached to the solid phase that is fenced with keratin. With spiked hair samples, drugs reside on the surface of hair, and can easily be washed off using any suitable solvent. Therefore, SOHT took the initiative to develop a hair pool that can be used for QC during drug analysis in hair. For this purpose, several laboratories participated in developing the hair pool via two methods: soaking the hair in a drug solution and measuring the contamination, and powdering the hair collected from drug addicts for a quantitative comparison of the results. The obtained results proved helpful for QC purposes for the analysis of drugs such as acetylmorphine, morphine, codeine, and benzoylecgonine because the results matched the reference laboratories' results. However, for cocaine, it was concluded that more analysis needs to be done to match the results with the reference laboratories (Sachs, 1997). A similar kind of work was developed in Italy called HAIR-VEQ, which was an external QC scheme in which 23 clinical laboratories participated in drug analysis in hair. Variations were observed in the interlaboratory analysis results that highlighted the need for guidelines for hair analysis in laboratories. The study also emphasized the participation of forensic laboratories in such programs to increase the quality of forensic hair analysis (Pichini et al., 2004).

4.7 Inks

Ink samples are often analyzed from questionable documents collected from a crime scene. Such documents can be recovered in cases especially related to forgery. Spectroscopic techniques as well as chromatographic techniques play important roles in the analysis of inks present in the questionable documents. Apart from documents, inks are also used in tattoos, which can help in the identification of an individual or cadaver found at a crime scene.

As far as the analysis of writing inks is concerned, guidelines have been given in the standards prepared by ASTM such as ASTM 1789-04 (for the identification of writing inks) and ASTM 1422-05 (for the comparison of writing inks). The major aim of these standards is to maintain the quality of laboratory results working on the forensic analysis of ink samples. The QC of ink analysis has to be mostly maintained during the analytical process in order to get reproducible results. However, the standard ASTM 1789-04 always mentions the standard ASTM 1422-05 whenever ink analysis is concerned. There is no description about the calibration curve or ways to reduce the variability in results. Therefore, the standards need to be modified with more guidelines in order to reduce the variability in interlaboratory results (Neumann and Margot, 2010). Persechino et al. also developed a QC method for the analysis of tattoo inks. The study emphasized the hyphenation of high-performance thin layer chromatography (HPTLC) with Raman spectroscopy for the detailed analysis of all the components in tattoo inks for QC purposes. HPTLC separates all the components, which are later analyzed using Raman spectroscopy (Persechino et al., 2019).

5 Conclusion

Forensic science is an extremely crucial part of the justice system. Therefore, it is important that the results and service produced or provided in this field be of the highest quality. There are several steps to be followed when investigating a crime scene, from the collection of the evidence and examination of the body to the preparation of testimony for court. In order to ensure that high-quality results are produced, several ISO guidelines have been developed for quality management in forensic science. Quality management in forensic science is done through quality assurance and control. Quality assurance is a proactive method that provides confidence to the organization to produce reliable, valid, and credible results while quality control covers the operational protocols and activities to ensure the fulfillment of the quality requirements of a product or a service. Quality management in forensic science has seen significant progress in the past decades and several accreditation bodies based on the ISO guidelines that are specific to each country have been developed as well.

References

Aderjan, R.E., 2008. Aspects of quality assurance in forensic toxicology. In: Handbook of Analytical Separations. Elsevier, pp. 763–819. (Chapter 23). https://doi.org/10.1016/S1567-7192(06)06023-2.

Ansell, R., 2013. Internal quality control in forensic DNA analysis. Accred. Qual. Assur. 18, 279–289. https://doi.org/10.1007/s00769-013-0968-9.

Archer, M.S., Wallman, J.F., 2017. The development of forensic entomology in Australia and New Zealand: an overview of casework practice, quality control and standards. Aust. J. Forensic Sci. 49, 125–133. https://doi.org/10.1080/00450618.2015.1137972.

Bodner, M., Bastisch, I., Butler, J.M., Fimmers, R., Gill, P., Gusmão, L., Morling, N., Phillips, C., Prinz, M., Schneider, P.M., Parson, W., 2016. Recommendations of the DNA Commission of the International Society for Forensic Genetics (ISFG) on quality control of autosomal Short Tandem Repeat allele frequency databasing (STRidER). Forensic Sci. Int. Genet. 24, 97–102. https://doi.org/10.1016/j.fsigen.2016.06.008.

Butler, J.M., Devaney, J.M., Marino, M.A., Vallone, P.M., 2001. Quality control of PCR primers used in multiplex STR ampli®cation reactions. Forensic Sci. Int. 10.

Cheng, B.-C., Chen, H., 2007. Quality assurance for evidence collection in network forensics. In: Lee, J.K., Yi, O., Yung, M. (Eds.), Information Security Applications. In: Lecture Notes in Computer Science, Springer, Berlin, Heidelberg, pp. 121–132. https://doi.org/10.1007/978-3-540-71093-6_10.

Christophersen, A.S., Mørland, J., 1994. Drug analysis for control purposes in forensic toxicology, workplace testing, sports medicine and related areas. Pharmacol. Toxicol. 74, 202–210. https://doi.org/10.1111/j.1600-0773.1994.tb01099.x.

Cooper, G.A.A., Kronstrand, R., Kintz, P., 2012. Society of hair testing guidelines for drug testing in hair. Forensic Sci. Int. 218, 20–24. https://doi.org/10.1016/j.forsciint.2011.10.024.

Doyle, S., 2019a. Brief introduction to some important concepts and key terms. In: Quality Management in Forensic Science. Elsevier, pp. 1–24. https://doi.org/10.1016/B978-0-12-805416-1.00001-3.

Doyle, S., 2019b. Organizations exercising responsibility for quality management in forensic science. In: Quality Management in Forensic Science. Elsevier, pp. 25–61. https://doi.org/10.1016/B978-0-12-805416-1.00002-5.

Ferlan-Marolt, V., Balažic, J., 1999. Quality assurance in morphological diagnosis making sense of evaluation and comparison of results in clinical and forensic pathology. Forensic Sci. Int. 8.

Gaudry, E., Dourel, L., 2013. Forensic entomology: implementing quality assurance for expertise work. Int. J. Legal Med. 127, 1031–1037.

Gill, P., Evett, I.W., Woodroffe, S., Lygo, J.E., Millican, E., Webster, M., 1991. Databases, quality control and interpretation of DNA profiling in the Home Office Forensic Science Service. Electrophoresis 12, 204–209. https://doi.org/10.1002/elps.1150120217.

Graeme, H., 2020. Part 1: quality assurance mechanisms for digital forensic investigations: introducing the Verification of Digital Evidence (VODE) framework. Forensic Sci. Int. 2, 100038. https://doi.org/10.1016/j.fsir.2019.100038.

Gullberg, R.G., 2006. Estimating the measurement uncertainty in forensic breath-alcohol analysis. Accred. Qual. Assur. 11, 562–568. https://doi.org/10.1007/s00769-006-0176-y.

Horsman, G., 2020. Part 2: quality assurance mechanisms for digital forensic investigations: knowledge sharing and the Capsule of Digital Evidence (CODE). Forensic Sci. Int. 2, 100035. https://doi.org/10.1016/j.fsir.2019.100035.

Irwin, J.A., Saunier, J.L., Strouss, K.M., Sturk, K.A., Diegoli, T.M., Just, R.S., Coble, M.D., Parson, W., Parsons, T.J., 2007. Development and expansion of high-quality control region databases to improve forensic mtDNA evidence interpretation. Forensic Sci. Int. Genet. 1, 154–157. https://doi.org/10.1016/j.fsigen.2007.01.019.

Lennard, C., 2013. Principles of quality assurance. In: Encyclopedia of Forensic Sciences. Elsevier, pp. 509–514. https://doi.org/10.1016/B978-0-12-382165-2.00234-8.

Levin, B.C., Cheng, H., Reeder, D.J., 1999. A human mitochondrial DNA standard reference material for quality control in forensic identification, medical diagnosis, and mutation detection. Genomics 55, 135–146. https://doi.org/10.1006/geno.1998.5513.

Levy, S., Bergman, P., Frank, A., 1999. Quality assurance in forensic science. Accred. Qual. Assur. 4, 253–255.

Malcolm, M.J., Húdson, J.C., Proulx, J.G.F., Sharp, M.E., Whiting, C., 1995. Internal quality control of a general GC drug screen in forensic toxicology: experience, questions, proposals. J. Can. Soc. Forensic Sci. 28, 215–228. https://doi.org/10.1080/00085030.1995.10757485.

Morris, K.B., Law, E.F., Jefferys, R.L., Dearth, E.C., Fabyanic, E.B., 2017. An evaluation of the discriminating power of an Integrated Ballistics Identification System ® Heritage™ system with the NIST standard cartridge case (Standard Reference Material 2461). Forensic Sci. Int. 280, 188–193. https://doi.org/10.1016/j.forsciint.2017.09.004.

Neumann, C., Margot, P., 2010. Considerations on the ASTM Standards 1789-04 and 1422-05 on the Forensic Examination of Ink: CONSIDERATIONS ON THE ASTM STANDARDS. J. Forensic Sci. 55, 1304–1310. https://doi.org/10.1111/j.1556-4029.2010.01454.x.

Pandey, G., Rawtani, D., Agrawal, Y.K., 2017a. Future aspects of halloysite nanotubes in forensic investigations. J. Nanomed. Res. 6.

Pandey, G., Tharmavaram, M., Rawtani, D., Kumar, S., Agrawal, Y., 2017b. Multifarious applications of atomic force microscopy in forensic science investigations. Forensic Sci. Int. 273, 53–63. https://doi.org/10.1016/j.forsciint.2017.01.030.

Parson, W., Gusmão, L., Hares, D.R., Irwin, J.A., Mayr, W.R., Morling, N., Pokorak, E., Prinz, M., Salas, A., Schneider, P.M., Parsons, T.J., 2014. DNA Commission of the International Society for forensic genetics: revised and extended guidelines for mitochondrial DNA typing. Forensic Sci. Int. Genet. 13, 134–142. https://doi.org/10.1016/j.fsigen.2014.07.010.

Penders, J., Verstraete, A., 2006. Laboratory guidelines and standards in clinical and forensic toxicology. Accred. Qual. Assur. 11, 284–290. https://doi.org/10.1007/s00769-006-0131-y.

Persechino, S., Toniolo, C., Ciccola, A., Serafini, I., Tammaro, A., Postorino, P., Persechino, F., Serafini, M., 2019. A new high-throughput method to make a quality control on tattoo inks. Spectrochim. Acta A Mol. Biomol. Spectrosc. 206, 547–551. https://doi.org/10.1016/j.saa.2018.08.037.

Pichini, S., Ventura, M., Pujadas, M., Ventura, R., Pellegrini, M., Zuccaro, P., Pacifici, R., de la Torre, R., 2004. HAIRVEQ: an external quality control scheme for drugs of abuse analysis in hair. Forensic Sci. Int. 145, 109–115. https://doi.org/10.1016/j.forsciint.2004.04.025.

Purdy, D.C., 1985. Basic elements of a quality assurance system for forensic document examiners. Forensic Sci. Int. 29, 11–20. https://doi.org/10.1016/0379-0738(85)90029-5.

Quarino, L., Wilson, R., Ferrara, J., Baum, H., Shaler, R.C., 1994. Quality assurance in a forensic biology laboratory. Forensic Sci. Int. 68, 17–28. https://doi.org/10.1016/0379-0738(94)90375-1.

Rawtani, D., Tharmavaram, M., Pandey, G., Hussain, C.M., 2019. Functionalized nanomaterial for forensic sample analysis. TrAC Trends Anal. Chem. 120, 115661. https://doi.org/10.1016/j.trac.2019.115661.

Sachs, H., 1997. Quality control by the Society of Hair Testing. Forensic Sci. Int. 84, 145–150. https://doi.org/10.1016/S0379-0738(96)02057-9.

Song, J., Vorburger, T., Reneger, T., Rhee, H., Zheng, A., Ma, L., Libert, J., Ballou, S., Bachrach, B., Bogart, K., 2006. Correlation of topography measurements of NIST SRM 2460 standard bullets by four techniques. Meas. Sci. Technol. 17, 500–503.

Sutter, K., 2002. Determination of ethanol in blood: analytical aspects, quality control, and theoretical calculations for forensic applications. Chimia (Aarau) 56, 59–62. https://doi.org/10.2533/000942902777680775.

Szibor, R., Edelmann, J., Hering, S., Plate, I., Wittig, H., Roewer, L., Wiegand, P., Calì, F., Romano, V., Michael, M., 2003. Cell line DNA typing in forensic genetics—the necessity of reliable standards. Forensic Sci. Int. 138, 37–43. https://doi.org/10.1016/j.forsciint.2003.09.002.

Yao, Y.-G., Bravi, C.M., Bandelt, H.-J., 2004. A call for mtDNA data quality control in forensic science. Forensic Sci. Int. 141, 1–6. https://doi.org/10.1016/j.forsciint.2003.12.004.

CHAPTER 23
Concluding notes

1 Introduction

Forensic science is a branch of science that serves as a bridge between the crimes and the justice system. This field involves analyzing various physical, biological, and chemical evidence found at a crime scene in order to get information about the nature of the crime, the victim, and the perpetrator. Over the years, the evidence samples have been analyzed using various analytical techniques, including spectroscopic, chromatographic, microscopic, and emerging techniques such as nanotechnology. In this book, such analytical techniques for the analysis of various types of forensic evidence have been discussed. Also, the ethical and legal implications of forensic science and the accreditation processes of forensic science laboratories have been discussed.

2 Analytical techniques and forensic science

There are mainly three types of forensic evidence: physical, chemical, and biological evidence. Physical evidence includes samples such as fingerprints, tire marks, footprints, paint, glass, and building materials. Chemical evidence includes samples such as explosives, poisons, and chemical weapons while biological evidence includes samples such as blood, DNA, semen, saliva, urine, vomit, hair, and feces. The nature of these samples is highly different, due to which each sample needs to be analyzed through a specific analytical technique. Spectroscopic and chromatographic techniques not only analyze the nature of the sample, but they can be used in conjunction with derivative techniques to obtain confirmatory analysis of the samples. Microscopic techniques also provide a visual confirmation of the sample. Apart from such conventional techniques, emerging techniques such as nanotechnology, lab-on-chip technology, and X-ray diffraction have also seen an increase in their use for the detection of the samples. In order to ensure that the results produced from the analysis of the samples are trustworthy and reliable, it is important that the laboratories are properly accredited and appropriate ethical and legal issues are followed.

Handbook of Analytical Techniques for Forensic Samples
https://doi.org/10.1016/B978-0-12-822300-0.00023-9

3 Rereads of the book

A crucial aspect of analyzing forensic samples is first identifying the nature of the sample. As mentioned before, physical, chemical, and biological evidence have separate sample collection and preparation techniques. The collection of this evidence is done with utmost care in order to ensure that there is absolutely no contamination. Also, documentation is also done at every step and a chain of custody for the evidence is maintained. The sample preparation of such techniques is necessary so that they can be optimally analyzed through different analytical techniques.

The analytical techniques that are commonly used for the analysis of forensic evidence are spectroscopic, chromatographic, microscopic, and other emerging analytical techniques such as mass spectrometry, X-ray diffraction, lab-on-chip devices, and nanotechnology.

Spectroscopic techniques are mainly based on the analysis of evidence using the spectral regions of electromagnetic radiation. Electromagnetic radiation consists of two components: the electric and the magnetic vectors that propagate at right angles to each other. It has a dual nature characteristic and therefore possesses a particle- as well as a wave-like nature. Spectroscopic techniques such as UV-visible and fluorescence, Fourier transform infrared (FTIR), near-infrared (NIR), atomic absorption and emission, and Raman and nuclear magnetic resonance spectroscopy are most commonly used for the analysis of various forensic evidence. When electromagnetic radiation of a particular wavelength, frequency, and energy hits a sample, it either get absorbed, scattered, or transmitted through. Quite often, when the radiation energy is absorbed, the electrons in the sample get excited and move to a higher energy level. However, they do not stay in this state for a long time and emit a series of energies when transitioning back to their ground state. This serves as the basis of all the spectroscopic techniques with the exception of nuclear magnetic resonance, in which the intrinsic magnetic moment of a molecule in the presence of an external magnetic field is taken into account. Spectroscopic techniques are highly useful when performing both the preliminary and confirmatory analysis of the samples. Techniques such as FTIR and NIR are nondestructive in nature and can be used for the onspot analysis of any sample. Spectroscopic techniques have been used for the analysis of biological evidence such as blood, urine, and saliva as well as physical evidence such as paint and cement and chemical evidence such as explosives (Hofmann, 2010).

Apart from spectroscopic techniques, chromatographic techniques have also been used for the analysis of forensic evidence. Commonly used chromatographic techniques are high-performance liquid (HPLC), gas (GC), high-performance thin-layer (HP-TLC), and hyphenated chromatographic techniques. In such techniques, a solid adsorbent serves as the stationary phase and is exposed to a mobile phase that consists of the sample. Based on the affinity of the sample toward either the stationary phase or the mobile phase, the molecules are separated and can be easily detected. Chromatographic techniques are excellent in identifying the constituents of any sample if appropriate methods have been selected. The techniques are then coupled with structure detection techniques such as mass spectroscopy. Such hyphenation allows the separation of the constituents in the evidence along with their simultaneous detection (Bogusz, 1999; Hashimoto et al., 1988).

Apart from the spectroscopic and chromatographic techniques, microscopy techniques have also been used for the analysis of forensic evidence. A major advantage with microscopy techniques is that visual images of the evidence can be obtained, which means they can be easily produced as testimony in a court of law. There are mainly three types of microscopes used for the analysis of forensic evidence: the optical, electron, and scanning probe microscopes. Optical microscopes use light as their main source of visualization while electron microscopes, especially scanning and transmission electron microscopes, use an electron beam as their source of visualization. Scanning probe microscopes such as atomic force microscopes use a tip or probe to obtain a three-dimensional surface morphology of the sample. Microscopy techniques offer a confirmatory analysis of the sample and therefore are useful (Amelinckx et al., 1997; Basu, 2006; Haynes, 1984; Pandey et al., 2017b).

While spectroscopic, chromatographic, and microscopy techniques are conventionally and routinely used for the analysis of forensic evidence, emerging techniques such as X-ray diffraction, lab-on-chip devices, and nanotechnology have also been used. XRD itself is not a new technique, but its application in the analysis of forensic evidence is recent and is mainly used to obtain the crystalline information of a sample. Lab-on-chip devices are basically miniaturized full-fledged laboratory devices that are capable of carrying out independent reactions when provided with a sample. Lab-on-chip devices employ microfluidic technology and therefore even with micro-or nanovolumes of samples, valid results can be obtained. These devices are also often paired with nanotechnology in which nanosized materials are used to increase the efficiency of detection. Due to the nanosize of the materials, they have increased surface area and chemical reactivity as well as a tunable surface chemistry. Therefore, they can be used for the detection of numerous forensic samples (Pandey et al., 2017a,b; Rawtani et al., 2019).

As mentioned before, forensic laboratories conducting this analytical testing must be accredited by proper accreditation bodies to ensure that the results obtained are valid, credible, and trustworthy. Also, appropriate quality control and quality management also allow a laboratory to function smoothly. Various ethical and legal issues associated with the professionals in this field must also be dealt with the utmost care as this is the only scientific field that directly deals with the justice system.

4 Conclusion

Forensic science is a scientific field that deals with the analysis of forensic evidence to ensure smooth deliverance of justice to victims. Crime scene investigation is a highly crucial aspect of this field where evidence is collected and sent to laboratories for analysis. Spectroscopic, chromatographic, microscopy, and other emerging techniques have been employed for this analysis. While forensic scientists are responsible for the analysis of such evidence, it is also important that they maintain an impartial attitude toward this process in order to remove any bias from the entire process. Laboratories have been routinely employing such techniques and in the future, it is hoped that newer developments are made that enhance the efficiency of these techniques.

References

Amelinckx, S., Van Dyck, D., van Landuyt, J., van Tendeloo, G., 1997. Electron Microscopy: Principles and Fundamentals. VCH Verlagsgesellschaft mbH, Weinheim (Federal Republic of Germany), Germany.

Basu, S., 2006. Scanning electron microscopy in forensic science. In: Encyclopedia of Analytical Chemistry. American Cancer Society. https://doi.org/10.1002/9780470027318.a1122.

Bogusz, M.J., 1999. Hyphenated liquid chromatographic techniques in forensic toxicology. J. Chromatogr. B Biomed. Sci. Appl. 733, 65–91. https://doi.org/10.1016/S0378-4347(98)00520-9.

Hashimoto, Y., Kawanishi, K., Moriyasu, M., 1988. Forensic chemistry of alkaloids by chromatographic analysis. In: The Alkaloids: Chemistry and Pharmacology. Elsevier, pp. 1–77. https://doi.org/10.1016/S0099-9598(08)60215-1 (Chapter 1).

Haynes, R., 1984. Optical Microscopy of Materials. Springer US, Boston, MA. https://doi.org/10.1007/978-1-4757-6085-9.

Hofmann, A., 2010. Spectroscopic techniques: I Spectrophotometric techniques. In: Wilson, K., Walker, J. (Eds.), Principles and Techniques of Biochemistry and Molecular Biology.

Cambridge University Press, Cambridge, pp. 477–521. https://doi.org/10.1017/CBO9780511841477.013.

Pandey, G., Rawtani, D., Agrawal, Y.K., 2017a. Future aspects of halloysite nanotubes in forensic investigations. J. Nanomed. Res. 6, 00153.

Pandey, G., Tharmavaram, M., Rawtani, D., Kumar, S., Agrawal, Y., 2017b. Multifarious applications of atomic force microscopy in forensic science investigations. Forensic Sci. Int. 273, 53–63. https://doi.org/10.1016/j.forsciint.2017.01.030.

Rawtani, D., Tharmavaram, M., Pandey, G., Hussain, C.M., 2019. Functionalized nanomaterial for forensic sample analysis. TrAC Trends Anal. Chem. 120, 115661. https://doi.org/10.1016/j.trac.2019.115661.

Exercise 1: UV-visible spectroscopic analysis of forensic samples (chemical and biological)

1 Aim

To analyze the chemical (drug) and biological (DNA in blood and saliva) forensic samples using UV-visible spectrophotometry.

2 Principle

UV-visible spectrophotometry is a spectroscopic technique that works on the basis of the amount of light absorbed or transmitted by the sample. The main principle behind this technique is the Beer Lambert law, which is an amalgamation of the Beer and Lambert laws. Lambert's law states that the loss in the intensity of light when it passes through a sample is directly proportional to its intensity and path length. Beer's law states that as long as the concentration of the sample and the path length remain constant, the transmittance of the light also remains constant. The mathematical expression of the Beer Lambert law is as follows:

$$A = abc$$

where A is the absorbance, a is the molar absorptivity coefficient of the sample molecule, b is the path length, and c is the concentration of the sample. Therefore, by determining the absorbance of a sample, it is possible to directly determine the concentration of the sample present.

3 Sample preparation and working procedure

3.1 Chemical forensic sample

Drugs are among the most common types of chemical forensic samples encountered at a crime scene. This section describes the sample preparation and working procedure for the UV-visible spectrophotometric analysis of drugs seized from a crime scene.

- Weigh the seized drug sample and dissolve it in a suitable solvent (polar or nonpolar), depending upon the solubility of the drug.
- The concentration of the drug solution should not be very high or very low as these concentrations do not obey the Beer-Lambert law.

- A 1–1000 ppm concentration of the drug is generally considered to be optimum for UV-visible spectrophotometric analysis.
- Take the solvent used to prepare the drug solution as blank in the cuvette and measure the absorbance in the UV-visible range of 200–800 nm.
- Now take the drug solution and measure its absorbance in the same range.
- Subtract the absorbance of the blank from the drug solution to know the actual absorbance of the drug solution.
- In the case of a drug solution of unknown concentration, the linear equation obtained from the calibration curve of the known concentrations of the drug should be used to find the unknown concentration of the drug.

3.2 Biological forensic sample

Body fluids are generally obtained as biological evidence from a crime scene, and they are analyzed for the presence of DNA using UV-visible spectroscopy. This section describes the collection, preparation, and analysis of body fluids (blood and saliva) for estimating the amount of DNA present using UV-visible spectroscopy.

- Collect the saliva samples via sterile buccal swabs, and blood from a finger by pricking using a sterile lancet. Drop the blood on the cotton swab.
- Dry the swabs, and remove the heads of the swabs.
- Centrifuge the swab heads in a buffer solution with a neutral pH.
- Take the supernatant and perform DNA extraction using a commercially available kit or organic solvent, followed by alcoholic precipitation.
- Take the extracted DNA solution in the cuvette and measure the absorbance of the DNA sample at 260 and 280 nm.
- The buffer solution should be taken as the blank.
- Subtract the blank's absorbance from the DNA's absorbance to calculate the actual absorbance of DNA, and quantify its concentration using the calibration curve of DNA with known concentrations.

Exercise 2: Fluorescence spectroscopic analysis of forensic samples (biological)

1 Aim

To analyze biological forensic samples (saliva and blood) using fluorescence spectroscopy.

2 Principle

The fluorescence spectroscopic technique is based on the analysis of the fluorescence energy emitted by a sample. When the sample is irradiated with electromagnetic radiation of a particular wavelength, frequency, and energy, it either absorbs, scatters, or transmits the energy. When the electrons in the sample absorb the energy of the electromagnetic radiation, they get excited and move to a higher energy level. However, they do not stay in this state for a long time and tend to go back to their ground state. During this transition, a series of energies is emitted in which fluorescence is also included. Also, during this transition, the photons emitted are of a higher wavelength than the incident photons. The parameters that are looked into when performing an experiment are the intensity of the fluorescence, the emission wavelength, and the time spent by the molecule in the excited state.

3 Sample preparation and working procedure

3.1 Saliva

Saliva is a body fluid that can be recovered from the scene of crime, and can be analyzed using fluorescence spectroscopy for the presence of fluorescent compounds such as amino acids. This section describes the sample preparation and procedure for the analysis of saliva samples (liquid and dried) using fluorescence spectroscopy.

- Collect the liquid saliva in a vial and store it at low temperature before analysis.
- In the case of a dry saliva sample, such as saliva dried on glass, use a cotton swab dipped in a buffer solution of neutral pH to collect the saliva.
- Dip the cotton swab with the saliva sample in the buffer solution and store that solution with the saliva sample at low temperature.

- Take the dried and liquid saliva solutions in separate cuvettes and place them in the fluorescence spectrometer.
- Analyze the fluorescence of the saliva samples with excitation wavelengths in the range of 280–290 nm.
- Saliva samples with more amino acids such as tryptophan and tyrosine will have a higher intensity of fluorescence.

3.2 Blood

Blood is very commonly encountered biological evidence at a crime scene. It can be analyzed using fluorescence spectroscopy for confirming its presence or checking the presence of hemoglobin or DNA. This section describes the sample preparation and procedure for the analysis of blood using fluorescence spectroscopy.

- Blood collected at the crime scene can be directly analyzed for fluorescence.
- In order to enhance the fluorescence spectra of blood, it can be treated with a fluorescent dye.
- Take the blood collected from the crime scene and mix it with a solution of fluorescent dye such as leucocrystal violet.
- Take the resulting solution in a cuvette and keep it in the fluorescence spectrometer.
- Analyze the fluorescence of the blood with excitation wavelengths in the range of 630–640 nm.
- For comparison purposes, measure the fluorescence of untreated blood also and equate the fluorescence of both the samples (treated and untreated).
- An enhanced fluorescence intensity will be observed for the blood sample treated with the fluorescent dye.

Exercise 3: FTIR analysis of forensic samples (chemical, physical, and biological)

1 Aim

To analyze chemical (drugs and explosives), physical (paint and questionable documents), and biological (body fluids) forensic samples using FTIR spectroscopy.

2 Principle

FTIR is a type of vibrational spectroscopy that takes into account the vibrational frequencies of the molecule when it absorbs the energy of the incident photons in order to determine the structure of the molecule. Similar to other spectroscopic techniques, an incident IR radiation is imparted on the sample and when the energy of the incident photon corresponds to that of the electrons in the sample, it absorbs the energy and starts to vibrate. There are different types of vibrations such as rocking, scissoring, and wagging, and these vibrations are similar to the action of a spring tethered from one side. The vibrational frequencies are characteristic of the structure of a molecule and therefore serve as a highly functional technique to identify the structure of the sample.

3 Sample preparation and working procedure

3.1 Chemical forensic sample

Drugs and explosives are generally found at a crime scene in powdered form. These can be analyzed using FTIR spectroscopy for their chemical profiling. This section describes the sample preparation and procedure for the FTIR analysis of explosives and drugs.

- Take the sample (drug or explosive) and finely grind it using a mortar and pestle.
- Mix the powdered sample with potassium bromide (KBr) in approximately 1:100 ratio, and again grind the mixture to a fine powder.
- Make the pellet of the mixture using a hydraulic pellet press. A KBr pellet without a sample should be used as the background.

- Put the pellet in the sample holder and place the holder in the FTIR spectrometer.
- Analyze the background pellet followed by the sample pellet in the suitable wavenumber range (400–4000 cm^{-1}).

3.2 Physical forensic sample

Paint chips are recovered in burglaries and hit-and-run cases while questionable documents are found in cases of forgery and counterfeiting. These samples need no sample preparation when analyzed with ATR-FTIR spectroscopy. The steps involved in ATR-FTIR analysis are described in this section.

- Collect the sample (paint chip or questionable document) and clear any kind of dust or impurity attached on the surface of the sample.
- A small area of the sample can be cut and then placed on the crystal of the ATR assembly.
- Press the sample against the crystal and record the spectra in the desirable range (400–4000 cm^{-1}).
- The spectra of air (blank crystal without any sample) should be taken as the background.

3.3 Biological forensic sample

Body fluids such as blood, semen, saliva, sweat, etc., are readily recovered from crime scenes, and are analyzed using ATR-FTIR spectroscopy for their chemical profiling. ATR assembly with FTIR requires no sample preparation. The steps involved in the ATR-FTIR analysis are described in this section.

- Collect the body fluids (blood, saliva, semen, sweat) in suitable vials and store at low temperature before analysis to prevent their degradation.
- Take a small amount of the fluid using a micropipette and place a drop of it on the crystal of the ATR assembly.
- Analyze the body fluid in the spectral range as desired (400–4000 cm^{-1}).
- The spectra of air (blank crystal without any sample) should be taken as the background.

Exercise 4: NIR spectroscopic analysis of forensic samples (chemical and biological)

1 Aim

To analyze chemical (drugs and dyes) and biological (body fluids) forensic samples using NIR spectroscopy.

2 Principle

NIR spectroscopy is a type of vibrational spectroscopy that uses the near-infrared region of the electromagnetic spectrum to obtain a structural analysis of the sample. It fulfills this purpose by analyzing the molecular overtone and combination vibrations of the molecule. Typically, there are two types of models for a diatomic model: the classical mechanical model and the anharmonic model, in which it is assumed that the molecule is made of two spheres attached to each other via a string. The classical mechanical model also used in FTIR follows Hooke's law. However, this model does not serve well in NIR spectroscopy as it does not allow certain types of transitions. The anharmonic model takes into account some nonharmonic oscillations as well. This model considers the electrical properties of a molecule as well and provides a very plausible explanation for the overtones or combinations of two or more vibrations. This data is depicted in a spectra intensity of absorption versus wavelength.

3 Sample preparation and working procedure

3.1 Chemical forensic sample

Drugs as well as dyes present in fibers and questionable documents are often analyzed via NIR spectroscopy for their chemical profiling and identification based on their absorbance in the NIR region. This section describes the sample preparation and procedure involved in the analysis of drugs and dyes using NIR spectroscopy.

- For drugs, weigh the powder and dissolve it in a suitable solvent depending on the drug's solubility.
- In the case of dyes, extract them from fibers or documents using chromatographic techniques.

- The extracted dye should be solubilized in a suitable solvent.
- Take the drug or dye solution in a cuvette and put it in the NIR spectrometer.
- Measure the absorbance of drug and dye samples in the NIR region with a spectral range roughly between 750 and 1200 nm.
- The solvent used for dissolving the drug or dye sample should be taken as blank before taking the absorbance of the samples.
- The actual absorbance of the sample can be calculated by subtracting the blank's absorbance from the sample's absorbance.

3.2 Biological forensic sample

Body fluids such as semen, urine, saliva, and blood are often encountered at a crime scene, and their chemical profiling can be carried out using NIR spectroscopy. These samples can be directly used for NIR analysis; that procedure is described in this section.

- Body fluids should be preserved at low temperature before performing the analysis in order to prevent their degradation.
- Take the body fluid sample from the vial and put it into a cuvette.
- Place the cuvette in the NIR spectrometer.
- Measure the absorbance of the body fluid sample in the NIR region with a spectral range roughly between 750 and 1200 nm.
- Because water is a major part and solvent in body fluids, it should be taken as blank before taking the absorbance of the samples.
- The actual absorbance of the sample can be calculated by subtracting the blank's absorbance from the sample's absorbance.

Exercise 5: Atomic absorption spectrometric analysis of forensic samples (chemical and biological)

1 Aim

To analyze chemical (seized drug tablets) as well as biological (tissues) forensic samples with the help of atomic absorption spectroscopy (AAS).

2 Principle

Atomic absorption spectroscopy is a technique that takes into account the absorption of incident radiation by free gaseous atoms to perform elemental mapping of the sample. In this technique, the samples are volatilized from the liquid or solid state to gaseous states. Once the sample has been nebulized, it is subjected to incident radiation. Based on the difference in energy levels between the sample atoms and the incident radiation, they get absorbed. The amount of energy absorbed is directly proportional to the concentration of the atoms and therefore acts according to the principle of the Beer-Lambert law. Atomizers are used to nebulize the samples to free atoms. Just like in UV visible analysis, a standard curve needs to be plotted prior to sample analysis of unknown concentration. The radiation source is of higher energy with a higher spectral resolution.

3 Sample preparation and working procedure

3.1 Chemical forensic sample

Rave parties and smuggling cases often have illicit drugs that are recovered in different forms, especially tablets, during investigations. AAS analysis helps in understanding the elemental composition of such drugs. The sample preparation and procedure involved during the analysis of seized drug tablets is discussed in this section.

- Take the seized drug tablet and crush it using a mortar and pestle to a fine powder.
- Add nitric acid and hydrogen peroxide in approximately 5:1 ratio to let the drug sample get digested and ionized.

- Incubate the mixture for a night in an oven at high temperature, followed by centrifugation.
- Remove the supernatant and reconstitute the pellet formed using deionized water followed by centrifugation in order to remove any impurities.
- The final pellet formed needs to be reconstituted in deionized water.
- Insert the prepared sample in the AAS instrument and analyze the presence and concentration of different elements at different wavelengths, depending on the type of element to be analyzed.

3.2 Biological forensic sample

Whenever any cadaver is recovered in cases of drowning, poisoning, electric shock, etc., elemental composition analysis of the tissues become important. AAS analysis can solve this purpose for different types of biological samples, including tissue. This section discusses the sample preparation and procedure involved in the analysis of tissues using AAS.

- Take the tissue sample and wash it with deionized water in order to remove any impurity present on the surface.
- Cut the tissue into small pieces and store in a refrigerator at low temperature until analysis.
- Add nitric acid and hydrogen peroxide in approximately 5:1 ratio to the tissue pieces to let them get digested.
- Put the mixture in a microwave for some time, followed by centrifugation.
- Take the supernatant, which is the extract of tissue, for the AAS analysis.
- Put the sample in the AAS instrument and analyze the presence and concentration of different elements at different wavelengths, depending on the type of element to be analyzed.

Exercise 6: Optical microscopic analysis of physical and biological forensic samples

1 Aim

To study the morphology of physical (paint chips) and biological (developmental stages of flies) forensic samples using optical microscopy.

2 Principle

Optical microscopes use light as the visualization source and are used to obtain the visualization of any sample. There are mainly two types of optical microscopes namely, the simple and the compound microscope. The simple microscope consists of a single lens such as a magnifying glass. The compound microscope is composed of two or more lenses, namely the objective and condenser lenses. The objective lens is responsible for gathering the light scattered from the specimen while the condenser lens is responsible for collecting the light from the light source and then focusing it on the specimen. Commonly used optical microscopes for forensic sample analysis are confocal, phase contrast, bright-field, stereomicroscope, and polarizing.

3 Sample preparation and working procedure

3.1 Physical forensic sample

Paint chips are collected as physical evidence by investigators in cases of hit and run and burglaries. The optical microscopic analysis of such samples helps in understanding the structure of paint layers, their color, and their texture. The sample preparation and procedure involved in the analysis of paint chips through optical microscopy is discussed in this section.

- Collect the paint chips carefully by avoiding any kind of contamination such as soil or any other foreign particles.
- In the case of a liquid paint sample, dip a clean glass slide into the paint container and allow it to dry. After drying, collect the paint chip from the slide.
- Take the paint chip and produce thin sections of it with the help of a microtome.

- Place the section on the glass slide and apply any suitable mounting medium.
- Place the cover slip above the sample and make sure that the sample is completely sealed with no air bubbles inside the cover slip.
- Visualize the glass slide under the microscope using different lenses. Use an oil immersion technique for viewing under the 100 × lens.

3.2 Biological forensic sample

There are many flies belonging to the Diptera order that have forensic importance because they feed and lay eggs in the cadaver. The microscopic analysis of the developmental stages (instars) of such flies can give an idea about the time of death. The sample preparation and procedure involved in the analysis of instars of flies through optical microscopy is discussed in this section.

- Collect all instars from the F_1 progeny if possible.
- Relocate the collected instars in rearing trays, followed by sacrificing them in boiling water for some minutes.
- In the first and second instars, sections are directly made on their body at different positions depending on the requirements.
- In the third instar, it needs to be placed in a high alkaline solution, preferably potassium hydroxide, for a few minutes in order to digest the body tissue before proceeding with the sectioning.
- Place the sectioned instar on the glass slide and apply a few drops of mounting medium.
- Cover the specimen with the coverslip, keeping in mind that there is no air bubble inside the cover slip.
- Visualize the glass slide under the microscope using different lenses. For 100 × lens, use an oil immersion technique for viewing.

Exercise 7: SEM analysis of physical and biological forensic samples

1 Aim

To study the morphological details of physical (soil) and biological (hair) forensic samples using scanning electron microscopic (SEM) analysis.

2 Principle

A scanning electron microscope is a type of electron microscope that uses an electron beam to visualize the surface morphology of any sample. In this technique, an electron beam is made to fall on the sample through a series of objective and condenser lenses. Once the electron beam hits the sample, several electrons are scattered, depending on the surface morphology of the sample. These scattered electrons are detected in order to obtain a two-dimensional (2D) visualization of the sample's surface. Because this technique highly depends on the interaction of the electron beam with the surface of the sample, it is important that the sample surface is conducting in nature. Therefore, prior to usage, they are often coated with a conducting element. This technique can also be paired with electron diffraction analysis to obtain elemental information of the sample.

3 Sample preparation and working procedure

3.1 Physical forensic sample

Soil samples are present on almost all the objects present at a crime scene, especially if it is an open crime scene. The morphological analysis of soil samples becomes important because their texture and morphology can give an idea about the location to which the soil belongs. This section describes the sample preparation and procedure involved in the SEM analysis of soil particles.

- Collect the soil sample and grind it using a mortar and pestle, followed by drying the sample.
- Pass the ground soil sample through a sieve to remove large particles and other debris.

- Take the sample and put it on double-sided adhesive tape attached on the aluminum stub.
- If required, sputter the sample attached on the stub with gold in order to form a conductive layer on the surface of soil particles.
- Transfer the stub with the sample to the sample stage in the SEM instrument carefully in a dust-free environment.
- Analyze the soil sample at a desirable magnification to view different morphological details of the soil particles.

3.2 Biological forensic sample

Hair samples are commonly encountered at a crime scene. Because their shape and color vary with different races and ethnic groups, their morphological analysis becomes an important task in forensic investigations. This section describes the sample preparation and procedure involved in the SEM analysis of hair samples recovered from a crime scene.

- Collect the hair sample and cut it into smaller fragments of around 1 cm from the middle portion of the hair.
- Deposit the hair fragments on the aluminum stub with double-sided adhesive tape attached on its surface.
- Perform the sputtering of gold particles on the hair samples in order to enhance the conductive nature of their surface.
- Transfer the stub with the sample to the sample stage in the SEM instrument carefully in a dust-free environment.
- Analyze the structure of the hair sample at a desirable magnification to view the different morphological details.

Exercise 8: TEM analysis of physical and biological forensic samples

1 Aim

To analyze physical (gunshot residue) and biological (tissues) forensic samples using transmission electron microscopic (TEM) analysis.

2 Principle

A transmission electron microscope is a type of electron microscope that is used to analyze the internal morphology of any sample. This technique uses high-energy electron beams in order to ensure that it passes through the sample. Based on the type of internal structure, some of the electrons are reflected or scattered while the remainder pass through the sample to the detector. Because the electron beam has to pass through the sample, the samples are sectioned ultrathin and placed on specific sample holders. Similar to SEM, there are several magnetic lenses that serve as condensers in order to achieve high-resolution electron beams.

3 Sample preparation and working procedure

3.1 Physical forensic sample

Gunshot residue is recovered in cases where firearm incidents have taken place. It can be recovered from the hands of the shooter, the clothes, and the nearby vicinity of the exit of the bullet from the firearm and the entry of the bullet into some object. The sample preparation and procedure involved in the TEM analysis of gunshot residue to understand its morphological details are discussed in this section.

- Take the sample using a cotton swab, carefully avoiding the nearby dust and impurities.
- Put the cotton head of the swab in an organic solvent (such as *n*-propanol) and sonicate for a few minutes.
- Remove the cotton head and allow the solution to sonicate for a few more minutes.
- While sonication is in progress, take a small volume of solution using a micropipette and drop it on the carbon-coated copper TEM grid.

- Allow the solvent to get evaporated by mild heating on a hot plate. Avoid overheating.
- Place the grid on the sample holder and transfer it on the sample stage in the TEM instrument.
- Analyze the sample of gunshot residue at a desirable magnification in order to understand the morphological details.

3.2 Biological forensic sample

Tissue analysis is often required in cases of poisoning with toxic plants, animal-based products, or heavy metals because they alter the morphological details of tissues in different parts of the body. This section therefore discusses the sample preparation and procedure involved in the TEM analysis of tissues in order to understand the morphological changes they go through due to poisoning.

- Collect the tissue sample and cut it into small pieces using a sterile and sharp object.
- Take each piece and perform its fixation using glutaraldehyde and paraformaldehyde for a day.
- Fixation should be followed by dehydration using ethanol by constantly increasing its concentration.
- The dehydrated sample should be transferred to a mold with resin for the process of embedding.
- Perform the sectioning of the embedded tissue in resin using ultramicrotome.
- Transfer the ultrathin section on the carbon-coated copper grid, followed by staining with a suitable dye.
- After drying, transfer the grid to the sample holder and place it on the sample stage of the TEM instrument.
- Analyze the tissue samples at a desirable magnification in order to understand the morphological details.

Exercise 9: AFM analysis of physical and biological forensic samples

1 Aim

To analyze the morphological details of physical (fired cartridge cases) and biological (diatoms) forensic samples using atomic force microscopic (AFM) analysis.

2 Principle

AFM is a type of scanning probe microscopy that uses a specific tip or probe to obtain a three-dimensional (3D) surface morphology of the sample. The working of AFM is similar to that of a blind man walking in which he uses a white stick to analyze the path before him. There are mainly three types of modes in AFM: noncontact, contact, and tapping. A certain threshold limit is set and based on this, the changes in the forces are measured. In the contact mode, the tip is constantly in touch with the sample surface and based on the deflections of the tip, which is constantly monitored by a laser, the 3D surface morphology is obtained. In noncontact mode, the forces between the tip and the sample surface are taken into consideration. Tapping mode is another mode in which the tip scans the surface by tapping it intermittently.

3 Sample preparation and working procedure

3.1 Physical forensic sample

Fired cartridge cases are recovered from crime scenes that have had firearm incidents. The microscopic analysis of such cartridge cases gives an idea about the type of firearm used and its manufacturer. This section thus discusses the sample preparation and procedure involved in the AFM analysis of fired cartridge cases to understand their surface morphology.

- A replica molding of the case head of the fired cartridge case needs to be done for AFM analysis.
- Prepare a mixture of a polyorganosiloxane and its curing agent and degas it to remove any air bubbles.
- Pour the degassed mixture on the case head and cure it at room temperature for several hours.

- Peel the replica off the case head and rinse it with deionized water, followed by drying.
- Put the replica on the sample holder disc of AFM and mount it on the sample stage in AFM.
- Analyze the surface morphology of the replica at a desirable magnification, which will give an idea about the surface details of the fired cartridge case.

3.2 Biological forensic sample

The presence of diatoms inside a cadaver indicates a case of drowning. It is a basic test performed in drowning cases. These microscopic organisms are analyzed using AFM to understand the morphology of their surface. The sample preparation and procedure involved for such analysis is discussed in this section.

- Collect the sample and concentrate the cells in the form of a pellet through centrifugation.
- Resuspend the pellet of diatom cells in a small amount of sterile water.
- Place the suspension of cells of a coverslip attached on a petri dish.
- Fix the coverslip on the petri dish using petroleum jelly.
- Mount the petri dish on the metal disc with double-sided adhesive tape and transfer the disc onto the sample stage present in the AFM instrument.
- Analyze the surface morphology of the diatoms at a desirable magnification.

Exercise 10: Mass spectrometric analysis of forensic samples (chemical and biological)

1 Aim

To analyze chemical (drug) and biological (urine) forensic samples using mass spectrometry (MS).

2 Principle

A mass spectrometer is a highly versatile technique used for the structural determination of any sample. The structural determination is done by separating the fragmented ions of a molecule according to their mass-to-charge ratio. An ionization source fragments the molecules into daughter ions and other metabolites. Once ionized, they are passed through a mass analyzer where they are separated according to their mass-to-charge ratio. Once analyzed, the detector detects the molecule present. A mass spectrum therefore displays the intensity of the daughter ion along with different ions of the sample separated according to their mass-to-charge ratio. Mass spectrometry is often used in conjunction with chromatographic techniques such as liquid and gas chromatography. It has also been used with high-performance thin-layer chromatography. This technique can also be used for the direct analysis of samples as long as they have been prepared in a suitable manner.

3 Sample preparation and working procedure

3.1 Chemical forensic sample

Illicit and counterfeit drugs are often analyzed during forensic investigations for their identification. MS helps in their identification on the basis of their mass-to-charge ratio. MS is often attached with liquid chromatography (LC) for drug analysis. The sample preparation and procedure involved in the LC-MS analysis of drugs is discussed in this section.

- Prepare stock solutions of illicit drugs by dissolving the drugs in alcohol.
- If the suspected samples are present in free powder form, directly suspend them in alcohol, and if present in any heterogenous blend such as cigarettes or coffee, the drug can be extracted using alcohol.

- Homogenize the drug samples and dissolve them in alcohol for some time. Centrifuge the mixture and remove the supernatant.
- Dry the pellet and reconstitute it in an organic solvent. Inject the sample in the instrument for further analysis.
- Initially, inject the solution in LC and confirm the retention time of the molecule with the existing literature.
- Identify the molecular structure of the drug sample in the MS spectra by comparing it with existing libraries.

3.2 Biological forensic sample

Urine analysis is an important investigation in forensic science because urine contains the metabolites of many drugs and alcohol, and therefore, such analysis becomes important in drug and alcohol abuse cases. The sample preparation and procedure involved in the LC-MS analysis of drugs is discussed in this section.

- Take the urine sample and add some volume of an internal standard solution to a small portion of this to make it optimal for analysis with LC. The standard solution typically consists of a buffer to maintain the pH and mobile phase.
- Homogenize the solution and centrifuge it.
- Take the supernatant for further analysis.
- Inject the solution in LC and confirm the retention time of the molecule with the existing literature.
- Identify the molecular structure of the metabolites present in the urine sample in the MS spectra by comparing it with existing libraries.

Index

Note: Page numbers followed by *f* indicate figures and *t* indicate tables.

A

AAS. *See* Atomic absorption spectroscopy (AAS)
Absorbance detector, 135
Absorption volume, 77–78
Accreditation
  crime scene investigators, 402–404
  digital forensics, 401–402
  DNA analysis, 404–405
  in forensic science, 397–406
  forensic toxicology, 405–406
  history and need, 395–397
  process of, 399–401
Accreditation body (AB), 397, 400–401
Acetylcholinesterase (AChE), 370
Active pharmaceutical ingredient (API), 103
AFM. *See* Atomic force microscopy (AFM)
Age determination, of bloodstains, 264–267
Alcohol
  detection, biosensors for, 369
  gas chromatography, 155–158
  quality control, for forensic samples, 415
Alcohol dehydrogenase (ADH), 369, 415
Alcohol oxidase, 369
Alkaloids, high-performance liquid chromatography, 142–143
American Academy of Forensic Science (AAFS), 389–390
American Association for Laboratory Accreditation (A2LA), 406
American Board of Forensic Toxicology (ABFT), 406
American National Standards Institute (ANSI), 406
American Society of Crime Laboratory Directors/Laboratory Accreditation Board (ASCLD/LAB), 396–397, 406
Ammonium nitrate fuel oil, 180–181
Analog detector, 325–326
Analytical column, 135
Analytical techniques, 425
ANSI-ASQ National Accreditation Board (ANAB), 398–399
Anthropogenic evidence
  electron microscopy, 248–249
  X-ray diffraction, 327–331
Anticounterfeiting, lab-on-chip devices, 350–351
Anti-Stokes shift, 110
Atmospheric pressure ionization, 313–314
Atomic absorption spectroscopy (AAS), 5, 289–290
  biological samples, 81–82, 440
  chemical forensic sample, 439–440
  drugs, 82–83
  explosives, gunshot residue, and ammunition, 83–84
  glass analysis, 85–86
  ignitable materials, 86
  inks and paper, 86–87
  instrumentation, 77–80, 79*f*
  principle, 75–77, 77*f*, 439
  sample preparation, 439–440
  soil, 87
  textile fibers, 84–85
  working procedure, 439–440
Atomic emission spectroscopy (AES), 5, 289–290
  biological samples, 81–82
  drugs, 82–83
  explosives, gunshot residue, and ammunition, 83–84
  glass analysis, 85–86
  ignitable materials, 86
  inks and paper, 86–87
  instrumentation, 79–80, 81*f*
  principle and theory, 75–77, 77*f*
  soil, 87
  textile fibers, 84–85

Atomic force microscopy (AFM), 9–10, 259–260
- age determination of bloodstains, 264–267, 268*f*
- biological forensic sample, 30, 448
- chemical forensic samples, 27
- components, 260–261
- contact mode, 261–263, 263*f*
- explosive analysis, 271–272
- fingermark examination, 272–273
- for forensic investigations, 263–274, 265–266*t*
- gunshot residue analysis, 271–272
- hair examination, 267–269
- instrumentation of, 260–263, 261*f*
- modes of, 260–263, 263*f*
- noncontact mode, 261–262, 263*f*
- physical forensic sample, 25, 447–448
- plastic wrapping material analysis, 269–270
- principle, 260, 447
- questionable document analysis, 273–274
- sample preparation, 447–448
- SIM card analysis, 271
- tapping mode, 261–262, 263*f*
- textile fiber examination, 270
- working procedure, 447–448

Atomic spectroscopy, 75
- biological forensic samples, 29

Atomizers, 439

Autonomous underwater vehicle (AUV), 349–350

B

Backscattered electrons, 283

Ball milling, 362

Barium sulfate, 292–294

Barn, 281–282

Beer-Lambert law, 76, 323, 431

Benzodiazepine, 176–179

Beryllium, 284–285

Biological evidence
- electron microscopy, 243–248
- X-ray diffraction, 333–335

Biological fluids, lab-on-chip devices, 348–349

Biological samples
- chromatographic analyses, 29
- classification, 21
- collection, preservation, and handling, 21–22
- DNA samples, 30
- energy dispersive X-ray-coupled microscopic analysis, 286–287
- lab-on-chip devices, 30
- mass spectrometry, 30, 307–309
- microscopic analyses, 29–30
- spectroscopic analyses, 28–29
- X-ray diffraction, 30

Biopyrellia bipuncta, scanning electron microscopy, 246–248, 247*f*

Biosensors, 368–370
- for forensic markers, 369
- for marine toxins, microbes, and alcohol, 369
- for warfare agents, 370

Blood, 434

Bloodstains, age determination of, 264–267, 268*f*

Body fluids, 21–22, 43–45, 432, 436
- identification, 102

Botanical forensic samples, 22

Bragg-Brentano parafocusing system, 325

Bragg's diffraction, 323

Bragg's law, 323

Bremsstrahlung X-rays, 281–282

Bright-field microscope, 214–216

C

California Association of Criminalists (CAC), 389–390

Cannabis, 137–140

Capillary electrophoresis (CE), 346

Carbon nanotubes (CNT), 365

Cartridges, 19, 415–416

Case management, 404

Cathinone drugs, 327–328, 329*f*

Cathodoluminescence, 240

Cell lysis, 346

Chain of custody, 384

Chemical analytical techniques
- chromatographic techniques, 5–8
- emerging analytical techniques, 10–13
- microscopic techniques, 8–10

spectroscopic techniques, 1–5
Chemical desorption, 303–304
Chemical forensic samples
chromatographic analyses, 26–27
classification, 20
collection, preservation, and handling, 20–21
lab-on-chip devices, 28
mass spectrometry, 27–28
microscopic analyses, 27
nanotechnology, 28
spectroscopic analyses, 26
X-ray diffraction, 27–28
Chemical ionization technique, 303–304
Chemical shift, 92–93
Chemical vapor deposition (CVD), 362–363
Chiral detector, 136
Chlamydia trachomatis, 369
Chromatographic techniques, 427
biological forensic samples, 29
chemical forensic samples, 26–27
gas chromatography, 6–7
high performance liquid chromatography, 6
high-performance thin-layer chromatography, 7
hyphenated techniques, 7–8
physical forensic samples, 24
Circular dichroism (CD), 136
Cliff-Lorimer method, 283–284
^{13}C NMR, 93–94
Code of Ethics and Conduct, 389–390
Codes of ethics
in fields of forensic science, 391–393
models of, 389–391
College of American Pathologists (CAP), 406
Committee on Conformity Assessment (CASCO), 397–398
Compound microscope, 441
Computer forensics, 386–387
and legal issues, 386–387
Concrete, X-ray diffraction, 333
Conductivity detector, 136
Confocal microscope, 219, 220*f*
Constant pressure pump, 133
Continuous-flow PCR chip, 346
Convergence angle, 241
Correlational spectroscopy (COSY), 94
Counterfeit beverages, 45
Counterfeit drugs, 350–351
Fourier transform infrared spectroscopy, 67, 68*f*
identification, 102–103
prevention, 372–373
X-ray diffraction, 331
Crime investigation procedure, 382–383
Crime scene
examination, 403
investigators, 402–404
preliminary assessment of, 403
recording, 404
supervision, 403
Crime Scene Accreditation Committee (CSAC), 402–403
CuKα, 328–335
Cyclonite (RDX), 180–181

D

Daly detector, 306
Data processing unit, 136
Detector, 135–136
Dichlorvas, 372–373
Dicing, 344
Digital detectors, 325–326
Digital evidence, 385–386
Digital forensics, 401–402
quality assurance in, 412–413
Dinitrobenzene (DNB), 365
Dinitrotoluene (DNT), 349–350
DNA Advisory Board (DAB), 404–405
DNA analysis, 404–405
ethics in, 392
lab-on-chip devices, 344–348
legal aspects in, 387–388
legal issues, 388
nanomaterials, 371–372
quality control in, 416–417
Drugs, 20, 45–46
atomic absorption and emission spectrometry, 82–83
FTIR analysis of, 435–436

Drugs *(Continued)*
high-performance liquid chromatography, 137–141
high-performance thin-layer chromatography, 176–179
hyphenated techniques, 198–200, 199*f*
mass spectrometry, 309–310
quality control, 417–418
Dry etching, 343
Dyes, 117–119

E

Ecstasy, 101–102
EDX. *See* Energy dispersive X-ray (EDX)
Elastic scattering, 323
Electrochemical detector, 136
Electromagnetic radiation, 426, 433
Electron beam, 237–238, 240
Electron guns, 238–240
Electron ionization, 303
Electron microscopes, 8–9, 235
anthropogenic evidence, 248–249
biological evidence, 243–248
forensic investigations, 243–253
geological evidence, 249–253
instrumentation of, 238–243, 239*f*
modes of, 241–243
principle, 236–238
sample preparation, 240–241
scanning electron microscopy, 237–238
scattering of electrons, 236, 237*f*
theory of, 236–238
transmission electron microscopy, 238
Electron multipliers, 306
Electrospray ionization, 311–313, 313*f*
Emerging analytical techniques
lab-on-chip devices, 12–13
mass spectrometry, 10–11
nanotechnology, 12
X-ray diffraction, 11–12
Energy dispersive X-ray (EDX), 10, 281
coupled microscopic analysis, 285–296
detectors in, 284–285
instrumentation of, 284–285
principle of, 281–284
SEM, 283
TEM, 283–284
theory of, 281–284
Energy dispersive X-ray (EDX)-coupled microscopic analysis, 285–296
biological samples, 286–287
explosives, 287–288, 288*f*
glass, 289–290
gunshot residue, 290–292, 291*f*
paints, 292–294, 293*f*
soil, 294–295, 296*f*
toxins, 295–296, 297*f*
Entomological samples, 22
Etching, 362
Ethics, 381, 388–391
in DNA analysis, 392
in forensic examination, 391–392
in forensic psychiatry, 392–393
theory of, 388–391
Ethyl glucuronide, 101–102
NCI-MS spectra of, 307–309, 308*f*
European DNA Profiling Group (EDNAP), 416
European Network of Forensic Science Institutes (ENFSI), 416
Evaporative light-scattering detector (ELSD), 136
Evidence
biological, 425
chemical, 425
defined, 383–384
interpretation, 403
physical, 425
rules and legal aspects, 383–384
Exhibits, collection of, 404
Expert witnesses, 385
Explosives, 21
atomic absorption and emission spectrometry, 83–84
atomic force microscopy analysis, 271–272
detection, of nanomaterials, 365–368
energy dispersive X-ray-coupled microscopic analysis, 287–288, 288*f*
Fourier transform infrared spectroscopy, 65–67, 66*f*, 435–436
gas chromatography, 160–161
high-performance thin-layer chromatography, 180–181

hyphenated techniques, 200–201, 202*f*
lab-on-chip devices in detection of, 349–350
mass spectrometry, 311–312
optical microscopy, 227–228
Raman spectroscopy, 119–120

F

Fabrication techniques, of lab-on-chip devices, 342–344, 345*f*
Faraday cup, 306
Fibers, 19–20, 46–48
high-performance thin-layer chromatography, 182–183
hyphenated techniques, 201–203, 203*f*
mass spectrometry, 312–313, 313*f*
optical microscopy, 226–227
Raman spectroscopy, 120–121, 122*f*
X-ray diffraction, 331
Field emission (FE) guns, 238–240
Field ionization, 303–304
Fingermarks
atomic force microscopy, 272–273
latent, visualization of, 370–371
Fingerprints, 48–49, 283, 370
gas chromatography, 161
latent, 272–273
Firearm, 19, 416
Flammable liquids, 20–21
Flesh fly. *See Ravinia belforti*
Fluorescence detector, 135
Fluorescence microscope, 216–218, 218*f*
Fluorescence spectroscopy, 2–3
blood, 434
body fluid, 43–45, 45*f*
chemical forensic samples, 26
drugs, 45–46
fibers, 46–48
fingerprints, 48–49
forensic samples analysis, 43–51
instrumentation, 40–43, 42*f*
paint, 49–50
petroleum products, 50
physical forensic samples, 23
principle, 40, 433
questionable documents, 50–51
saliva, 433–434
sample preparation, 433–434
working procedure, 433–434
Fluorophores, 371–372
Food products
hyphenated techniques, 204
mass spectrometry, 309–310
Forensic biology, quality assurance in, 413–414
Forensic markers, biosensors for, 369
Forensic sample
biological forensic samples, 21–22
chemical forensic samples, 20–21
demonstrative evidence, 17
physical forensic samples, 18–20
preparation techniques, 22–30
real evidence, 17
types, 17–22
Forensic science, 301, 395, 409
analytical techniques, 1
definition, 1
legal issues in, 382–388
quality assurance in, 409–414
samples collection, 18–22
samples preparation techniques, 22–30
Forensic toxicology, 405–406, 413–414
Forensic urine drug testing (FUDT), 406
Fourier transform infrared spectroscopy (FTIR), 426
bank notes, 68
biological sample, 62–64, 62*f*, 436
chemical forensic sample, 435–436
explosives and gunshot residue, 65–67, 66*f*
illicit materials and counterfeit drugs, 67, 68*f*
inks and questionable documents, 64–65, 64*f*
instrumentation, 58–59, 58*f*
paint, 68–69, 70*f*
physical forensic sample, 436
principle, 56–57, 435
sample preparation, 435–436
spectrum analysis, 61
working procedure, 435–436
Fourier transform ion cyclotron resonance (FTICR), 305

G

Gas chromatography (GC), 6–7
  alcohol, 155–158
  chemical forensic samples, 27
  explosives and ignitable materials, 160–161
  fingerprints, 161
  forensic science, 149
  illicit drugs, 158–160, 159*f*
  inks, 161–162, 162*f*
  instrumentation, 152–154, 155*f*
  paints, 162–163, 163*f*
  physical forensic samples, 24
  principle and theory, 149–151, 152*f*
  toxins, 163–164
  volatile organic compounds, 155–158, 156–157*f*
Gas chromatography-mass spectrometry (GC-MS)
  hyphenated techniques, 190, 191*f*
Geological evidence
  electron microscopy, 249–253
  X-ray diffraction, 332–333
Glass, 19–20
  atomic absorption and emission spectrometry, 85–86
  energy dispersive X-ray-coupled microscopic analysis, 289–290
  fragments, 289
  optical microscopy, 228–229
Grazing incidence XRD (GIXRD), 333
Guard column, 134
Gunshot residue (GSR), 19
  atomic absorption and emission spectrometry, 83–84
  atomic force microscopy analysis, 271–272
  energy dispersive X-ray-coupled microscopic analysis, 290–292, 291*f*
  Fourier transform infrared spectroscopy, 65–67, 66*f*
  mass spectrometry, 311–312
  nanomaterials, 365–368
  Raman spectroscopy, 119–120, 120*f*
Gyromagnetic ratio, 92–93

H

Hair, 267–269
  atomic force microscopy, 267–269
  quality control for, 415–416
HAIR-VEQ, 419–420
Hall method, 283–284
Hemoglobin, 264
High-performance liquid chromatography (HPLC), 6
  applications, 136–143
  biological forensic samples, 29
  chemical forensic samples, 26–27
  classification, 131–132
  drugs, 137–141
  instrumentation, 132–136, 133*f*
  pesticides, 141–142
  physical forensic samples, 24
  plant toxins and alkaloids, 142–143
  principle, 130
High-performance liquid chromatography-Fourier transform infrared (HPTLC-FTIR), 195–198, 197*f*
High-performance liquid chromatography-mass spectrometry (HPTLC-MS), 195–197, 198*f*
High-performance thin-layer chromatography (HPTLC)
  chemical forensic samples, 27
  drugs, 176–179
  explosives, 180–181
  fibers, 182–183
  forensic samples analysis, 176–184, 177–178*t*
  inks, 181–182
  instrumentation, 172–176, 174–175*f*
  principle and theory, 169–172, 170*f*
  toxins, 183–184
High-throughput microfluidic immunosensors (HTMI), 349–350
Hollow cathode lamp (HCL), 78–79
Homeland security, 372–373
Horseradish peroxidase (HRP), 369
HPLC. *See* High-performance liquid chromatography (HPLC)
Hyperspectral imaging (HSI), 264
Hyphenated techniques, 7–8
  drugs, 198–200, 199*f*

explosives, 200–201, 202*f*
fibers, 201–203, 203*f*
food products, 204
GC-MS, 190, 191*f*
HPTLC-FTIR, 195–198, 197*f*
HPTLC-MS, 195–197, 198*f*
IMS-MS, 194–195, 196*f*
inks and questionable documents, 204–205, 205*f*
LC-IR, 192–193
LC-MS, 190, 192*f*
LC-NMR, 193–194, 194*f*
toxins, 206–207, 207*f*

I

Ignitable materials
atomic absorption and emission spectrometry, 86
gas chromatography, 160–161
Illicit drugs
lab-on-chip devices, 351–353
nanomaterials, 371–372
Impression marks, 19
Inductively coupled plasma AES (ICP-AES), 81–82
Inductively coupled technique, 79–80
Inelastic scattering, 323
Infrared spectroscopy, 3
Infrared (IR) spectroscopy, 55
biological forensic samples, 28
chemical forensic samples, 23, 26
Inks
atomic absorption and emission spectrometry, 86–87
gas chromatography, 161–162, 162*f*
hyphenated techniques, 204–205, 205*f*
quality control (QC) for, 420–421
Integrated Ballistics Identification System (IBIS), 416
Integrating detectors, 325–326
Internal Laboratory Accreditation Guide (ILAC), 410–411
International Electrotechnical Commission (IEC), 396–397
International Laboratory Accreditation Cooperation (ILAC), 397–398
International Organization for Standardization (ISO), 396–397, 410–411
International Society of Forensic Genetics (ISFG), 404–405
Ionization energy, 281–282
Ion mobility spectroscopy (IMS-MS), 194–195, 196*f*
Ion trap analyzers, 305
ISO 17025, 399

J

J coupling constant, 92–93

K

Kastle-Mayer test, 348–349

L

Lab-on-chip devices, 12–13, 339
anticounterfeiting, 350–351
biological fluid identification, 348–349
biological forensic samples, 30
chemical forensic samples, 28
components of, 340–342, 342*f*
DNA analysis, 344–348
explosive detection, 349–350
fabrication techniques of, 342–344, 345*f*
illicit drug detection, 351–353
injectors, 340–341
Lactate dehydrogenase (LDH), 369
The Larmor Frequency, 92–93
Laser-ablation, 362
Laue method, 324–325
Legal aspects
in computer and digital forensics, 386
crime investigation procedure, 382–383
in DNA forensics, 387–388
evidence, 383–384
of reports issued by forensic laboratories, 384–385
subpoena, 383
testimony by expert witnesses, 385
Legal issues, in forensic science, 382–388
Leucocrystal violet (LCV), 43–44
Ligand exchange reactions, 363–364
Light microscopes, 235
Liquid chromatography, 129

Liquid chromatography-infrared (LC-IR) spectroscopy, 192–193
Liquid chromatography-mass spectrometry (LC-MS), 190, 192*f*
Liquid chromatography-nuclear magnetic resonance (LC-NMR), 193–194, 194*f*
Listeria monocytogenes, 369
Localized surface plasmon resonances (LSPR), 110–111

M

Magnetic microbeads, 341
Magnetogyric ratio. *See* Gyromagnetic ratio
Marine toxins, biosensors for, 369
Mass analyzers, 304–305
Mass spectrometry (MS), 10–11, 301
  biological sample analysis, 29–30, 307–309, 450
  chemical forensic sample, 27–28, 449–450
  detectors, 305–306
  drugs and food examination, 309–310
  explosives analysis, 311–312
  fiber analysis, 312–313, 313*f*
  gunshot residue analysis, 311–312
  instrumentation of, 302–303, 302*f*
  ionization sources, 303–304
  mass analyzers, 304–305
  oil spills, investigation of, 314–316
  petroleum products, investigation of, 314–316
  physical forensic samples, 24–25
  principle, 301–302, 449
  questionable documents, inks detection in, 313–314
  sample preparation, 449–450
  theory of, 301–302
  working procedure, 449–450
Matrix-assisted laser desorption ionization (MALDI), 174–176, 304
Metaethics, 388–389
Methemoglobin, 264
Methomyl, 295–296
Micellar electrokinetic capillary chromatography (MECC), 351–353
Microbes, biosensors for, 369
Microbial samples, 22
Microfluidic devices, 339
  for DNA extraction and amplification, 347*f*
  for sex determination, 346–348
Microscopic techniques
  atomic force microscope, 9–10
  biological forensic samples, 29–30
  chemical forensic samples, 27
  electron microscopes, 8–9
  energy dispersive X-ray, 10
  optical microscopes, 8
  physical forensic samples, 24–25
Microtaggant, 351, 352*f*
Mitochondrial DNA (mtDNA) analysis, 417
Molecular beam epitaxy (MBE), 343, 362–363
Molecular ions, 301–302
Morphologically directed RS (MDRS), 121–123

N

Nanomaterials, 359
  bottom-up approach, 361–362
  characterization techniques, 360–361
  fluorescent, 370–371
  in forensic science, 366–367*t*
  functionalization strategies of, 363–364
  one-dimensional (1D), 360
  synthetic approaches of, 361–363, 364*f*
  three-dimensional (3D), 360
  top-down approach, 361–362
  two-dimensional (2D), 360
  types of, 360–361
Nanotechnology, 12
  biosensors, 368–370
  chemical forensic samples, 28
  counterfeiting prevention, 372–373
  DNA analysis, 371–372
  explosive, detection of, 365–368
  gunshot residues, detection of, 365–368
  homeland security, 372–373
  illicit drug analysis, 371–372
  latent fingermarks, visualization of, 370–371
  nanomaterials, 360–364

physical forensic samples, 25
tracking devices, 373
Nanotrackers, 373
National Accreditation Board for Testing and Calibration Laboratories (NABL), 398–399
National Accreditation of Measurement and Sampling (NAMAS), 406
National Association of Testing Laboratories (NATA), 406
National Institute of Standards and Technology (NIST), 416
Near-field scanning microscopy (NSOM), 360–361
Near-field scanning optical microscopy (NSOM), 259
Near-infrared (NIR) spectroscopy, 3, 55, 426
bank notes, 68
biological samples, 62–64, 438
chemical forensic sample, 437–438
explosives and gunshot residue, 65–67, 66*f*
illicit materials and counterfeit drugs, 67, 68*f*
inks and questionable documents, 64–65, 64*f*
instrumentation, 59–61, 60*f*
paint, 68–69, 70*f*
principle, 56, 437
sample preparation, 437–438
spectrum analysis, 61
working procedure, 437–438
Negative chemical ionization (NCI), 307–309
NMR. *See* Nuclear magnetic resonance (NMR) spectroscopy
Nonbody fluids, 22
Nondestructive techniques, 273–274
Normative ethics, 388–389
Nuclear magnetic resonance (NMR) spectroscopy, 4
advantages, 91
biological forensic samples, 29
body fluids, 102
chemical forensic samples, 26
counterfeit products, 102–103
drug analysis, 96–99
explosives, 99–100
forensic application, 96, 97*f*, 98*t*
instrumentation, 92–95, 96*f*
postmortem changes, 100–101
principle, 92–95
types, 93–94
xenobiotics examination, 101–102
Nuclear overhauser effect spectroscopy (NOESY), 94

O

1D NMR, 93–94
On-wafer testing, 344
Ophyra capensis, 246–248
Opiates, 140–141
Optical compound microscopes
bright-field microscope, 214–216
confocal microscope, 219, 220*f*
fluorescence microscope, 216–218, 218*f*
phase contrast microscope, 216, 217*f*
polarizing microscope, 219–221, 222*f*
stereomicroscope, 221–222, 222*f*
Optical microscopy, 8, 427
biological samples, 30, 223–225, 442
chemical forensic samples, 27
explosives, 227–228
fibers, 226–227
forensic science, 213, 223–231
fundamentals, 214, 215*f*
glass, 228–229
paint, 229
physical samples, 24, 441–442
principle, 441
questionable documents, 229, 230*f*
sample preparation, 441–442
soil and minerals, 230–231, 231*f*
working procedure, 441–442
Optical rotary dispersion (ORD), 136
Organophosphates (OPs), 370

P

Paints, 19–20, 49–50
energy dispersive X-ray-coupled microscopic analysis, 292–294, 293*f*
Fourier transform infrared spectroscopy, 68–69, 70*f*

Paints *(Continued)*
FTIR analysis of, 436
gas chromatography, 162–163, 163*f*
optical microscopic analysis, 441–442
optical microscopy, 229
Raman spectroscopy, 117–119
Para-hydrogen induced polarization (PHIP) technique, 97–99
PeakForce Quantitative Nanomechanical Mapping (PF QNM), 272–273
Pentaerythritoltetranitrate (PETN), 180–181, 311
Perchlorate, 331
Pesticides, 141–142
Petroleum products, 21, 50, 314–316
Phase contrast microscope, 216, 217*f*
Photolithography, 343
Photomultiplier tube (PMT), 240
Photon counting detectors, 325–326
Physical forensic samples
chromatographic analyses, 24
classification, 18
collection, preservation, and handling, 18–20
mass spectrometry, 25
microscopic analyses, 24–25
spectroscopic analyses, 23–24
X-ray diffraction, 25
Physical vapor deposition (PVD), 362–363
Piezoelectric detection, 260–261
Plastic wrapping material analysis, 269–270
^{31}P NMR, 93–94
Polarizing microscope, 219–221, 222*f*
Polydimethylsiloxane (PDMS), 249, 346–348
Polyethylene (PE), 269
Polymerase chain reaction (PCR), 417
Polymeric coatings, 363–364
Polymers, 343–344, 363–364
Polymethyl methacrylate (PMMA), 346–348
Polyvinyl chloride (PVC), 269
Postmortem interval (PMI), 333, 335
Powder diffraction method, 325
Precession, 92–93
Prostate specific antigen (PSA), 369
Psychoactive drugs, 327–328
Pumps, 133

Q

QR code, 351
Quadrupole analyzers, 305
Quality assurance, 409–412
aspects of, 411–412
clinical and forensic pathology, 414
in digital forensics, 412–413
in forensic biology and toxicology, 413–414
in questionable document analysis, 414
standard development organizations, 410–411
Quality control (QC)
alcohol, 415
cartridge cases, 415–416
DNA analysis, 416–417
drugs, 417–418
entomological evidence, 418–419
hair, 419–420
inks, 420–421
Quantitative XRD (QXRD), 332–333
Quartz crystal microbalance (QCM), 351–353
Questionable document
atomic force microscopy analysis, 273–274
quality assurance in analysis, 413–414
Questionable documents, 50–51, 248–249, 313–314

R

Raman scattering, 4–5, 110
Raman spectroscopy (RS), 4–5
biological samples, 114–116, 115–116*f*
chemical forensic samples, 26
dyes, 117–119
explosive analysis, 119–120
fibers, 120–121, 122*f*
gunshot residue, 119–120, 120*f*
instrumentation, 111–113, 112*f*
paint analysis, 117–119
physical forensic samples, 23–24
pigments, 117–119
principle, 110–113, 111*f*
questionable documents, 123–124

recreational drugs, 116–117, 118*f*
soil, 121–123
Ravinia belforti, 246–248
Real-time polymerase chain reaction (RT-PCR), 371–372
Reciprocal discovery, 383
Reciprocating piston pump, 133
Recreational drugs, 116–117, 118*f*
Refractive index detector, 136

S

Saliva, 44–45, 45*f*, 433–434
Sample injector, 134, 134*f*
Scanning analyzers, 304–305
Scanning electron microscopy (SEM)
biological forensic samples, 30, 444
of *Biopyrellia bipuncta*, 246–248, 247*f*
of Caucasian and Afro hair after thermal and bleaching treatment, 243–245, 244*f*
of coal particles, 253, 254*f*
energy dispersive X-ray, 283
firearm cartridges, analysis of, 249, 251*f*
of gum binders, 292–294, 294*f*
instrumentation of, 238–243, 239*f*
modes of, 241–243
nanomaterials, characterization, 360–361
physical forensic samples, 24–25, 443–444
of polyethylene terephthalate fibers, 249, 252*f*
principle, 443
red sealing layered under gel pen ink, 248–249, 250*f*
sample preparation, 443–444
theory of, 237–238
working procedure, 443–444
Scanning probe microscopy (SPM), 259, 427
Scanning tunneling microscopy (STM), 259
Scientific Working Group on Digital Evidence (SWDGE), 402
Scintillator detector, 240
Secondary ion mass spectrometry (SIMS), 304
Self-incrimination, 383
SEM. *See* Scanning electron microscopy (SEM)
Semen-stained objects, 21–22
Semiconductor detectors, 240
Short tandem repeats (STRs), 417
Silanization, 363–364
Si-Li detector, 286–287
SIM card analysis, 271
Simple microscope, 441
Small angle X-ray scattering (SAX), 326
Society of Hair Testing (SOHT), 418
Soils, 19–20
energy dispersive X-ray-coupled microscopic analysis, 294–295, 296*f*
optical microscopy, 230–231, 231*f*
Solid-phase extraction (SPE), 141–142
Solid phase extraction chromatography (SPEC), 96–97
Solid phase microextraction (SPME), 141–142
Solvent delivery system, 132
Sonochemical technique, 362–363
Spatial sensitive detectors, 325–326
Specific Criteria for Accreditation of Forensic Science Laboratories, 398–399
Spectroscopic techniques, 426
atomic emission and absorption spectroscopy, 5
biological forensic samples, 28–29
chemical forensic samples, 26
electromagnetic radiation, 1–2
fluorescence spectroscopy, 2–3
infrared and near-infrared techniques, 3
nuclear magnetic resonance spectroscopy, 4
physical forensic samples, 23–24
Raman spectroscopy, 4–5
UV-visible spectroscopy, 2–3
Spin-spin splitting, 92–93
Standard development organizations (SDOs), 410–411
Standard operating procedures (SOPs), 411–412
Standard reference materials (SRMs), 415–416
Stereomicroscope, 221–222, 222*f*
Stokes shift, 110
Subpoena, 383

Surface enhanced Raman scattering (SERS), 110–111
Surface-enhanced Raman spectroscopy (SERS), 349–353
Surface plasmon resonance (SPR) biosensors, 369
Swabing, 21–22
Synchrotrons, 322
Syringe pumps, 133

T
TEM. *See* Transmission electron microscopy (TEM)
Tetrodotoxin, 369
Textile fibers
  atomic absorption and emission spectrometry, 84–85
  atomic force microscopy, 270
Thomson scattering, 323
Total correlated spectroscopy (TOCSY), 94
Toxins
  energy dispersive X-ray-coupled microscopic analysis, 295–296, 297*f*
  gas chromatography, 163–164
  high-performance thin-layer chromatography, 183–184
  hyphenated techniques, 206–207, 207*f*
Transmission electron microscopy (TEM)
  biological forensic samp, 446
  energy dispersive X-ray, 283–284
  instrumentation of, 238–243, 239*f*
  modes of, 241–243
  nanomaterials, characterization, 360–361
  physical forensic sample, 445–446
  principle, 445
  sample preparation, 445–446
  theory of, 238
  working procedure, 445–446
Triacetone triperoxide (TATP), 311
Trinitrobenzene (TNB), 349–350
Trinitrotoluene (TNT), 180–181, 349–350, 365
  sensing via molecularly imprinted carbon dots, 365, 368*f*
2D NMR, 93–94

U
Ultraviolet (UV) light, 343
Uranium dioxide (UO_2), 333
Urine analysis, 450
UV-visible detector, 135
UV-visible spectroscopy, 2–3
  biological forensic sample, 432
  body fluid, 43–45, 45*f*
  chemical forensic samples, 26, 431–432
  counterfeit beverages, 45
  drugs, 45–46
  fibers, 46–48
  fingerprints, 48–49
  instrumentation, 40–43, 40*f*
  paint, 49–50
  petroleum products, 50
  physical forensic samples, 23
  principle, 38–39, 431
  questionable documents, 50–51
  sample preparation, 431–432
  working procedure, 431–432

V
Variable number tandem repeats (VNTRs), 387–388
Verification of digital evidence (VODE) framework, 412–413
Volatile organic compounds (VOCs), 155–158, 156–157*f*

W
Warfare agents, biosensors for, 370
Wave-particle duality, 37
Weapons of mass destruction (WMDs), 372–373
Wet etching, 343
Wide-angle X-ray diffraction (WXRD), 326

X
Xenobiotics, 101–102
X-ray diffraction (XRD), 11–12, 321, 427–428
  anthropogenic evidence, 327–331
  biological evidence, 333–335
  biological forensic samples, 30
  chemical forensic samples, 27–28

detectors in, 325–326
geological evidence, 332–333
instrumentation of, 324–326, 325*f*
nanomaterials, characterization, 360–361
physical forensic samples, 25
theory of, 323
types of, 326
X-rays, 235, 281–282, 321
absorption, 323
Bremsstrahlung, 281–282
hard, 322
Kα, 281–282
Kβ, 281–282
soft, 322
theory of, 322–323
XRD. *See* X-ray diffraction (XRD)

Kolibri 360 GmbH
Ferdinand-Jühlke-Straße 7
99095 Erfurt, Deutschland
produktsicherheit@kolibri360.de